T0305983

CHEMICAL REACTION ENGINEERING

Beyond the Fundamentals

CHEMICAL REACTION ENGINEERING

Beyond the Fundamentals

L.K. Doraiswamy
Deniz Üner

CRC Press is an imprint of the
Taylor & Francis Group, an **informa** business

CRC Press
Taylor & Francis Group
6000 Broken Sound Parkway NW, Suite 300
Boca Raton, FL 33487-2742

Version Date: 20130514

ISBN 13: 978-1-4398-3122-9 (hbk)

Library of Congress Cataloging-in-Publication Data

Doraiswamy, L. K. (Laxmangudi Krishnamurthy)
Chemical reaction engineering : beyond the fundamentals / authors, L.K. Doraiswamy and Deniz Uner.
pages cm
Includes bibliographical references and index.
ISBN 978-1-4398-3122-9 (hardback)
1. Chemical engieering. 2. Chemical reactions. I. Title.

TP155.D67 2013
660--dc23 2013003677

Visit the Taylor & Francis Web site at
http://www.taylorandfrancis.com

and the CRC Press Web site at
http://www.crcpress.com

Dedicated to my former students and colleagues at Iowa State University (USA) and National Chemical Laboratory (India), from whom I learned so much.

L.K. Doraiswamy

Dedicated to L.K. Doraiswamy (1927–2012) and his continuing legacy in Chemical Reaction Engineering.

D. Uner

Contents

Preface . . . . **xxv**

Notations . . . . **xxxi**

Overview . . . . **xxxvii**

PART I FUNDAMENTALS REVISITED

Objectives 1
Introduction 1
The essential minimum of chemical reaction engineering 2
The skill development 2
Getting started 2
Warm-up questions 3
  Qualitative 3
  Quantitative 3

1 Reactions and reactors: Basic concepts **5**

Chapter objectives 5
Introduction 5
Reaction rates 5
  Different definitions of the rate 6
  Basic rate equation 8
Stoichiometry of the rate equation 9
  Basic relationships 9
  Conversion–concentration relationships 10
  Variable-density reactions 11
  Reactors 12
  Batch reactor 13
    Reactions without volume change 14
    Reactions with volume change 14
  Nonisothermal operation 16
    Optimal operating policies 18

Plug-flow reactors 19
Basic PFR equation 20
Design equations 21
Nonisothermal operation 21
Perfectly mixed flow reactor (MFR) 23
Basic CSTR equation 23
Nonisothermal operation 24
Multiple steady states 26
MSS in a CSTR 26
Adiabatic CSTR 27
Stability of the steady states 28
Comparison of BR, PFR, and MFR 29
Explore yourself 30
References 31
Bibliography 31

2 Complex reactions and reactors 33

Chapter objectives 33
Introduction 33
Reduction of complex reactions 34
Stoichiometry of simple and complex reactions 34
Mathematical representation of simple and complex reactions 35
Independent reactions 36
Rate equations 38
The concept of extent of reaction 38
Determination of the individual rates in a complex reaction 39
Selectivity and yield 39
Definitions 40
Analytical solutions 40
Maximizing selectivity in a complex reaction: Important considerations 43
Multistep reactions 46
Definitions 46
Yield versus number of steps 47
Reactor design for complex reactions 48
Batch reactor design based on number of components 48
Use of extent of reaction or reaction coordinates 50
Plug-flow reactor 52
Continuous stirred tank reactor 53
Reactor choice for maximizing yields/selectivities 56

Parallel reactions (nonreacting products) 56
The general case 56
Effect of reaction order 58
One of the reactants undergoes a second reaction 59
Parallel–consecutive reactions 59
Plug-flow reactor with recycle 60
The basic design equation 60
Optimal design of RFR 62
Use of RFR to resolve a selectivity dilemma 64
Semibatch reactors 64
Constant-volume reactions with constant rates of addition and removal: Scheme 1 64
Variable-volume reactor with constant rate of inflow: Scheme 2 66
Variable-volume reactor with constant rate of outflow of one of the products: Scheme 3 67
General expression for an SBR for multiple reactions with inflow of liquid and outflow of liquid and vapor: Scheme 4 70
Nonisothermal operation 71
Optimum temperatures/temperature profiles for maximizing yields/selectivities 71
Optimum temperatures 72
Optimum temperature and concentration profiles in a PFR 72
Parallel reactions 72
Consecutive reactions 73
Extension to a batch reactor 73
Explore yourself 74
References 75
Bibliography 76

Interlude I 77
Reactive distillation 77
Membrane reactors 78
Inorganic membranes for organic reactions/synthesis 78
Potentially exploitable features of membranes 79
Equilibrium shift in membrane reactors 79
Controlled addition of reactants 80
Preventing excess reactant "slip" in reactions requiring strict stoichiometric feeds 80

Mimicking trickle-bed operation with improved performance 80
Coupling of reactions 80
Hybridization 80
Phase transfer catalysis 81
References 82

3 Nonideal reactor analysis 85

Chapter objectives 85
Introduction 85
Two limits of the ideal reactor 86
Plug-flow reactors with recycle 86
Tanks-in-series model 87
Nonidealities defined with respect to the ideal reactors 88
Nonidealities in tubular reactors 88
Axial dispersion model 89
Nonidealities in MFR 90
Residence time distribution 91
Theory 91
Types of distribution 93
Concept of mixing 95
Regions of mixing 95
Fully segregated flow 97
Micromixing policy 98
Models for partial mixing 99
Axial dispersion model 99
Tanks-in-series model 100
Models for partial micromixing 100
Degree of segregation defined by the age of the fluid at a point 101
Turbulent mixing models 101
Characteristic timescales 102
Engulfment-deformation diffusion model 104
Interaction by exchange with a mean 104
Zone model 105
Joint PDF 105
Practical implications of mixing in chemical synthesis 106
General considerations 106
Dramatic illustration of the role of addition sequence of reagents 108
Explore yourself 109
References 110
Bibliography 110

Interlude II **111**

Limits of mean field theory 111
The predator–prey problem or surface mixing 111
Mixing problem addressed 113
Short contact time reactors 113
Microfluidic reactors 114
Passive devices for mixing and pumping 115
Knudsen pump 115
Mixing 115
Slug flow as a mixer 115
Dean flow as a static mixer 115
Elastic turbulence 116
References 116

PART II BUILDING ON FUNDAMENTALS

Introduction 117
The different tools of the trade 117
Relationship between thermodynamics and chemical reaction engineering 118
Relationship between transport phenomena and chemical reaction engineering 118
Relationship between chemical reaction engineering and kinetics 118
Chemical reaction engineering as an experimental and theoretical science 118

4 Rates and equilibria: The thermodynamic and extrathermodynamic approaches **121**

Chapter objectives 121
Introduction 121
Basic thermodynamic relationships and properties 122
Basic relationships 122
Heats of reaction, formation, and combustion 122
Implications of liquid phase reactions 124
Free energy change and equilibrium constant 124
Standard free energy change and equilibrium constant 124

Equilibrium compositions in gas phase reactions 126
Accounting for condensed phase(s) 126
Complex equilibria 128
Simultaneous solution of equilibrium equations 128
Extension to a nonideal system 129
Minimization of free energy 130
Thermodynamics of reactions in solution 131
Partial molar properties 131
Medium and substituent effects on standard free energy change, equilibrium constant, and activity coefficient 132
General considerations 132
Solvent and solute operators 133
Comments 134
Extrathermodynamic approach 134
Basic principles 134
Group contributions or additivity principle 135
Extrathermodynamic relationships between rate and equilibrium parameters 136
Polanyi and Brønsted relations 136
Hammett relationship for dissociation constants 137
Extrathermodynamic approach to selectivity 138
Theoretical analysis 138
Thermodynamics of adsorption 139
Henry's law 141
Langmuir isotherm 141
Inhomogeneities expressed in terms of a site-energy distribution 142
Two-dimensional equations of state and their corresponding adsorption isotherms 143
Appendix 144
Derivation of chemical equilibrium relationships for simple reactions 144
Reactions in gas phase 146
Reactions in liquid phase 146
Explore yourself 147
References 147
Bibliography 148

Interlude III 149

Reactor design for thermodynamically limited reactions 149
Kinetics 149

Optimization of temperatures and pressures 150
References 152

5 Theory of chemical kinetics in bulk and on the surface 153

Chapter objectives 153
Chemical kinetics 153
Collision theory 154
Transition state theory 155
Proposing a kinetic model 158
Brief excursion for the classification of surface reaction mechanisms 159
Langmuir–Hinshelwood–Hougen–Watson models 159
Langmuir isotherm 159
Rate-determining step 160
Basic procedure 160
Eley–Rideal mechanism 165
Mars–van Krevelen mechanism 166
Michelis–Menten mechanism 168
Influence of surface nonideality 168
Paradox of heterogeneous kinetics 169
Microkinetic analysis 169
Postulate a mechanism 171
Determine the kinetic parameters 171
Simplify the mechanism 171
Compare the model predictions with the kinetic data 172
Explore yourself 174
References 175
Bibliography 175

6 Reactions with an interface: Mass and heat transfer effects. 177

Chapter objectives 177
Introduction 177
Diffusivity 178
Diffusivities in gases 178
Diffusivities in liquids 179
Effective diffusivity 179
Transport between phases 180
General remarks 180
Film theory 182

Penetration theory 183
Surface renewal theory 184
Characteristic times for diffusion, reaction, and mass transfer 185
Two-film theory of mass and heat transfer for fluid–fluid reactions in general 185
Mass transfer 185
Heat transfer 186
Mass transfer across interfaces: Fundamentals 187
Solid catalyzed fluid reactions 189
Overall scheme 189
Role of diffusion in pellets: Catalyst effectiveness 189
First-order isothermal reaction in a spherical catalyst 191
Weisz modulus: Practical useful quantity 196
Nonisothermal effectiveness factors 197
Multicomponent diffusion 199
Miscellaneous effects 199
Extension to complex reactions 200
Noncatalytic gas–solid reactions 200
Gas–liquid reactions in a slab 204
Two-film theory 205
Slow reactions 205
Instantaneous reactions 206
Effect of external mass and heat transfer 209
External effectiveness factor 209
Combined effects of internal and external diffusion 209
Relative roles of mass and heat transfer in internal and external diffusion 210
Gas phase reactants 210
Liquid phase reactants 211
Regimes of control 212
Explore yourself 213
References 214

7 Laboratory reactors: Collection and analysis of the data 217

Chapter objectives 217
Chemical reaction tests in a laboratory 217
A perspective on statistical experimental design 218
Batch laboratory reactors 220
Rate parameters from batch reactor data 221

From concentration data 221
From pressure data 223
Flow reactors for testing gas–solid catalytic reactions 225
Differential versus integral reactors 226
Eliminating or accounting for transport disguises 229
Eliminating the film mass transfer resistance 229
Eliminating the pore diffusion resistances 230
Eliminating axial dispersion effects 231
Koros–Nowak criterion 231
Catalyst dilution for temperature uniformity 231
Gradientless reactors 231
Transport disguises in perspective 231
Guidelines for eliminating or accounting for transport disguises 234
Analyzing the data 235
Modeling of solid catalyzed reactions 235
The overall scheme 235
LHHW models 236
Selection of the most plausible model 236
Influence of surface nonideality 239
Explore yourself 239
References 240

PART III BEYOND THE FUNDAMENTALS

Objectives 243
Introduction 244
The different tools of the trade 244
Process intensification 246
Microfluidics 247
Membrane reactors 248
Combo reactors 249
Homogeneous catalysis 249
Phase-transfer catalysis 249
References 249

8 Fixed-bed reactor design for solid catalyzed fluid-phase reactions 251

Chapter objectives 251
Introduction 251

Effect of catalyst packing in a tubular reactor 251
Fixed-bed reactor 252
Nonisothermal, nonadiabatic, and adiabatic reactors 254
Design methodologies for NINA-PBR 256
Quasi-continuum models 257
Cell model 257
Models based on the pseudo-homogeneous assumption 257
Homogeneous, pseudo-homogeneous, and heterogeneous models 257
1D pseudo-homogeneous nonisothermal, nonadiabatic flow 260
Reduction to isothermal operation 261
Momentum balance 261
The basic model: 2D pseudo-homogeneous nonisothermal, nonadiabatic with no axial diffusion 262
Extension to nonideal models with and without heterogeneity 265
Adiabatic reactor 266
The approach 266
A unique conversion–temperature relationship 266
Single-bed reactor 268
Multiple-bed reactor 270
A simple graphical procedure 273
Strategies for heat exchange 273
Choice between NINA-PBR and A-PBR 274
Some practical considerations 275
Backmixing or axial dispersion 275
Nonuniform catalyst distributions between tubes 275
Scale-up considerations 276
Alternative fixed-bed designs 276
Radial-flow reactors 276
Material, momentum, and energy balances 279
Material balance 279
Mass balance 279
Momentum balance 279
Some important observations 279
Catalytic wire-gauze reactors 280
Explore yourself 281
References 282
Bibliography 282

9 Fluidized-bed reactor design for solid catalyzed fluid-phase reactions 285

Chapter objectives 285
General comments 285
Fluidization: Some basics 286
Minimum fluidization velocity 286
Two-phase theory of fluidization 287
Geldart's classification 287
Classification of fluidized-bed reactors 288
Velocity limits of a bubbling bed 289
Fluid mechanical models of the bubbling bed 291
Complete modeling of the fluidized-bed reactor 291
Bubbling bed model of fluidized-bed reactors 292
Bubbling bed 292
Bubble rise velocity 293
Main features 293
Mass transfer between bubble and emulsion 294
Solids distribution 294
Estimation of bed properties 295
Heat transfer 295
Calculation of conversion 297
End region models 297
Dilute bed region 297
Grid or jet region 298
Practical considerations 299
Recommended scale-up procedure 300
Strategies to improve fluid-bed reactor performance 302
Packed fluidized-bed reactors 303
Reactor model for packed fluidized beds 303
Staging of catalyst 305
Extension to other regimes of fluidization types of reactors 306
Turbulent bed reactor 307
Fast fluidized-bed reactor 307
Transport (or pneumatic) reactor 308
Circulation systems 309
Deactivation control 310
Heat transfer controlled 312
Reactor choice for a deactivating catalyst 312
Basic equation 313
Fixed-bed reactor 314

Fluidized-bed reactor 315
Moving bed reactor 315
Some practical considerations 318
Slugging 318
Defluidization of bed: Sudden death 318
Gulf streaming 318
Effects of fines 318
Start-up 319
Fluidized-bed versus fixed-bed reactors 319
Explore yourself 320
References 321

10 Gas–solid noncatalytic reactions and reactors 325

Chapter objectives 325
Introduction 325
Modeling of gas–solid reactions 326
Shrinking core model 327
Volume reaction model 329
Zone models 331
The particle–pellet or grain models 332
Other models 334
Extensions to the basic models 334
Bulk-flow or volume-change effects 334
Effect of temperature change 335
Models that account for structural variations 336
Effect of reaction 336
Effect of sintering 338
A general model that can be reduced to specific ones 338
Gas–solid noncatalytic reactors 339
Fixed-bed reactors 340
Moving-bed reactors 343
Fluidized-bed reactors 345
References 345

11 Gas–liquid and liquid–liquid reactions and reactors 347

Chapter objectives 347
Introduction 347
Diffusion accompanied by an irreversible reaction of general order 350
Diffusion and reaction in series with no reaction in film: Regimes 1 and

2 (very slow and slow reactions), and regimes between 1 and 2 350
Regimes 1 and 2: Very slow and slow reactions 350
Regimes between 1 and 2 352
Diffusion and reaction in film, followed by negligible or finite reaction in the bulk: Regime 3 (fast reaction), and regime covering 1, 2, and 3 352
Reaction entirely in film 352
Reactions both in film and bulk (regimes 1–2–3) 353
Measurement of mass transfer coefficients 354
Microfluidic devices 354
Reactor design 355
A generalized form of equation for all regimes 356
Regime 1: Very slow reaction 356
Regime 2 and regime between 1 and 2: Diffusion in film without and with reaction in the bulk 356
Regime 3: Fast reaction 357
Regime between 2 and 3 357
Regime 4: Instantaneous reaction 357
Classification of gas–liquid contactors 358
Classification-1 (based on manner of phase contact) 358
Classification-2 (based on the manner of energy delivery) 359
Mass transfer coefficients and interfacial areas of some common contactors 359
Role of backmixing in different contactors 359
Reactor design for gas–liquid reactions 361
The overall strategy 361
Calculation of reactor volume 361
Case 1: Plug gas, plug liquid, and countercurrent steady state 363
Case 2: Same as case 1 but with cocurrent flow 363
Case 3: Plug gas, mixed liquid, and steady state 364
Case 4: Mixed gas, mixed liquid, and steady state 364
Case 5: Mixed gas, batch liquid, and unsteady state 364
Comments 365
Reactor choice 369
The criteria 369

Volume minimization criterion 369
General discussion 369
Limitations of volume minimization 371
Steps in volume minimization 371
Energy minimization criteria 372
Criterion 2(a): Homogeneous regime (regime 1) 372
Criterion 2(b): Heterogeneous regime (regimes 2–4) 374
Comparison of criteria 374
Liquid–liquid contactors 375
Classification of liquid–liquid reactors 375
Values of mass transfer coefficients and interfacial areas for different contactors 376
Calculation of reactor volume/reaction time 377
Stirred tank reactor: Some practical considerations 379
References 380

12 Multiphase reactions and reactors 383

Chapter objectives 383
Introduction 383
Design of three-phase catalytic reactors 383
The approach 383
Semibatch reactors: Design equations for (1,0)- and (1,1)-order reactions 384
Continuous reactors 385
Types of three-phase reactors 387
Mechanically agitated slurry reactors 389
Mass transfer 389
Minimum speed for complete suspension 390
Gas holdup 390
Controlling regimes in an MASR 390
Bubble column slurry reactors 390
Regimes of flow 391
Mass transfer 391
Minimum velocity for complete solids suspension 392
Gas holdup 392
Loop slurry reactors 396
Types of loop reactors 396
Mass transfer 397
Trickle bed reactors (TBRs) 397
Regimes of flow 397
Mass transfer 398
Controlling regimes in TBRs 398

Collection and interpretation of laboratory data for three-phase catalytic reactions 398
Experimental methods 398
Effect of temperature 398
Interpretation of data 399
Three-phase noncatalytic reactions 401
Solid slightly soluble 402
Negligible dissolution of solid in the gas–liquid film 402
Significant dissolution of solid in the gas–liquid film 403
Solid insoluble 403
References 408
Bibliography 409

13 Membrane-assisted reactor engineering. 411
Introduction 411
General considerations 411
Major types of membrane reactors 411
Modeling of membrane reactors 414
Packed-bed inert selective membrane reactor with packed catalyst (IMR-P) 414
Model equations 415
Extension to consecutive reactions 420
Fluidized-bed inert selective membrane reactor (IMR-F) 420
Catalytic selective membrane reactor (CMR-E) 421
Model equations 422
Main features of the CMR-E 422
Packed-bed catalytic selective membrane reactor (CMR-P) 423
Catalytic nonselective membrane reactor (CNMR-E) 424
Catalytic nonselective hollow membrane reactor for multiphase reactions (CNHMR-MR) 424
Immobilized enzyme membrane reactor 425
Operational features 425
Combining exothermic and endothermic reactions 425
Controlled addition of one of the reactants in a bimolecular reaction using an IMR-P 426
Effect of tube and shell side flow conditions 428

Comparison of reactors 428
Effect (1) 429
Effect (2) 429
Combined effect 429
Examples of the use of membrane reactors in organic technology/synthesis 429
Small- and medium-volume chemicals 430
Vitamin K 430
Linalool (a fragrance) 431
Membrane reactors for economic processes (including energy integration) 431
References 432

14 Combo reactors: Distillation column reactors 435

Distillation column reactor 436
Enhancing role of distillation: Basic principle 436
Batch reactor with continuous removal of product 436
Case 1: Accumulation of S 437
Case 2: S is not completely vaporized 438
Packed DCR 439
Overall effectiveness factor in a packed DCR 441
Residue curve map (RCM) 442
Design methodology 443
Generating residual curve maps 445
Distillation–reaction 447
Dissociation–extractive distillation 448
Basic principle 448
Theory 448
References 451

15 Homogeneous catalysis 453

Introduction 453
General 453
Formalisms in transition metal catalysis 453
Uniqueness of transition metals 453
Oxidation state of a metal 455
Coordinative unsaturation, coordination number, and coordination geometry 457
Ligands and their role in transition metal catalysis 457
Electron rules ("electron bookkeeping") 459
18-electron rule 459
16–18-electron rule 460

Operational scheme of homogeneous catalysis 460
Basic reactions of homogeneous catalysis 461
Reactions of ligands (mainly replacement) 461
Elementary reactions (or activation steps) 462
Coordination reactions 462
Addition reactions 462
Main reactions 463
Insertion 463
Elimination 464
Main features of transition metal catalysis in organic synthesis: A summary 464
A typical class of industrial reactions: Hydrogenation 465
Hydrogenation by Wilkinson's catalyst 465
Wilkinson's catalyst 465
The catalytic cycle 466
Kinetics and modeling 466
A general hydrogenation model 467
General kinetic analysis 468
Intrinsic kinetics 468
Multistep control 468
Role of diffusion 469
Complex kinetics—Main issue 469
Reactions involving one gas and one liquid 470
Regimes 1 and 2 470
Regime between 1 and 2 (reaction in bulk) 470
Regime 3 (reaction in film) 472
Two gases and a liquid 473
References 473

16 Phase-transfer catalysis. 475

Introduction 475
What is PTC? 475
Fundamentals of PTC 476
Classification of PTC systems 476
Phase-transfer catalysts 477
Mechanism of PTC 478
Liquid–liquid PTC 478
Solid–liquid PTC 480
Solid-supported PTC or triphase catalysis (TPC) 481
Modeling of PTC reactions 482
LLPTC models 483
SLPTC models 484

Interpretation of the role of diffusion: A cautionary note 486
Supported PTC (TPC) 487
Kinetic mechanism of TPC systems 488
Methodology for modeling solid-supported PTC reactions 489
Supported PTC with LHHW kinetics 490
"Cascade engineered" PTC process 494
References 495

17 Forefront of the chemical reaction engineering field 497
Objective 497
Introduction 497
Resource economy 497
Carbon and hydrogen 498
Bio-renewables 498
Energy economy 499
Heat integration in microreactors 500
Sonochemical reaction triggering 500
Photochemical or photocatalytic systems 500
Electrochemical techniques 500
Microwaves 501
Chemical reaction engineer in the twenty-first century 501
In Closing 502

Subject Index 503

Author Index. 515

Preface

> When one sees Eternity in things that pass away and infinity in finite things, then one has pure knowledge.
>
> But if one merely sees the diversity of the things, with their divisions and limitations, then one has impure knowledge.
>
> And if one selfishly sees a thing as if it were everything, independent of the ONE and the many, then one is in the darkness of ignorance.
>
> **Bhagavad Gita XVIII, 20–22**

Academically oriented books can broadly be placed under three categories: undergraduate texts, advanced books in broad areas/disciplines, and monographs in specific areas. Chemical engineering is no exception. In this important area, more specifically its most expansive component, chemical reaction engineering (CRE), what is missing is a "textbook" for graduate-level teaching to twenty-first-century graduate students. The question arises: Should there be a text at all for any graduate course? One can argue both for and against following a single textbook for a graduate course. So we would like, at the outset, to state our purpose in writing this book. Our objective is to create a book that spans the extremes of an undergraduate text and a highly advanced book that does not even remotely revisit the undergraduate material. Stated differently, it may be regarded as a ramp connecting the two levels, increasingly encompassing the higher level but stopping appropriately short. It should prepare graduate students for the needs of twenty-first-century chemical engineering research and technology.

To accomplish this objective, several factors had to be seriously considered: language and format; striking a balance between crossing the t's and dotting the i's, on the one hand, and letting the student do most of it, on the other hand; giving problems as home assignments with precise solutions along with a solution manual (assumed to be miraculously out of the students' reach); and including some recent additions to the well-traversed repertoire of CRE as well as a few nontraditional areas as fresh infusions into this highly absorptive area. We shall see how a consideration of these factors resulted in the structure and the format in which this book has finally emerged.

Undergraduate texts are for the complete beginner. They must drive home the basic principles unambiguously with no room for interpretation or mistaken understanding. Every detail must be explained and computational methods clearly illustrated. Only the most important books and publications need to be referred, with emphasis on language and illustration that would ensure utter clarity of the foundational building blocks of the area. Problems worked out in the text or assigned as homework should mostly, if not always, have only a single solution. We make a point of this because, as one moves up the complexity scale, alternative solutions depending on different physical pictures and use of different correlations increasingly come into play. Books to address this situation belong to a different category and must conform to the level and needs of students cutting their first teeth at the graduate level.

Undergraduate texts in CRE are strong in homogeneous reactions and gas–solid reactions where the solid is a catalyst, with rapidly decreasing attention to gas–liquid, gas–liquid–solid (slurry), and gas–solid reactions where the solid is a reactant. Varying degrees of attention are given to the question of nonideality and its role in reactor performance. Experimental methods for collecting basic kinetic data and simple statistical methods for analyzing them also form part of the undergraduate curriculum. Rigorous methods of treating complex reactions are avoided. Also, due to the limitations in context, very little attention is given to emerging methods and newer concepts.

So, as students get their first taste of a graduate course in CRE, they are normally well grounded in the elements of it, *thanks to the excellent books that are available.* A survey of the literature shows that there are many good books on certain advanced aspects of CRE as well, which have been catering to the needs of a graduate program in CRE. We would like to mention, in particular, the 1985 book *Chemical Reactor Analysis* by Gilbert Froment and Kenneth Bischoff and its 3rd edition in 2010, *Chemical Reactor Analysis and Design* with Juray De Wilde as the third author. This new edition addresses computational fluid dynamics (CFD) and a few other related tools of CRE. *Chemical Reactor Analysis and Design Fundamentals* by J.B. Rawlings and J.G. Ekerdt entered the field with well-presented emphasis on the use of computers in the design and analysis of chemical reactors but still essentially at the undergraduate level. During the many years that we have taught CRE at both the undergraduate and graduate levels in the United States, India, and Turkey, we noticed that no book particularly suited as a first year graduate text in CRE is available. The requirements of such a book, which may loosely be regarded as intermediate between an undergraduate text and an advanced book that takes all undergraduate material for granted, may be stated as follows:

- There should be a distinct connectivity with what the students learned as undergraduate juniors/seniors.
- The format of the book should be different from that of a typical undergraduate text, in that spoonfeeding must be minimized, and full solutions to a problem should not always be given—leaving it open in some cases for alternative solutions to be suggested by the students.
- Attenuated explanations should be adequate in many, if not all, cases, leaving it to the students to supply the rest.
- Derivations for important equations should be fully given; for other associated equations, only the final solutions need be presented, with all the equations for a given class of situations, such as for gas–liquid reactions under different conditions, consolidated in tabular form. Students should be encouraged to derive those equations (as home assignments).
- Chapter-end exercises should span the whole range of Bloom's taxonomy of educational objectives: knowledge, understanding, application, analysis, synthesis, evaluation, and even valuation. These problems will give the student a feeling for the stage of his/her learning. As desirable in a graduate level text, we emphasize the higher-level skills associated with analysis, synthesis, and evaluation in a limited number of open-ended chapter-end "Explore yourself" problems. We decided to provide up-to-date problems and self-paced learning modules at the website of the book at www.metu.edu.tr/~uner in a continuously updated fashion.
- Students are seldom exposed to analyses/equations for cases not described in the book. This situation should be specifically addressed by including representative equations as exercises in derivation under on-line learning modules.

The format of the book has been designed to accommodate all these features. The material is divided into three parts, comprising a total of 17 chapters.

Part I: Fundamentals Revisited is devoted to a recapitulation of the salient features of the undergraduate course in CRE. The material is recast, wherever needed, in a format that would easily dovetail into the more advanced chapters to follow. Still, we introduce the concepts of mixing, unsteady-state operations, multiple steady states, and complex reactions in this part as they are fundamental to the design of reactors in a world driven by emphasis on high selectivity, raw material economy, and green engineering.

Part II: Building on Fundamentals is dedicated, if we may, to skill building, especially in the area of catalysis and catalytic reactions. This part covers chemical thermodynamics with special emphasis on the thermodynamics of adsorption and complex reactions; a brief section on the fundamentals of chemical kinetics, with special emphasis on microkinetic analysis, as we believe that good literacy in CRE requires a clear understanding of mechanisms postulated and tested by microkinetic analysis; and heat and mass transfer effects in catalysis from a classical point of view. Finally, we devote a full chapter to giving graduate students the tools of the trade for making accurate kinetic measurements and analyzing the data obtained.

Part III: Beyond the Fundamentals is concerned, as the name implies, with material not commonly covered in present-day textbooks. This part begins with the treatment of solid catalyzed fluid-phase reactions and proceeds to reactions involving at least one liquid phase as separate chapters. It is addressed to the advanced learner trying to find out aspects of reactors involving more than one phase and intending to go in the direction of innovation in terms of process intensification and sustainable engineering. The fundamental background available in the literature is succinctly summarized to equip the advanced learner with the concepts and the wherewithal to deal with a variety of situations, including such little explored territories as the cell as a chemical reactor.

While writing the book, we deliberated at length on whether to focus on a single computational tool, or opt for more. The final decision was to keep the learner, and also the instructor, free to choose any computational tool of their preference and not to limit them to one of our choice. This might tend to somewhat limit the span of some of our examples, but we have remained loyal to the central philosophy of the book: to point out the vast territory and prod the learner to learn more—and explore.

We attempt in this book a unique approach toward examples. Instead of giving many brief examples embedded in texts within chapters, we give elaborate accounts of technologies and designs as INTERLUDES between chapters. Some of the interludes are laden with plenty of questions for the learner to answer—and learn. Some interludes are fully solved design problems. This, we believe, will give the reader and the instructor great flexibility between covering the background on a subject and covering an example in sufficient detail.

As with any other book, this too comes, no doubt, with its own share of merits and demerits; we worked hard to have more of the former and less of the latter, but still the final judgment must rest with the user. We would be glad to have any errors brought to our attention. And finally, the broad objective of this book has been to provide the student, particularly the advanced learner, with a new perspective of *Chemical Reaction Engineering: Beyond the Fundamentals*—coupled with the incentive and the wherewithal to transition from solving close-ended problems to exploring open-ended ones. The extent to which this has happened would be a fitting measure of our success.

Acknowledgments

This book has its origin in Doraiswamy's 2001 *Organic Synthesis Engineering* published by Oxford University Press (OUP). When it was decided to completely revise and shorten the book with the objective of providing a textbook for a first graduate course in CRE, and OUP showed no interest in this effort, we asked them for permission to use it as the basis for writing the proposed book for publication by Taylor & Francis. We are thankful to OUP, in particular to senior editor Jeremy Lewis, with whom Doraiswamy was dealing with in this matter, for permission to do so. Numerous examples were covered from Doraiswamy's earlier publications, especially from Doraiswamy and Sharma's "Heterogeneous Reactions: Analysis, Examples and Reactor Design," and the massive chapter by Joshi and Doraiswamy on "Chemical Reaction Engineering" in *Albright's Encyclopaedia of Chemical Engineering.* The incredible support from the team at Taylor & Francis, in particular by our editor Barbara Glunn, is deeply appreciated.

L.K. Doraiswamy records the following: I am particularly thankful to Surya Mallapragada, chair of the Department of Chemical and Biological Engineering, Iowa State University, for her continued support. I thank Linda Edson, a staff member of CBE, for her help in many ways, Vivek Renade, B.D. Kulkarni, and Imran Rehman for fruitful discussions and illustrative examples. I also join Deniz Uner in thanking her many students for their enthusiastic help with many of the illustrative problems that appear in the book, particularly to Ibrahim Bayar for redrawing the figures in the manner required by the publisher. I was plagued by stretches of ill health during the entire period of this writing, but without the affectionate care of my daughter Sandhya and son Deepak, this book would not have been possible. I cannot adequately thank them and their spouses, Sankar and Kelly, for all they have done.

Deniz Uner records the following: Many thanks are due to Ibrahim Bayar for his patient handling of the figures, Necip Berker Uner for his illustrative examples and careful editing of the equations, and Cihan Ates for his patient proofreading and corrections. Special thanks are due to the graduate students enrolled in my research group between 2010 and 2012, particularly Mustafa Yasin Aslan, Hale Ay, Atalay Çalışan, and Arzu Kanca for their contributions at various stages of writing. The active participation of the students enrolled in ChE 510 Advanced Chemical Reaction Engineering during the Fall semester of 2011 and the Fall semester of 2012 at METU (Middle East Technical University) in the implementation of a course with the draft document shaped this book in the final structure. Our editor, Barbara Glunn, is deeply appreciated for her patience throughout and for her personal support, especially during the impeded progress after the unfortunate passing of L.K. Doraiswamy. The final text benefited immensely from patient and meticulous editing in the hands of Deepak Doraiswamy. My gratitude extends further to Deepak Doraiswamy and Sandhya Raghavan for their friendship, support, and encouragement along the way. Finally, I would like to express my love and deep appreciation to my children Imre, Sedef, Esen, and Firat. They supported me in many ways that cannot be expressed in words, while they transformed to independent adults as this book was being written.

Notations

$[A]$	Concentration (moles per volume) of species A
A	Reactant
a	Area per unit volume, 1/m
A_h	Heat transfer surface, m^2
A_m	Area of membrane, m^2
A_p	Area of particle, m^2
Ar	Archimedes number
Bd	Bond number
Bi	Biot number
C_p	Heat capacity
C_{pA}	Heat capacity of reactant, kcal/mol K
Da	Damköhler number
D_{AB}	Diffusion coefficient of a binary system
d_e	Equivalent reactor diameter
D_j	Diffusivity of species j, m^2/s or mol/m atm s
d_M	Molecular diameter, m
Dn	Dean number
d_t	Tube diameter
E	Total energy
F_i	Molar flow rate of i, mol/s
F_j	Flow rate of species j, mol/s
Fo	Fourier number
Fr	Froude number
F_j^T, F_j^S	Flow rates of species j in inner tube and outer shell, respectively, mol/s

Ga	Galileo number
G_j^T, G_j^S	Flow rates per unit area of species j in inner tube and outer shell, respectively, mol/m^2 s
H	Enthalpy
H	Product (hydrogen)
$\Delta \underline{H}$	Change in molar enthalpy
I	Inert (sweep) gas, usually argon
K	Reaction equilibrium constant
k	General symbol for rate constant (m^3/mol)$^{n-1}$(1/s)
k_B	Boltzmann constant
KE	Kinetic energy
Kn	Knudsen number
k^o	Arrhenius frequency factor, same units as the rate constant
k_{per}	Percolation rate constant, appropriate units (usually 1/s)
L	Reactor length, m
m	Mass; order of reaction
M_{H}	Hatta modulus
M_j	Molecular weight of species j
n	Order of reaction
N_i	Number of moles of species i
N_0	Avogadro number
P	Total pressure
PE	Potential energy
Pe	Peclet number
P^S, P^T	Pressures on the shell and tube sides
p_j^S, p_j^T	Shell and tube side partial pressures of species j
q	Heat transferred
Q_A	Volumetric flow rate of A
Q_j	Volumetric flow rate of species j, m^3/s
R	Reactant; radius
R_1	Inner radius of membrane tube, m
R_2	Outer radius of membrane tube, m

R_3	Inner radius of shell, m
Re	Reynolds number
R_g	Gas constant, kcal/mol k
r_i	Rate of reaction of species i
r_j	Rate of reaction of species j, mol/m^{3}s
S	Product
Sc	Schmidt number
Sh	Sherwood number
S_R	Selectivity of R (moles of R formed/moles of A converted)
T	Temperature, °C or K
t	Time
$\bar{t}$	Residence time (volume/volume per unit time) (V/Q_0), s
t_m	Thickness of membrane wall, m
T_w	Wall temperature, K
U	Internal energy; overall heat transfer coefficient, kcal/m^2 K s
u	Velocity, m/s
V	Reactor volume, m^3
v_i	Stoichiometric coefficient of species i
W	Work
X_A	Conversion of species A
X_e^S, X_e^T	Equilibrium conversion on the shell and tube sides
y_A	Mole fraction of species A
Y_R	Yield of R (moles of R formed/moles of A fed)

Greek

α	Ratio of sweep gas to feed gas flow rates at inlet (Chapter 13)
α	Coefficient of thermal expansion
β	Ratio of ethane to oxygen
γ_{bu}	Efficiency of bulk liquid utilization
δ	Molar change parameter; constrictivity
ε	Effectiveness factor

ε_A	Volume change parameter of species A
η	Enhancement factor
θ	Surface coverage of adsorbed species
λ	Thermal conductivity
λ	Mean free path, m (Chapter 13)
ℓ	Length parameter, m
μ	Viscosity
$\Delta\nu$	Volumetric difference between products and reactants
ν_i	Stoichiometric coefficient of species i
ν_j	Stoichiometric coefficient of species j
ξ	Extent of reaction
ρ	Density
$\sigma_{A,B}$	Mean molecular diameter of A and B
τ	Tortuosity
ϕ	Thiele modulus
ϕ_p	Porosity
ψ_i	Feed stoichiometry, $[i]/[A]_0$, $i \neq A$
Π_j	Specific rate of permeation of species j, mol/m^2 s
Ω	Activity
Γ	Heat transfer rate, $UA_m/C_{pA}F_{A0}$

Subscripts/superscripts

Subscripts

0	Initial/input, initial/entrance conditions (Chapter 13)
1	Output/exit
i	For/belonging to species i
rxn	Reaction
s	Shaft
SS	Steady state
t	Total
w	Wall

+	Generation
–	Removal

Superscripts

S	Shell side
T	Tube side

Overview

Chemical reactions all around us

Chemical reaction engineering (CRE) embraces a wide variety of chemical reactions involving all the three states of matter: gas, liquid, and solid. Till recently, it was confined to reactor sizes encountered in typical inorganic and organic conversions, such as those in the fertilizer, petrochemical, and organic chemicals industries. In Table O.1, we summarize the classes of reactions with respect to the states of aggregation of the contacting phases and an example. They are considered in varying degrees of detail, appropriate to their importance and the state of development, in the subsequent chapters of the book.

Enzymes are increasingly being used as catalysts due to their extraordinarily high selectivities and amenability to low-temperature operation. They are used both in the liquid phase and anchored to solid substrates. Some flowers and seeds act as nanoscale natural reactors, and emulating them in the laboratory is one of the challenges faced by CRE today. Another emerging class of reactors known as microfluidic reactors and the whole concept of miniaturization will almost certainly come to the fore as future generation realities of the chemical industry.

Table O.1 Chemical Reactions Classified According to the Contacting Phases and an Example from Industry

Solid catalyzed vapor-phase reactions	Ammonia synthesis Xylene isomerization
Gas–liquid reactions	Absorption of gases to produce acids such as nitric acid or sulfuric acid
Gas–solid reactions	Smelting of ores
Solid–liquid reactions	Hydration of lime
Liquid–liquid reactions	Hydrolysis of oil, production of biodiesel
Solid–liquid–gas (slurry) reactions	Hydrogenation of glucose to sorbitol
Solid–solid reactions	Catalytic oxidation of diesel soot
Solid catalyzed gas–solid reactions	Preparation of methylchlorosilanes

Example O.1: A typical conventional reaction: Synthesis of nitric acid

As a classical example, we will look at the industrial synthesis of nitric acid, a compound whose production revolutionized the fertilizer industry, and the world population growth is strongly correlated with the amount of nitric acid (or ammonia) synthesized.

The primary ingredients of nitric acid are nitrogen, oxygen, and hydrogen, but it is very difficult to synthesize nitric acid starting from the elementary ingredients due to the energy barriers associated with them. The primary reaction in nitric acid synthesis is the oxidation of nitrogen given by

$$N_2 + O_2 \rightarrow 2NO, \quad \Delta G = 173.2 \text{ kJ/mol } N_2 \qquad \text{(R1)}$$

Since highly positive Gibbs free energy changes make the reaction highly unfavorable, this reaction proceeds spontaneously only at temperatures in excess of 3000 K. On the other hand, if the oxidation is carried out from a reduced nitrogen compound, such as ammonia, the Gibbs free energy changes become negative, and these reactions become favorable even at low temperatures as shown below:

$$N_2 + 3H_2 \rightarrow 2NH_3, \quad \Delta G = -33 \text{ kJ/mol } N_2 \qquad \text{(R2)}$$

$$2NH_3 + \tfrac{5}{2}O_2 \rightarrow 2NO + 3H_2O, \quad \Delta G = -479 \text{ kJ/mol } N_2 \qquad \text{(R3)}$$

the net reaction being

$$N_2 + 3H_2 + \tfrac{5}{2}O_2 \rightarrow 2NO + 3H_2O, \quad \Delta G = -512 \text{ kJ/mol } N_2 \qquad \text{(R4)}$$

Industrial fertilizer synthesis starts from ammonia synthesis, and ammonia is then easily oxidized in a separate reactor to nitric oxide over PtRh wire gauze catalyst. Formation of nitric acid requires further oxidation of nitric oxide to nitrogen dioxide (NO_2) and absorption of the nitrogen dioxide in water. Overall, three different chemical process plants are used for the synthesis of nitric acid. The ammonia synthesis reaction takes place in a high-temperature, high-pressure reactor that requires recycling of products due to the thermodynamic limitations of chemical conversion. The ammonia oxidation reaction is very fast and takes place over a very small reactor length. Finally, nitric acid synthesis takes place in absorption columns.

The ammonia synthesis reaction is rightfully called "The World's Greatest Fix" (Lehigh, 2004), creating a major breakthrough in agriculture. Developments in the reaction have involved three Nobel Prize winners: the first, Fritz Haber in 1918 for finding the catalyst and the process parameters; the second, Carl Bosch in 1931 for developing the high-pressure process; and the third, Gerhard Ertl in 2007 for developing the surface reaction

Figure O.1 (a) Fritz Haber, (b) Carl Bosch, and (c) Gerhard Ertl.

mechanism (Figure O.1). Despite all the development, the industrial process still lags behind the nitrogen-fixation bacteria, which carry out the same conversion at room temperature and more importantly at atmospheric pressure.

One of the most innovative designs of the ammonia synthesis reactor is from the Haldor-Topsøe company. We give in Figure O.2 the flow cross-sectional view of one of the earlier designs of the so-called radial flow reactor (US Patent 3 372 988, filed September 18, 1964) and ask you to figure out the answers to the following questions:

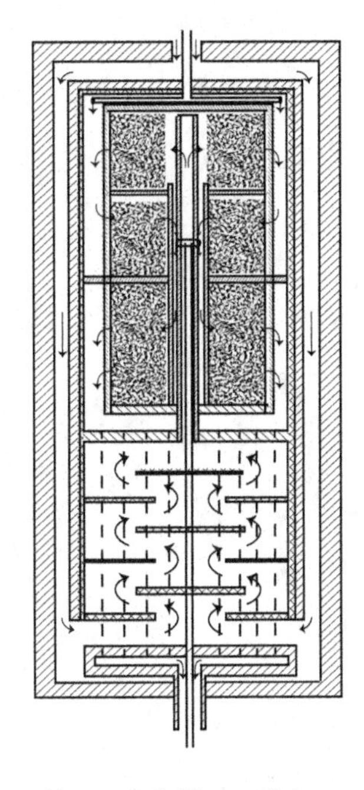

Figure O.2 The radial flow reactor of Topsøe. (Hansen, H.J., US Patent 3 372 988, 1968.)

> Questions to ponder: What are the salient radial flow features of this reactor? Where does the gas enter? Where does the gas leave? What are the advantages of the given design over the straight flow packed-bed reactor with intermediate cooling?

As an exercise, compare the design shown in Figure O.2 with more contemporary designs. Comment on the major technology developments in the flow design of the two systems over a time span of 40 years.

Example O.2: Chemical conversions on nanoscale: The multi-reactor sunflower seed

This gives us an idea of the immense variety of chemical reactors in which reactions are combined with selective transport processes. Selectivities offered by enzyme catalysts are yet to be matched by selectivities in the inorganic catalysts designed and produced in the laboratory. The secret in enzyme selectivities lies not only in the structure and site of the catalysts in the enzymes, but also in the rates at which the reactants and products are transported to/from the catalytic sites. The fundamentals of such processes will perhaps remain a mystery for a long time but it will be instructive for us to go through such a reaction system.

An outstanding example is the sunflower seed. The industrial ammonia synthesis process mentioned under Example O.1 is still technologically too primitive in comparison to the millimeter-sized sunflower seed. This seed, after being planted and exposed to sufficient nutrients and solar energy, establishes an oil factory, a cellulose factory, and a paint factory within six months. It has several process control units, but the most explicit one makes the sunflower follow the sun. It is the chemist/chemical engineer's dream to emulate this unique feature of the sunflower and be able to synthesize on demand chemicals, as much as, when and where needed. As we lamented earlier, it is still a dream.

Photosynthesis is one of the key reactions that constitute the present-day holy grail of energy and chemical conversion research.

In the example below, we will scratch the surface of the chemical conversion legend of the photosynthetic plant cell.

Example O.3: Looking at the photosynthetic plant cell with the eyes of a chemical reaction engineer

The photosynthetic plant cell, shown in Figure O.3, has several compartments for various functions involved in the chemical conversion of CO_2 and H_2O to glucose and related chemicals. We will focus our attention on the so-called granum to begin with. The granum is the compartment where the water oxidation reaction takes place. The CO_2-related reactions are limited to the stroma side of the chloroplast, while the water oxidation reactions take place in the granum, at a one-reaction-per-granum rate. The isolation of CO_2 and H_2O reactions helps increase the selectivity of the reaction as we will presently see.

The outer shell of the granum, called the thylakoid membrane, creates a concentration gradient and is separated from the stroma as far as the water oxidation reaction is concerned. The photocatalytic splitting of water takes place in two steps, commensurate with the energy content of the visible light photons and the chemical potentials required to produce an oxygen molecule and a hydrogen atom. The oxygen molecule is synthesized by photosystem II. Protons and electrons released during the reaction are transported to photosystem I for further reduction. As a result, a four-proton, four-electron reaction takes place. It is important to keep the oxygen molecule separated from the rest of the chemical conversion compartments. This is because the Gibbs free energies dictate that oxidation is the preferred route of chemical conversion. Once formed, the oxygen molecule is released from the cell through the semipermeable membranes. Oxygen, as much as it is needed for the sustenance of life, is a harmful by-product of the reaction as far as the photosynthesis is concerned. Therefore, the formation reaction is isolated in the cell and is discarded from the

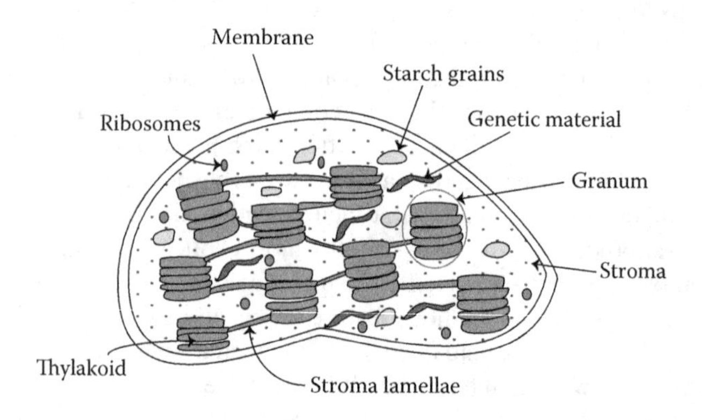

Figure O.3 Schematic of the chloroplast photosynthetic plant cell.

cell upon formation. The electrochemical potential gradients assist the easy removal of oxygen from the plant cells.

The electrochemical potential gradients in the plant cells serve more functions than just the isolation of the oxygen molecule. The transport of the protons and electrons formed during the *S* cycle, the academic name of the four-proton, four-electron oxygen evolution reaction, across the membrane requires chemical intermediates that are responsible for transporting the redox intermediates across membranes without disturbing the electrochemical gradients. There are two energy carrier molecules across the pH-polarized membranes: one is NADP–NADPH (nicotinamide adenine dinucleotide phosphate) and the other is ADP–ATP (adenosine diphosphate and adenosine triphosphate). These pairs simply serve for the transport of protons and electrons within and through the cell walls.

For a chemical reaction engineer, the fundamental question in this problem is about the relative rates of transport and of chemical conversions. The reaction system is one that we call a complex reaction system comprised of parallel and series reaction pathways. A change in the reaction direction occurs through these parallel and series reactions, and this is evident from our mere observation of the natural processes: photosynthetic plants converting to respiration in the absence of light. Thus, once the rates of reaction and transport are at a critical balance, the photosynthesis proceeds. If you do not water the plants, the plants may maintain life for a while, but they stop growing. In an abundance of CO_2, the rate of plant growth increases. While we may be perceived as making such statements rather lightly, it should be noted that each of these conclusions required carefully designed, controlled experiments. Thus, it is imperative that we understand the rates of mass transfer and of chemical conversions, and sometimes it is also necessary that we supply the reaction system with intermediate steps that assist the reaction progress with milder activation barriers toward the desired product.

Example O.4: Chemical conversions in microscale via microfluidic reactors

Process intensification is a very important concept that entered our vocabulary for improved raw material and energy economy. When the design constraints of chemical processes started to involve penalties for pollution, for excess use of energy, and very recently for CO_2 footprints, they complicated the optimization processes and changed our approach to design significantly. At such a complicated level of optimization, the "art" component of design seems as important as the science component—never mind the increased cost. It is tantamount to judging mathematical equations for their "beauty" as a measure of their accuracy when all else fails. Thus, we are moving at an unprecedented pace toward designing efficient reactors with high selectivity, incorporating both the "science" and the "art" parts in the design portfolio. Part of the intentions of this book is to give the reader the "art" of chemical reactor design, especially in the parts where we discuss process intensification where the emerging technologies are difficult to judge based on science only. Among the many different technologies developed so far for process intensification, an example from microfluidics is selected.

In microfluidic reactors, the high-pressure, high-temperature requirements of many industrial reactions are combined with the need to carry out the reactions on a small scale to obtain enhanced selectivities. The advances in microscale processing have made it possible to design and construct systems with micron-sized diameters.

Although microreactors are still in their infancy, the controlled transport versus kinetic processes via controlling bulk and surface forces make them a very attractive choice as selectivity enhancement tools for low-throughput processes. Mixing is a key parameter in chemical conversions. Sometimes it is necessary to mix the reacting streams thoroughly. In some other situations, such as explored by us in photosynthesis, mixing is detrimental to high selectivities. Understanding the role of mixing in chemical conversions and exploiting it to achieve higher conversions and selectivities as such constitute an important problem of today's CRE. Experimental as well as computational tools enabled the design of elaborate mixing schemes to improve the selectivities in fine chemical syntheses.

The control of mixing in microscopic scale was both very important and a challenge until the development of microreactor technology, owing its existence largely to developments in microelectronics-related manufacturing processes. Microreactor technologies, albeit still in their infancy, made it possible to control mixing almost at the molecular level. The same technology also made it possible to control mixing by creating uniform bubbles in more than one phase (Figure O.4).

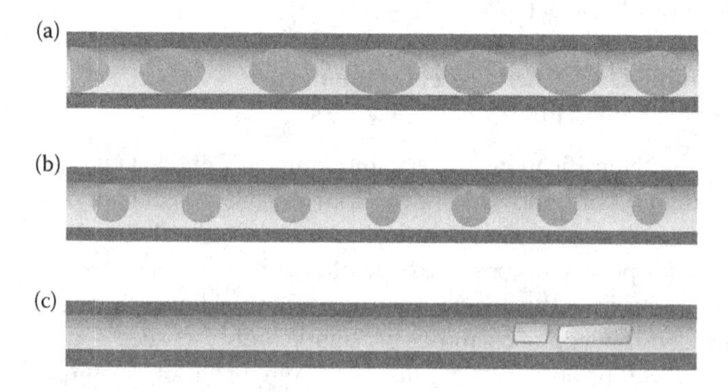

Figure O.4 Observed flow regimes in the capillary microreactor (Y-junction ID = 1 mm, capillary ID = 1 mm). (a) Slug flow, (b) drop flow, and (c) deformed interface flow. (Adapted from Kashid, M.N. and Agar, D.W., *Chem. Eng. J.* **131**, 1, 2007.)

What is ahead?

The introduction given above is just a brief glimpse of the endless horizons of CRE. The span of the topic starts from molecular conversions and ends in mega million tons scales of manufacture. The conscientious engineer of the twenty-first century is required to accomplish the needed chemical conversion from the most basic raw materials in the smallest reactor requiring the smallest amount of energy. The ability to do so will transform our economies in the same way as the petroleum refining and fertilizer manufacturing technologies developed by the chemical engineers of the late nineteenth and early twentieth centuries did. The wisdom required for sustainable development and sustainable manufacture has its roots in the knowledge basis of CRE comprising the following fundamental components:

1. *A strong foundation of chemistry*: Without understanding the chemical interactions and conversions, it is not possible to scale up the processes, replace harmful ingredients with benign ones, and apply novel reaction triggering mechanisms to already existing manufacturing methods. We limit our coverage of the chemistry here to the kinetics and, to a certain extent, thermodynamics, but urge the reader to explore the immense knowledge base that the chemical fields offer us.
2. *A strong foundation of transport phenomena*: Transport phenomena in the toolbox of a chemical engineer provide a unique set of capabilities for understanding the chemical conversions and tuning them to requirements. In most situations, we consider transport phenomena as a disguise limiting the kinetics. But resistances due to transport in complex reaction schemes become a very useful aid in tuning selectivity toward desired products. Tuning of these resistances is becoming easier with the development of new and novel materials that can be tailored at the nano or molecular scale.
3. *A strong foundation of mathematics*: A sound mathematical basis is very important for the successful application of CRE in the design of the most efficient reactors using the cheapest raw materials. The mathematical foundations are not limited to the ability to solve algebraic and differential equations. The ability to use statistics to extract workable equations including rate expressions needed in the design and the ability to use computers to solve complex mathematics are all included in the reaction engineers' expanding repertoire of capabilities. CFD is finding a strong niche in CRE where mixing and turbulence play a very vital role in conversions and selectivity.
4. *A sound basis of engineering*: The engineering foundations required in CRE should not exclude the social responsibility and ethics of the engineer. The engineer should be able to foresee beyond-the-immediate impact of expected/predicted

outcomes. The cost-optimization-based engineering era "when environment had not been discovered either by the industry or by the public at large" (Cropley, 2004) was based on different constraints. With increasing social responsibilities of the chemical industry, new technologies should take the environmentally benign solutions carefully into account, nonfossil-based fuels should provide the energy input, and process intensification should be a must.

5. *Seamless CRE*: These four pillars, listed above, provide the fundamental basis for the three parts of the text that follow. In closing this introductory chapter, we attempt to illustrate this emerging internal integration of the CRE domain by considering a particularly relevant class of reactions: fluid-phase reactions catalyzed by solids.

   A typical catalytic reaction involves:

   i. Reaction on the catalyst surface: *The surface field problem*
      - Understanding the chemistry of the reaction
      - Development of the most selective and robust catalyst
      - Analysis of the interplay between physics, chemistry, and mathematics on the surface of the catalyst, which in a single term we might call the chemical physics of catalysis.

   ii. Interaction between happenings at the surface and in the interior of the catalyst, and the role of the fluid microenvironment immediately outside the catalyst—in other words, the combined effect of reaction and transport phenomena within the catalyst and its microenvironment: *The internal field problem.*

   iii. Conveyance of events inside the internal field of the catalyst to the flowing fluid in the reactor—in other words, the interparticle events in the reactor outside the catalyst: *The external field problem.*

The three fields mentioned above are schematically illustrated in Figure O.5. We have tried to place equal emphasis on all these, without explicitly talking about the three fields in classifying the total design. It is hoped that such an approach would provide the graduate student with an opportunity to appreciate the seamless continuity in the development of CRE.

In closing this overview, it is important to emphasize that the three fields mentioned above need not be the only fields. As CRE encompasses more and more chemistry-based disciplines into its fold, the number of field equations is likely to increase. For instance, when reaction engineers looked at electrochemical processes, they found the need to include a set of equations defining the electrochemical field in the analysis and design of electrochemical reactors. Similarly, when sonochemical reactions were added, a new set of equations defining the sonochemical field had to be added, and so on.

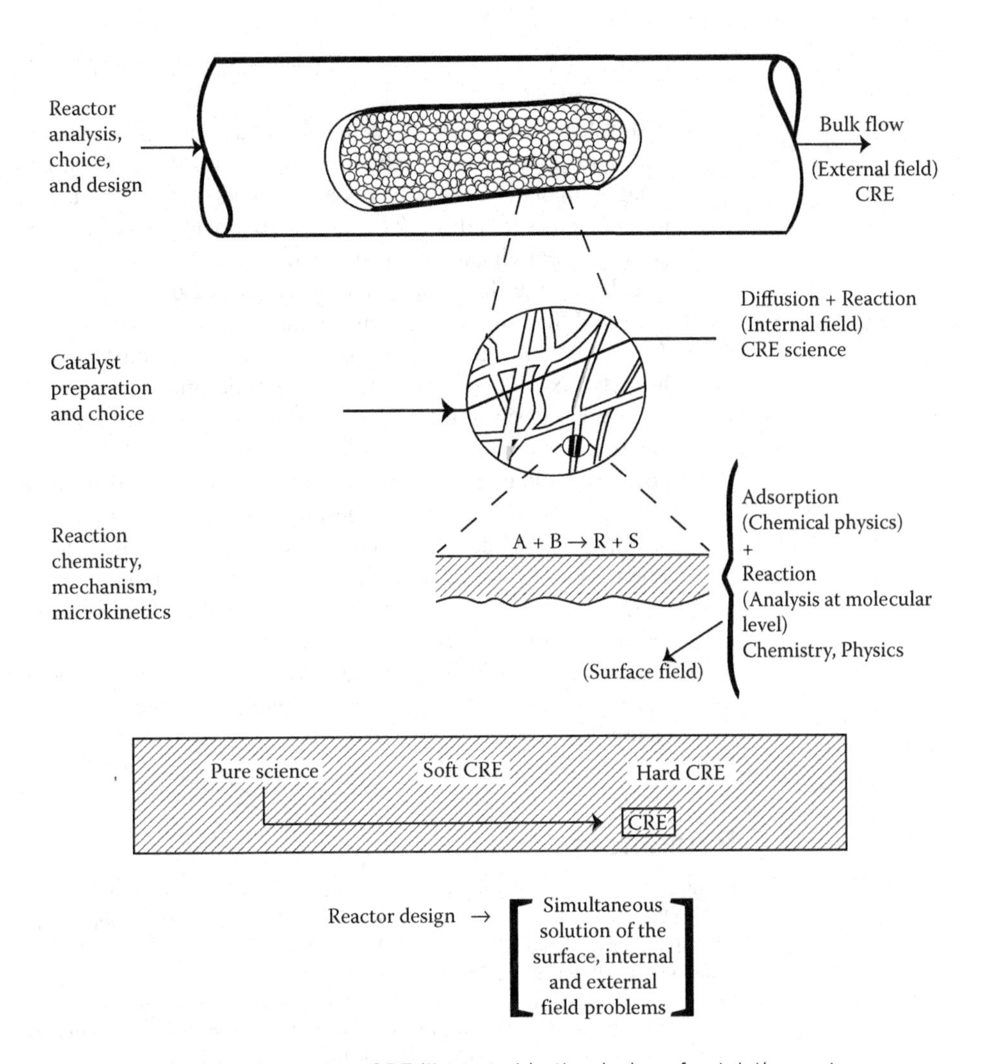

Figure 0.5 The principle of seamless CRE illustrated in the design of catalytic reactors.

Explore yourself

1. Select an enzymatic reaction system from the human body. Identify the active site, and show the relevant heat- and mass-transport process across the enzyme peripherals as well as the rate-controlling step. Discuss the relative roles of chemical conversion and transport rates and determine the order of magnitude of the overall rate of conversion.
2. Search the patent literature and find recent patents on hydro desulfurization reaction schemes. Comment on how developments in catalysis changed the plant design, temperatures, and pressures.
3. From the most recent literature, identify a reaction system where the microreactor technology has been commercialized.

Comment on the advantages and disadvantages of this microreactor application.

4. List at least four cases of chemical manufacture that can benefit from microreactor technology. What are the pitfalls, and what technology developments are needed to drive the use of microreactors in industry? Base your comments on conversion, activity/selectivity, and cost considerations. What would you do to enhance the throughput if you had to design a reactor and were told that mixing was detrimental?
5. The petrochemicals industry uses a variety of reactor designs. Identify six processes that use different designs, comment on why those designs were chosen, and describe a few salient details.
6. Focus on fluid catalytic cracking (FCC) reactors used in refineries. Answer the following questions after reading the earlier patents on FCC design. Why are two catalytic beds needed? What types of fluid and solid transport are used? How is the thermal energy needed provided? What are the advantages and disadvantages of using such a reactor?
7. Find examples of activity and selectivity improvement in chemical reactors when the product is separated from the reaction chamber.

References

Cropley, J.B., in *CRC the Engineering Handbook*, Second edition, R.C. Dorf ed. *Chapter 80: Scale Up of Chemical Reaction Systems from Laboratory to Plant*, CRC Press, Boca Raton, FL, 2004.

Hansen, H.J., US Patent 3 372 988, 1968.

Kashid, M.N. and Agar, D.W., *Chem. Eng. J.* **131**, 1, 2007.

Leigh, G.J., *The World's Greatest Fix: A History of Nitrogen and Agriculture*, Oxford University Press, New York, NY, 2004.

Prinz F.B., and Grossman A.R., Photosynthetic bioelectricity, 2005, http://gcep.stanford.edu/research/factsheets/photosynthesis_bioelectricity.html (accessed August 10, 2012).

Part I
Fundamentals revisited

There will come a time when you believe that everything is finished. That will be the beginning.

Louise A'Mour

Objectives

This part is dedicated to be a review for graduate students with a chemical engineering background and an introduction to those new to the field. After completion of this part, the successful student must be able to

- Design reactors for simple reactions
- Design reactors for complex reactions
- Choose the most suitable ideal reactor for a given reaction
- Perform energy balance analyses around isothermal and adiabatic ideal reactors
- Identify nonidealities in chemical reactors and mathematically analyze the nonidealities for improved design protocols
- Choose the best operation mode after differentiating the advantages and disadvantages of steady-state and nonsteady-state operations of chemical reactors

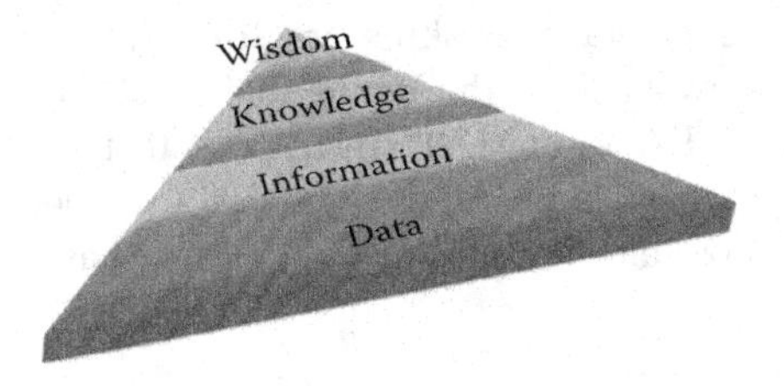

Introduction

The growth rate of information in the twenty-first century is exponential. Thus, it is utterly important to develop skills to differentiate the important and the relevant. It is one of the intentions of this book to teach the students how to distill data to extract wisdom.

In the pyramid of wisdom, this part serves at the data level. The more a student grasps from the contents of this part, the better would be the final wisdom we hope to develop in each individual after reading through the material.

The essential minimum of chemical reaction engineering

The data level of chemical reaction engineering has the following basics:

- The ability to write steady- and nonsteady-state material balances involving chemical conversions
- The ability to write steady- and nonsteady-state energy balances involving chemical conversions
- The ability to solve the differential or algebraic equations resulting from the conservation laws
- The ability to make approximations when necessary with the awareness of the limitations of such approximations

These are the skills that chemical engineers develop at the undergraduate level. However, a good review is necessary at the graduate level for several reasons. First of all, we have to acknowledge that if not used, the information stored even in the long-term memories need some assistance to retrieve. Second, this part serves as a warm-up for the rest of the book. Finally, even if we label the part as fundamentals, we still look at the bigger picture from the framework of an advanced learner, thus, we present a higher level of analysis of the fundamentals.

The skill development

We anticipate that the students will come with a good background in math and in computations. Those who feel a little bit behind in terms of computational skills are strongly encouraged to download tutorials available on the World Wide Web for using EXCEL, MATHCAD, MATLAB®, or MATHEMATICA that will come in very handy in solving complex problems. We leave the student and the instructor free to choose any or none of the above-listed available software.

Getting started

A good review of what you know so far would help. As with any personal development, learning must be internally driven, and if you do not feel the need to learn, no tool will be sufficient to transfer the necessary information.

Warm-up questions

Qualitative

1. Does the conversion increase, decrease, or remain constant as the reactor volume increases for an ideal PFR or ideal CSTR? Speculate and defend against the opposite of your answer.
2. Does the conversion increase, decrease, or remain constant as the temperature increases? Speculate and defend against the opposite of your answer.
3. What happens to the temperature in an adiabatic reactor during reaction? What additional information do you need to give a clear answer?
4. Do we need heat exchange to keep a reactor isothermal? Why?
5. What is the difference between the clock time of a batch reactor (BR) and the space time of the plug-flow reactor (PFR) and continuously stirred tank reactor (CSTR)?
6. Compare and contrast batch reactor and PFR design equation mathematically.
7. The variation in the reaction enthalpy, $-\Delta H_r$, with respect to temperature is frequently neglected. How good an assumption is this? How can we eliminate this assumption?
8. What is the reason for multiple steady states in CSTRs? Do we have them in PFRs and BRs?

Quantitative

1. For the elementary liquid phase reaction

$$A + B \rightarrow C$$

   i. Derive the design equation, list any assumptions that you make
   ii. Construct a stoichiometric table on the basis that only the reactants are initially present
      a. For the batch reactor
      b. For the PFR
      c. For the CSTR
      d. What are the main differences between the design equations in parts a, b, and c?
2. Redo Problem 1 for an elementary gas phase reaction. List all of the assumptions.
3. Derive the final equations for (space) time it takes to achieve conversion, X_A for all cases presented in Problem 1 for the following situations:
   a. $[A_0] = [B_0]$
   b. $[A_0] = 2[B_0]$, no inerts
   c. $[A_0] = [B_0] = [I_0]$

d. $[A_0] = [B_0] = \frac{1}{2}[I_0]$

e. Comment on the role of inerts, I, in achieving high conversions in a given (space) time

4. Redo Problem 3 for an elementary gas phase reaction. Is the effect of presence of inerts more or less pronounced in the gas phase? Discuss.
5. Design a reactor for the hypothetical cracking reaction.

$$A_n \rightarrow nA$$

a. Choose your reactor type. What criteria did you use as a basis of your choice?

b. List all the possible questions that you have for an accurate choice of reactor system.

c. What information do you need for an accurate design analysis?

6. Set up and solve energy balances for the following situations:

a. Determine the adiabatic flame temperature of CH_4 and C_2H_6. In order to calculate the adiabatic flame temperature, you will use stoichiometric amount of air and the fuel to determine the final temperature of combustion products in a hypothetical adiabatic chamber.

b. Provided that the feed enters the combustor at stoichiometric amounts and at room temperature, determine the T_f at the exit of a combustor if

- CH_4 is burned with air.
- CH_4 is burned with O_2.

c. What is your opinion about the relative sizes of a combustor if,

- The oxidizer is air?
- The oxidizer is O_2?
- The oxidizer is NO?

d. Discuss the role of inerts in chemical reactions based on your answers to the questions above.

7. Determine the volume of a CSTR for a hypothetical reaction $A + B \rightarrow$ products, for the following situations:

a. $-r_A = k[A]$

b. $-r_A = k[A]^2$

c. $-r_A = k[A][B]$, $[A]_0 = [B]_0$

d. $-r_A = k[A]/[B]$, $[A]_0 = [B]_0$. What happens if $[B] \rightarrow 0$?

8. Determine the volume of a PFR for the situations given in Problem 7.

Chapter 1 Reactions and reactors

Basic concepts

Chapter objectives

This chapter is intended as a review for students with chemical engineering background and as a quick introduction to chemical engineering graduate students with interdisciplinary backgrounds.

Upon successful completion of this chapter, the students should be able to

- Differentiate between batch, semibatch, and continuous operations around ideal reactors.
- Specify the flow characteristics and isotropy around the ideal reactors called plug-flow reactors (PFRs) and continuously stirred tank reactors (CSTRs).
- Set up and solve material balance equations around an ideal chemical reactor.
- Set up and solve energy balances around an ideal chemical reactor.
- Derive equations that lead to multiple steady states (MSS) in continuous reactors.
- Analyze and quantify the parameters that lead to MSS.

Introduction

Choosing a reactor for a given reaction is based on several considerations and combines *reaction analysis* with *reactor analysis*. Thus, we consider in this chapter the following aspects of reactions and reactors, much of which should serve as an introduction to chemists and a refresher to chemical engineers: reaction rates, stoichiometry, rate equations, and the basic reactor types.

Reaction rates

The first step in any consideration of reaction rates is the definition of reaction time. This depends on the mode of reactor operation: batch or continuous (see Figure 1.1). For the batch reactor (BR), the reaction

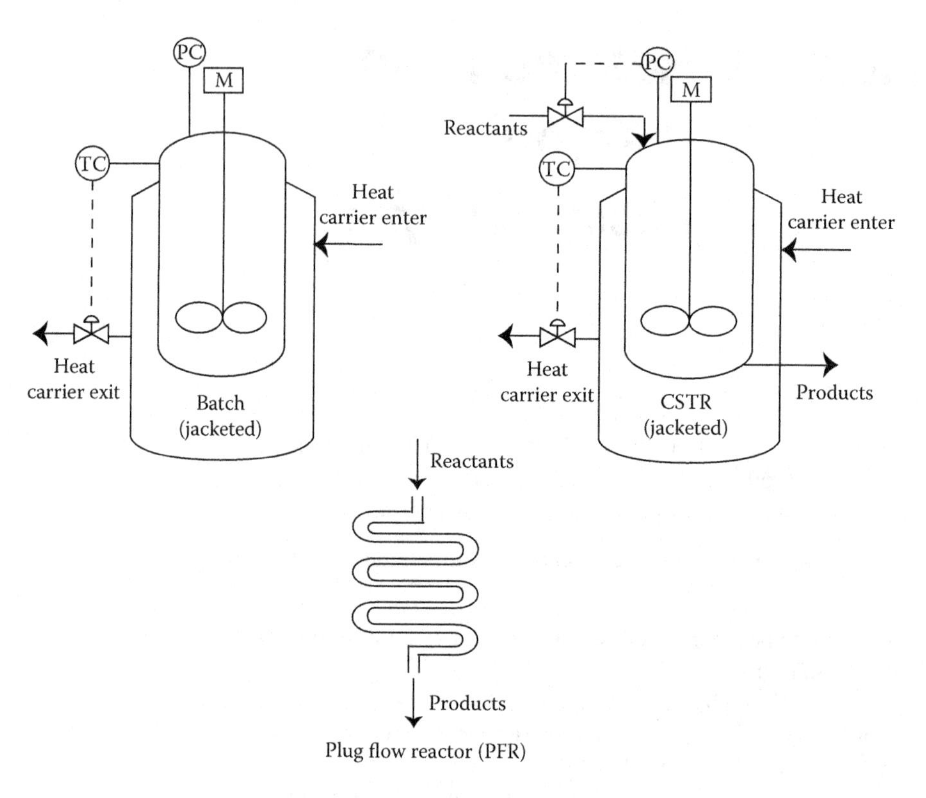

Figure 1.1 Types of ideal reactors: CSTR, PFR, and BR.

time is the elapsed time, whereas for the continuous reactor, it is given by the time the reactant spends in the reactor, referred to as the *residence time*, and is measured by the ratio of reactor volume to flow rate (volume/volume per unit time, with units of time). An equally important consideration is the concept of *reaction space* (which can have units of volume, surface, or weight), leading to different definitions of the reaction rate. We begin this section by considering different ways of defining the reaction rate based on different definitions of reaction time and space.

Different definitions of the rate

The basis of all reactor designs is an equation for the reaction rate. The rate is expressed as

$$\begin{bmatrix} \text{Rate of consumption of a} \\ \text{reactant or formation of a} \\ \text{product} \end{bmatrix} = \frac{\text{Moles consumed or formed}}{(\text{Time})(\text{Reaction space})} \tag{1.1}$$

Different definitions are possible depending on the definition of reaction space. This in turn depends on whether the reaction is homogeneous or heterogeneous.

Consider a BR that contains a fixed amount of reactant in a single phase. If the reactant is a gas, it occupies the entire reactor space. If it is a liquid, then the reaction space is less than the reactor volume, usually about two-thirds. This volume is generally indicated by the symbol V (or sometimes specifically by V_r). Thus, for any component i, the rate is defined as

$$\text{Rate, } r_i = \frac{1}{V}\left(\frac{dN_i}{dt}\right) \quad (1.2)$$

In the case of a catalytic reactor, the reaction space can be the weight, volume, or surface of the catalyst. It can also be the total reactor volume (catalyst + voids). Thus, four definitions are normally used:

$$r_i = \frac{1}{V}\left(\frac{dN_i}{dt}\right)$$ Moles per unit time per unit volume of the reactor (1.3a)

Rate per volume of reactor

$$r_{Wi} = \frac{1}{W}\left(\frac{dN_i}{dt}\right)$$ Moles per unit time per unit weight of the catalyst (1.3b)

Rate per weight of catalyst

$$r_{Vi} = \frac{1}{V_{\text{catalyst}}}\left(\frac{dN_i}{dt}\right)$$ Moles per unit time per unit volume of the catalyst (1.3c)

Rate per volume of catalyst

$$r_{Si} = \frac{1}{S}\left(\frac{dN_i}{dt}\right)$$ Moles per unit time per unit surface of the catalyst (1.3d)

Rate per unit surface of catalyst

In the case of gas–liquid or liquid–liquid reactions, Equation 1.3a, based on total reactor volume, which is identical to Equation 1.2 for homogeneous reactions, may be used to give r_i. Alternatively, the rate can be expressed as r_{Si} given by Equation 1.3d for catalytic reactions but based on gas–liquid or liquid–liquid (instead of the catalyst) interfacial area. In this case, we generally use the symbol r_i'.

There could be situations where all the three phases would be present: gas, liquid, and solid (as catalyst or reactant). The most common example of this is the slurry reactor used in reactions such as hydrogenation. Here, the rate is sometimes expressed as in the case of catalytic reactions, that is, per unit catalyst volume, weight, or surface, but more commonly in terms of the total reactor volume (liquid + gas + catalyst).

Various representations of the rates are summarized in Table 1.1 along with interrelationships between them. Depending on the nature of the

Table 1.1 Units of Rates/Rate Constants (for a First-Order Reaction) for Different Classes of Reactions

Reaction	Reaction Rate	Rate Constant
Any reaction	r'_{VA} (or r'_A)$[=]\dfrac{\text{mol}}{\text{m}^3\ \text{reactor}\cdot\text{s}}$	k (or k_V)$[=]\dfrac{1}{\text{s}}$
Gas–liquid, liquid–liquid reactions	r'_{VA} (or r'_A)$[=]\dfrac{\text{mol}}{\text{m}^2\ \text{interfacial area}\cdot\text{s}}$	$k'[=]\dfrac{m}{\text{s}}$
Catalytic reactions	$r_{VA}[=]\dfrac{\text{mol}}{\text{m}^3\ \text{cat}\cdot\text{s}}$	$k_V[=]\dfrac{1}{\text{s}}$
	$r_{SA}[=]\dfrac{\text{mol}}{\text{m}^2\ \text{surface area}\cdot\text{s}}$	$k_S[=]\dfrac{m}{\text{s}}$
	$r_{WA}[=]\dfrac{\text{mol}}{\text{kg cat}\cdot\text{s}}$	$k_W[=]\dfrac{\text{m}^3}{\text{kg cat}\cdot\text{s}}$

Interrelationships

k_V (or k) $= (1 - f_B)k_V = (1 - f_B)S_g\rho_c k_s$

k_V (or k) $= ak'$, $k_W = k_V/\rho_c$

ρ_c: catalyst density, g-cat/m^3

f_B: volume fraction of the voids, m^3 voids/m^3 reactor

S_g: specific surface area, m^2/g-cat

reaction, any of these definitions may be used. Since the present chapter is largely restricted to homogeneous reactions, rate based on total volume (r_i) is used throughout the chapter.

Basic rate equation

Any rate equation can be written in its most general form as

Reaction rate depends on temperature and concentration

$$\text{Rate} = f\begin{bmatrix}\text{temperature,}\\ \text{concentration,}\\ \text{mode of contact,}\\ \text{and history of catalyst}\end{bmatrix} \tag{1.4}$$

In this equation, no parameter is independent of the other and hence a single functionality is assigned. This is referred to as a *nonseparable kinetic equation*. For homogeneous reactions, this usually reduces to

$$\text{Rate} = f\,[\text{temperature, concentration}] \tag{1.5}$$

which can be further simplified to

$$\text{Rate} = f_1(\text{temperature})\,f_2(\text{concentration}) \tag{1.6}$$

In this *separable rate equation*, temperature and concentration are treated as independent functions and the effects of the two are separated. Our treatment will be based on separable kinetics.

The temperature dependence is expressed in terms of the Arrhenius equation as

$$k = k_0 \, e^{-E/RT} \tag{1.7}$$

where E is the activation energy for the reaction, k_0 the preexponential term, which may be regarded as the temperature-independent rate constant, and R the ideal law gas constant. The concentration dependence is usually expressed as a power of the concentration, and this power is referred to as the *reaction order.* Thus, for the reactions

Swante Arrhenius (1859–1927), Nobel Prize in Chemistry, 1903

$$A \rightarrow R \tag{R1}$$

$$A + B \rightarrow R \tag{R2}$$

the rate equations, referred to as *power law models*, are

$$-r_A = k[A]^n \tag{1.8}$$

$$-r_A = k[A]^n \, [B]^m \tag{1.9}$$

where m and n are the reaction orders with respect to A and B, respectively. If the reaction order is equal to the *stoichiometric coefficient* of the concerned reactant, that is, the number of molecules of the reactant taking part in the reaction, then it is referred to as an *elementary reaction.*

Reactions R1 and R2 involve only a single step and are called simple reactions. Rate equations similar to Equations 1.8 and 1.9 can be written for a variety of such simple reactions and solved to obtain the concentration of any component as a function of time as the reaction progresses.

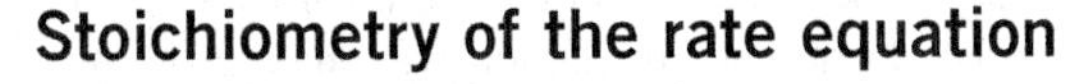

Stoichiometry of the rate equation

Basic relationships

It is clear from the previous section that the rate of reaction depends on the concentrations of reactants. In the case of reversible reactions, it depends additionally on the concentrations of products. Often, in a laboratory experiment, one of the reactants is selected and its concentration monitored as a function of time or flow rate, depending on whether the reactor is batch or continuous. The results are then expressed in terms of the conversion of that reactant (say A) defined as

$$X_A = \frac{\text{Moles of } A \text{ converted}}{\text{Moles of } A \text{ fed}} \tag{1.10}$$

It is often necessary to convert the rate of reaction of any component to that of another. Thus, a stoichiometric relationship between the rates is required. To obtain such a relationship, consider the general reaction

General reaction

$$\nu_A A + \nu_B B \rightarrow \nu_R R + \nu_S S \tag{R3}$$

From its stoichiometry, we can easily write

Relative rates according to the stoichiometry

$$\frac{r_A}{\nu_A} = \frac{r_B}{\nu_B} = \frac{r_R}{\nu_R} = \frac{r_S}{\nu_S} \tag{1.11}$$

Conversion–concentration relationships

Consider Reaction R3. No stoichiometric relationship is needed for calculating the conversion, which is given simply by Equation 1.10. But the rate equation contains other concentrations as well, which can be related to X_A through the stoichiometry of the reaction. These relations will depend on whether or not there is a volume change accompanying the reaction. Volume change can occur due to a change in the number of moles (when $\nu_A + \nu_B \neq \nu_R + \nu_S$), temperature, or pressure, or a combinations of these. Reactions in which no volume change occurs due to any of these factors are referred to as *constant-density* (*constant-volume*) reactions. Those in which one or more of these occur are referred to as *variable-density* (*variable-volume*) reactions. The effect of volume change is important in gas phase reactions but is negligible in homogeneous liquid phase reactions.

For a BR of volume V, the concentration of A is given by

Concentration in a BR

$$\text{Batch reactor:} \quad [A] = \frac{N_A}{V} \tag{1.12}$$

where the volume V is constant and equal to the initial volume V_0. For a continuous flow reactor with a volumetric feed rate of Q_A liters per second and a molar rate of F_{A0} moles per second, the concentration is given by

Concentration in a continuous reactor

$$\text{Continuous reactor:} \quad [A] = \frac{F_{A0}}{Q_0} \tag{1.13}$$

where Q_A is constant and equal to the initial value Q_{A0}. We also define the feed stoichiometry:

Feed stoichiometry

$$\psi_i = \frac{N_{i0}}{N_{A0}} = \frac{[i]_0}{[A]_0} = \frac{F_{i0}}{F_{A0}} \tag{1.14}$$

and the mole change parameter δ_A as

Mole change parameter

$$\delta_A = \frac{\text{Change in total number of moles at complete conversion of } A}{\text{Number of moles of } A \text{ reacted}} \tag{1.15}$$

which is important for the gas phase reactions. For the liquid phase reactions where the density remains constant during the course of the reaction, and for the gas phase reactions where the net reaction stoichiometry is equal to zero, the volume change parameter is zero. Considering Reaction R3 with A as the key component, we can now write the following expressions for the number of moles of each component remaining after a certain conversion X_A:

$$N_A = N_{A0}(1 - X_A) \quad (1.16)$$

Conversion

$$N_i = N_{A0}\left(\psi_i - \frac{\nu_i}{\nu_A}X_A\right), \quad i \neq A \quad (1.17)$$

In writing Equation 1.18, we make sure that the sign of the stoichiometric coefficient ν_i is positive for products and negative for reactants:

$$N_t = N_{t0} + \delta_A N_{A0} X_A \quad (1.18)$$

where

$$\delta_A = \frac{\sum \nu_i}{|\nu_A|} \quad (1.19)$$

Note again that the sign for ν_i is negative for the reactant and positive for the product. For reactions with no volume change, we can combine Equations 1.16 through 1.18 with Equation 1.12 to give

$$[A] = [A]_0(1 - X_A) \quad (1.20)$$

Mole balance for no volume change

$$[i] = [A]_0\left(\psi_i - \frac{\nu_i}{\nu_A}X_A\right) \quad (1.21)$$

Identical concentration equations can be derived for a flow system, using F in place of N in Equations 1.16 through 1.18 and Equation 1.13 for the concentration.

Variable-density reactions

The general case of variable-density reactions is applicable mostly to gas phase reactions and seldom to liquid phase reactions. Since the gas law gives the precise relationship between P, V, T, and N, we start with that equation. Based on the ideal gas law, we can write

$$V = V_0\left(\frac{P_0}{P}\right)\left(\frac{T}{T_0}\right)\left(\frac{N_t}{N_{t0}}\right) = V_0\left(\frac{P_0}{P}\right)\left(\frac{T}{T_0}\right)(1 + \delta_A y_{A0} X_A) \quad (1.22)$$

The ideal gas law to account for the changes in T and P

where

$$y_{A0} = \frac{N_{A0}}{N_{t0}} \quad (1.23)$$

is the initial mole fraction of A. Our concern here is with the change in total volume when A has reacted completely in relation to the total volume present initially. Thus, we define a volume change parameter ε_A as

Volume change parameter ε_A

$$\varepsilon_A = \frac{V|_{X_A=1} - V|_{X_A=0}}{V|_{X_A=0}} \tag{1.24}$$

Since ε_A represents the volume change at complete reaction per mole of A reacted (see Equation 1.19), we have

$$\varepsilon_A = y_{A0}\delta_A \tag{1.25}$$

Combining Equations 1.22 and 1.25, we obtain

$$V = V_0\left(\frac{P_0}{P}\right)\left(\frac{T}{T_0}\right)(1 + \varepsilon_A X_A) \tag{1.26}$$

which is the basic equation to account for volume/density change. The concentration equations now readily follow:

Accounting for volume/density changes for ideal gases

$$[A] = [A]_0\left(\frac{1 - X_A}{1 + \varepsilon_A X_A}\right)\left(\frac{T_0}{T}\right)\left(\frac{P}{P_0}\right) \tag{1.27}$$

$$[i] = [A]_0\left[\frac{\psi_i + (v_i/v_A)X_A}{(1 + \varepsilon_A X_A)}\right]\left(\frac{T_0}{T}\right)\left(\frac{P}{P_0}\right) \tag{1.28}$$

The analysis can be extended to a flow system by expressing the volume change in terms of change in Q_A. The following expression similar to Equation 1.26 results:

Volumetric flow rate in its most general form

$$Q_A = Q_{A0}\left(\frac{P_0}{P}\right)\left(\frac{T}{T_0}\right)(1 + \varepsilon_A X_A) \tag{1.29}$$

Combining the above with Equation 1.13, equations similar to Equations 1.27 and 1.28 can be written.

Reactors

Before we go into the details of reactors, let us first write down the general mole balance under unsteady-state conditions around an arbitrary control volume shown in Figure 1.2:

Mole balance for a flow reactor

$$\frac{dN_A}{dt} = \sum_i F_{Ai_{\text{in}}} - \sum_i F_{Ai_{\text{out}}} + r_A V \tag{1.30}$$

Here, N_A represents the total number of moles contained within the control volume, $F_{A,i}$ is the molar flow rate of species A (mol/time) through

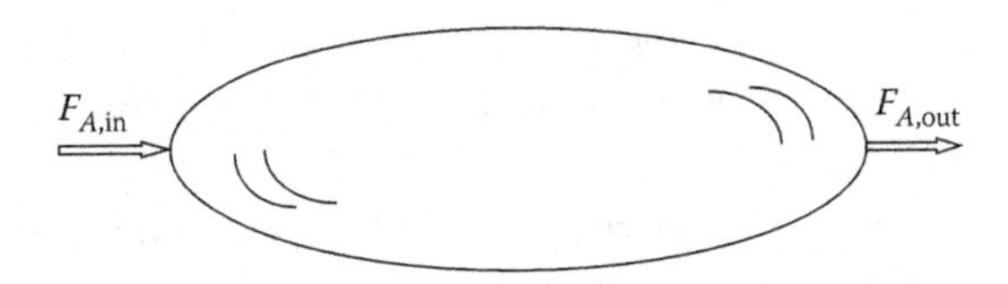

Figure 1.2 The control volume and the flow streams.

stream i, and subscripts in and out refer to the incoming and outgoing streams, respectively. The term $r_A V$ signifies the moles converted per unit time given that the reaction takes place in a volume of V. In the subsequent sections, we will first describe the BR most commonly used by chemists and its continuous counterpart, the PFR. These represent one extreme characterized by the total absence of backmixing from fluid elements "downstream" in time or space. On the other extreme, we have the mixed flow reactor (MFR), which is called the CSTR, when perfectly mixed. The three reactors mentioned above are referred to as *ideal reactors*. Effects due to departures from these limits of perfect and no mixing can lead to the so-called *partially mixed reactors*. The mixing effects will be covered in detail in Chapter 3.

Batch reactor

The BR is essentially a reactor in which a batch of reactants is allowed to react under predetermined conditions. The reactor is continuously stirred to maintain a uniform composition at any time during the reaction. This composition will of course change with time.

BRs are most common in the pharmaceutical, perfumery, essential oil, and other fine chemicals industries. They are also extensively used in the pesticides industry. For relatively small-scale productions, it is customary not to have a BR exclusively for a single reaction. The scheduling of its use can be done in such a way as to ensure maximum utilization. Consider the simple irreversible reaction $A \rightarrow R$, carried out in a BR. The mole balance equation given in Equation 1.30 is simplified for the absence of continuous input or output of material to give

$$-\frac{dN_A}{dt}\frac{1}{V} = (-r_A) \tag{1.31}$$

Mole balance for a BR

Since $N_A = N_{A0}(1 - X_A)$, we have

$$t = N_{A0}\int_{X_{A0}}^{X_{Af}} \frac{dX_A}{V(-r_A)} \tag{1.32}$$

Equation 1.32 is written in terms of the number of moles and hence independent of volume. If, however, the moles have to be expressed as concentrations, volume comes into the picture, and different expressions are needed for reactions with and without volume change.

Reactions without volume change Since V is constant, $[A] = N_A/V$ and Equation 1.31 becomes

Constant-volume BR design equation

$$-\frac{d[A]}{dt} = (-r_A) \tag{1.33}$$

Integrating between the limits of initial and final concentrations gives

$$t = -\int_{[A]_0}^{[A]} \frac{d[A]}{(-r_A)} = [A]_0 \int_0^{X_A} \frac{dX_A}{(-r_A)} \tag{1.34}$$

The time t for achieving a stated conversion can be obtained by introducing the appropriate equation for the rate in Equation 1.34 and solving it. For the simple first-order case ($-r_A = k[A]$), the solution is

$[A(t)]$ for a first-order reaction in a BR

$$\frac{[A]}{[A]_0} = e^{-kt} \quad \text{or} \quad \ln\frac{[A]_0}{[A]} = kt \tag{1.35}$$

For higher order reactions the design equations are given in Table 1.2. A common type of reaction is the second-order reversible reaction represented by

$$A + B \leftrightarrow R + S \tag{R4}$$

Esterification reactions such as

$$C_2H_5OH + CH_3COOH \leftrightarrow CH_3COOC_2H_5 + H_2O \tag{R5}$$

are typical of this class. The rate equation for this second-order reversible reaction is given by

$$-r_A = k_+[A][B] - k_-[R][S] \tag{1.36}$$

Detailed tables for the time required to achieve a specific conversion are given in Doraiswamy (2001) along with solutions to a few other reversible reactions. Equation 1.34 can also be solved by plotting $-1/r_A$ versus X_A or $[A]$ between the limits 0 and X_{Af} or $[A]_0$ and $[A]_f$, as shown in Figure 1.3.

Reactions with volume change Let us first consider the volume change resulting from a change in the number of moles, such as in the reaction

$$A \rightarrow R + S \tag{R6}$$

To account for volume change, we express V in terms of Equation 1.26 and modify Equation 1.32 to give

Table 1.2 Analytical Solutions (Design Equations) for Simple Reactions in a BR

Reaction	Rate Equations	Analytical Solution[a]
1. $A \to R$	$-r_A = k[A]$	$kt = \ln\left[\frac{1}{1-X_A}\right], \quad \frac{[A]}{[A]_0} = e^{-kt}$
2. $2A \to R$	$-r_A = k[A]^2$	$k[A]_0 t = \left[\frac{1}{(1-X_A)^2} - 1\right], \quad \frac{[A]}{[A]_0} = \frac{1}{1+k[A]_0 t}$
3. $3A \to R$	$-r_A = k[A]^3$	$k[A]_0^2 t = \frac{1}{2}\left[\frac{X_A}{1-X_A}\right], \quad 2kt = \frac{1}{[A]^2} - \frac{1}{[A]_0^2}$
4. $A \to R$	$-r_A = k[A]^n$	$k[A]_0^{n-1} t = \frac{1}{(n-1)}\left[(1-X_A)^{1-n} - 1\right], \quad n \neq 1$
5. $A + B \to R$, $\psi_B = 1$	$-r_A = k[A][B]$	$k[A]_0 t = \left[\frac{X_A}{1-X_A}\right], \quad t = \frac{1}{k}\left(\frac{1}{[A]} - \frac{1}{[A]_0}\right)$
6. $A + B \to R$, $\psi_B \neq 1$	$-r_A = k[A][B]$	$k[A]_0 t = \frac{1}{(\psi_B - 1)}\ln\left[\frac{\psi_B - X_A}{\psi_B(1-X_A)}\right]$
7. $v_A A + v_B B \to R$, $\psi_B = v_B/v_A$	$-r_A = k[A][B]$	$k[A]_0 t = \frac{1}{\psi_B}\left[\frac{X_A}{1-X_A}\right]$
8. $A + 2B \to R$, $\psi_B = 2$	$-r_A = k[A][B]^2$	$k[A]_0^2 t = \frac{1}{8}\left[\frac{X_A}{(1-X_A)^2} - 1\right]$
9. $A + B \to R$, $\psi_B = 1$	$-r_A = k[A][B]^2$	$k[A]_0^2 t = \frac{1}{2}\left[\frac{1}{(1-X_A)^2} - 1\right]$
10. $A + 2B \to R$, $\psi_B \neq 2$	$-r_A = k[A][B]^2$	$k[A]_0 t = \frac{1}{(2-\psi_B)^2}\left[\ln\left(\frac{\psi_B - 2X_A}{\psi_B(1-X_A)}\right) + \frac{2X_A(2-\psi_B)}{\psi_B(\psi_B - 2X_A)}\right]$
11. $v_A A + v_B B \to R$, $\psi_B \neq v_B/v_A$	$-r_A = k[A][B]$	$k[A]_0 t = \frac{1}{\left[\psi_B - \frac{v_B}{v_A}\right]}\ln\left[\frac{\psi_B - (v_B/v_A)X_A}{\psi_B(1-X_A)}\right]$
12. $v_A A + v_B B \to R$, $\psi_B = v_B/v_A$	$-r_A = k[A]^n[B]^m$	$k[A_0]^{m+n-1} t = \frac{1}{\psi_B(m+n-1)}\left[\frac{1}{(1-X_A)^{m+n-1}} - 1\right]$

Note: The same equations are valid for PFR when time is replaced with space time.

[a] LHS = $k[A_0]^{n-1}t$, where k has the units of an nth [or $(m+n)$th]-order reaction, $(m^3/mol)^{n-1}$ (1/s); $\psi_B = [B]_0/[A]_0$.

$$t = [A]_0 \left(\frac{P}{P_0}\right)\left(\frac{T_0}{T}\right)\int_0^{X_A} \frac{dX_A}{(-r_A)(1+\varepsilon_A X_A)} \tag{1.37}$$

Accounting for volume change

This can be graphically integrated as shown in Figure 1.3 to give the time for accomplishing a given conversion.

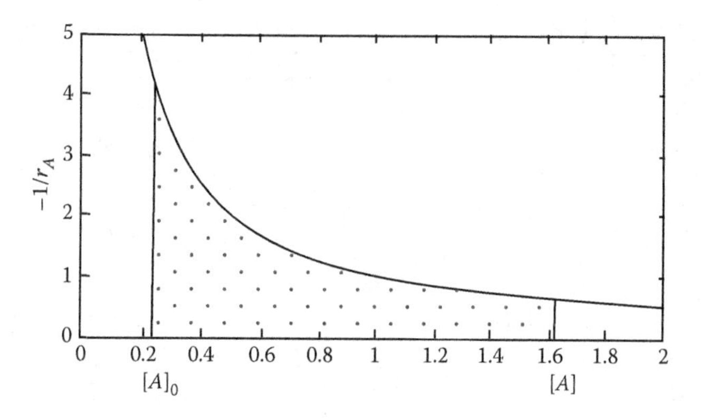

The graphical integration

Figure 1.3 The graphical integration method.

Nonisothermal operation

Most reactions are characterized by reasonable heats of reaction and hence it is not always possible to operate them under isothermal conditions with a constant heat exchange rate. It is necessary in such cases to write the energy balance in addition to the mass balance and solve them simultaneously.

Some kind of heat exchange is necessary to control the temperature, and this is usually achieved by circulating a heat exchange fluid through the reactor jacket or an immersed coil as the case may be. Three major parameters are involved in accounting for heat exchange between the fluids: heat transfer area, temperature difference or driving force between the phases, and the rate of heat transfer expressed through a heat transfer coefficient.

We will begin by writing the general energy balance for an open system:

The first law of thermodynamics

$$\frac{dE}{dt} = \dot{m}_0 \hat{E}_0 - \dot{m}_1 \hat{E}_1 + \dot{q} + \dot{W} \tag{1.38}$$

where E represents the total energy, q heat exchange, and W the work terms. Work term is composed of the sum of the flow work, the shaft work, and the surface boundary work:

$$\dot{W} = \dot{W}_f + \dot{W}_s + \dot{W}_b = Q_0 P_0 - Q_1 P_1 + \dot{W}_s - P\frac{dV}{dt} \tag{1.39}$$

The total energy in the system is represented as the sum of the kinetic, potential, and internal energies:

$$E = U + KE + PE \tag{1.40}$$

Substitution of this along with the definition of enthalpy, $H = U + PV$, in Equation 1.38 results in

$$\frac{d(U+KE+PE)}{dt} = \dot{m}_0(\hat{H}_0 + \widehat{KE}_0 + \widehat{PE}_0) - \dot{m}_1(\hat{H}_1 + \widehat{KE}_1 + \widehat{PE}_1) + \dot{q} + \dot{W}_s - P\frac{dV}{dt} \quad (1.41)$$

Change of variable to enthalpy

For a BR, there are no flow streams and hence the equation becomes

$$\frac{d(U + KE + PE)}{dt} = +\dot{q} + \dot{W}_s - P\frac{dV}{dt} \quad (1.42)$$

Energy balance for a BR

We will now focus our attention on the situation where there is no shaft work and the kinetic and potential energy terms are negligible:

$$\frac{dU}{dt} + P\frac{dV}{dt} = \dot{q} \quad (1.43)$$

If we substitute the definition of enthalpy, $H = U + PV$ in the equation above, then we obtain

$$\frac{dH}{dt} - V\frac{dP}{dt} = \dot{q} \quad (1.44)$$

Enthalpy can be expressed as a function of T, P, and number of moles N_j in differential form as

$$dH = \left(\frac{dH}{dT}\right)_{P,N_j} dT + \left(\frac{dH}{dP}\right)_{T,N_j} dP + \sum\left(\frac{dH}{dN_j}\right)_{T,P,N_{k\neq j}} dN_j \quad (1.45)$$

The first term on the right is the definition of heat capacity C_p. The second term can be expressed as (see Sandler, 2006 for a detailed derivation)

$$\left(\frac{\partial H}{\partial P}\right)_{T,N_j} = V - T\left(\frac{\partial V}{\partial T}\right)_{P,N_j} = V(1 - \alpha T) \quad (1.46)$$

where $\alpha = (1/V)(\partial V/\partial T)_{P,N_j}$ is the coefficient of thermal expansion.

The derivative in the summation term is the definition of partial molar enthalpy for component j, $\bar{H}_j$, such that

$$dH = m\hat{C}_p dT + (1 - \alpha T)V\,dP + \Sigma\bar{H}_j\,dN_j \quad (1.47)$$

When this definition of enthalpy is substituted in the energy balance for the BR (Equation 1.44), the resulting expression is

$$m\hat{C}_p\frac{dT}{dt} - \alpha TV\frac{dP}{dt} + \sum\bar{H}_j\frac{dN_A}{dt} = \dot{q} \quad (1.48)$$

From the reaction stoichiometry, it is possible to write $dN_j/dt = (v_j/v_i)(dN_i/dt)$ and we assume that the partial molar enthalpy is just equal to the pure component-specific enthalpy, that is, $\bar{H}_i = \underline{H}_i$ such that we have

Energy balance in chemically reacting systems

$$m\hat{C}_p \frac{dT}{dt} - \alpha TV \frac{dP}{dt} = -\Delta \underline{H}_{rxn} \frac{dN_A}{dt} + \dot{q} \tag{1.49}$$

Finally, we substitute for the reaction rate from the material balance for a BR to obtain

$$m\hat{C}_p \frac{dT}{dt} - \alpha TV \frac{dP}{dt} = -\Delta \underline{H}_{rxn} Vr_A + \dot{q} \tag{1.50}$$

For reaction systems where the reaction enthalpy is a weak function of temperature, the pressure is constant, and the rate of heat transfer is represented by an overall heat transfer coefficient U_o, such that the rate of heat transfer is

Rate of heat transfer

$$\dot{q} = U_o A_h (T_w - T) \tag{1.51}$$

Upon substitution of the heat transfer term, Equation 1.50 takes the form

$$N_{A0}\underline{C}_p \frac{dT}{dt} = -\Delta \underline{H}_{rxn} N_{A0} \frac{dX_A}{dt} + U_o A_h (T_w - T) \tag{1.52}$$

Noting that $T = T(t)$, but for the sake of simplicity, ignoring that and integration of this equation between inlet (X_{A0}, T_0) and outlet (X_A, T) gives

$$N_{A0}\underline{C}_p (T - T_0) - (-\Delta \underline{H}_{rxn}) N_{A0} (X_A - X_{A0}) = U_o A_h (T_w - T)t \tag{1.53}$$

For the special case of adiabatic operation, no heat abstraction or addition occurs. Hence, the last term of Equation 1.53 vanishes, and the following unique relationship between temperature and conversion results:

Conversion–temperature relationship in an adiabatic reactor

$$(X_A - X_{A0}) = \alpha_H (T - T_0) \tag{1.54}$$

where

$$\alpha_H = \frac{C_p}{-\Delta H_r} \tag{1.55}$$

Note that no such relationship exists for nonadiabatic operation.

Optimal operating policies While the design of a BR considered above gives the batch time for a given duty, this time is not necessarily the optimum reaction time for maximum profit. Aris (1965, 1969) suggests a method for calculating the optimum time for maximizing profit at a given temperature. It is more important, however, to compute a time–temperature policy for maximizing performance. Thus, since the reaction rate for a simple reaction always increases with temperature, the optimum temperature policy for a simple reaction is merely the maximum temperature possible. This is fixed by considerations such as material of construction of the reactor, catalyst deactivation (in catalytic reactions), and so on.

On the other hand, different time–temperature policies are optimal for different classes of complex reactions and these are considered in Chapter 2. Although the reversible reaction is also a complex reaction in the sense that two reactions occur, it is equally true that no additional species are involved in the second (reverse) reaction. Hence, the reversible reaction can also be regarded as a simple reaction. If the reaction is endothermic, its reversible nature makes no difference since both the reaction rate constant and the equilibrium constant increase with temperature, and the maximum practicable temperature continues to be the optimal temperature. But if the reaction is exothermic, an increase in temperature has opposite effects: it lowers the equilibrium constant but raises the rate constant. Hence, a thermodynamic optimum temperature exists. For any reaction such as $A \leftrightarrow R$ with the rate equation $-r_A = k([A] - [R]/K)$, this optimum can be found by integrating the expression

$$t = [A]_0 \int_{X_{A0}}^{X_{Af}} \frac{dX_A}{[-r_A(X_A,T)]} \tag{1.56}$$

for different constant values of T (and hence of k and K and the rate) and finding the optimum temperature for minimum reaction time.

Optimum temperature

Plug-flow reactors

The characteristic feature of the PFR is that there is no feedback from downstream to upstream. This kind of ideal behavior eliminates many complications associated with fluid flow and leads to relatively simple reactor equations.

The PFR is usually a long tube, straight or coiled, a set of straight tubes connected in series at their ends, or a bank of independent tubes. The diameter of the tube is usually not more than 4–5 cm. For the PFR assumption to be valid, the length-to-diameter ratio should be very high, at least 30. These reactors are common for solid catalyzed vapor-phase reactions. Where no solids are present, the reactor tube is sometimes coiled to accommodate high residence times. An example of this is the coiled reactor for the production of ethylenediamine by reaction between ethylene dichloride and aqueous ammonia sketched in Figure 1.4 (Venkitakrishnan and Doraiswamy, 1982).

A serious drawback of PFR is the variation of temperature with length, leading to significant temperature gradients within the reactor. If a

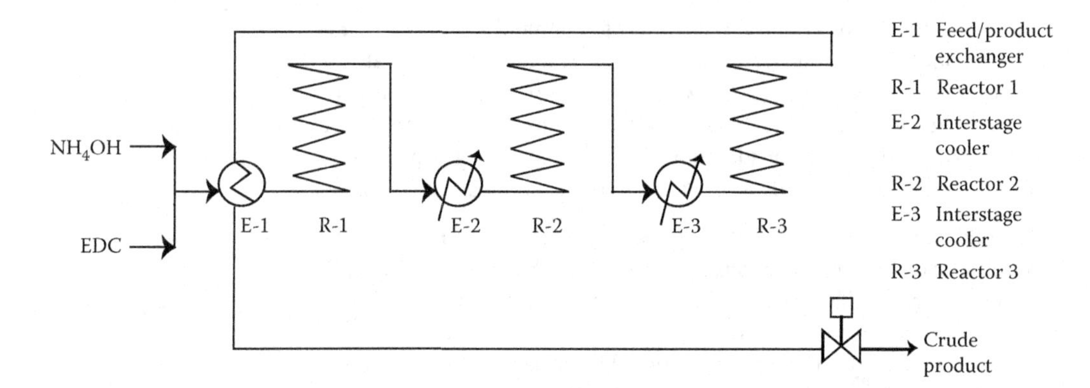

Figure 1.4 Coiled reactor for ethylenediamine manufacture.

reaction is highly temperature-sensitive, has a large heat of reaction, and operation at an optimum temperature (determined from laboratory experiments) is essential, a PFR cannot be used. Either a BR or an MFR (to be described later in this chapter) would appear to be a more suitable choice. However, a PFR with controlled heat exchange to give a favorable temperature profile can also be used—and may often be the preferred candidate. In general, liquid phase reactions are carried out in batch (or semibatch, to be described later) reactors and large-volume vapor-phase catalytic reactions in PFR of the heat exchanger type.

Basic PFR equation A sketch of a PFR along with the inlet and outlet concentrations and flow rates is shown in Figure 1.5. The material balance of Equation 1.1 holds equally for a differential element dV of this reactor over which the rate is assumed to be constant. However, there is a finite flow into the element and a finite flow out of it now and as a result of the steady-state assumption there is no accumulation in the reactor. In terms of the nomenclature of Figure 1.5, this equation can therefore be recast as

Steady-state material balance for a PFR

$$0 = (F_A + dF_A) - F_A + (-r_A)dV \tag{1.57}$$

or

$$-dF_A = (-r_A)dV$$

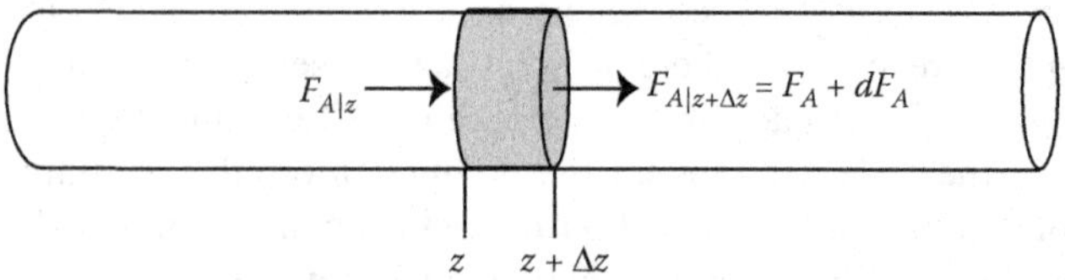

$F_{A|z}$ → $F_{A|z+\Delta z} = F_A + dF_A$

z $z + \Delta z$

Figure 1.5 PFR geometry and the differential control volume.

Since $F_A = F_{A0}(1 - X_A)$, we have

$$F_{A0}\,dX_A = (-r_A)dV \qquad (1.58)$$

Plug-flow design equation in differential form

Solution of this equation with boundary conditions ($V = 0, X_A = X_{A0}$) and ($V = V, X_A = X_{Af}$) gives

$$\frac{V}{F_{A0}} = \int_{X_{A0}}^{X_{Af}} \frac{dX_A}{-r_A} \qquad (1.59)$$

Plug-flow design equation in integral form

This is the basic design equation for PFR.

It is useful to modify Equation 1.59 so that it becomes equivalent to Equation 1.34 for BR, particularly since the two are identical if time is replaced by residence time V/Q or z/u. For this purpose, we define a space time

$$\bar{t} = \frac{V[A]_0}{F_{A0}} = \frac{V}{Q_0} \qquad (1.60)$$

Space time

to represent the time needed to treat one reactor volume of the feed stream. This is equal to the residence time for a constant-density system.

Equation 1.59 thus becomes

$$\bar{t} = [A]_0\left(\frac{P}{P_0}\right)\left(\frac{T_0}{T}\right)\int_0^{X_{Af}} \frac{dX_A}{(1+\varepsilon_A X_A)(-r_A)}, \quad \varepsilon_A = \text{any value} \qquad (1.61)$$

$$\bar{t} = \int_{[A]_f}^{[A]_0} \frac{d[A]}{-r_A}, \quad \varepsilon_A = 0 \qquad (1.62)$$

Design equations It will be noticed that Equations 1.59 and 1.61 are general and valid, irrespective of volume change. Integrated forms for constant-density systems are identical to those for BR. Graphical integration is straightforward and gives the reactor volume directly, as shown in Figure 1.6. Reciprocal rate is plotted as a function of either X_A (Equation 1.61) or $[A]$ (Equation 1.62). Alternatively, any of the several numerical integration methods can be used, and this is perhaps the most attractive.

Nonisothermal operation In a PFR, where time is not usually a parameter, nonisothermicity is reflected in a temperature profile in the reactor from the inlet to the outlet. Nonisothermal operation of a PFR is, therefore, an inherent feature of the reactor.

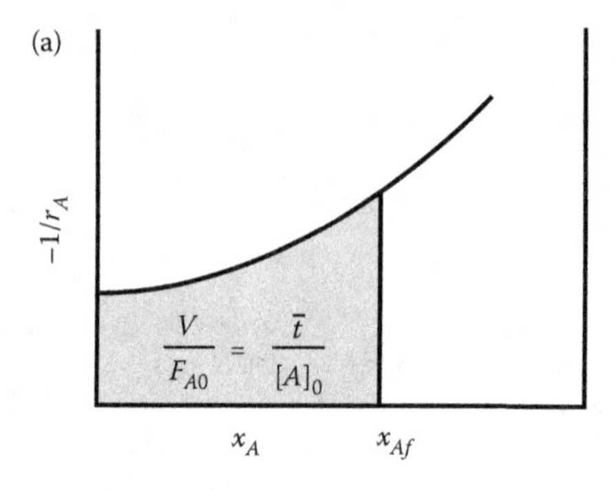

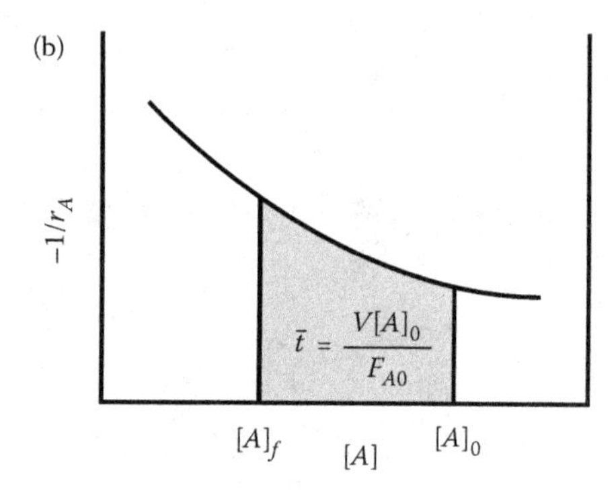

Figure 1.6 Graphical integration of (a) Equation 1.61 and (b) Equation 1.62.

Consider reaction 1.1 (with first-order kinetics) conducted in a nonisothermal PFR. The mass balance for the PFR given by Equation 1.58 must now be supplemented by the energy balance. Restricting our attention to adiabatic constant-pressure operation, the energy balance may be written as

Nonisothermal reactors

$$F_{A0}\, dX_A\left(-\Delta H_r\right) = F_t C_{p_m} dT \tag{1.63}$$

or

$$dX_A = \frac{F_t C_{p_m}}{F_{A0}\left(-\Delta H_r\right)} dT \tag{1.64}$$

Mean heat capacity

where C_{p_m} is the mean heat capacity defined as $\Sigma C_{p_i} v_i$.

Integration of Equation 1.64—assuming that $(-\Delta H_r)$ is independent of temperature—leads to Equation 1.54 with α modified as

$$\alpha = \frac{F_t C_{p_m}}{F_{A0}\left(-\Delta H_r\right)} \tag{1.65}$$

Then, assuming first-order kinetics, no volume change, that is, $\varepsilon_A = 0$, and with Arrhenius dependence of k on T, Equation 1.59 can be written as

Accounting for the temperature change in the design equation

$$\frac{V}{F_{A0}} = \frac{1}{k_0[A]_0} \int_{T_0}^{T} \frac{\alpha\, e^{E/RT}}{\{1 - [X_{A0} + \alpha(T - T_0)]\}} \left(\frac{P}{P_0}\right)\left(\frac{T_0}{T}\right) dT \tag{1.66}$$

A fully analytical solution to this equation is not possible, but a semi-analytical solution has been given by Douglas and Eagleton (1962). However, numerical solution is quite straightforward, as shown below, and appears to be the method of choice. The objective is to establish the temperature and concentration profiles in the reactor. For this purpose, we express V as the product $A_c L$ (where A_c is the cross-sectional area and

L is the length of the reactor) and use Equations 1.58 and 1.64 to write the following:

$$\Delta X_A = \beta' \Delta Z \tag{1.67}$$

$$\Delta T = \alpha' \Delta Z \tag{1.68}$$

where

$$\alpha' = \frac{(-\Delta H_r) A_c (-r_A)}{C_{p_m} F_t}, \quad \beta' = \frac{A_c (-r_A)}{F_{A0}} \tag{1.69}$$

Choosing small increments of ΔZ would provide meaningful concentration and temperature profiles along the reactor.

Perfectly mixed flow reactor (MFR)

The MFR is a continuous reactor with a constant volumetric inflow of reactants and outflow of products, but the fluid (usually liquid) within the reactor is in a state of perfect mixing. As a result, the composition of this liquid is spatially uniform within the reactor. Thus, the outflow from the reactor will be at the same composition as the liquid within. To differentiate between the two cases, we will call the MFR under perfect mixing a CSTR.

Basic CSTR equation A sketch of a continuous CSTR is shown in Figure 1.7. The inflow and outflow rates and compositions in the two streams as well as within the reactor are clearly marked. Again, the material balance of Equation 1.30 holds, leading to

Mole balance for a CSTR

$$Q_0[A]_0 - Q_f[A]_f - (-r_{Af})V = 0 \tag{1.70}$$

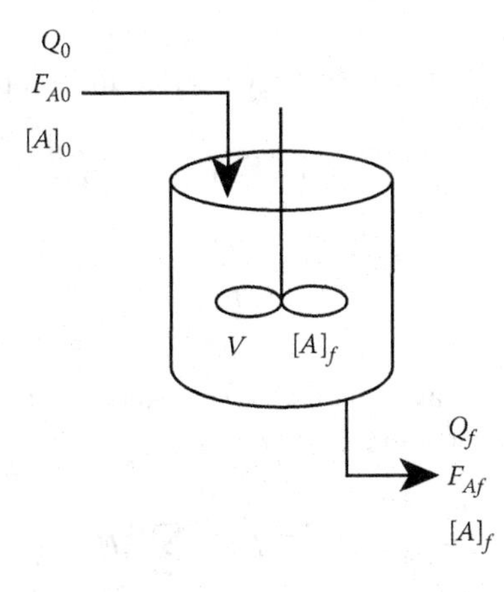

Figure 1.7 MFR, also called CSTR, under perfect mixing conditions.

for component A. Notice that the rate has been written in terms of the outlet conditions. This is a distinctive feature of a CSTR, one that enables algebraic equations to be written instead of differential equations as in the case of a PFR in which the rate and concentration change from the inlet to the outlet.

Space time

For liquid phase systems and gas phase systems without volume change, that is, $\varepsilon_A = 0$, we can assume that $Q_0 = Q_f$, and Equation 1.70 simplifies to

$$[A]_0 - [A]_f = (-r_{Af})\bar{t} \quad \text{or} \quad \bar{t} = \frac{[A]_0 - [A]_f}{-r_A} \tag{1.71}$$

From the relation $X_A = [([A]_0 - [A])/[A]_0]$, and the definition of residence time given by Equation 1.60, the above equation can be recast as

$$\frac{V}{F_{A0}} = \frac{X_{Af}}{-r_{Af}} \quad \text{or} \quad \bar{t} = \frac{[A]_0 X_{Af}}{-r_{Af}} \tag{1.72}$$

which for a first-order reaction becomes

Conversion for a first-order reaction in a CSTR

$$X_{Af} = \frac{k\bar{t}}{1 + k\bar{t}} \tag{1.73}$$

Nonisothermal operation

The heat effect in a stirred reactor is quite different from that in a PFR. Unlike in a PFR, there is no spatial variation of temperature or concentration in a stirred reactor because it is fully mixed. Hence, the effect of the energy balance is restricted to raising or lowering the single temperature at which the reactor is operating. As a result, the analysis is simpler. Thus, assuming first-order kinetics, and using T as the general notation for reactor temperature instead of T_f, we now write the energy balance.

We will start from the general energy balance for an open system:

General energy balance for an open system

$$\frac{d(U + KE + PE)}{dt} = m_0(\hat{H}_0 + \widehat{KE}_0 + \widehat{PE}_0) - m_1(\hat{H}_1 + \widehat{KE}_1 + \widehat{PE}_1) + \dot{q} + \dot{W}_s - P\frac{dV}{dt} \tag{1.74}$$

Neglecting the kinetic and potential energy terms and taking the different components into account

$$\frac{d(U)}{dt} + P\frac{dV}{dt} = \sum_{\text{in}} F_{i_0}\bar{H}_i - \sum_{\text{out}} F_i\bar{H}_i + \dot{q} + \dot{W}_s \tag{1.75}$$

or in enthalpy form

$$\frac{dH}{dt} - V\frac{dP}{dt} = \sum_{\text{in}} F_{i_0}\bar{H}_i - \sum_{\text{out}} F_i\bar{H}_i + \dot{q} + \dot{W}_s \qquad (1.76)$$

In enthalpy form

Similar to the BR case, we can write the LHS of Equation 1.76 as Equation 1.48. Since $F_i = F_{i_0} - F_{A_0}X_A\nu_i/\nu_A$,

$$m\hat{C}_p\frac{dT}{dt} - \alpha TV\frac{dP}{dt} + \sum\bar{H}_j\frac{dn_j}{dt} = \sum_{\text{in}} F_{i_0}\bar{H}_i - \sum_{\text{out}} F_{A_0}\left(\psi_i - \frac{\nu_i}{\nu_A}X_A\right)\bar{H}_i + \dot{q} + \dot{W}_s \qquad (1.77)$$

where $\psi_i = F_{i_0}/F_{A_0}$. Now the summation on the far right can be separated:

$$m\hat{C}_p\frac{dT}{dt} - \alpha TV\frac{dP}{dt} + \sum\bar{H}_j\frac{dn_j}{dt} = \sum_{\text{in}} F_{i_0}\bar{H}_i - \sum_{\text{out}} F_{i_0}\bar{H}_i + \frac{F_{A_0}X_A}{\nu_A}\sum\bar{H}_i\nu_i + \dot{q} + \dot{W}_s \qquad (1.78)$$

The enthalpy of reaction is defined as $\Delta\bar{H}_{rxn} = \Sigma\bar{H}_i\nu_i$. In and out summations enthalpy on the RHS can be combined and Equation 1.78 can be changed into

$$m\hat{C}_p\frac{dT}{dt} - \alpha TV\frac{dP}{dt} + \sum\bar{H}_j\frac{dn_j}{dt} = \sum F_{i_0}(\bar{H}_{\text{in}} - \bar{H}_{\text{out}}) + \frac{F_{A_0}X_A}{\nu_A}\sum\bar{H}_i\nu_i + \dot{q} + \dot{W}_s \qquad (1.79)$$

and can then be finalized as

$$m\hat{C}_p\frac{dT}{dt} - \alpha TV\frac{dP}{dt} + \sum\bar{H}_j\frac{dn_j}{dt} = m_{\text{in}}C_{p_m}(T_{\text{in}} - T_{\text{out}}) + \frac{F_{A_0}X_A}{\nu_A}\sum\bar{H}_i\nu_i + \dot{q} + \dot{W}_s \qquad (1.80)$$

Note that the average heat capacity in the energy balance depends on the inlet feed only. For constant-pressure, steady-state operation, the terms on the RHS of the equation disappear. Under negligible shaft work, the W_s term also disappears. Substitution of the reaction rate for first-order kinetics, using the definition of enthalpy as C_pT and the definition of $\dot{q}$ from Equation 1.51 leaves us with

$$Q_0\rho C_p(T_0 - T) + k(T)[A]V(-\Delta H_r) - U_oA_h(T - T_w) = 0 \qquad (1.81)$$

Rearranging gives

$$Q_0\rho C_p(T_0 - T) - U_oA_h(T - T_w) = -V(-\Delta H_r)k(T)[A] \quad (1.82)$$

or

$$J[(T - T_0) - \kappa(T - T_w)] = \frac{\bar{t}k}{1 + \bar{t}k} \quad (1.83)$$

where

$$J = \frac{\rho C_p}{(-\Delta H_r)[A]_0}, \quad \kappa = \frac{U_oA_h}{Q_0\rho C_p} \quad (1.84)$$

and $k(T)$ is given by the Arrhenius equation $k_0\, e^{-E/RT}$. Equation 1.83 gives the temperature at which the reactor would operate at steady states. For adiabatic operation, the heat transfer term (second on the LHS) vanishes.

This equation is important not only in the context of establishing the temperature of the reactor, but also because it can have more than one solution, leading to operation at more than one steady state.

Multiple steady states

We noted in the previous section that Equation 1.83 was very important in the context of nonisothermal operation of a CSTR. This algebraic equation has more than one solution, leading to the concept of MSS. On the other hand, the differential equation characterizing a PFR has only one solution, that is, the PFR operates at a single steady state. MSS are of particular concern to us because they can occur in the physically realizable range of variables, that is, between zero and infinity, and not at some absurd values such as a negative concentration or temperature (which would then be no more than a mathematical artifact).

MSS in a CSTR

We examine this concept of MSS further by reconsidering Equation 1.83: all the terms on the LHS, which we shall collectively designate q_-, represent heat removal (for an exothermic reaction). The terms on the RHS, designated q_+, represent heat generation. Thus, Equation 1.82 can be more expressively recast as

Heat generation and heat removal terms

$$q_-(T) = q_+(T) \quad (1.85)$$

From the nature of the terms, we see that q_- is a linear function of T, whereas q_+ is a nonlinear function of T. Figure 1.8a shows the generation curve specifically for a first-order irreversible reaction, along with the heat-removal curves with different initial temperatures, T_0, whereas Figure 1.8b shows a first-order irreversible reaction with the heat-removal

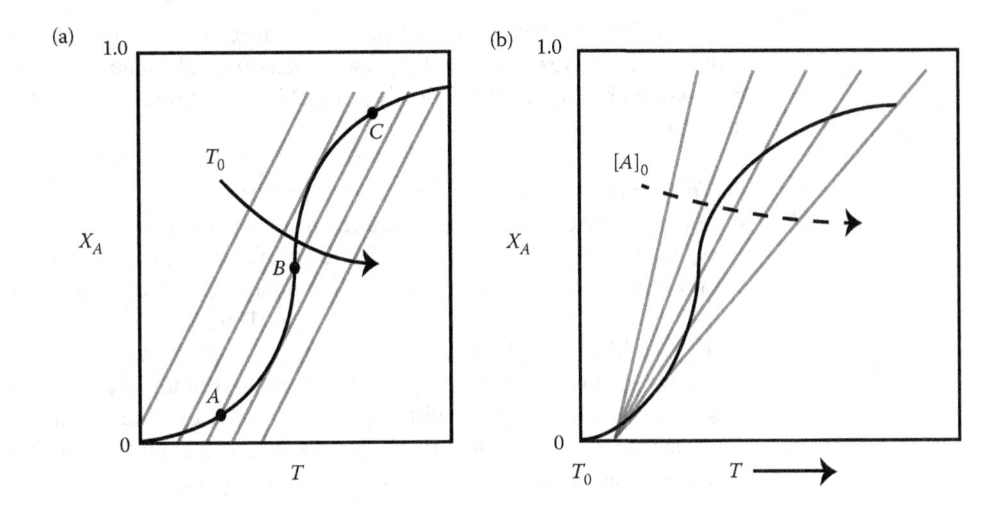

Figure 1.8 The effect of (a) initial temperature T_0 and (b) initial concentration $[A]_0$ on MSS in an adiabatic CSTR.

curves showing the effect of the initial concentration $[A]_0$. As can be seen from the figures, the heat-generation and -removal curves can intersect at one point or at three points. Operation at the temperatures corresponding to these points is referred to as *autothermal operation*, and the three states are referred to as *multiple steady states* (see the original paper by van Heerden, 1953). Among many books, reviews, and leading articles on the general subject of instabilities and oscillations in chemical reactions and reactors that have since appeared, we particularly recommend the book by Varma et al. (1999) examining various aspects of the parametric sensitivity in chemical reactors.

Adiabatic CSTR

Adiabatic CSTR

When a CSTR is operated adiabatically, the heat transfer term in Equation 1.83 vanishes and we obtain

$$J(T - T_0) = \frac{\bar{t}k(T)}{1 + \bar{t}k(T)} \tag{1.86}$$

Let us now prepare constant $\bar{t}$ plots. For this, we make use of Equation 1.73:

$$X_A = \frac{\bar{t}k(T)}{1 + \bar{t}k(T)} \tag{1.87}$$

This represents the RHS of Equation 1.86, which can be plotted as X_A versus T for different constant values of residence time $\bar{t}$. This is the sigmoidal material balance plot (Figure 1.8). We also plot X_A versus T (the LHS of Equation 1.86), which gives a straight line with slope J.

Theoretically, the two curves can intersect at three points. Points A and C are physically realizable and represent the two steady states of operation, while point B is not physically realizable and represents the (third) unstable steady state.

Stability of the steady states We have seen that there can be two stable steady states, one at a low temperature corresponding to quench conditions, and the other at a high temperature corresponding to ignition conditions. An important consideration is the approach to the steady state, for that would determine the fate of a reaction. Thus, if we started at any initial condition, will the reaction approach one of the steady states, and if so how quickly? Alternatively, if we started the reaction at a condition close to a stable steady state, will it approach that steady state? To answer these questions, we must modify the steady-state mass and heat and balance equations to include a time-dependent component. The resulting *transient equations* are

$$\bar{t}\frac{dX_A}{dt} = -X_A + \frac{\bar{t}(-r_{Af})}{[A]_0} \tag{1.88}$$

$$\bar{t}\frac{dT}{dt} = T_0 - T - k(T - T_c) + \frac{\bar{t}}{[A]_0}(-r_{Af}) \tag{1.89}$$

A stable steady state

A steady state is said to be stable when the system returns to it after a small perturbation. Alternatively, the perturbation can grow exponentially with time or the reaction can be quenched to extinction. To determine which one of these alternatives will prevail, we linearize the equation about the steady state and examine the behavior of the perturbations in conversion and temperature, that is, of

$$\alpha = X_A - X_{A,SS} \tag{1.90}$$

$$\beta = T - T_{SS} \tag{1.91}$$

These disturbances can be expressed in terms of the simple differential equations

$$\bar{t}\frac{d\alpha}{dt} = a\alpha + b\beta \tag{1.92}$$

$$\bar{t}\frac{d\beta}{dt} = c\alpha + d\beta \tag{1.93}$$

which can then be combined to give

$$\frac{d^2\alpha}{dt^2} - (a + d)\frac{d\alpha}{dt}(ad + bc)\alpha = 0 \tag{1.94}$$

where a, b, c, and d are constant coefficients defined by lengthy expressions involving the steady-state values of X_A and T, activation energy of the reaction, rate constant, and residence time. The general solution to this equation is

$$\alpha(t) = C_1\, e^{\lambda t/\bar{t}} + C_2\, e^{-\lambda t/\bar{t}} \tag{1.95}$$

where the *eigenvalues* $\lambda_{\pm}$ are related to the coefficients by the expression

$$\lambda = \frac{1}{2}(a + d) \mp \frac{1}{2}[(a - d)^2 + 4bd]^{1/2} \tag{1.96}$$

The solution is stable if the real part of λ is negative, that is, if $[(a-d)^2 + 4bc]^{1/2} > 0$. On evaluating this term using actual values of the coefficients a, b, c, and d, we find that (i) when there is only one steady state, it is a stable one, and (ii) when there are three steady states, only the extreme states are stable and the middle one is unstable. Further analysis of MSS in an adiabatic CSTR reveals some interesting features (see, e.g., Schmidt, 1998), which we summarize in the form of qualitative statements below.

1. When the initial temperature T_0 in an adiabatic CSTR is varied slowly, different heat-removal lines are obtained as shown in Figure 1.8a. As expected, either one or three steady states can be obtained.
2. When the initial concentration $[A]_0$ is varied, the resulting situation for an adiabatic CSTR is sketched in Figure 1.8b. The number of steady states varies from a single low-conversion steady state at low $[A]_0$ to a single high-conversion steady state at high $[A]_0$ and MSS in between.
3. An important consideration in reactor operation is the time dependence of MSS, that is, the transients that develop during startup or shutdown. Any apparent approach to the middle unsteady steady state is deceptive for it quickly turns to either of the extreme steady states. The behavior is identical with respect to temperature.
4. When a CSTR is operated nonadiabatically, a heat removal term (for an exothermic reaction) must be added to the system equations. This complicates the situation and increases the likelihood of MSS.

Comparison of BR, PFR, and MFR

We derived above the performance equations for the three ideal reactors: batch, plug flow, and mixed flow. The BR and PFR are exactly comparable, with the reaction time t in BR related to the residence time $\bar{t}$ at the corresponding axial position in PFR by

$$t = \bar{t} = \frac{V}{Q} \quad \text{or} \quad \bar{t} = \frac{z}{u} \text{ for a constant area PFR} \tag{1.97}$$

where z/u indicates the fractional volume transversed by the fluid and u is the linear velocity of the fluid. An MFR, on the other hand, operates on a different principle, that of complete backmixing, and therefore no direct relationship between the MFR and BR or PFR is possible.

Explore yourself

1. Identify systems with chemical conversions in your immediate environment. Can you classify the reactor types as one (or more) of the ideal reactors. For the situations where you had to use more than one reactor type, explain clearly why you needed to do so. Explain clearly the improvement in accuracy you achieved after adding a second (or more) reactor.
2. Browse through the Internet for the most recent investments in chemical manufacturing. Choose a particular plant or chemical manufacturing technology and answer the following questions:
   a. Can you identify the reactor type they use?
   b. How old is the design/technology?
   c. Do they have a patent on the reactor? If they do not have a patent on the reactor technology, can you identify why?
   d. When was the technology first put in place? What improvements were made between the time it was first installed and now?
   e. Can you draw a map of the technology development for this process?
3. Internal combustion engines (ICE) are a special class of reactors. There are a large number of ICE. In this question, we ask you to go a little bit deep into the subject.
   a. List as many types of ICE as possible.
   b. What is the fuel?
   c. What is the “advertised” engine efficiency? Why are some of the engines more efficient than the others?
   d. Can you model the ICE as one or more of ideal reactors? Prepare a table with the information you gathered so far, and also list whether the ICE can be classified as a BR, a PFR, or a CSTR.
   e. Choose one of them and describe the features as a chemical reactor. Is it isothermal, adiabatic? Is it constant volume, constant pressure? Is it an open system or a closed system?
   f. What can you say about the CO_2 footprint of an ICE operation?
4. A very small amount of NO is oxidized during combustion in the ICE discussed in the previous problem.
   a. Suggest as many solutions as possible to inhibit NO_x formation.

   b. What is the present-day technological solution to NO_x emission problem? What type of reactor is used? Is it universal?
   c. List the salient features of a diesel engine and a gasoline engine "after-treatment" technology for NO_x abatement from the point of view of a reactor analysis. Compare and contrast them from the point of view of CRE.
5. Relate the chaos theory to CRE in one of the following areas:
   a. Multiple steady states in adiabatic reactors
   b. Kinetic phase transitions
   c. Parametric sensitivity of chemical reactors
   d. Vapor–liquid equilibria
   e. Oscillatory reactions

References

Aris, R., *Introduction to the Analysis of Chemical Reactors*, Prentice-Hall, Englewood Cliffs, NJ, 1965.
Aris, R., *Elementary Chemical Reactor Analysis*, Prentice-Hall, Englewood Cliffs, NJ, 1969.
Doraiswamy, L.K., *Organic Synthesis Engineering*, Oxford University Press, NJ, 2001.
Douglas, J.M. and Eagleton, L.C., *Ind. Eng. Chem. Fundam.*, **1**, 116, 1962.
Sandler, S.I., *Chemical, Biochemical and Engineering Thermodynamics*, 4th edition, Wiley, NY, 2006.
Schmidt, L.D., *The Engineering of Chemical Reactions*, Oxford University Press, NY, 1998.
Van Heerden, C., *Ind. Eng. Chem.*, **45**, 1242, 1953.
Venkitakrishnan, G.R. and Doraiswamy, L.K., *National Chemical Laboratory Report*, Pune, India, 1982.

Bibliography

For the detailed analysis of multiple steady states, refer to the following:

Aris, R., *Problems in the Dynamics of Chemical Reactors*, Paper Presented at the International Chemical Reaction Engineering Conference, Pune, 1984.
Gray, P. and Scott, S.K., *Chemical Oscillations and Instabilities: Non-Linear Chemical Kinetics*, Clarendon Press, Oxford, 1990.
McGreavy, C. and Dunbobbin, B.R., In *Frontiers in Chemical Reaction Engineering* (Eds. Doraiswamy, L.K., and Mashelkar, R.A.), Wiley Eastern, New Delhi, India, 1984.
Razon, L.F. and Schmitz, R.A., *Chem. Eng. Sci.*, **42(5)**, 1005, 1987.
Slin'ko, M.M. and Jaeger, N.I., *Oscillating Heterogeneous Catalytic Systems*, Elsevier Science, Amsterdam, 1994.
Varma, A., Morbidelli, M., and Wu, H., *Parametric Sensitivity in Chemical Systems*, Cambridge University Press, Cambridge, UK, 1999.

Chapter 2 Complex reactions and reactors

Chapter objectives

In this chapter, we will review complex reactions and the concepts of selectivity and yield around such reactions. Upon successful completion of this chapter, you should be able to

- Differentiate between multiple reactions and multistep reactions.
- Define and use the concepts of yield and selectivity.
- Select the best reactor type for the highest selectivity in a multiple reaction scheme.
- Select the most suitable reactor operation mode, that is, batch, continuous, or semibatch, to improve selectivity and yield.
- Compare and contrast batch, continuous, and semi-batch reactors from the point of view of backmixing.

Introduction

When a reactant or a set of reactants undergoes several reactions (at least two) simultaneously, the reaction is said to be a complex reaction. The total conversion of the key reactant, which is used as a measure of reaction in the case of simple reactions, has little meaning in complex reactions, and what is of primary interest is the fraction of reactant converted into the *desired product*. Thus, the more pertinent quantity is *product distribution* from which the conversion to the desired product can be calculated. This is usually expressed in terms of the *yield* or *selectivity* of the reaction with respect to the desired product.

From the design point of view, an equally important consideration is the analysis and quantitative treatment of complex reactions. A common example is the dehydration of alcohol represented by the following reactions:

$$C_2H_5OH \rightarrow C_2H_4 + H_2O \qquad \text{(R1.1)}$$

$$C_2H_5OH + C_2H_4 \rightarrow C_2H_5OC_2H_5 \qquad \text{(R1.2)}$$

Complex multiple reactions

The net reaction is the sum of the reactions given by

$$2C_2H_5OH \rightarrow C_2H_5OC_2H_5 + H_2O \qquad (R1.3)$$

We refer to such a set of simultaneous reactions as complex *multiple* reactions.

It is also important to note that many chemical syntheses involve a number of steps, each carried out under different conditions (and sometimes in different reactors), leading to what we designate as *multistep reactions* (normally referred to by chemists as a *synthetic scheme*). This could, for example, be a sequence of reactions such as dehydration, oxidation, Diels–Alder, and hydrogenation. The purpose of this chapter is to outline simple procedures for the treatment of complex multiple and multistep reactions and to explain the concepts of selectivity and yield.

Reduction of complex reactions

Stoichiometry of simple and complex reactions

Consider a "simple" reaction such as the chlorination of methane to methyl chloride:

$$CH_4 + Cl_2 \rightarrow CH_3Cl + HCl \qquad (R2)$$

Understanding the kinetics and mechanism of this reaction would involve understanding the elementary steps leading to the ultimate reaction represented by R2. These steps are

$$Cl_2 \rightarrow 2Cl^{\cdot} \qquad (R2.1)$$

$$Cl^{\cdot} + CH_4 \rightarrow HCl + CH_3^{\cdot} \qquad (R2.2)$$

$$CH_3^{\cdot} + Cl_2 \rightarrow CH_3Cl + Cl^{\cdot} \qquad (R2.3)$$

Consider now the set of reactions

$$CH_4 + Cl_2 \rightarrow CH_3Cl + HCl \qquad (R3.1)$$

$$CH_3Cl + Cl_2 \rightarrow CH_2Cl_2 + HCl \qquad (R3.2)$$

$$CH_2Cl_2 + Cl_2 \rightarrow CHCl_3 + HCl \qquad (R3.3)$$

$$CHCl_3 + Cl_2 \rightarrow CCl_4 + HCl \qquad (R3.4)$$

The difference between reactions R2 and R3 is that the intermediates in reaction R2 are of a transitory nature, whereas those in reaction R3 are stable compounds present in finite quantities in the final product. Thus, reaction R2 can be treated as a "single" reaction, but reaction R3 cannot. It is a multiple reaction. Note that the single reaction is also constituted of many steps. These steps are often referred to as elementary steps or reactions. Our primary concern in this chapter is with reactions of the type represented by R3.

Mathematical representation of simple and complex reactions

The reaction rate expression given by

$$A + 2B \rightarrow R \tag{R4}$$

can be equally represented as

$$-A - 2B + R = 0 \tag{R5}$$

which is often mathematically more convenient. In this equation, we bring all the constituents to one side and set the other side equal to zero. Note that the reactants are denoted by a negative sign and the products by a positive sign. The equation is then further modified to read as

$$\sum_{i=A,B,R}^{N} v_i A_i = 0 \tag{2.1}$$

where the term within the summation sign represents $(-A - 2B + R)$. For the most general case of a single reaction involving a large number of components $A_1, A_2, \ldots, A_j, \ldots A_N$, we can write

$$\sum_{i=1}^{N} v_i A_i = 0 \tag{2.2}$$

where i represents any species from 1 to N.

If a reaction system consists of a number of components reacting with one another in more than one reaction, the result is a complex reaction network. Mathematically, a complex reaction consisting of N components and M reactions can be represented as

$$\sum_{i=1}^{N} v_{ij} A_i = 0, \quad j = 1,2,\ldots,M \tag{2.3}$$

where v_{ij} is the stoichiometric coefficient of A_i in the jth reaction.

Mathematical representation of a complex reaction network

We shall now see how a complex reaction network can be conveniently represented in matrix form. Thus, consider a simple reaction

$$A \rightarrow R \tag{R6}$$

with the rate equation

$$-\frac{d([A]V)}{dt} = -r_A V \tag{2.4}$$

Clearly, a single rate equation is all that is needed to kinetically describe the system. But when extended to a complex reaction represented by Equation 2.3, a set of N ordinary differential equations, one for each component, must be written to describe the system. These may be expressed concisely in the language of matrix mathematics as

Reaction rates in a complex scheme

$$\frac{d([\mathbf{A}]V)}{dt} = \boldsymbol{\nu}\boldsymbol{r}V \qquad (2.5)$$

where $[\mathbf{A}]$ is a vector ($N \times 1$ matrix) of component concentrations, $\boldsymbol{\nu}$ an ($M \times N$) matrix of stoichiometric coefficients, and $\boldsymbol{r}$ a vector ($M \times 1$ matrix) of reaction rates.

Independent reactions

A typical complex organic reaction usually consists of a number of reactions, some of which can be obtained by algebraic addition of two or more reactions of the network. Thus in Scheme 2.1 describing the dehydration of alcohol, reaction R1.3 can be obtained by the addition of reactions R1.1 and R1.2, and hence is not an independent reaction. This can be stated more formally as follows: For a set of reactions to be *independent*, no reaction from the set shall be obtainable by algebraic additions of other reactions (as such or in multiples thereof) and each member shall contain one new species exclusively.

Mathematically, if a set of complex reactions is represented in matrix form, then the number of independent reactions is given by the rank of the matrix, as illustrated in the example below. It can also be found by a simple stepwise manipulation of the matrix (see Aris, 1969).

Example 2.1: Number of independent reactions in the reactions of propylene glycol in the cyclization of ethylenediamine and propylene glycol, and in the ethylation of aniline

The cyclization of ethylenediamine (EDA) and propylene glycol (PG) over a mixture of zinc and chromium oxides to 2-methylpyrazine (MP) is a basic step in the synthesis of 2-amidopyrazine, a well-known antitubercular drug. This is a highly complex reaction in which EDA and PG each react independently to give a variety of products, as shown below (Forni and Miglio, 1993). It is desired to find the number of independent reactions from this set.

$$CH_3\text{–}CHOH\text{–}CH_2\text{–}OH + H_2 \rightarrow CH_3\text{–}CH_2\text{–}OH + CH_3\text{–}OH$$

$$CH_3\text{–}CH_2\text{–}OH \rightleftharpoons CH_3\text{–}CHO + H_2$$

$$CH_3\text{–}CHOH\text{–}CH_2\text{–}OH \rightarrow H_2O + CH_3\text{–}CO\text{–}CH_3$$

$$CH_3\text{–}CHOH\text{–}CH_2\text{–}OH \rightarrow H_2O + CH_3\text{–}CH_2\text{–}CHO$$

$$CH_3\text{–}CH_2\text{–}CHO + H_2 \rightarrow CH_3\text{–}CH_2\text{–}CH_2\text{–}OH$$

$$CH_3\text{–}CHOH\text{–}CH_2\text{–}OH \rightarrow CH_2{=}CH\text{–}CH_2\text{–}OH + H_2O$$

$$CH_3\text{–}CHOH\text{–}CH_2\text{–}OH \rightarrow CH_2{=}CH\text{–}CHO + H_2O + H_2$$

$$CH_3\text{–}CH_2\text{–}OH + CH_3\text{–}CH_2\text{–}OH \rightarrow CH_3\text{–}CO\text{–}CH_2\text{–}CH_3 + CO + 3H_2$$

$$2CH_3\text{–}CH_2\text{–}OH \rightarrow CH_3\text{–}CO\text{–}CH_3 + CO + 3H_2$$

$$2CH_3\text{–}CH_2\text{–}CH_2\text{–}OH \rightarrow CH_3\text{–}CH_2\text{–}CO\text{–}CH_2\text{–}CH_3 + CO + 3H_2$$

$$2CH_3\text{–}CO\text{–}CH_3 \rightarrow (CH_3)_2C{=}CH\text{–}CO\text{–}CH_3 + H_2O$$

$$2CH_3\text{–}CH_2\text{–}CHO \rightarrow CH_3\text{–}CH_2\text{–}CH{=}C(CH_3)\text{–}CHO + H_2O$$

SOLUTION

The reaction scheme consists of 12 reactions involving 16 species. Thus, a (12 × 16) matrix of stoichiometric coefficients can be written as shown below where the reactions are marked R1, R2,. . ., R12.

	$CH_3CHOHCH_2OH$	H_2	CH_3CH_2OH	CH_3OH	CH_3CHO	H_2O	CH_3COCH_3	CH_3CH_2CHO	$CH_3CH_2CH_2OH$	$CH_2{=}CHCH_2OH$	$CH_2{=}CHCHO$	$CH_3COCH_2CH_3$	CO	$CH_3CH_2COCH_2CH_3$	$(CH_3)_2{=}CHCOCH_3$	$CH_3CH_2CH{=}C(CH_3)CHO$
R1	1	1	−1	−1	0	0	0	0	0	0	0	0	0	0	0	0
R2	0	−1	1	0	−1	0	0	0	0	0	0	0	0	0	0	0
R3	1	0	0	0	0	−1	−1	0	0	0	0	0	0	0	0	0
R4	1	0	0	0	0	−1	0	−1	0	0	0	0	0	0	0	0
R5	0	1	0	0	0	0	0	1	−1	0	0	0	0	0	0	0
R6	1	0	0	0	0	−1	0	0	0	−1	0	0	0	0	0	0
R7	1	−1	0	0	0	−1	0	0	0	0	−1	0	0	0	0	0
R8	0	−3	1	0	0	0	0	0	1	0	0	−1	−1	0	0	0
R9	0	−3	2	0	0	0	−1	0	0	0	0	0	−1	0	0	0
R10	0	−3	0	0	0	0	0	0	2	0	0	0	−1	−1	0	0
R11	0	0	0	0	0	−1	2	0	0	0	0	0	0	0	−1	0
R12	0	0	0	0	0	−1	0	2	0	0	0	0	0	0	0	−1

We now make use of the fact that the rank of the matrix is the number of independent rows in the matrix, which in turn is the number of independent reactions. We use MATLAB® to determine the rank of the matrix.

Rank of a matrix gives the number of independent reactions

$$\text{Rank} = \text{Number of independent reactions} = 12$$

In other words, all the reactions in the set are independent. A useful conclusion from this illustration is that it is not always possible to reduce the number of reactions to be considered from a given complex reaction sequence.

Rate equations

There are two aspects to a rate equation: its formulation from laboratory kinetic data and its use in reactor design. We shall consider in this section a procedure for formulating rate expressions for the independent reactions of a complex set and defer the question of reactor design to later sections in this chapter.

The concept of extent of reaction

Consider a simple reaction

$$\nu_A A + \nu_B B \rightarrow \nu_R R \tag{R7}$$

with no restriction regarding volume change. In such a situation, the amounts of A and B converted and R formed can be expressed in terms of the actual number of moles before and after reaction, since these are independent of volume change:

The extent of a reaction

$$\frac{N_A - N_{A0}}{\nu_A} = \frac{N_B - N_{B0}}{\nu_B} = \frac{N_R - N_{R0}}{\nu_R} = \xi \tag{2.6}$$

where ξ is the *extent of reaction* or *reaction coordinate.* Note that ξ has the units of moles, while the conversion X_A is dimensionless. The rate and extent of reaction are obviously related.

Thus, let us consider the reaction

$$2A + 3B \rightleftharpoons R + 2S \tag{R8}$$

The rate of this reaction can only be understood in terms of the rates of formation or disappearance of the various components. The rate of a reaction as such is difficult to define unless it is postulated that it is based on the rate of formation of a product or disappearance of a reactant with a specified stoichiometric coefficient. *Usually it is the rate of formation of a product with a stoichiometric coefficient of unity.* Thus, in this case we choose the rate of formation of R. The rate of disappearance of A is then twice this rate, that of B is three times this rate, and the rate of formation of S is twice this rate.

Let us now extend the concept to each of the reactions comprising a complex set, such as

$$2A + 3B \rightleftharpoons 2R \tag{R9}$$

$$2C + D \rightleftharpoons S$$

The rates of formation or disappearance of the different components of the reaction can then be expressed as multiples of this reaction coordinate ξ. For example, for reaction R9 we can write

$$\frac{N_A - N_{A0}}{-2} = \frac{N_B - N_{B0}}{-3} = \frac{N_R - N_{R0}}{2} = \xi_1 \tag{2.7}$$

$$\frac{N_C - N_{C0}}{-2} = \frac{N_D - N_{D0}}{-1} = \frac{N_S - N_{S0}}{1} = \xi_2 \tag{2.8}$$

The compositions of A, B, C, D, R, and S can then be expressed in terms of the reaction coordinates ξ_1 and ξ_2 defined by Equations 2.7 and 2.8, respectively. Hence, the number of equations would be (a) six if written in terms of the rates of formation/disappearance of the individual components or (b) two if written in terms of the extent of reaction in each step.

Extent of reaction decreases the number of independent variables

Determination of the individual rates in a complex reaction

The method depends on whether the reactor is operated as a PFR or an MFR. If operated as an MFR, the experimentally determined product composition directly gives the rates. Thus, considering A, the rate is given by

$$-r_{Af} = \frac{[A]_0 X_{Af}}{\bar{t}} \tag{2.9}$$

where $\bar{t} = [A]_0 V_r / F_{A0}$. Note that r_{Af} is the rate corresponding to the final composition. By varying the initial composition and flow rate, the rates corresponding to different compositions can be obtained.

Space time

On the other hand, for a PFR the outlet conversion (or concentration) of any component i must be plotted as a function of $\bar{t}$, and the rate at any value of $\bar{t}$ is determined by measuring the slope of the curve at that value (Figure 2.1), for example,

$$\text{Measured slope} = r_R = \frac{d[R]}{d\bar{t}} \tag{2.10}$$

This rate corresponds to the composition at that value of $\bar{t}$. Reactions carried out on a catalyst in a tubular reactor (which conforms to plug flow) can also be treated in a similar way as described in Chapter 1, with this difference that we now plot conversion to i as a function of W/F_{A0} where W is the weight of the catalyst. The dimensions of the rate in this case would be moles per unit weight of catalyst per unit time.

Selectivity and yield

We consider below the concepts of yield and selectivity as applied to multiple as well as multistep reactions. The reaction engineering

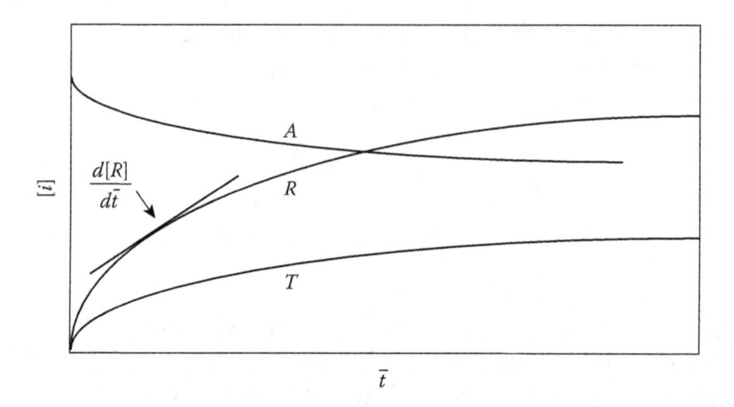

Figure 2.1 Obtaining the rate from the concentration versus time, or space time, data.

literature uses the two terms "multiple" and "multistep" interchangeably. However, when dealing with an overall organic synthesis comprising many separate steps, it is desirable to distinguish between the two.

Definitions

Consider the reaction

$$A + B \rightarrow R \qquad \text{(R10)}$$

$$C + B \rightarrow S$$

Conversion, yield, and selectivity based on this example are then defined as

Conversion

$$\text{Conversion, } X_A = \frac{\text{moles } A \text{ converted}}{\text{moles } A \text{ fed}} = \frac{[A]_0 - [A]}{[A]_0}$$

Yield

$$\text{Yield, } Y_R = \frac{\text{moles } R \text{ formed}}{\text{moles } A \text{ fed}} = \frac{[R] - [R]_0}{[A]_0}$$

Selectivity

$$\text{Selectivity, } S_R = \frac{\text{moles } R \text{ formed}}{\text{moles } A \text{ converted}} = \frac{[R] - [R]_0}{[A]_0 - [A]}$$

Note that

$$\text{Yield} = \text{conversion} \times \text{selectivity}$$

Analytical solutions

To illustrate the procedures used in the analysis of multiple reactions, we considered relatively complex cases in the earlier sections. Such complex schemes do not normally admit analytical solutions. The simplest are the *parallel* and the *consecutive* (or series) reactions:

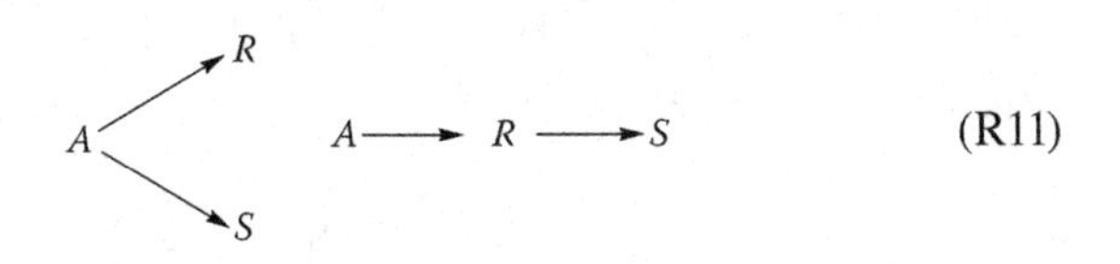

Parallel Series (consecutive)

Three other important schemes commonly encountered are

$$A + B \longrightarrow R \qquad A \longrightarrow R \longrightarrow S \qquad A \longrightarrow R, \; A \longrightarrow S \qquad \text{(R12)}$$
$$R + B \longrightarrow S \qquad A \longrightarrow T, \; R \longrightarrow U \qquad R \longrightarrow S$$

Series–parallel Denbigh Triangular

Example 2.2: Concentration versus space time in series reactions

For the series reaction scheme given in R11, derive the concentration versus space-time expression for both PFR and CSTR given the rate constants for the elementary steps as k_A and k_R. Qualitatively plot the concentrations of all of the species as functions of space time. Comment on the reactor operation scheme if the desired product is R or S.

SOLUTION

Optimum space time in a CSTR:

$$A \xrightarrow{k_A} R \xrightarrow{k_R} S$$

The rates are given by

$$r_A = -k_A[A]$$

$$r_R = k_A[A] - k_R[R]$$

$$r_S = k_R[R]$$

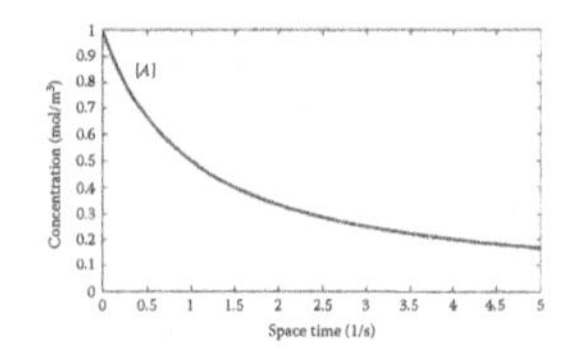

The mole balance for A in a CSTR:

$$Q[A]_0 - Q[A] + r_A V = 0$$

$$Q[A]_0 - Q[A] - k_A[A]V = 0$$

is solved to obtain

$$[A] = \frac{[A]_0}{1 + \bar{t}k_A} \quad \text{where } \bar{t} = \frac{V}{Q}$$

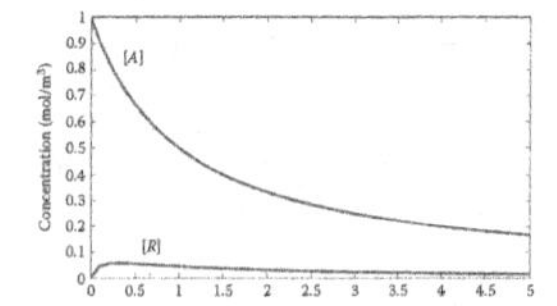

The mole balance for [R]:

$$-Q[R] + r_R V = 0$$

$$-Q[R] + (k_A[A] - k_R\,[R])V = 0$$

yields

$$[R] = \frac{\bar{t}k_A[A]_0}{(1 + \bar{t}k_A)(1 + \bar{t}k_R)}$$

The mole balance for [S]:

$$-Q[S] + r_S V = 0$$

$$-Q[S] + k_R[R]V = 0$$

gives the exit concentration of S as

$$[S] = \frac{\bar{t}^2 k_A k_R[A]_0}{(1 + \bar{t}k_A)(1 + \bar{t}k_R)}$$

It is clear from the equations that [A] decreases, [S] increases, and [R] goes through a maximum with increasing $\bar{t}$. As a result, it is possible to choose an optimum space time to maximize R if it is the desired product. If [S] is the desired product, longest possible time for the reaction (determined based on the relative rate constants) is optimum.

Optimum space time in a PFR:

We will again start from the design equation for a PFR for A:

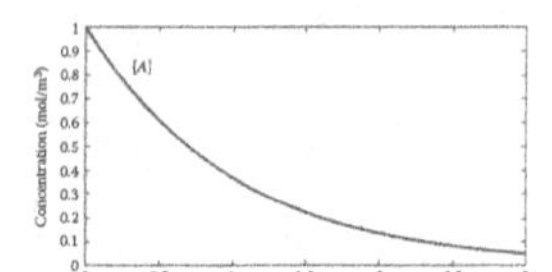

$$\frac{d[A]}{d\bar{t}} = r_A$$

$$\frac{d[A]}{d\bar{t}} = -k_A[A]$$

Solved with the initial condition when $\bar{t} = 0$ and $[A] = [A]_0$ to yield

$$\ln\frac{[A]}{[A]_0} = -k_A\bar{t} \quad \text{or} \quad [A] = [A]_0 e^{-k_A\bar{t}}$$

Balance on R:

$$\frac{d[R]}{d\bar{t}} = r_R = k_A[A] - k_R[R] = k_A[A]_0 e^{-k_A\bar{t}} - k_R[R]$$

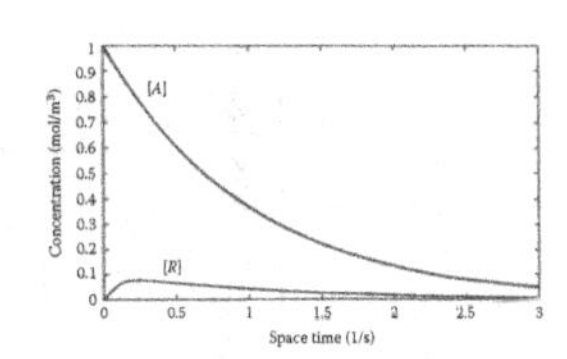

This is a first-order linear differential equation with the initial condition at $\bar{t} = 0$ $[RR] = 0$, giving $[R]$ as

$$[R] = \frac{k_A}{k_A + k_R}(e^{-k_A\bar{t}} - e^{-k_R\bar{t}})[A_0]$$

Balance on S:

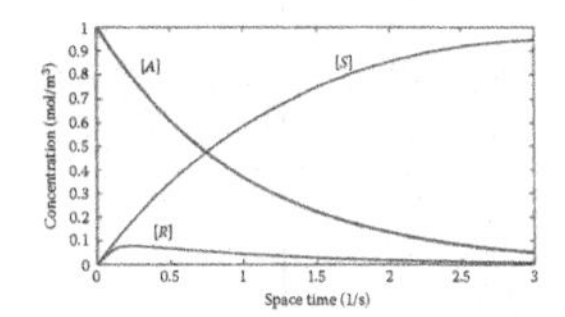

$$\frac{d[S]}{d\bar{t}} = r_S = k_R[R]$$

One can substitute for $[R]$ and integrate, or use simple material balance for $[S]$:

$$[S] = [A]_0 - [A] - [R]$$

In addition to the tuning of the reactor for maximum selectivity, this type of information is also useful for the identification of the intermediates and reaction mechanism. The evolution of concentration with time provides a clear indication of whether a particular component is an end product or an intermediate product which undergoes further reaction even as it is formed.

Exercise

The rate equations and solutions for all the five schemes under plug flow or fully mixed conditions are given in Table 2.1. Derive the design equations for the complex reaction schemes given above. Compare your results with the results presented in Table 2.1.

An important feature of the concentration–time (t or $\bar{t}$) profiles for the series scheme mentioned above for an intermediate product is that it shows a maximum at a specific time t_{max}. If that compound happens to be the desired product, it is best to operate the reactor at t_{max}.

Relationships for $[R]_{max}$ in terms of the kinetic parameters of the reactions are important in maximizing production. They are summarized in Table 2.2 for the reaction schemes given in 2.11 and 2.12.

Maximizing selectivity in a complex reaction: Important considerations Choosing a reaction pathway for any reaction, simple or complex, is always a difficult task, more so for a new, untried product. Even so, several considerations are common to both new and old products.

Table 2.1 Integrated Forms of Rate Equations for the More Common Complex Scheme

Reaction	Rate Equations	Solutions for BR/PFR (Use t for BR, for PFR)	Solutions for MFR
$A \overset{1}{\nearrow} R$, $A \overset{2}{\searrow} S$ (Parallel)	$-\frac{d[A]}{dt} = (k_1 + k_2)[A]$ $\frac{d[R]}{dt} = k_1[A]$ $\frac{d[S]}{dt} = k_2[A]$	$\ln\frac{[A]_0}{[A]} = (k_1 + k_2)t$ $[R] = [R]_0 + \frac{k_1[A]_0}{k_1 + k_2}\{1 - \exp[-(k_1 + k_2)t]\}$ $[S] = [S]_0 + \frac{k_2[A]_0}{k_1 + k_2}\{1 - \exp[-(k_1 + k_2)t]\}$	$\frac{[A]}{[A]_0} = \frac{1}{1 + (k_1 + k_2)\bar{t}}$ $\frac{[R]}{[A]_0} = \frac{[R]_0}{[A]_0} - \frac{k_1\bar{t}}{1 + (k_1 + k_2)\bar{t}}$ $\frac{[S]}{[A]_0} = \frac{[R]_0}{[A]_0} - \frac{k_2\bar{t}}{1 + (k_1 + k_2)\bar{t}}$
$A \xrightarrow{1} R \xrightarrow{2} S$ (Series)	$-\frac{d[A]}{dt} = k_1[A]$ $\frac{d[R]}{dt} = k_1[A] - k_2[R]$ $\frac{d[S]}{dt} = k_2[R]$	$\frac{[A]}{[A]_0} = \exp(-k_1 t)$ $\frac{[A]}{[A]_0} = \frac{k_1}{k_2 - k_1}[\exp(-k_1 t)$ $- \exp(-k_2 t)] + \frac{[R]_0}{[A]_0}\exp(-k_2 t)$ $[S] = [A]_0 + [R]_0 + [S]_0 - [A] - [R]$	$\frac{[A]}{[A]_0} = \frac{1}{1 + k_1\bar{t}}$ $\frac{[R]}{[A]_0} = \frac{k_1\bar{t}}{(1 + k_1\bar{t}) + (1 + k_2\bar{t})}$ $+ \frac{[R]_0}{[A]_0}\frac{1}{1 + k_2\bar{t}}$ $[S] = [A]_0 + [R]_0 + [S]_0 - [A] - [R]$
$A + B \xrightarrow{1} R$ $R + B \xrightarrow{2} S$ (Series–parallel)	$-\frac{d[A]}{dt} = k_1[A][B]$ $\frac{d[R]}{dt} = k_1[A][B] - k_2[R][B]$ $\frac{d[S]}{dt} = k_2[R][B]$ $-\frac{d[B]}{dt} = k_1[A][B] + k_2[R][B]$	$\frac{d[A]}{dt} = k_1[A][A]_0$ $\times\left\{\frac{[B]_0}{[A]_0} - \frac{k_2}{k_1 - k_2}\left[\frac{[A]}{[A]_0} - 1\right]\right.$ $\left.+ \frac{k_2}{k_1 - k_2}\left[\left(\frac{[A]}{[A]_0}\right)\frac{k_2}{k_1} - 1\right] + \frac{[A]}{[A]_0} - 1\right\}$ $[A] = f(t)$ (numerical soln.) $\frac{[B]}{[A]_0} = \frac{[B]_0}{[A]_0} - \frac{k_2}{k_1 - k_2}\left[\frac{[A]}{[A]_0} - 1\right]$ $+ \frac{k_1}{k_1 - k_2}\left[\left(\frac{[A]}{[A]_0}\right)\frac{k_2}{k_1} - 1\right] + \frac{[A]}{[A]_0} - 1$	$\frac{[A]}{[A]_0} = 1 - \frac{\bar{t}}{[A]_0}k_1[A][B]$ $\frac{[R]}{[A]_0} = \frac{[R]_0}{[A]_0} + \frac{\bar{t}}{[A]_0}$ $(k_1[A][B] - k_2[R][B])$ $\frac{[B]}{[A]_0} = \frac{[B]_0}{[A]_0} - \frac{\bar{t}}{[A]_0}$ $(k_1[A][B] + k_2[R][B])$ $[S] = [A]_0 + [R]_0 + [S]_0 - [A]$

		$\frac{[R]}{[A]_0} = -\frac{[A]}{[A]_0}\left(\frac{k_1}{k_1 - k_2}\right) + \frac{k_1}{k_1 - k_2}\left(\frac{[A]}{[A]_0}\right)^{\frac{k_2}{k_1}}$ $\frac{[S]}{[A]_0} = \frac{[S]_0}{[A]_0} + \frac{k_2}{k_1 - k_2}\left[\frac{[A]}{[A]_0} - 1\right] - \frac{k_1}{k_1 - k_2}\left[\left(\frac{[A]}{[A]_0}\right)^{\frac{k_2}{k_1}} - 1\right]$	
A, 1, 3 (Triangular)	$-\frac{d[A]}{dt} = (k_1 + k_3)[A]$ $\frac{d[R]}{dt} = k_1[A] - k_2[R]$ $\frac{d[S]}{dt} = k_3[A] + k_2[R]$	$\frac{[A]}{[A]_0} = \exp[-(k_1 + k_3)t]$ $\frac{[R]}{[A]_0} = \frac{k_1}{k_2 - k_1 - k_3}\{\exp[-(k_1 + k_3)t] - \exp(-k_2 t)\}$ $\frac{[S]}{[A]_0} = 1 - \frac{k_3}{k_1 + k_3}\exp[-(k_1 + k_3)t] - \frac{k_2 k_1}{(k_1 + k_3)(k_2 - k_1 - k_3)} \times \exp[-(k_1 + k_3)t] + \frac{k_1}{k_2 - k_1 - k_3}\exp(-k_2 t)$	$\frac{[A]}{[A]_0} = \frac{1}{1 + (k_1 + k_3)\bar{t}}$ $\frac{[R]}{[A]_0} = \frac{[R]_0/[A]_0 + k_1\bar{t}/[1 + (k_1 + k_3)\bar{t}]}{1 + k_2\bar{t}}$ $[S] = [A]_0 + [R]_0 + [S]_0 - [A] - [R]$
$A \xrightarrow{1} R \xrightarrow{2} S$, $A \xrightarrow{3} T$, $R \xrightarrow{4} V$ (Denbigh)	$-\frac{d[A]}{dt} = (k_1 + k_3)[A]$ $\frac{d[R]}{dt} = k_1[A] - (k_2 + k_4)[R]$ $\frac{d[S]}{dt} = k_2[R]$ $\frac{d[T]}{dt} = k_3[A]$ $\frac{d[V]}{dt} = k_4[R]$	$\frac{[A]}{[A]_0} = \exp[-(k_1 + k_3)t]$ $\frac{[R]}{[A]_0} = \frac{k_1}{k_2 + k_4 - k_1 - k_3}\{\exp[-(k_1 + k_3)t] - \exp(-(k_2 + k_4)t)\} + \frac{[R]_0}{[A]_0}\exp(-(k_2 + k_4)t)$ $\frac{[S]}{[A]_0} = \frac{k_1 k_2}{k_2 + k_4 - k_1 - k_3}\left\{\frac{\exp[-(k_2 + k_4)t]}{k_2 + k_4} - \frac{\exp[-(k_1 + k_3)t]}{k_1 + k_3}\right\} + \frac{k_1 k_2}{(k_1 + k_3)(k_2 + k_4)} + \frac{[R]_0}{[A]_0}\frac{k_2}{k_2 + k_4}\{1 - \exp(-(k_2 + k_4)t)\} + \frac{[S]_0}{[A]_0}$ $\frac{[T]}{[A]_0} = \frac{k_3}{k_1 + k_3}\{1 - \exp[-(k_1 + k_3)t]\} + \frac{[T]_0}{[A]_0}$ $\frac{[U]}{[A]_0}$, same as for $\frac{[S]}{[A]_0}$ but with $k_2 \leftrightarrow k_4$ and $[S]_0 \leftrightarrow [V]_0$	$\frac{[A]}{[A]_0} = \frac{1}{1 + (k_1 + k_3)\bar{t}}$ $\frac{[R]}{[A]_0} = \frac{k_1\bar{t}}{[1 + (k_1 + k_3)\bar{t}] + (1 + (k_2 + k_4)\bar{t})} + \frac{[R]_0}{[A]_0}\frac{k_2\bar{t}}{1 + (k_2 + k_4)\bar{t}}$ $\frac{[S]}{[A]_0} = \frac{k_1 k_3 \bar{t}^2}{[1 + (k_1 + k_3)\bar{t}][1 + (k_2 + k_4)\bar{t}]} + \frac{[R]_0}{[A]_0}\frac{k_2\bar{t}}{[1 + (k_1 + k_3)\bar{t}]} + \frac{[T]_0}{[A]_0}$ $\frac{[T]}{[A]_0} = \frac{k_3\bar{t}}{[1 + (k_1 + k_3)\bar{t}]} + \frac{[T]_0}{[A]_0}$ $\frac{[U]}{[A]_0}$, same as for $\frac{[S]}{[A]_0}$ but with $k_2 \leftrightarrow k_4$ and $[S]_0 \leftrightarrow [V]_0$

Table 2.2 Equations for $[R]_{max}/[A]_0$ for Different Complex Reaction Schemes, No Product in Feed

Class of Reaction	Batch Reactor (or PFR)	Mixed Reactor
Series reaction $A \xrightarrow{k_1} R \xrightarrow{k_2} S$	$\left(\frac{k_1}{k_2}\right)^{k_2/(k_2-k_1)}$	$\frac{1}{\left[(k_2/k_1)^{1/2}+1\right]^2}$
Series–parallel reaction $A+B \xrightarrow{k_1} R$; $R+B \xrightarrow{k_2} S$; $n = 2$ for both	$\left(\frac{k_1}{k_2}\right)^{k_2/(k_2-k_1)}$	$\frac{1}{\left[(k_2/k_1)^{1/2}+1\right]^2}$
Denbigh reaction $A \xrightarrow{k_1} R \xrightarrow{k_2} S$; $A \xrightarrow{k_3} T$; $R \xrightarrow{k_4} V$	$\left(\frac{k_1}{k_1+k_3}\right)\left(\frac{k_1+k_3}{k_2+k_4}\right)^{(k_2+k_4)/((k_2+k_4)-(k_1+k_3))}$	$\left(\frac{k_1}{k_1+k_3}\right)\frac{1}{\left[((k_2+k_4)/(k_1+k_3))^{1/2}+1\right]^2}$

Note: All steps are first order except when noted otherwise.

Conventional wisdom dictates that higher economic worth per unit quantity corresponds to lower capacity and fewer materials and operational cost restrictions. Other considerations are cost and economic recovery of solvents, minimization of side reactions, use of relatively mild reaction conditions, and minimization and efficient disposal of wastes. In a multistep (as opposed to multiple) process, considered in the next section, minimization of the number of steps is particularly important.

Multistep reactions

Multistep reactions

Definitions Multistep reactions may be classified as *simple multistep* and *complex multistep* reactions. In the single multistep scheme, each step of the synthesis is a simple reaction with no side products (this can often be assumed if reaction conditions for each step are so chosen that side products, if any, are negligibly small). A general example of such a scheme would be

Step 1: $A \rightarrow B + C$

Step 2: $C + D \rightarrow E$

Step 3: $E + F \rightarrow G + H$

Step 4: $H \rightarrow I + J$ (R13)

If, on the other hand, each step of a synthetic strategy happens to be complex, then we have a complex multistep reaction, for example,

Step 1:

$$A + B \rightarrow C$$

$$C + B \rightarrow D$$

Step 2:

$$C \rightarrow E$$

$$C \rightarrow F$$

Step 3:

$$E + G \rightarrow H$$

$$H + G \rightarrow I$$

$$H + E \rightarrow J$$

Step 4:

$$H + K \rightarrow L$$

$$L \rightarrow M \tag{R14}$$

with L as the final product. For such a scheme, we define overall conversion, yield, and selectivity as follows:

Overall conversion

Overall yield

Overall selectivity

$$\text{overall conversion, } X_{ov} = \prod_n (\text{conversion in each step})$$

$$\text{overall yield, } Y_{ov} = \prod_n (\text{yield in each step})$$

$$\text{overall selectivity, } S_{ov} = \prod_n (\text{selectvity in each step}) \tag{2.11}$$

where n is the number of steps.

Yield versus number of steps

It would be instructive to elaborate on the relationship between yield and the number of steps. For this, we consider the reaction

$$A + \mathrm{B} \rightarrow C$$

$$C + D \rightarrow E$$

$$E + F \rightarrow G$$

$$G + H \rightarrow I \tag{R15}$$

Assuming the same yield for each step, the overall yield is plotted in Figure 2.2 as a function of the number of steps for different values of the individual yield. It will be seen that for a five-step reaction, quite common in organic synthesis, the overall yield is only 58% for individual yields of 90% in each step. If the individual yields can be raised to 95% (an increase of just 5%), the overall yield goes up to 77%, an increase of about 20%. This underscores the importance of maximizing the yield of each step in a multistep reaction.

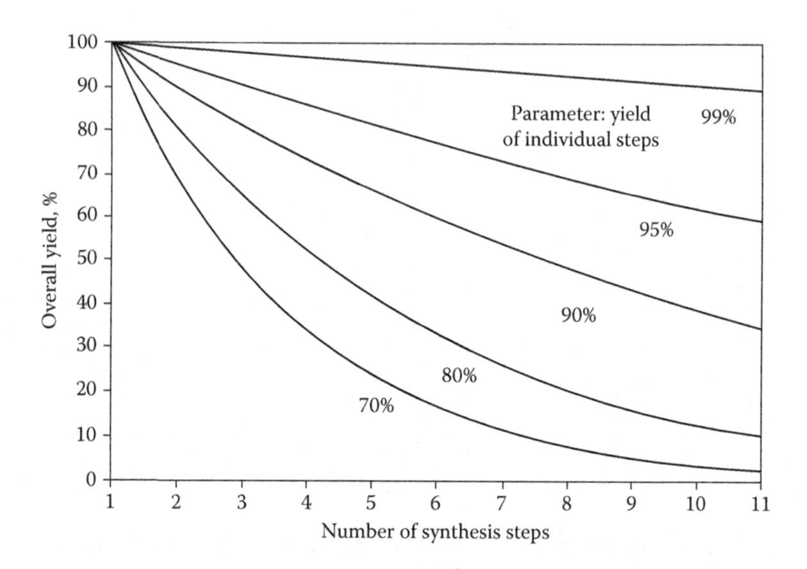

Figure 2.2 Overall final yield as a function of the number of synthesis steps at selected values of the individual yield.

Reactor design for complex reactions

We now turn to the design of reactors for complex reactions. We will focus on the ethylation reaction, using the following less formal nomenclature: A = aniline, B = ethanol, C = monoethylaniline, D = water, E = diethylaniline, F = diethyl ether, and G = ethylene. The four independent reactions then become

$$C_6H_5NH_2 + C_2H_5OH \rightarrow C_6H_5NHC_2H_5 + H_2O \qquad \text{(R16.1)}$$

$$A \quad + B \quad \rightarrow C \quad + D$$

$$C_6H_5NHC_2H_5 + C_2H_5OH \rightarrow C_6H_5N(C_2H_5)_2 + H_2O \qquad \text{(R16.2)}$$

$$C \quad + B \quad \rightarrow E \quad + D$$

$$2C_2H_5OH \rightarrow C_2H_5OC_2H_5 + H_2O \qquad \text{(R16.3)}$$

$$2B \quad \rightarrow F \quad + D$$

$$C_2H_5OH \rightarrow C_2H_4 + H_2O \qquad \text{(R16.4)}$$

$$B \quad \rightarrow G \quad + D$$

Using this set of equations as the basis, we now formulate design equations for various reactor types in the vapor phase over a solid catalyst. Detailed expositions of the theory are presented in a number of books, in particular, Aris (1965, 1969) and Nauman (1987).

Batch reactor design for complex reactions

Batch reactor design based on number of components

Consider a reaction network consisting of N components and M reactions. A set of N ordinary differential equations, one for each component, would be necessary to mathematically describe this system. They may be concisely expressed in the form of Equation 2.5, that is,

$$\frac{d([\mathbf{A}]V)}{dt} = \boldsymbol{\nu}\mathbf{r}V$$

The use of this equation in developing batch reactor equations for a typical complex reaction is illustrated in Example 2.3.

Example 2.3: Batch reactor equations based on number of components: Ethylation of aniline

Applying Equation 2.5 in more explicit form to reaction R16, and assuming constant volume and first-order dependence of the rate on each concentration, we obtain

$$\frac{d}{dt}\begin{pmatrix}[A]\\ [B]\\ [C]\\ [D]\\ [E]\\ [F]\\ [G]\end{pmatrix} = \begin{pmatrix}-1 & 0 & 0 & 0\\ -1 & -1 & -2 & -1\\ 1 & -1 & 0 & 0\\ 1 & 1 & 1 & 1\\ 0 & 1 & 0 & 0\\ 0 & 0 & 1 & 0\\ 0 & 0 & 0 & 1\end{pmatrix}\begin{pmatrix}r_1\\ r_2\\ r_3\\ r_4\end{pmatrix} \tag{2.12}$$

where $r_1 = k_1[A][B]$, $r_2 = k_2[B][C]$, $r_3 = k_3[B]^2$, and $r_4 = k_4[B]$.

We now express the rate of formation/disappearance of each component by accounting for its rates of formation and disappearance by the different steps comprising the reaction. For example, A is consumed by reaction R16.1 with a stoichiometric coefficient of −1 and is not involved in any of the other three reactions. A similar analysis of all components leads to the following set of ordinary differential equations:

$$\begin{aligned}
\frac{d[A]}{dt} &= -k_1[A][B]\\
\frac{d[B]}{dt} &= -k_1[A][B] - k_2[B][C] - 2k_3[B]^2 - k_4[B]\\
\frac{d[C]}{dt} &= k_1[A][B] - k_2[B][C]\\
\frac{d[D]}{dt} &= k_1[A][B] + k_2[B][C] + k_3[B]^2 + k_4[B]\\
\frac{d[E]}{dt} &= k_2[B][C]\\
\frac{d[F]}{dt} &= k_3[B]^2\\
\frac{d[G]}{dt} &= k_4[B]
\end{aligned} \tag{2.13}$$

Note that there are as many equations as the number of components. Also, the rate constants would be known. Hence, these equations can be solved to obtain the product distribution by any convenient numerical method.

Extent of reaction

Use of extent of reaction or reaction coordinates

Consider a simple reaction

$$\nu_A A + \nu_B B \rightarrow \nu_R R \tag{R17}$$

with no restriction regarding volume change. Referring to Equation 2.6, the material balance relations for this reaction can be written as

$$\begin{aligned} N_A - N_{A0} &= \nu_A \xi \\ N_B - N_{B0} &= \nu_B \xi \\ N_R - N_{R0} &= \nu_R \xi \end{aligned} \tag{2.14}$$

These can be recast in the form

$$\begin{pmatrix} N_A \\ N_B \\ N_R \end{pmatrix} - \begin{pmatrix} N_{A0} \\ N_{B0} \\ N_{R0} \end{pmatrix} = \begin{pmatrix} \nu_A \\ \nu_B \\ \nu_R \end{pmatrix} \xi \tag{2.15}$$

When extended to a complex reaction of N components and M reactions, this becomes

$$\begin{pmatrix} N_A \\ N_B \\ N_R \\ \vdots \end{pmatrix} - \begin{pmatrix} N_{A0} \\ N_{B0} \\ N_{R0} \\ \vdots \end{pmatrix} = \begin{pmatrix} \nu_{A1} & \nu_{A2} & \cdots & \nu_{AM} \\ \nu_{B1} & \nu_{B2} & \cdots & \nu_{BM} \\ \nu_{R1} & \nu_{R2} & \cdots & \nu_{RM} \\ \vdots & \vdots & \vdots & \vdots \end{pmatrix} \begin{pmatrix} \xi_1 \\ \xi_2 \\ \xi_3 \\ \vdots \end{pmatrix} \tag{2.16}$$

or

$$N - N_0 = \nu\xi \tag{2.17}$$

where N and N_0 are $(N \times 1)$ matrices, respectively, of the final and initial moles of each component, ν is the $(M \times N)$ matrix of stoichiometric coefficients, and ξ is the $(M \times 1)$ reaction coordinate matrix.

Since the units of ξ and r are moles and moles/(time) (volume), respectively, we can also write

$$\frac{d\xi}{dt} = Vr \tag{2.18}$$

This is the basic equation for a complex reaction and expresses the rates of the individual reactions in terms of the corresponding reaction coordinates. It can be solved by expressing the rates r of the individual elements in terms of the number of moles according to Equation 2.2. Thus, for the complex reaction

$$\begin{aligned} 2A + 3B &\rightarrow R, \quad r_1 = k_1 [A][B] \\ B + D &\rightarrow S, \quad r_2 = k_2 [B][D] \end{aligned} \tag{R18}$$

the following relationships are obtained:

$$\begin{aligned} N_A - N_{A0} &= -2\xi_1 \\ N_B - N_{B0} &= -3\xi_1 - \xi_2 \\ N_D - N_{D0} &= -\xi_2 \\ N_R - N_{R0} &= \xi_1 \\ N_S - N_{S0} &= \xi_2 \end{aligned} \tag{2.19}$$

Equation 2.18 can now be expanded to give

$$\frac{d}{dt}\begin{pmatrix} \xi_1 \\ \xi_2 \end{pmatrix} = V\begin{pmatrix} k_1[A][B] \\ k_2[B][D] \end{pmatrix} = \begin{pmatrix} k_1 N_A N_B / V \\ k_2 N_B N_D / V \end{pmatrix} \tag{2.20}$$

and the Ns then eliminated by combining with Equations 2.19 to give expressions for $d\xi/dt$ in terms of ξ and N_0.

The method is demonstrated below for the same complex reaction considered earlier: ethylation of aniline.

Example 2.4: Batch reactor equations based on the extent of reaction for the ethylation of aniline

The four independent reactions of this network are given by reaction, Equation 2.16. Applying Equation 2.18 to these reactions and resolving the matrices into their elements gives

$$\frac{d}{dt}\begin{pmatrix} \xi_1 \\ \xi_2 \\ \xi_3 \\ \xi_4 \end{pmatrix} = V\begin{pmatrix} k_1[A][B] \\ k_2[B][C] \\ k_3[B]^2 \\ k_4[B] \end{pmatrix} = \begin{pmatrix} k_1 N_A N_B / V \\ k_2 N_B N_C / V \\ k_3 N_B^{\,2} / V \\ k_4 N_B \end{pmatrix} \tag{2.21}$$

We now express the individual Ns in terms of the corresponding N_0s and ξs by writing Equation 2.19 for the individual components:

$$\begin{aligned} N_A - N_{A0} &= -\xi_1 \\ N_B - N_{B0} &= -\xi_1 - \xi_2 - 2\xi_3 - \xi_4 \\ N_C - N_{C0} &= \xi_1 - \xi_2 \\ N_D - N_{D0} &= \xi_1 + \xi_2 + \xi_3 + \xi_4 \\ N_E - N_{E0} &= \xi_2 \\ N_F - N_{F0} &= \xi_3 \\ N_G - N_{G0} &= \xi_4 \end{aligned} \tag{2.22}$$

Combining Equations 2.21 and 2.22 leads to

$$\begin{aligned} \frac{d\xi_1}{dt} &= \frac{k_1}{V}\left[(N_{A0} - \xi_1)(N_{B0} - \xi_1 - \xi_2 - 2\xi_3 - \xi_4)\right] \\ \frac{d\xi_2}{dt} &= \frac{k_2}{V}\left[(N_{B0} - \xi_1 - \xi_2 - 2\xi_3 - \xi_4)(N_{B0} + \xi_1 - \xi_1)\right] \\ \frac{d\xi_3}{dt} &= \frac{k_3}{V}(N_{B0} - \xi_1 - \xi_2 - 2\xi_3 - \xi_4)^2 \\ \frac{d\xi_4}{dt} &= k_4(N_{B0} - \xi_1 - \xi_2 - 2\xi_3 - \xi_4) \end{aligned} \tag{2.23}$$

Since the values of N_0 and k are expected to be known in a design calculation, the set of ordinary differential equations 2.23 can be solved to give the individual reaction coordinates (ξ) as functions of time. Thus, the product distribution at the end of a stipulated time for a reactor of known volume can be found. Note that only four equations are required (although the number of components is seven).

Plug-flow reactor design for complex reactions

Plug-flow reactor

As explained in Chapter 1, the plug-flow reactor differs from the batch reactor only with respect to the time coordinate. For the batch reactor, time elapsed since the commencement of reaction is directly used as a measure of this coordinate, whereas in the plug-flow reactor it is replaced by the time required to traverse a given distance in the tubular reactor: $t = z/u$, where z is the distance and u the average velocity. Thus, the rate equation now becomes

$$-r_A = -u\frac{d[A]}{dz} \tag{2.24}$$

Rate equations can be written for all the components of a complex reaction. For the ethylation reaction considered in the above example, for instance, these will be identical to Equations 2.21 except that $(d[i]/dt)$

will be replaced by ($u\ d[i]/dz$). This set of equations can be solved by any of the well-known numerical methods.

Continuous stirred tank reactor

CSTR for complex reactions

The performance equation for a CSTR was developed in Chapter 1. We use the same equation now but with a complex rate equation replacing the simpler one of the earlier chapter. We describe the method below for any complex reaction consisting of N components N in M reactions. The following material balances can be written for the different constituents of the complex reaction at hand (Figure 2.3):

$$\begin{aligned} Q_0[A]_0 - Q_f[A]_f + \nu_A(r_{Af})V &= 0 \\ \vdots \qquad\qquad & \vdots \\ Q_0[M]_0 - Q_f[M]_f + \nu_M(r_{Mf})V &= 0 \end{aligned} \tag{2.25}$$

where subscript f represents the final condition. In compact matrix form, this becomes

$$Q_0\boldsymbol{c}_0 - Q_f\boldsymbol{c}_f + \nu rV = 0 \tag{2.26}$$

where $\boldsymbol{c}_0$ and $\boldsymbol{c}_f$ are the initial and final concentration matrices, respectively. Thus, we have N simultaneous equations and N unknown concentrations plus one unknown outlet flow rate Q_f (i.e., a total of $N+1$ unknowns). If we assume the volumetric flow rate to be approximately constant (which is true for liquid systems, but only very approximately so for gaseous systems), then $Q_0 = Q_f$. Thus, we would have only N unknowns, and an equation of state for the system which would relate Q_f to Q_0 would not be needed. If the reactions are assumed to be first order, further simplification results, and Equation 2.25 can be solved for the N unknown concentrations. Even with these simplifications, the solution becomes quite difficult as the number of reactions increases.

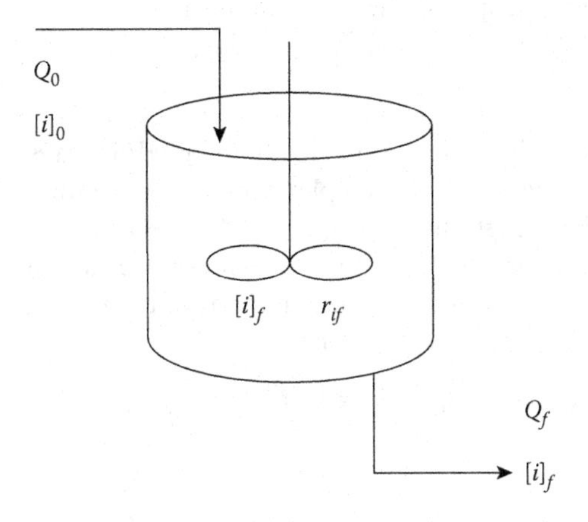

Figure 2.3 Material flow in a CSTR for a complex reaction.

Hence, as in the case of the batch reactor, it is convenient to reduce the number of equations to the number of independent reactions M. For this purpose, we define a reaction coordinate ξ' with units of moles/time, which is equivalent (but not equal) to the reaction coordinate ξ for the batch reactor with units of moles. Using this coordinate, we obtain (without the assumption, $Q_f = Q_0$)

$$Q_0\mathbf{c}_0 - Q_f\mathbf{c}_f + \nu\xi' = 0 \tag{2.27}$$

where ξ' is a modified extent of reaction vector whose elements ξ_1', $\xi_2',\ldots, \xi_M'$ represent the values for reactions 1, 2,. . ., M.

A comparison of Equations 2.26 and 2.27 shows that

$$\nu\,(\mathbf{r}V - \xi') = 0 \tag{2.28}$$

which can be recast as

$$\Sigma\nu_{ij}[r_{jf}V - \xi_j] = 0 \tag{2.29}$$

Since the ν_{ij} are nonzero, the only way that Equation 2.29 to be true is

$$\xi_1' - r_{1f}V = 0 \tag{2.30}$$

This equation gives the basic rate of reaction 1, which can be used to express the rates of formation or disappearance of the other components of that reaction as multiples thereof. Similar equations can be written for the other reactions of the system, leading to the following full set of equations:

$$\begin{aligned} \xi_1' - r_{1f}V &= 0 \\ \xi_2' - r_{2f}V &= 0 \\ &\vdots \\ \xi_M' - r_{Mf}V &= 0 \end{aligned} \tag{2.31}$$

These can now be expressed in compact form as

$$\xi' = \mathbf{r}V \tag{2.32}$$

Equation 2.32 is the basic design equation for MFR and is applicable to any number of reactions in a complex scheme. To obtain the rate for use in this equation, recall that in all MFR calculations the rate is based on outlet concentrations. Since these are unknown, we use the material balance Equation 2.25 to express them in terms of the inlet concentrations. Thus, for component A we can write

$$Q_f[A]_f - Q_0[A]_0 = \nu_{A1}\xi_1' + \nu_{A2}\xi_2' + \nu_{A3}\xi_3'\cdots \tag{2.33}$$

or, if $Q_f = Q_0$,

$$Q_0([A]_f - [A]_0) = \nu_{A1}\xi_1' + \nu_{A2}\xi_2' + \nu_{A3}\xi_3'\cdots \tag{2.34}$$

We now illustrate the use of Equation 2.32 in the continuing example of this chapter: ethylation of aniline.

Example 2.5: Design of a mixed–flow reactor based on extent of reaction for the ethylation of aniline

Considering the four independent reactions of this system (reaction 2.16) and assuming the rates to be first order in each reactant, we obtain

$$\begin{aligned}
\xi_1' &= r_{1f}V = k_1[A]_f[B]_fV \\
\xi_2' &= k_2[B]_f[C]_fV \\
\xi_3' &= k_3[B]_f^2V \\
\xi_4' &= k_4[B]_fV
\end{aligned} \tag{2.35}$$

The next step is to express the unknown outlet concentrations in terms of the initial concentrations. For this, we use the material balance equations for each of the components. For component A, for example, this is given by

$$Q([A]_f - [A]_0) = \nu_{A1}\xi_1' + \nu_{A1}\xi_1' \cdots = \sum_{j=1}^{4} \nu_{Aj}\xi_j' \tag{2.36}$$

Thus, we obtain the following equations for the seven components:

$$\begin{aligned}
[A]_f &= -\frac{\xi_1'}{Q_0} + [A]_0 = \left(-\xi_1' + [A]_0Q_0\right)\frac{1}{Q_0} \\
[B]_f &= -\frac{\xi_1' + \xi_2' + 2\xi_3' + \xi_4'}{Q_0} + [B]_0 \\
[C]_f &= \frac{\xi_1' - \xi_2'}{Q_0} + [C]_0 \\
[D]_f &= \frac{\xi_1' + \xi_2' + \xi_3' + \xi_4'}{Q_0} + [D]_0 \\
[E]_f &= \frac{\xi_2'}{Q_0} + [E]_0 \\
[F]_f &= \frac{\xi_3'}{Q_0} + [F]_0 \\
[G]_f &= \frac{\xi_4'}{Q_0} + [G]_0
\end{aligned} \tag{2.37}$$

The final step in the formulation of the design equations is to replace the outlet concentrations appearing in Equation 2.35 by the equations in Equation 2.36. This leads to the following set of four simultaneous equations corresponding to the four independent reactions of the system:

$$\xi_1' = k_1\bar{t}(-\xi_1' + Q_0[A]_0)\{Q_0[B]_0 - (\xi_1' + \xi_2' + 2\xi_3' + \xi_4')\}\frac{1}{Q_0}$$

$$\xi_2' = k_2\bar{t}\{Q_0[B]_0 - (\xi_1' + \xi_2' + 2\xi_3' + \xi_4')\}(\xi_1' - \xi_2' + Q_0[A]_0)\frac{1}{Q_0}$$

$$\xi_3' = k_3\bar{t}\{Q_0[B]_0 - (\xi_1' + \xi_2' + 2\xi_3' + \xi_4')\}^2\frac{1}{Q_0}$$

$$\xi_4' = k_4\bar{t}\{Q[B]_0 - (\xi_1' + \xi_2' + 2\xi_3' + \xi_4')\} \tag{2.38}$$

where $\bar{t} = V/Q_0$. Notice that these are algebraic equations as against the differential equations that characterize PFR. Solution can be cumbersome but is relatively straightforward.

Reactor choice for maximizing yields/selectivities

In reactions where the product does not react further (i.e., parallel reactions), yields and selectivities can be easily calculated from the ratios of the rates. Where a product reacts further, no such simple analysis is possible, and resort to numerical solution is often necessary. As a general rule, however, whenever an intermediate product is the desired product, PFR is the preferred reactor.

Parallel reactions (nonreacting products)

The general case For a two-step parallel reaction represented by

$$A \xrightarrow{1} R, \quad A \xrightarrow{2} S \tag{R19}$$

we can define a point selectivity as

Point selectivity

$$S_p = \frac{d[R]/dt}{-d[A]/dt} = -\frac{d[R]}{d[A]} \tag{2.39}$$

S_p can be calculated either from knowledge of the rates of formation and disappearance of R and A, respectively, or directly from a plot of $[R]$ versus $[A]$ (see Figure 2.4).

The questions now are: What type of reactor would be best suited for a given parallel scheme? Is the choice to be made on the basis of conversion or yield? The answers to these questions depend largely on the nature of the S_p versus [A] curve and whether A can be separated from the product and recycled at a relatively low additional cost. This recycle is to be distinguished from the recycle of the recycle flow reactor (RFR; to be discussed in Chapter 3) where part of the exit stream is recycled as such, without separation of the reactant from the product.

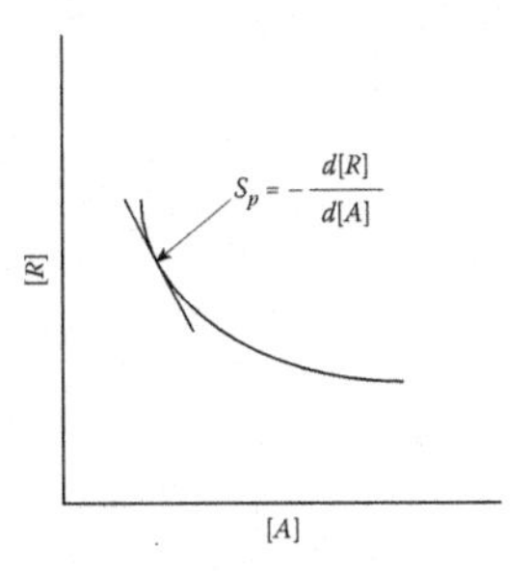

Figure 2.4 Obtaining the selectivity of R in the parallel scheme $A \rightarrow R$, $A \rightarrow S$ in a tubular reactor.

Let us first examine the question of reactor choice for maximizing [R]. Consider Figure 2.4 which shows a plot of S_p versus [A]. The values of [R] corresponding to plug- and mixed-flow modes of operation are given by

$$\text{PFR: } [R] = \int_{[A]_f}^{[A]_0} S_p d[A] \quad (2.40)$$

$$\text{MFR: } [R] = S_{pf}([A]_0 - [A]_f) \quad (2.41)$$

Note that S_{pf} is the overall selectivity in MFR at the same time.

It should also be noted, however, that different types of selectivity curves are possible, as shown in Figure 2.5. In Figure 2.5a, S_p increases with [A].

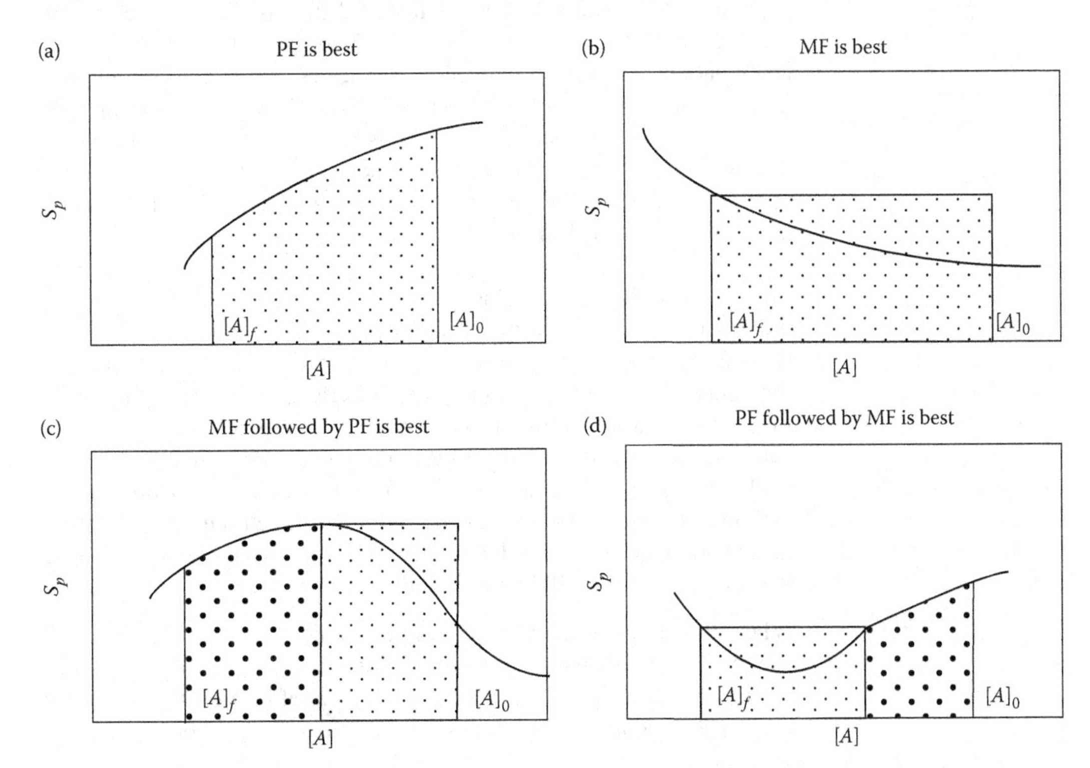

Figure 2.5 Reactor choice for different forms of S_p–[A] curves for the parallel reaction $A \rightarrow R$, $A \rightarrow S$.

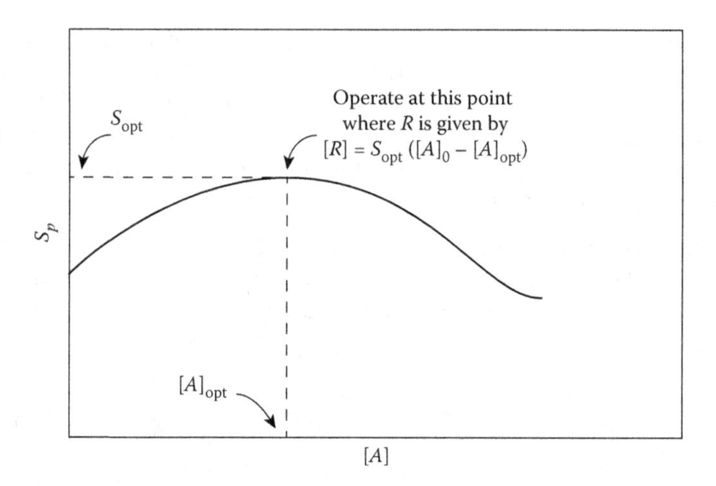

Figure 2.6 Optimum operating strategy for a parallel reaction $A \rightarrow R$, $A \rightarrow S$ where unreacted A is recycled.

The value of $[R]$ corresponding to given initial and final values of $[A]$ is seen to be higher for plug flow than for mixed flow. In other words, for a curve of this type, plug flow is the preferred mode of operation. By the same reasoning, it can be seen that mixed flow is the preferred mode for a curve of the type shown in Figure 2.5b. Parts (c) and (d) indicate combinations of plug- and mixed-flow reactors as the preferred modes. The design principles of these reactors have already been outlined in Chapter 1. In cases where the reactant is recycled, the extent of conversion is unimportant and the reactor can be operated at $[A]$ corresponding to S_{max}, as shown in Figure 2.6. If the selectivity curve does not show a maximum, no such clear-cut decision is possible.

Effect of reaction order Choice of reactor becomes relatively easy if the orders of the two reactions of Scheme R19 are known. If reaction R1 is first order and R2 second order in A, then the concentration environment of the reaction will have a major effect on the selectivity of R. Recall that conversion in an MFR occurs at the greatly reduced outlet concentration, while that in a PFR is the cumulative value determined by the changing concentration environment of the reactor starting from a high initial value. Thus, the selectivity for R will be higher in the MFR. On the other hand, if the order of reaction R1 is R2 and of R2 is R1, then the selectivity for R will be lower in the MFR.

Where the orders of the two steps are the same, it is clear that the yield would be the same irrespective of the concentration profiles within the reactor, since any profile would have identical effects on the two reactions. The yield and selectivity for such a situation would be given by

Yield and selectivity

$$Y_R = \frac{[R]}{[A]_0}, \quad S_R = \frac{[R]}{[R]+[S]} \tag{2.42}$$

One of the reactants undergoes a second reaction Let us now consider the scheme

$$A + B \rightarrow R$$
$$B \rightarrow S \tag{R20}$$

Polymerization of reactants is a common occurrence in many reactions. Although this is also a parallel scheme, it will be noticed that high concentrations of A combined with low concentrations of B will favor the desired product R. Thus, a semibatch reactor (SBR) would be the preferred candidate since the above condition is met in this reactor. We will see the design equations and principles of operation of SBRs later in this chapter. On the other hand, the common BR, PFR, and MFR would all give lower selectivities because they all allow the second reaction to proceed without hindrance.

Parallel–consecutive reactions

Consider first the reaction

$$A + B \rightarrow R$$
$$R + B \rightarrow S \tag{R21}$$

The selectivity, yield, and conversion are given by

$$S_R = \frac{[R]}{[R]+[S]}, \quad Y_R = \frac{[R]}{[A]_0}, \quad X_A = \left(1 - \frac{[A]}{[A]_0}\right) \tag{2.43}$$

The selectivity of R will be high if [A] and $[B]$ are high, and $[B]$ is low at high $[R]$. Since this pattern holds for BR and PFR, these are the preferred reactors for maximizing R. On the other hand, for an MFR, $[R]$ within the reactor is uniformly high, being equal always to the exit concentration. Hence, the first (desired) reaction is not favored, resulting in a lower selectivity for R than in a PFR or BR.

Scheme R21 can be extended to include a number of intermediates, namely

$$A + B \rightarrow R$$
$$R + B \rightarrow S$$
$$S + B \rightarrow T$$
$$T + B \rightarrow U \tag{R22}$$

Industrially important examples are listed in Table 2.3. Based on a detailed study of this scheme (Russell and Buzzelli, 1969), the following observations are important:

1. For a reaction in which the activation energies of the different steps are approximately equal, product distribution is a function

Table 2.3 Industrially Important Examples of Scheme R22

Reactants		Products		
A	*B*	*R*	*S*	*T*
Water	Ethylene oxide	Ethylene glycol	Diethylene glycol	Triethylene glycol
Ammonia	Ethylene oxide	Monoethanolamine	Diethanolamine	Triethanolamine
Ammonia	Ethylene dichloride	Ethylenediamine	Diethylenetriamine	Triethylenetramine
Methyl, ethyl, or butyl alcohol	Ethylene oxide	Monoglycol ether	Diglycol ether	Triglycol ether
Benzene	Chlorine	Monochlorobenzene	Dichlorobenzene	Trichlorobenzene
Methane	Chlorine	Methyl chloride	Dichloromethane	Trichloromethane (chloroform)

only of two variables: mole ratio of reactants and fraction of product recycled.

2. It is sometimes convenient to use a *secondary* reactor instead of a recycle to the first (*primary*) reactor. In such a case, while the input ratio (this time to two reactors) continues to be important, the more critical variable is the allocation of B between the primary and secondary reactors.
3. The choice between recycle to a single (primary) reactor and a secondary reactor is often dictated by cost considerations.

Consider next the scheme

$$\begin{aligned} A + B &\rightarrow R \\ 2A &\rightarrow S \end{aligned} \tag{R23}$$

If R is the desired product, its yield can be maximized by maintaining a low concentration of A throughout the reactor. This can be done by feeding A at various points along a tubular reactor with an inlet feed of (B + some A), or by distributing B in the individual reactors of a CSTR sequence. For further details on this and other similar schemes, reference may be made to Van de Vusse and Voetter (1961) and Denbigh and Turner (1971).

Finally, consider the special consecutive–parallel scheme

$$\begin{aligned} A \rightarrow R &\rightarrow S \\ 2A &\rightarrow T \end{aligned} \tag{R24}$$

This poses an interesting problem in that a PFR would favor R by the first reaction, whereas a CSTR would suppress the undesired second reaction (Van de Vusse, 1964). This problem can be resolved by using a recycle reactor (with its partial mixing).

Plug-flow reactor with recycle

The basic design equation

The principle of the RFR is sketched in Figure 2.7. The single parameter that distinguishes it from PFR is the recycle flow ratio R,

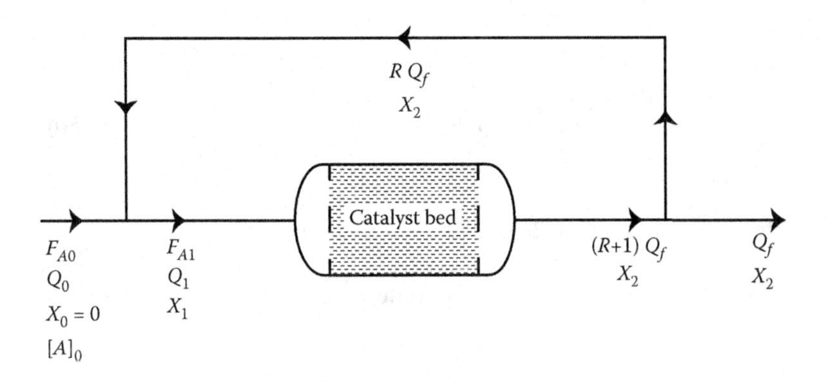

Figure 2.7 Recycle reactor.

$$R = \frac{\text{Volume of product recycled}}{\text{Volume of product leaving the reactor}} \tag{2.44}$$

Recycle ratio

With this recycle, the conditions at R corresponding to the reactor inlet are: $[A] = [A]_R$, $F_A = F_{AR}$. Thus, the PFR equation given in Chapter 1 becomes

$$\frac{V}{F_{A0}} = (1 + R)\int_{X_1}^{X_2} \frac{dX_A}{-r_A} \tag{2.45}$$

Material balance across the reactor gives

$$X_1 = \frac{X_{A0} + RX_2}{1 + R} \tag{2.46}$$

Since usually there is no conversion initially,

$$X_1 = \frac{RX_2}{1 + R} \tag{2.47}$$

Equation 2.46 can also be written in terms of concentration as

$$[A]_2 = \frac{F_{A1}}{Q_1} = [A]_0\left(\frac{1 + R - RX_2}{1 + R + R\varepsilon_A X_1}\right) \tag{2.48}$$

Using Equation 2.47 for X_{A1} as the inlet boundary condition, Equation 2.45 can be solved to give

$$\bar{t} = \frac{[A]_0 V}{F_{A0}} = (1 + R)[A]_0 \int_{X_1}^{X_2} \frac{dX_A}{(-r_A)}, \quad \text{for any } \varepsilon_A \tag{2.49}$$

or

$$\bar{t} = \frac{[A]_0 V}{F_{A0}} = (1+R)\int_{[A]_2}^{[A]_1} \frac{d[A]}{-r_A}, \text{ for } \varepsilon_A = 0 \tag{2.50}$$

Integration of Equation 2.50 for a first-order reaction

$$A \rightarrow \text{Products} \tag{R25}$$

gives

$$\tau = k\bar{t} = (1+R)\ln\left[\frac{[A]_0 + R[A]_2}{(1+R)[A]_2}\right] \tag{2.51}$$

Optimal design of RFR

It may be noted from Equations 2.46 and 2.48 that the recycle ratio R is given by

$$R = \frac{[A]_0 - [A]_1}{[A]_1 - [A]_2} \quad \text{for any } \varepsilon_A \tag{2.52}$$

$$R = \frac{X_1 - X_{A0}}{X_2 - X_1} \quad \text{for } \varepsilon_A = 0 \tag{2.53}$$

Also, for the common shape of the curve shown as A in Figure 2.8, RFR can never have a volume less than PFR. The RFR shows up at its best when the $(1/{-r_A})$ versus X_A curve is continuously falling, but that seldom happens. It can, however, exhibit a minimum (curve B of Figure 2.8),

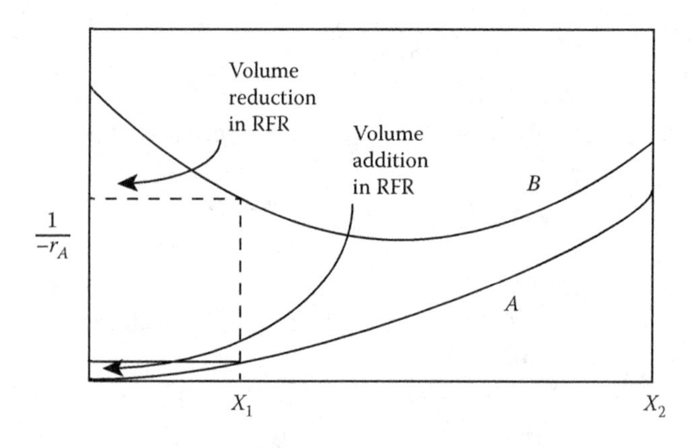

Figure 2.8 Plots of $1/{-r_A}$ versus XA for common reactions (curve A) and for autocatalytic and adiabatic reactions (curve B).

as in the case of an autocatalytic or adiabatic reactor. Clearly, in such a case, the recycle reactor can be superior to the plug-flow reactor under certain conditions, as shown by the lower reactor volume for curve B in the figure.

The central problem in the design of RFR is the determination of the optimum value of X_{AR} for minimizing the reactor volume. This can be obtained by setting

$$\frac{d\{\bar{t}/[A]_0\}}{dR} = 0,$$

$$\text{i.e.,} \quad \frac{d\{\bar{t}/[A]_0\}}{dR} = \frac{d\int_{X_1}^{X_2}(1+R)(dX_A/-r_A)}{dR} = 0 \tag{2.54}$$

with the result (Levenspiel, 1993)

$$\left.\frac{1}{-r_A}\right|_{X_1} = \frac{\int_{X_1}^{X_2} dX_A/-r_A}{X_2 - X_1} \tag{2.55}$$

Expressed in words, this means

$$(1/\text{rate}) \text{ at } X_1 = (1/\text{rate}) \text{ average in the reactor} \tag{2.56}$$

Equation 2.55 can be solved by iteration to find that the value of X_{AR} satisfies both sides. It can also be solved graphically as illustrated in Figure 2.9. The recycle is introduced in such a way that the rate corresponding to the value of X_{AR} is exactly equal to the average rate in the reactor. This is easily found by locating X_{AR} such that the areas M and N are equal.

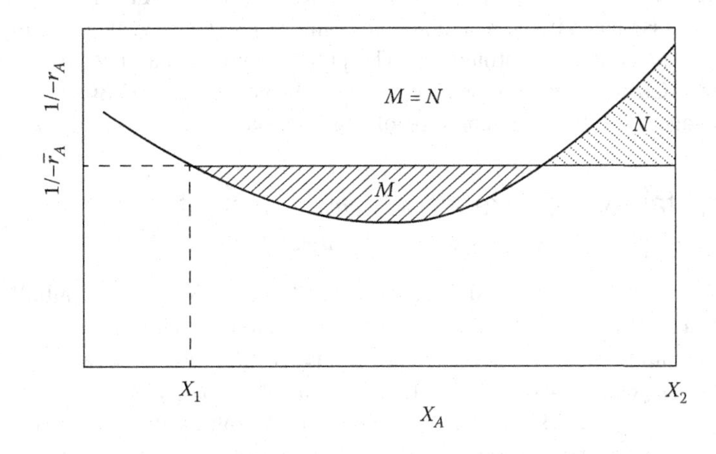

Figure 2.9 Minimization of the reactor volume in an RFR.

Use of RFR to resolve a selectivity dilemma

Role of backmixing in selectivity

The effect of mixing on the yield or selectivity of a desired product is an important consideration in reactor choice as we will see in Chapter 3. RFR is well suited to impose a controlled level of mixing to maximize selectivity. In this connection, we note the following two important effects of mixing in an isothermal reaction:

1. Mixing is detrimental to the yield of an intermediate product, favoring the formation of the final product.
2. Mixing favors the lowest order reaction in a system involving reactions of varying orders and has no effect when all orders are the same.

These effects, which can be used to advantage in many complex reactions, lead to contradictory reactor choices when applied to the scheme,

$$A \xrightarrow{k_1} R \xrightarrow{k_2} S \quad \text{(main reaction)} \qquad \text{(R26.1)}$$

$$A + A \xrightarrow{k_3} T \quad \text{(side reaction)} \qquad \text{(R26.2)}$$

where R is the desired product. Thus, while conclusion 1 calls for a PFR for maximizing R in the main reaction, conclusion 2 would require a CSTR to minimize the loss of A via the side reaction. It can be shown (Van de Vusse, 1964; Gillespie and Carberry, 1966) that neither of the extremes, PFR or CSTR, is the best for this reaction and that a reactor with an intermediate level of mixing as determined by the value of the recycle ratio R is optimal.

Semibatch reactors

Semibatch reactor design

SBRs are very common in industrial practice. The basic principle is that a reactant is placed in the reactor and the same or a second reactant, usually the latter, is added continuously. The product may or may not be withdrawn. Clearly, several modes of such semibatch operation (SBO) can be envisaged, and the more common of these are sketched in Figure 2.10.

Constant-volume reactions with constant rates of addition and removal: Scheme 1

In this constant-volume SBO, a stirred tank reactor is charged initially with a reactant or reactants (B). One reactant is fed to the reactor at a constant volumetric flow rate Q_0, and the product is removed at the same rate. This is a general description of the reactions of scheme 1 in Figure 2.10. In analyzing these schemes, we make the following reasonable assumptions to simplify the equations: there is no volume change upon mixing of the two liquids or upon reaction, and the reactor is perfectly mixed.

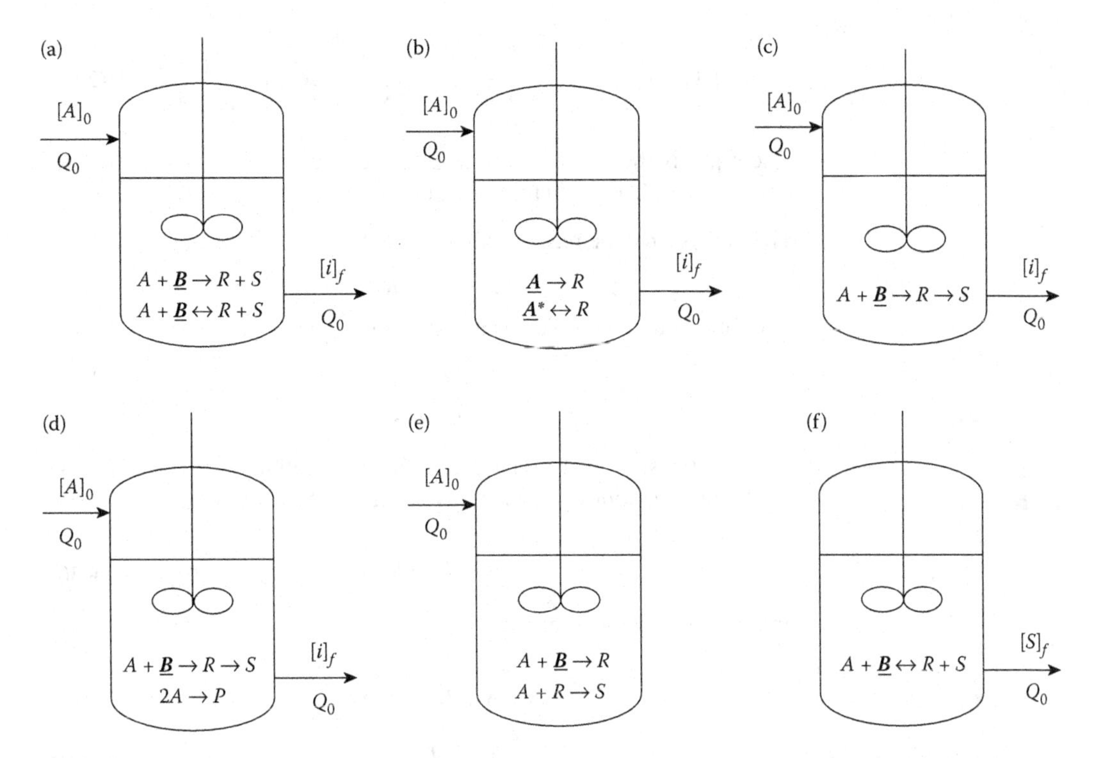

Figure 2.10 Representative modes of SBO for different reaction schemes. (a) Constant-volume SBO, scheme 1A, (b) constant-volume SBO, scheme 1B, (c) constant-volume SBO, scheme 1C, (d) constant-volume SBO, scheme 1D, (e) variable-volume SBO, scheme 2, and (f) variable-volume SBO, scheme 3.

Consider first the general case of an irreversible second-order reaction

$$A + B \xrightarrow{k_1} R + S \tag{R27}$$

As shown in Figure 2.10, the tank is initially charged with reactant B, and reactant A is added at a rate Q_0. The outlet stream contains both reactants A and B and products R and S. A general material balance can be written as

$$\text{In-out-disappearance} + \text{Generation} = \text{Accumulation} \tag{2.57}$$

For reactant A, this becomes

$$Q_0[A]_0 - Q_0[A] - (-r_A)V = \frac{d(V[A])}{dt} = V\frac{d[A]}{dt} \tag{2.58}$$

which can be recast as

$$\frac{d[A]}{dt} = -(-r_A) + \frac{[A]_0 - [A]}{\bar{t}} \tag{2.59}$$

and solved for $[A]$. Since generally $[A] \ll [B]$ in this mode of operation, the rate can be written as $(-r_A) = k_1[A]$, leading to the general solution

$$[A] = [A]_0\left(\frac{1}{k\bar{t} + 1}\right) + \left([A]_i - \frac{[A]_0}{k\bar{t} + 1}\right)\exp\left[-\left(\frac{k\bar{t} + 1}{\bar{t}}\right)t\right] \quad (2.60)$$

where $[A]_i$ is the concentration of A initially charged in the tank. Note, however, that $[A]_i = 0$ in the present case.

The analysis can be readily extended to the reversible reaction

$$A + B \rightarrow R + S \quad (R28)$$

The following other schemes are also important:

$$nA \rightarrow R \quad (R29)$$

$$nA \leftrightarrow R$$

in which the species A is charged to the tank initially at a concentration of $[A]_i$, and the following reactions occur: the series reaction

$$A + B \xrightarrow{k_1} R \xrightarrow{k_2} S \quad (R30)$$

and the van de Vusse reaction

$$A + B \xrightarrow{k_1} R \xrightarrow{k_2} S$$

$$2A \xrightarrow{k_3} P \quad (R31)$$

Exercise

Derive the expressions to show that the semibatch mode of operation of the van de Vusse scheme gives results similar to those of the recycle reactor. Thus, higher yields and selectivities for product R can be realized than in a PFR or MFR when $k_3[A]_0 \gg k_2$.

Variable-volume reactor with constant rate of inflow: Scheme 2

Variable-volume reactor design

Here, a tank is charged with an initial volume V_o of reactant B. Beginning at time zero, reactant A is added at a rate Q_o to the reactor, and no tank products are withdrawn [Figure 2.10(2)]. This SBO scheme is useful in several situations. If the reaction is highly exothermic, the rate of addition of one reactant can be manipulated to control the amount of heat evolved. Furthermore, if undesirable side reactions occur, it is possible to control the selectivity for the desired product. For example, consider the scheme

$$A + B \xrightarrow{k_1} R \quad (R32.1)$$

$$A + R \xrightarrow{k_2} S \quad (R32.2)$$

We can assume that $[A] \ll [B]$ but not that $[R] \ll [A]$, giving the following rate equations: $(-r_A)_1 = k_1[A]$ and $(-r_A)_2 = k_2[A][R]$. The resulting mass balance for A is

$$d\frac{(V[A])}{dt} = V\frac{d[A]}{dt} + [A]\frac{dV}{dt} = Q_0[A]_0 - Q_0[A] - (-r_A)_1 V - (-r_A)_2 V \quad (2.61)$$

Assuming linear variation of V with time, that is,

$$V = V_0 + Q_0 t \quad (2.62)$$

or

$$\frac{dV}{dt} = Q_0 \quad (2.63)$$

we obtain

$$\frac{d[A]}{dt} = \frac{Q_0([A]_0 - 2[A])}{V_0 + Q_0 t} - k_1[A] - k_2[A][R]$$
$$\frac{[A]_0 - 2[A]}{\bar{t}_0 + t} - k_1[A] - k_2[A][R] \quad (2.64)$$

where

$$\bar{t}_0 = \frac{V_0}{Q_0} \quad (2.65)$$

Similar expressions can be written for B, R, and S and are included in Table 2.4.

Variable-volume reactor with constant rate of outflow of one of the products: Scheme 3

In this scheme [Figure 2.10(3)],

$$A + B \rightarrow R + S \quad (R33)$$

the tank is initially charged with one of the reactant species as shown in reaction R33 (or with both), and one of the products is withdrawn at a constant rate. This operating mode is particularly useful when the reaction is highly reversible, so that the removal of one of the products causes a favorable shift in equilibrium. For example, in esterification reactions, water can easily be removed through boiling or under vacuum to shift the equilibrium. The following material balance can be written for this situation:

$$d\frac{(V[S])}{dt} = V\frac{d[S]}{dt} + [S]\frac{dV}{dt} = V\frac{d[S]}{dt} - Q_0[S] = r_s V - Q_0[S]_f \quad (2.66)$$

Table 2.4 List of Terms That Can Be Methodically Combined to Give the Design Equations for Several Schemes of Semibatch Operation[a]

Scheme	Reaction[b]	$\frac{d[A]}{dt}, r$	$\frac{d[B]}{dt}, r$	$\frac{d[R]}{dt}, r$	$\frac{d[S]}{dt}, r$	$\frac{d[P]}{dt}, r$	Exact Solutions for [*A*], [*B*], [*R*], [*S*] (See Below)
Part 1. Matrix for Identification of Terms for Constructing Design Equations							
1A-1	$A+\underline{B}\xrightarrow{k_1}R+S$	$\alpha 1, r1$	$\beta 1, r1$	$\gamma 1, r1$	$\sigma 1, r1$	—	1*a*, 1*b*, 1*r*, 1*s*
1A-2	$A+\underline{B}\overset{k_1}{\leftrightarrow}R+S$	$\alpha 1, r2$	$\beta 1, r2$	$\gamma 1, r2$	$\sigma 1, r2$	—	No
1B-1	$\underline{A}^n\xrightarrow{k_1}R$	$\alpha 1, r3$	—	$\gamma 1, r3$	—	—	1*a*, 1*r* if $n=1$
1B-2	$\underline{A}^n \leftrightarrow R$	$\alpha 1, r4$	—	$\gamma 1, r4$	—	—	No
1C	$A+\underline{B}\xrightarrow{k_1}R\xrightarrow{k_2}S$	$\alpha 1, r1$	$\beta 1, r1$	$\gamma 1, r5$	$\sigma 1, r6$	—	1*a*, 1*b*, equations get messy for *R* and *S*
1D	$A+\underline{B}\xrightarrow{k_1}R\xrightarrow{k_2}S$ $2A\xrightarrow{k_3}P$	$\alpha 1, r7$	$\beta 1, r1$	$\gamma 1, r5$	$\sigma 1, r6$	$\rho 1, r8$	No
2	$A+\underline{B}\xrightarrow{k_1}R$ $A+R\xrightarrow{k_2}S$	$\alpha 2, r9$	$\beta 2, r1$	$\gamma 2, r10$	$\sigma 2, r11$	—	No
3	$A+\underline{B}\leftrightarrow R+S$	$\alpha 3, r12$	$\beta 3, r12$	$\gamma 3, r12$	$\sigma 3, r12$	—	No

Part 2. Exact Solutions[c]

1*a*

$$[A]=\left(\frac{[A]_0}{1+k_1\bar{t}}\right)+\left([A]_i-\frac{[A]_0}{1+k_1\bar{t}}\right)\exp\left[\left(-\frac{1+k_1\bar{t}}{\bar{t}}\right)t\right]$$

1*b*

$$[B]=-\frac{k_1\bar{t}\,[A]_0}{1+k_1\bar{t}}+([A]_0-[A]_i+[B]_i)\exp\left(-\frac{t}{\bar{t}}\right)+\left([A]_i-\frac{[A]_0}{1+k_1\bar{t}}\right)\exp\left[\left(-\frac{1+k_1\bar{t}}{\bar{t}}\right)t\right]$$

1*r*

$$[R]=\frac{k_1\bar{t}\,[A]_0}{1+k_1\bar{t}}+\left(\frac{[A]_0}{1+k_1\bar{t}}-[A]_i\right)\exp\left[-\left(\frac{1+k_1\bar{t}}{\bar{t}}\right)t\right]+([A]_i-[A]_0+[R]_i)\exp\left(-\frac{t}{\bar{t}}\right)$$

1s

$$[S] = \frac{k_1\bar{t}\,[A]_0}{1+k_1\bar{t}} + \left(\frac{[A]_0}{1+k_1\bar{t}} - [A]_i\right)\exp\left[-\left(\frac{1+k_1\bar{t}}{\bar{t}}\right)t\right] + ([A]_i - [A]_0 + [S]_i)\exp\left(-\frac{t}{\bar{t}}\right)$$

Part 3. Equations for α, β, . . ., ρ

$$\alpha 1 = \frac{d[A]}{dt} = -r + \frac{[A]_0 - [A]}{\bar{t}} \qquad \beta 1 = \frac{d[B]}{dt} = -r - \frac{[B]}{\bar{t}}$$

$$\gamma 1 = \frac{d[R]}{dt} = r - \frac{[R]}{\bar{t}} \qquad \delta 1 = \frac{d[S]}{dt} = r - \frac{[S]}{\bar{t}}$$

$$\rho 1 = \frac{d[P]}{dt} = \frac{r}{2} - \frac{[P]}{\bar{t}}$$

$$\alpha 2 = \frac{d[A]}{dt} = -r + \frac{[A]_0 - [A]}{\bar{t}_0 + t} \qquad \beta 2 = \frac{d[B]}{dt} = -r - \frac{[B]}{\bar{t}_0 + t}$$

$$\gamma 2 = \frac{d[R]}{dt} = r - \frac{[R]}{\bar{t}_0 + t} \qquad \delta 2 = \frac{d[S]}{dt} = r - \frac{[S]}{\bar{t}_0 + t}$$

$$\alpha 3 = \frac{d[A]}{dt} = -r + \frac{[A]}{\bar{t}_0 - t} \qquad \beta 3 = \frac{d[B]}{dt} = -r - \frac{[B]}{\bar{t}_0 - t}$$

$$\gamma 3 = \frac{d[R]}{dt} = r - \frac{[R]}{\bar{t}_0 - t}$$

Part 4. Equations for Reaction Rates, r_1, r_2, . . ., r_{12}.[d]

$r_1 = k_1[A]$ $\qquad$ $r_2 = k_+[A] - k_-[R][S]$

$r_3 = k_1[A]^n$ $\qquad$ $r_4 = k_+[A]^n - k_-[R]$

$r_5 = k_1[A] - k_2[R]$ $\qquad$ $r_6 = k_2[R]$

$r_7 = k_1[A] + k_3[A]^2$ $\qquad$ $r_8 = k_3[A]^2$

$r_9 = k_1[A] + k_2[A][R]$ $\qquad$ $r_{10} = k_1[A] - k_2[A][R]$

$r_{11} = k_2[A][R]$ $\qquad$ $r_{12} = k_+[A][B] - k_-[R][S]$

[a] Locate scheme in Part 1 and read the exact solution, if available, in Part 2. If not available, construct ODE for numerical solution from Parts 3 and 4.

[b] $\underline{A}$, $\underline{B}$ indicate that are initially present in the reactor.

[c] Finite values of $[A]_i$, $[B]_i$, $[R]_i$, $[S]_i$ are assumed in Part 2 solutions; each is zero in the absence of initial charge.

[d] All rates are denoted by r_i, and not $(-r_i)$ for disappearance; thus, $(-r_A)$ in the text is equivalent to r_A in the table.

where V is given by Equation 2.62, and $[S]_f$ is the concentration of S in the outlet stream. Note that we assume the exit stream is pure S with no loss or gain of A, B, or R except by reaction. Equation 2.66 can be recast in the form

$$\frac{d[S]}{dt} = \frac{Q_0([S]-[S]_f)}{V_0 - Q_0 t} + k_+[A][B] - k_-[R][S] \tag{2.67}$$

Mass balance on A gives

$$\frac{d[A]}{dt} = \frac{Q_0[A]}{V_0 - Q_0 t} - k_+[A][B] + k_-[R][S] \tag{2.68}$$

General expression for an SBR for multiple reactions with inflow of liquid and outflow of liquid and vapor: Scheme 4

The schemes considered so far were all single or relatively simple two-step reactions with clear specification of inlet and outlet flows of liquid wherever such flows were present. To generalize the approach, we remove these restrictions and write equations for the most general isothermal case that would include multiple reactions, inflow of liquid, and outflow of liquid and vapor. Thus, consider N species ($i = 1, 2,\ldots, N$) undergoing M reactions ($j = 1, 2,\ldots, M$) with the following continuity equation:

$$\frac{d([i]V)}{dt} = Q_0[i]_0 - Q_f[i] - Q_{Vf}[i]_V - V\sum_{j=1}^{M} \nu_{ij} r_j \tag{2.69}$$

where Q_{Vf} is the volumetric vapor removal rate and V is the total volume of the liquid in the reactor at any time. In analogy with Equation 2.63 for a single (inlet) flow stream, we can write

$$\frac{dV}{dt} = Q_0 - Q_f - Q_{Lf} = Q_n \tag{2.70}$$

Note that, since this equation is for liquid flow, we use the liquid flow rate equivalent Q_{Lf} of the vapor flow rate Q_{Vf}, the two being related by the expression

$$Q_{Lf} = P\left(\frac{273}{T}\right)\left(\frac{Q_{Vf}}{22.4}\right)\left(\frac{1}{[L]_f}\right) \tag{2.71}$$

where $[L]_t$ is the total liquid phase concentration of all constituents.

Assuming linear variation of V with time, the following equivalent form of Equation 2.62 can be written as

$$V = V_0 + Q_n t \qquad (2.72)$$

Equation 2.69 can now be solved by substituting Equation 2.72 for V (see Froment and Bischoff, 1990, for further details).

Nonisothermal operation

Nonisothermal operation for semibatch reactors

As mentioned earlier in this section, SBO can be very advantageous for highly exothermic reactions since the rate of heat generation can be controlled by the rate of reactant addition. Specifically, what needs to be controlled is the evolving temperature progression with time. To do this for any of the schemes considered above, we must write the energy balance corresponding to the mass balance of that scheme and solve the two equations simultaneously. Since the mass balance given by Equation 2.69 is general, we shall write the energy balance corresponding to this equation. Thus,

$$\rho C_p \frac{d(VT)}{dt} = Q_0 C_{p\,\text{mean}} \rho T_0 - Q_f C_{p\,\text{mean}} \rho T - Q_{Vf} \sum_{i=1}^{N} [i]_V \Delta H_v + \dot{q} + V \sum_{j=1}^{M} r_j (-\Delta H_j) \qquad (2.73)$$

where ΔH_v is the heat of vaporization and $\dot{q}$ is the amount of heat exchanged by the control fluid.

The term $\dot{q}$ can be estimated from

$$\dot{q} = W_c C_{pc} (T_{c0} - T_{cf}) = U A_h \frac{(T_{c0} - T) - (T_{cf} - T)}{\ln(T_{c0} - T / T_{cf} - T)} \qquad (2.74)$$

where T_{c0} and T_{cf} are the inlet and outlet temperatures, respectively, of the control fluid, W_C the flow rate of the control fluid (kg/s), C_{pc} the heat capacity of the control fluid, and U the overall heat transfer coefficient whose value depends on whether the reactor is heated by the jacket or the coil.

Since all the parameters in Equations 2.69 and 2.73 are now known or can be estimated, the two equations can be solved numerically to obtain the various concentrations and temperature as functions of time.

Optimum temperatures/temperature profiles for maximizing yields/selectivities

In the case of MFR, the temperature is uniform within the reactor, and hence one can conceive of a single optimum temperature for maximizing the yield of a product in a complex reaction. On the other hand, in the case of PFR, there can be (and often is) a temperature profile within the reactor. The question therefore arises: Can one impose an optimum temperature profile for maximizing the yield in a tubular reactor? We examine both the cases below.

Optimum temperatures

Optimum temperatures for complex reactions

Consider the parallel scheme R19 with R as the desired product. If $E_1 > E_2$, the highest practical temperature should be used. For $E_1 < E_2$, an optimum temperature T_{opt} exists below which the rate would be too low (requiring a huge reactor) and above which the yield of R would be too low. An approximate expression for T_{opt} can be obtained by speculating on the largest allowable reactor size, and therefore τ_{max}, for given $[A]_0$ and F_{A0}, that is, by fixing

$$\tau_{max} = \frac{[A]_0 V_{max}}{F_{A0}} = \text{fixed} \tag{2.75}$$

This is quite practical since often the reaction is desired to be carried out in an available CSTR. Using this value of τ_{max}, T_{opt} can be obtained from

$$T_{opt} = \frac{E_2}{R \ln[k_2^{\circ} \tau_{max}((E_2 - E_1)/E_1)]} \tag{2.76}$$

Optimum temperature and concentration profiles in a PFR

This problem has been analyzed quite extensively over the years (see, Doraiswamy and Sharma, 1984), but we restrict the treatment here to a brief qualitative discussion followed by a presentation (without derivation) of equations for selected reaction schemes.

Take any two-step scheme in which product R of the first reaction is the desired product, while product S of the second reaction is the undesired product, irrespective of whether the scheme is parallel or consecutive. Let the activation energies of the two reactions be E_1 and E_2, with $E_1 < E_2$. The basic principle used in maximizing the yield of R in this situation is that the rate of reaction R1 is lower than that of reaction R2 at higher temperatures, and higher at lower temperatures. This is illustrated in Figure 2.11.

Parallel reactions Consider now the parallel scheme $A \rightarrow R$, $A \rightarrow S$, in which $E_1 > E_2$. Clearly, the yield of R is highest at the highest temperature that can be practically used. On the other hand, if $E_1 < E_2$, the temperature must be lowered to increase the yield of R. But there is a limit to which the temperature can be reduced consistent with the need to maintain a reasonably high rate of reaction. So we employ an increasing temperature profile, in which we start at the minimum feasible temperature (T_{min}) and

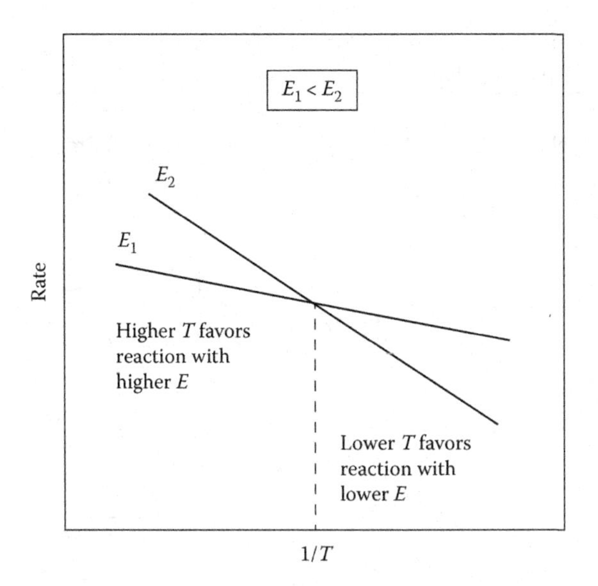

Figure 2.11 Effect of temperature on reactions with different activation energies.

complete as much reaction as possible at that temperature. In view of the high initial concentration of A, the concentration effect will offset much of the negative temperature effect due to low T. But as reaction progresses and $[A]$ falls, both T and $[A]$ will have a negative effect, and hence the temperature should be progressively raised to increase the overall rate.

Consecutive reactions For a consecutive reaction such as $A \rightarrow R \rightarrow S$ with $E_1 < E_2$, reflection will show that a decreasing temperature profile is optimal. Since R is not present initially, we start with the highest possible temperature and complete as much reaction as possible at that temperature. Since $E_1 < E_2$, the rate of decomposition of R is higher than its rate of formation at higher temperatures, and hence the temperature should be reduced as the reaction progresses in order to maintain a high yield of R.

Extension to a batch reactor In a batch reactor, instead of a temperature profile we vary the temperature with time. It is quite easy to divide the time cycle into different operating time zones, each corresponding to a particular predetermined temperature. It is often sufficient to have a qualitative knowledge of the nature of the profile. Operation at 3–4 discrete temperatures can lead to substantial improvements in conversion. The experiments can often be done quite easily in a chemist's laboratory, and the procedure implemented on a larger scale without the need for a detailed engineering analysis.

We will conclude this chapter by giving a summary of the temperature profiles for maximizing the selectivity for complex chemical reaction schemes (Table 2.5).

Table 2.5 Temperature–Time Profiles for Various Types of Complex Reactions in a BR (Also Applicable to PFR with Length Replacing Time)

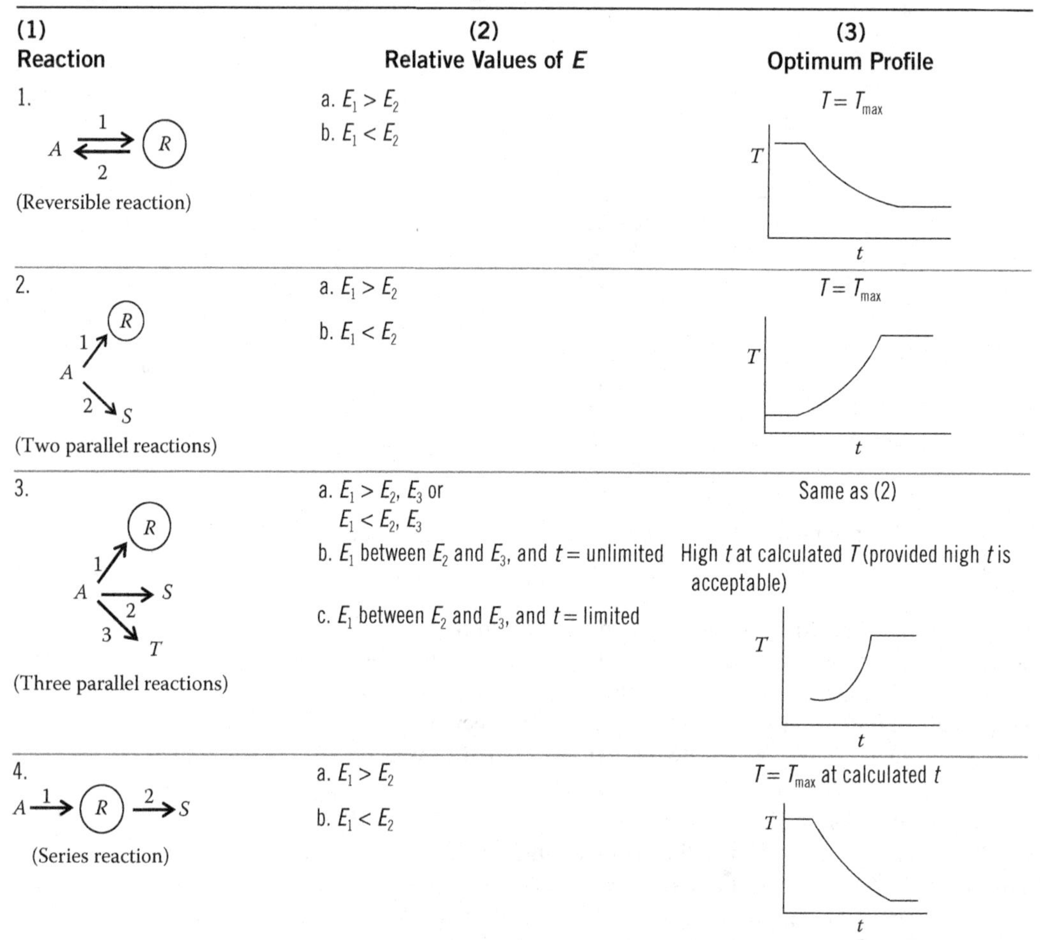

(1) Reaction	(2) Relative Values of E	(3) Optimum Profile
1. $A \underset{2}{\overset{1}{\rightleftarrows}} (R)$ (Reversible reaction)	a. $E_1 > E_2$ b. $E_1 < E_2$	$T = T_{max}$ [plot of T vs. t]
2. $A \xrightarrow{1} (R)$; $A \xrightarrow{2} S$ (Two parallel reactions)	a. $E_1 > E_2$ b. $E_1 < E_2$	$T = T_{max}$ [plot of T vs. t]
3. $A \xrightarrow{1} (R)$; $A \xrightarrow{2} S$; $A \xrightarrow{3} T$ (Three parallel reactions)	a. $E_1 > E_2, E_3$ or $E_1 < E_2, E_3$ b. E_1 between E_2 and E_3, and t = unlimited c. E_1 between E_2 and E_3, and t = limited	Same as (2) High t at calculated T (provided high t is acceptable) [plot of T vs. t]
4. $A \xrightarrow{1} (R) \xrightarrow{2} S$ (Series reaction)	a. $E_1 > E_2$ b. $E_1 < E_2$	$T = T_{max}$ at calculated t [plot of T vs. t]

Explore yourself

1. Analyze the combustion reaction as a complex reaction scheme. You can find the free radical reaction mechanisms of many combustion reactions in the NIST database or in the open literature. Answer the following questions:
   a. How many independent reaction steps are involved in the combustion of hydrogen?
   b. How many independent reaction steps are involved in the combustion of methane?
   c. What is the driving force behind the interest in identifying the detailed mechanism and the existence of such short-lived intermediates? How can we use such detailed information in the design and operation of the reactors?

d. Recall the internal combustion engine question at the end of Chapter 1. Do you think that the design (and the efficiency) can be improved if the detailed reaction mechanism is known?

2. Search for the reaction mechanism of the production of aspirin, a good example for a complex reaction.
   a. What are the reactants? What are the products?
   b. What does chirality mean?
   c. Do the pharmaceutical companies purify the product before sale?
   d. How can you improve the selectivity toward the desired product?
3. Inquire about the microfluidic reactor technology used for a complex reaction scheme, the reasons for choosing that particular reactor type, and the prospects of adopting a new design.
4. Find the optimal operational policy for a given reaction system for a partially emptying reactor.
   a. Suggest as many alternatives as possible for the type of reactions that would benefit from such an operation.
   b. What are the design parameters for a partially emptying reactor?
   c. How would you control selectivity in this type of reactor?
   d. Can you shift equilibrium conversions of reversible reactions in partially emptying reactors?
5. Describe the strategies for the following cases using a semi-batch reactor:
   a. To control the temperature of a highly exothermic reaction
   b. To control the temperature of a highly endothermic reaction
   c. Increase the selectivity of a complex reaction scheme
   d. How would you increase the equilibrium conversion of a reversible reaction?

References

Aris, R., *Introduction to the Analysis of Chemical Reactors*, Prentice-Hall, Englewood Cliffs, NJ, 1965.

Aris, R., *Elementary Chemical Reactor Analysis*, Prentice-Hall, Englewood Cliffs, NJ, 1969.

Denbigh, K.G. and Turner, J.C.R., *Chemical Reactor Theory*, Cambridge University Press, NY, 1971.

Doraiswamy, L.K. and Sharma, M.M., *Heterogeneous Reactions: Analysis, Examples, and Reactor Design,* Wiley, NY, 1984.

Forni, L. and Miglio, R., Catalytic synthesis of 2-methylpyrazine over Zn–Cr–O/Pd a simplified kinetic scheme, *Heterogeneous Catalysis and Fine Chemicals* III, 329–336, 1993.

Froment G.F. and Bischoff, K.B., *Chemical Reactor Analysis and Design*, Wiley, NY, 1990.

Gillespie, B.M. and Carberry, J.J., Reactor yield at intermediate mixing levels—An extension of Van De Vusse's analysis, *Chem. Eng. Sci.*, **21**, 472–475, 1966.

Levenspiel, O., *The Chemical Reactor Omnibook,* OSU Bookstore, Corvallis, OR, 1993.

Nauman, E.B., *Chemical Reactor Design*, Wiley, NY, 1987.

Russell, T.W.F. and Buzzelli, D.T., Reactor analysis and process synthesis for a class of complex reactions, *Ind. Eng. Chem. Proc. Des. Dev.* **8**, 2–9, 1969.

Van de Vusse, J.G., Plug-flow type reactor versus tank reactor, *Chem. Eng. Sci.*, **19**, 994–996, 1964.

Van de Vusse, J.G. and Voetter, H., C2. Optimum pressure and concentration gradients in tubular reactors, *Chem. Eng. Sci.,* **14**, 90–98, 1961.

Bibliography

For a more detailed treatment of multiple reactions:

Aris, R., *Elementary Chemical Reactor Analysis*, Prentice-Hall, Englewood Cliffs, NJ.

Nauman, E.B., 1987, *Chemical Reactor Design*, Wiley, NY, 1969.

Selectivity and yield:

Levenspiel, O., *The Chemical Reactor Omnibook,* OSU Bookstore, Corvallis, OR, 1993.

Interlude I

Selectivity and its relation to sustainability

Enhancing the selectivity in complex reaction schemes may be the biggest challenge on the shoulders of the chemists and chemical engineers. Sustainable production of desired chemicals requires improved material and energy economies. After the conclusion of Chapter 2, it must have been obvious that low selectivity results in wasting valuable raw materials. Additional energy penalties are due in terms of separation costs. Hence, selective manufacture of the desired chemicals is imperative in a world where sustainability is the primary concern.

In this interlude, we will cover some novel approaches toward improved selectivity via the product-removal strategy through recent examples in the literature. We will limit the coverage of information by only giving the framework descriptions of the methodologies. In-depth treatment of these subjects is postponed to Chapters 13 and beyond.

Reactive distillation

Reactive distillation

Looking back to the progress achieved in the areas of conventional separation and reactor design, it seems that major advances have now been made. This has led to increasing research into methods in which reaction and separation are combined in a single unit. The equipment in which this dual function is carried out is sometimes referred to as the *combo reactor.*

Combo reactors can be of two types: (1) reaction-oriented and (2) separation-oriented. In the first, distillation is used to enhance conversion beyond the equilibrium value, such as in esterification reactions. In the second, reaction is used to effect efficient separation, such as in the separation of *p*-cresol from its mixture with *m*-cresol.

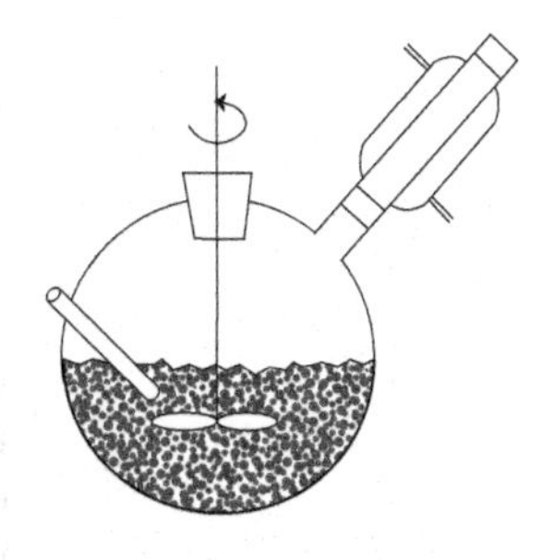

Figure I.1 A simple laboratory setup for reaction with distillation.

The principle behind the reaction-oriented strategy is the chemist's apparatus in which a reflux column condenser is connected to a batch reactor (usually a round-bottomed flask) as shown in Figure I.1. The product and the heat of reaction are continuously removed, and the reactant is returned to the reactor. Modeling of type 2 systems leads to equations for the so-called *Separation Factor* of a mixture enhanced by reaction. For type 1 systems, on the other hand, equations are obtained for the conversion or yield enhanced by separation. If the component separated is the desired product, it is really immaterial which definition is used. The reaction itself is unimportant in type 1 systems.

Irrespective of whether reaction or separation is of primary concern, three types of combo reactors are commonly used: *reaction–extraction, reaction–distillation,* and *reaction–crystallization.* Each of these can, in theory, be either reaction- or separation-oriented. Among other, less conventional methods of combining reaction with separation are biphasing and the use of membranes. Photochemistry, micelles, ultrasound, and microphases offer additional techniques/agents for enhancing the rate of a reaction, and a survey of the analysis and design of combo reactors involving these methods can be found in Doraiswamy (2001).

Membrane reactors

Membrane reactors

Like zeolites that combine shape selectivity with catalysis, membranes combine separation with catalysis to enhance reaction rates. The dual functionality of zeolites derives from the nature of the *catalytic material,* while that of membranes derives from the nature of *the reactor material.* The catalyst in the membrane reactor can be a part of the membrane itself, or be external to it (i.e., placed inside the membrane tube). The chief property of a membrane is its ability for selective *permeation* or *permselectivity* with respect to certain compounds.

Organic membrane reactions are best carried out in reactors made of *inorganic membranes,* such as from palladium, alumina, or ceramics. A recent trend has been to develop polymeric–inorganic composite-type membranes formed by the deposition of a thin dense polymeric film on an inorganic support. Another class of membranes under development for organic synthesis is the *liquid membrane.* The permselective barrier in this type of membrane is a liquid phase, often containing a dissolved "carrier" or "transporter" which selectively reacts with a specific permeate to enhance its transport rate through the membrane.

Inorganic membranes for organic reactions/synthesis

Inorganic membranes can be grouped under two broad classes, dense and porous. Membranes made of dense palladium or its alloys and of porous glass are the most commonly used membranes for organic reactions. For reactions involving hydrogen, the dense membranes are typically metallic and are made of Pt or Pd (Gryaznov et al., 1986). These can be in the form of hollow tubes or foils, or thin films deposited on porous supports prepared by various methods (Gryaznov et al., 1993; Li et al., 1993; Shu et al., 1993). These membranes have low permeability but high permselectivity.

For Knudsen diffusion see Chapter 6

Porous membranes are usually characterized by mesopores, with transport in the Knudsen regime. Typical examples are Vycor glass and γ-Al_2O_3. Thin-film ceramic membranes also belong to this category and are prepared by sol–gel coating techniques (Leenaars et al., 1984, 1985; Uhlhorn et al., 1987, 1992a,b) or chemical vapor deposition (CVD) (Gavalas et al., 1989; Lin and Burggraaf, 1992). Porous membranes have high permeability but relatively low permselectivity.

Polymeric–inorganic composites have been developed mainly for the separation of organic compounds through pervaporation (PV) and vapor permeation (VP). However, they can also be advantageously used for reactions such as esterification and condensation.

Potentially exploitable features of membranes

Several attractive features of membranes derive from their unique characteristics and amenability to novel modes of operation. These are sketched in Figure I.2 and briefly described below.

Equilibrium shift in membrane reactors The most important feature of a membrane reactor that gives it a pronounced advantage over other reactors is its ability to remove a product selectively by letting it permeate

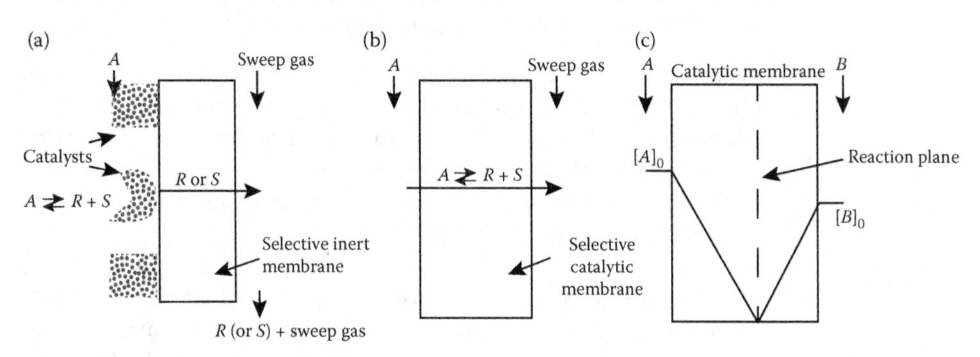

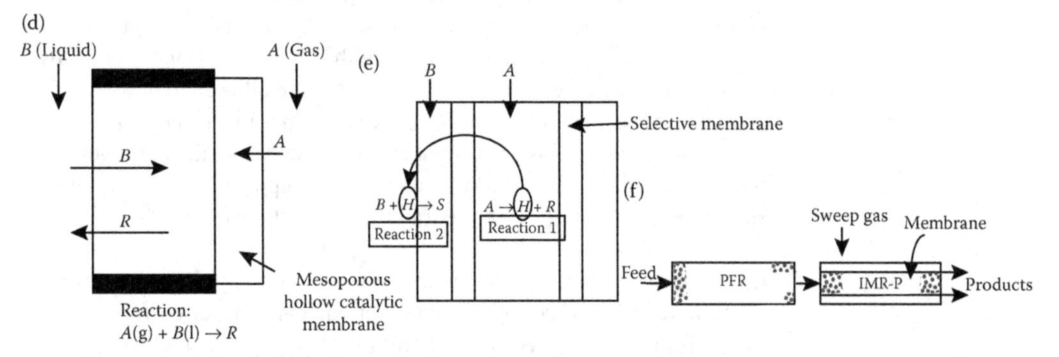

Figure I.2 Exploitable features of membrane reactors. (a) Enhancing the conversion of a reversible reaction in a packed-bed inert membrane reactor. (b) Enhancing the conversion of a reversible reaction in a catalytic membrane reactor. (c) Preventing slip in a reaction requiring stoichiometric feeds. (d) Enhancing the rate of a multiphase reaction. (e) Energetic, thermodynamic, or kinetic coupling of two reactions run on opposite sides of a membrane. (f) Hybrid of fixed-bed reactor (PFR) and selective inert membrane reactor (IMR-P) in series.

out of the reactor through its inert (Figure I.2a) or catalyst-containing (Figure I.2b) membrane wall. Thus, one can "beat the equilibrium" and achieve conversions beyond the limits of equilibrium.

Controlled addition of reactants Controlled dosing of one of the reactants (usually hydrogen or oxygen) is often an important consideration in partial hydrogenation and oxidation reactions. The lowering of selectivity in the inlet region of a fixed-bed reactor can be avoided by maintaining a uniformly low reactant concentration by controlled supply of the second reactant across the membrane wall of the entire reactor length.

Preventing excess reactant "slip" in reactions requiring strict stoichiometric feeds Certain reactions involving more than one reactant require introduction of the reactants in precisely stoichiometric proportions. Thus in a reaction such as

$$\nu_A A + \nu_B B \rightarrow \nu_R R + \nu_S S \qquad \text{(R1)}$$

any feed of A (or B) in excess of the ratio ν_A/ν_B (or ν_B/ν_A) "slips" into the product stream. Such a slip, generally undesirable, would be particularly unacceptable if the reactant concerned happened to be a toxic pollutant.

If A and B are allowed to diffuse into the membrane from the opposite sides and react instantaneously and completely at a plane whose location is determined by stoichiometry (Figure I.2c), then there would be no slip on either side (Zaspalis et al., 1991; Sloot et al., 1992). Two other noteworthy features of this concept are that permselectivity of the membrane is not essential, and permeability is usually low.

Mimicking trickle-bed operation with improved performance Another use of nonpermselective membranes is in multiphase organic reactions involving trickle-bed-type reactors (Harold et al., 1989, 1994; Cini and Harold, 1991). The reactor consists essentially of a hollow macroporous membrane tube coated on the inside with a hollow mesoporous catalyst layer, as shown in Figure I.2d. Liquid and gas are allowed to flow on opposite sides of the membrane. Because the gas comes into *direct contact* with the liquid-filled catalyst, it resembles a trickle-bed reactor. However, as there is no separate liquid film to hamper the supply of gas to the catalyst sites, it performs better than the traditional trickle-bed reactor.

Coupling of reactions A particularly attractive application of membrane reactors is in coupling two reactions carried out on the opposite sides of a membrane. A product from reaction 1 on one side can serve as a reactant for reaction 2 on the other side, while the exothermic heat from reaction 2 supplies the endothermic heat for reaction 1 (Gryaznov et al., 1986). This is illustrated in Figure I.2e.

Hybridization An operational variation reminiscent of combined fully mixed reactor followed by a tubular reactor (MT) and a tubular reactor followed by a fully mixed reactor (TM) is the use of a plug-flow fixed-bed reactor followed in series by a packed inert membrane reactor, as shown in Figure I.2f (Wu and Liu, 1992).

Phase transfer catalysis

Phase transfer catalysis

Phase transfer catalysis can be defined as a means to facilitate transport of reagents across phase boundaries in a multiphase reaction system, be it liquid–liquid, solid–liquid, or even gas–(solid)–liquid. As a result, reactions such as nucleophilic substitution, which would otherwise be hindered because the reactants are located in different phases and would therefore be inaccessible to each other, are greatly facilitated. Rather than homogenizing the system by adding a polar solvent, a phase transfer agent achieves this interphase interaction by causing the transfer of the nucleophile, typically in the form of anions, to the organic phase, where the two reagents, now in the same phase, can react to give the desired product. In the aqueous phase, anions that are left behind either form ion pairs with the cations of the catalyst (represented in general as Q^+) or form complexes with it, depending on the type of catalyst used. Charles Starks introduced the term "Phase Transfer Catalysis" for such a catalyst or agent in a landmark paper in 1971. The PT agent is considered to be a catalyst, since it is regenerated in the organic phase, once the anion reacts with the organic reagent and can be shuttled back to the aqueous phase to continue the cycle; also, only small amounts of the PT agent are required for effective phase transfer action. Hence the name "phase transfer catalysis" survives although it is not a catalyst in the traditional sense of the word. Given that the concentration of the active PT agent changes during the PT cycle, one can perhaps consider it as a catalytic process with changing catalytic activity.

In general, PTC involves a wide body of reactions in heterogeneous liquid–liquid or solid–liquid systems in which inorganic anions (or organic anions generated through deprotonation with an aqueous phase base) react with organic substrates through the mediation of a phase transfer catalyst. The reactive anions are introduced into the organic phase in the form of lipophilic ion pairs or complexes that they form with the PT catalyst (reaction 2.1). This ion-pair partitions between the organic and aqueous phases due to its lipophilic nature and once the anion is transferred to the organic phase, it can react with the organic substrate, yielding the desired product (reaction 2.2), with high yields and often with high selectivity. In the absence of the PT catalyst, the anions and the organic phase cannot react as they are physically isolated from each other in two different mutually immiscible phases.

$$Q^+Y^- + M^+X^- \leftrightarrow M^+Y^- + Q^+X^- \quad \text{Ion exchange reaction} \quad \text{(R2.1)}$$

$$Q^+Y^- + R - X \leftrightarrow R - Y + Q^+X^- \quad \text{Organic phase reaction} \quad \text{(R2.2)}$$

It should be noted here that reactions in the presence of bases follow a slightly different mechanism involving deprotonation of moderately to weakly acidic organic compounds at the interface, as discussed further later in Chapter 16.

In addition, PTC has the advantage that easily recoverable solvents such as dichlormethane, toluene, and hexane can be used rather than polar solvents such as dimethyl formamide (DMF), dimethyl sulfoxide (DMSO), and hexamethylphosphoramide (HMPA), which are costlier. All these benefits lead to enhanced productivity with higher safety and lower environmental impact. Also, it must be mentioned that although a PT catalyst is mainly used to enhance reaction rates and yield, it can be a useful tool in many cases to selectively synthesize one product or significantly reduce an undesired by-product. One of the main concerns with PTC that has been a significant barrier to industrial adoption, especially in the pharmaceutical and food additives industry, is the issue of catalyst recovery from the final product stream.

Most early work on PT-catalyzed systems considers pseudo-first-order reaction kinetics for the organic phase reaction. However, being multiphase systems, phase interfaces, and transport between phases are an integral part of the PTC cycle. Depending on the relative rates of ion exchange and interphase transport steps, the organic phase reaction may not be the rate-controlling step. In general, although pseudo-first-order kinetics applies in many limiting cases, the relative rates of the reactions and interphase anion transfer steps determine the overall reaction kinetics. In the case of reactions in the presence of a strong base, the limited extractability of the OH^- ion into organic phase leads to a slow anion transfer step. The choice of the catalyst becomes critical in this situation and can be different from the optimal choice for substitution reactions under neutral conditions. Reaction kinetics is affected strongly by reaction variables such as catalyst amount and structure, anion type and degree of hydration, agitation, amount of water in the system, reaction temperature, and type of solvent used, and is discussed in Chapter 16.

References

Cini, P. and Harold, M.P., *AIChE J.*, **37**, 997, 1991.

Doraiswamy, L.K., *Organic Synthesis Engineering,* Oxford University Press, New York, 2001.

Gavalas, G.R., Megiris, C.E., and Nam, S.W., *Chem. Eng. Sci.*, **44**, 1829, 1989.

Gryaznov, V.M., Mishchenko, A.P., Smirnov, V.A., Kashdan, M.V., Sarylova, M.E., and Fasman, A.B., French Patent 2,595,092, 1986.

Gryaznov, V.M., Serebryannikova, O.S., Serov, Y.M., Ermilova, M.M., Karavanov, A.N., Mischenko, A.P., and Orekhova, N.V., *Appl. Catal.*, **96**, 15, 1993.

Harold, M.P., Cini, P., Patanaude, B., and Venkatraman, K., *AIChE Symp. Ser.*, **85**, 26, 1989.

Harold, M.P., Lee, C., Burggraaf, A.J., Keizer, K., Zaspalis, V.T., and de Lange, R.S.A., *Mater. Res. Soc. MRS Bull.*, **29**, 34, 1994.

Leenaars, A.F.M., Keizer, K., and Burggraaf, A.J., *J. Mater. Sci.*, **19**, 1077, 1984; *J. Coll. Inter. Sci.*, **105**, 27, 1985.

Li, Z.Y., Maeda, H., Kusakabe, K., Morooka, S., Anzai, H., and Akiyama, S.J., *Membr. Sci.*, **78**, 247, 1993.

Lin, Y.S. and Burggraaf, A.J., *AIChE J.*, **38**, 444, 1992.

Shu, J., Grandjean, B.P.A., Ghali, E., and Kaliaguine, S., *J. Membr. Sci.*, **77**, 181, 1993.

Sloot, H.J., Smolders, C.A., Van Swaaij, W.P.M., and Versteeg, G.F., *AlChE J.*, **38**, 887, 1992.

Starks, C.M., *J. Am. Chem. Soc.*, **93**, 195, 1971.

Uhlhorn, R.J.R., Huis in't Veld, M.B.H.J., Keizer, K., and Burggraaf, A.J., *Sci. Ceram.*,**14**, 55, 1987.

Uhlhorn, R.J.R., Keizer, K., and Burggraaf, A.J., *J. Membr. Sci.*, **66**, 259, 1992a; *J. Membr. Sci.*, **66**, 271, 1992b.

Wu, J.C.S., and Liu, P.K.T., *Ind. Eng. Chem. Res.*, **31**(1), 322, 1992.

Zaspalis, V.T. and Burggraaf, A.F., In *Inorganic Membranes Synthesis, Characteristics and Applications* (Ed., Bhave, R.R.), Van Nostrand Reinhold, New York, 1991.

Chapter 3 Nonideal reactor analysis

Chapter objectives

Upon successful completion of the chapter, one should be able to

- Explain the difference between PFR and tubular reactor.
- Explain the difference between CSTR and MFR.
- Identify and model the nonidealities in chemical reactors.
- Propose ideal reactor arrangements and flow schemes that offer best representations of the nonidealities.
- Model a chemical reactor with an awareness of the role of mixing in its performance.
- Define macromixing, micromixing, and segregated flow concepts.
- Use tanks-in-series, axial dispersion, and RFR concepts interchangeably in modeling chemical reactors with knowledge of the limitations of each concept.
- Choose between different theories for modeling nonidealities in chemical reactors such as residence time distribution (RTD), interaction by exchange with the mean (IEM), engulfment deformation (E), and probability density function (PDF).
- Apply different mixing models in chemical reactor design and analysis.

Introduction

In this chapter, we move to the situation where the reactors are no longer ideal and begin our description by showing how the ideal PFR and MFR can be represented in terms of each other. As a result, we expect to develop an understanding of how simple combinations of these ideal reactors can be used to describe the behavior of complex reaction/reactor geometries. Then we logically move on to the subject of nonidealities in chemical reactors and go into the considerations of channeling and bypassing in MFRs and axial and radial dispersion in PFRs. While discussing these, we explain the RTD theory and axial dispersion theory. The RTD theory is a very frequently employed method of treating nonidealities in chemical reactors. Thus, we focus a little more on this theory in the next section and give its merits and demerits. In the same section, we also discuss the concept of mixing and give four commonly used mixing models from the literature. We dedicate the next section to modern theories of mixing that include the turbulent theory, the so-called zone model, the IEM model, the engulfment-deformation models, and the joint PDF model. We conclude the treatment by giving a comparative analysis of the timescales of different processes and their relevance to mixing.

Actually, reactors can operate under conditions where there is an arbitrary distribution of residence times, leading to different degrees of mixing with consequent effects on the reactor performance. Also, multiple solutions can exist for equations describing certain situations and these can have an important bearing on the choice of operating conditions. We also briefly review these important aspects of reactors in this chapter. However, as the subjects are highly mathematical, the treatment will be restricted to simple formulations and qualitative discussions that can act as guidelines in predicting the reactor performance.

Two limits of the ideal reactor

The ideal PFR and MFR will be seen to converge to one another in the two limits that we are going to cover below. This section shows that it is possible to model nonideal behavior as a combination of PFR and CSTR. We present two common models, each of which can be used to define the plug-flow and mixed-flow limits of ideal behavior.

Plug-flow reactors with recycle

The principle of the RFR, which we have already discussed in Chapter 2, is sketched in Figure 3.1. Here, we will examine the limits of infinite and zero recycle. Recall that the recycle ratio is defined as

Recycle ratio

$$R = \frac{\text{Volume of product recycled}}{\text{Volume of product leaving the reactor}} \tag{3.1}$$

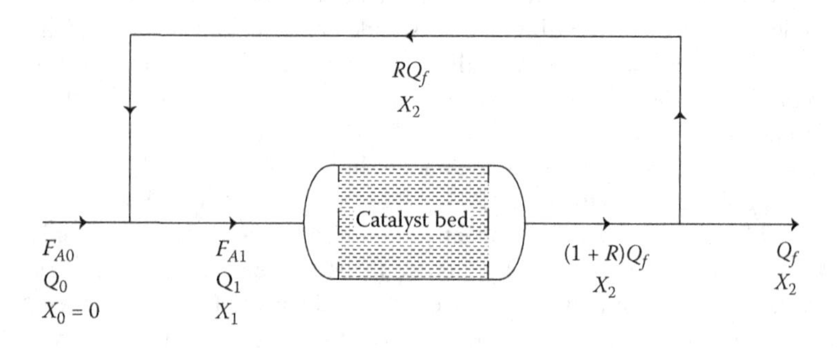

Figure 3.1 Recycle reactor.

The space time for the recycle reactor is

$$\frac{V}{F_{A0}} = (1 + R)\int_{X_1}^{X_2} \frac{dX_A}{-r_A} \tag{3.2}$$

Furthermore, the conversion after the mixing point is

Conversion in a recycle reactor

$$X_1 = \frac{RX_2}{1 + R} \tag{3.3}$$

Simple analysis of Equation 3.3 reveals that as $R \to \infty$, $X_1 \to X_2$, that is, the behavior of RFR approaches the behavior of MFR.

In other words, as R goes to infinity, that is, when the amount of effluent stream Q_f leaving the reactor is too small in comparison to the amount recycled RQ_f, the design equation converges to that for an MFR:

$$\frac{V}{F_{A0}} = \frac{X_f}{-r_{Af}} \tag{3.4}$$

On the other hand, in the limit of $R = 0$, the design equation is identical to that of a PFR. These two limits of PFR operation indicate that by adjusting the recycle ratio, it is possible to adjust the degree of *backmixing*, or it is possible to operate a PFR as an MFR.

Tanks-in-series model

The concept of using two or more CSTRs in series stems from the fact that mixed-flow performance can be made to approach plug-flow performance by increasing the number of CSTRs. In fact, the number of reactors (or tanks) can be regarded as a measure of the degree of mixing, and this description of partial mixing is commonly referred to as the *tanks-in-series model.*

Consider a sequence of stirred reactors in series as shown in Figure 3.2. The residence time for reactor 1 is given by

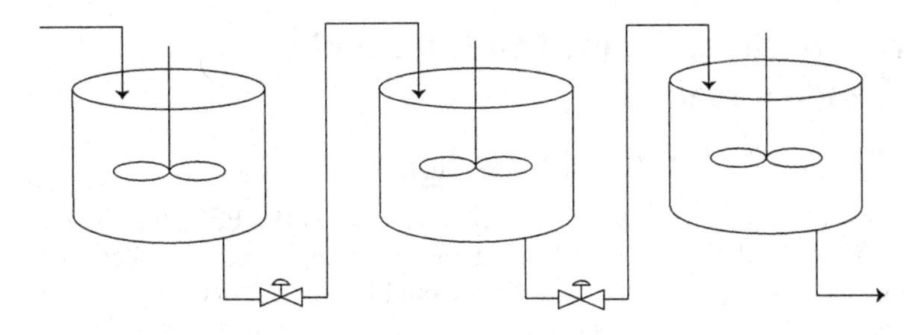

Figure 3.2 Cascade of CSTRs in series.

$$\bar{t}_1 = \frac{V_1[A]_0}{F_{A0}} \quad (3.5)$$

Combining this with the performance equation for a first-order reaction, we obtain

$$\bar{t}_1 = \frac{V_1[A]_0}{F_{A0}} = \frac{[A]_0 - [A]_1}{k[A]_1} \quad (3.6)$$

By writing similar equations for $\bar{t}_2, \ldots, \bar{t}_N$, we obtain following expression for conversion at the end of N reactors:

N tanks in series

$$\frac{[A]_N}{[A]_0} = (1 - X_A) = \left[\prod_{i=1}^{N}(1 + k\bar{t}_i)\right]^{-1} \quad (3.7)$$

This equation has been validated by the extensive experimental results of Eldridge and Piret (1950).

Now, if we assume $\bar{t}$ to be constant, that is, all the reactors to be of the same volume for a given volumetric flow rate, Equation 3.7 for a first-order irreversible reaction becomes

$$N\bar{t} = \frac{N}{k}\left[\left(\frac{[A]_0}{[A]_N}\right)^{1/N} - 1\right] \quad (3.8)$$

It is evident that PFR operation is approached as N approaches infinity. Under this condition, the PFR equation ($k\bar{t} = \ln\{[A]_0/[A]\}$) is recovered.

A similar analysis of nonfirst-order reactions leads to quite cumbersome equations for $[A]_N/[A_0]$. As the order increases, the telescoping functions involved in these equations become progressively more unwieldy so that a simple expression for $[A]_N/[A_0]$ for a series of N reactors similar to Equation 3.7 for a first-order reaction becomes impossible. In such cases, it is necessary to resort to step-by-step algebraic calculations.

Nonidealities defined with respect to the ideal reactors

Nonidealities in tubular reactors

For a tubular reactor, the ideal behavior limit is the PFR, that is, the velocity profile is flat, and in the radial direction, we have perfect mixing. In this scheme, the nonideality would be the absence of plug-flow behavior. Plug-flow behavior in a tubular reactor starts after the boundary layer formation is completed. Therefore, if the reactor is not long

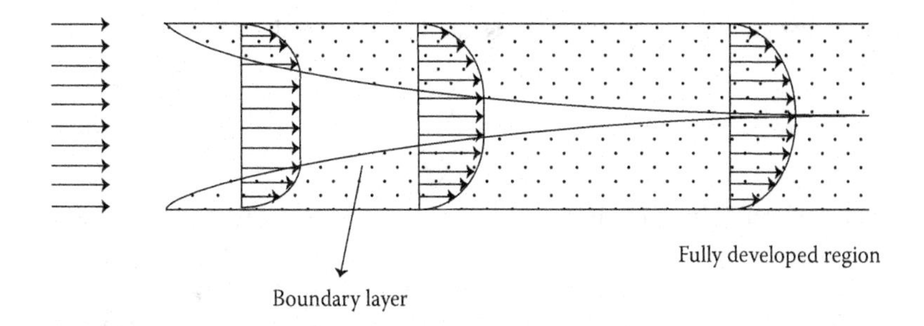

Figure 3.3 Nonidealities in a PFR: boundary layer development. The nonuniformities in the velocity fields cause mixing problems, giving rise to axial and/or radial dispersion effects.

enough, the boundary layer development will not be completed and severe velocity and concentration gradients will be formed (Figure 3.3). Another condition that ensures PFR behavior is a highly turbulent flow field. Thus, true PFR behavior can be safely assumed only if at least one of the following conditions is met:

Validity of PFR assumption

$$\text{PFR assumption is valid when } \frac{L}{D} > 50 \quad \text{or} \quad Re > 10{,}000$$

Axial dispersion model In the dispersion model, deviation from plug flow is expressed in terms of a dispersion or effective axial diffusion coefficient. The mathematical derivation is similar to that for plug flow except that a term is now included for diffusive flow in addition to that for convective flow. This term appears as $-D_{ez}$ $(d[A]/dz)$, where D_{ez} is the effective axial dispersion coefficient. The continuity equation in the absence of radial variations takes the form

Axial dispersion

$$\frac{\partial[A]_i}{\partial t} + u\frac{d[A]_i}{dz} = D_{iz}\frac{d^2[A]_i}{dz^2} + \sum_{j=1}^{R} V_{ji}r \tag{3.9}$$

For a simple first-order reaction, $-r_A = k[A]$, taking place under steady-state conditions, the equation simplifies to

$$D_{iz}\frac{d^2[A]}{dz^2} - u\frac{d[A]}{dz} - k[A] = 0 \tag{3.10}$$

This equation can be rendered dimensionless by introducing the dimensionless length $Z = z/L$ so that

$$\frac{d^2[A]}{dZ^2} - \frac{uL}{D_{Az}}\frac{d[A]}{dZ} - \frac{kL^2}{D_{Az}}[A] = 0 \tag{3.11}$$

Here, we define two dimensionless numbers, the Peclet number,*

Peclet number

$$Pe = \frac{uL}{D_{Az}} \tag{3.12}$$

and the Damköhler number

Damköhler number

$$Da = \frac{kL^2}{D_{Az}} \tag{3.13}$$

The boundary conditions for this case, also called Danckwerts or closed-boundary conditions, are

Danckwerts boundary conditions

$$-D_{ez}\frac{d[A]}{dz} = u([A]_0 - [A]) \text{ at } z = 0 \tag{3.14}$$

$$\frac{d[A]}{dz} = 0 \quad \text{at} \quad z = L \tag{3.15}$$

The solution of Equation 3.11 with the boundary conditions 3.14 and 3.15 is (Nauman, 1987)

$$\frac{[A]_{\text{out}}}{[A]_{\text{in}}} = \frac{4\sqrt{1 + 4k\bar{t}/Pe}\exp(Pe/2)}{[1 + \sqrt{1 + 4k\bar{t}/Pe}]^2\exp(\sqrt{1 + 4k\bar{t}/Pe}(Pe/2)) - [1 - \sqrt{1 + 4k\bar{t}/Pe}]^2\exp(-\sqrt{1 + 4k\bar{t}/Pe}(Pe/2))} \tag{3.16}$$

As $Pe \rightarrow 0$ MFR
As $Pe \rightarrow \infty$ PFR

We will leave it to the reader to show that as $Pe \rightarrow 0$, the equation approaches the MFR limit, whereas for $Pe \rightarrow \infty$, the PFR result is recovered.

Nonidealities in MFR In Chapter 1, we called an ideal MFR a CSTR. In the CSTR domain, the concentration and temperature are uniform throughout the reactor. This is ensured by good mixing: if the recirculation time is 100 times greater than the residence time, then we can safely take the CSTR condition to be valid. In addition to poor mixing in an MFR, we may also have stagnant dead zones and bypassing that can cause serious nonidealities as shown in Figure 3.4.

The CSTR assumption is valid when 100 × recirculation time > residence time. For more details, see Westerterp et al. (1983).

* A number of correlations for predicting the Peclet number both for liquids and gases in fixed and fluidized beds are available and have been reviewed by Wen and Fan (1975).

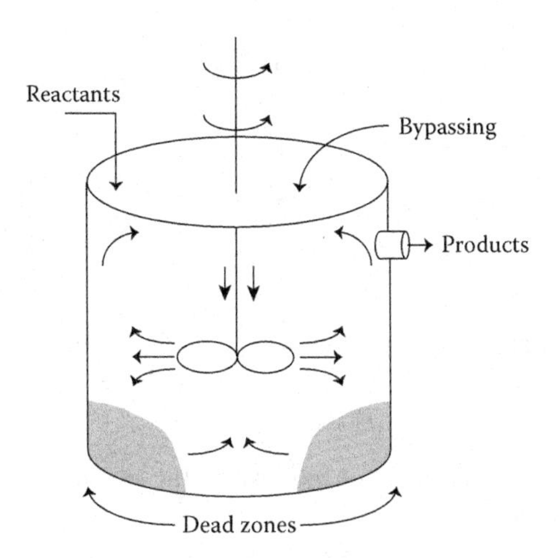

Figure 3.4 Nonidealities in an MFR. (From Fox, R.O. *Computational Models for Turbulent Reacting Flows*, Cambridge, UK, 2003.)

Residence time distribution

Theory

The theory of RTD was first enunciated by Danckwerts (1953). It is a useful technique for identifying the nonidealities in chemical reactors, but as with any theory, it comes with its own limitations. In this section, we describe the RTD theory with its merits and demerits. In the next section, we give examples of more recent theories of mixing and nonideal reactor modeling.

When a steady stream of fluid flows through a vessel, different elements of the fluid spend different times within it. The time spent by each fluid element can be identified by an inert tracer experiment, where a pulse or a step input of a tracer is injected into the flow stream, and the concentration of the pulse in the effluent is detected. As the reader may quickly infer, the tracer must leave the PFR undisturbed. On the other hand, a step pulse may give rise to an exponential distribution in a CSTR. In the beginning of this chapter, we already demonstrated that PFR behavior approaches that of a CSTR under infinite recycle. It follows that infinite CSTRs in series behave like a PFR. Thus, we conclude that any nonideal reactor can be represented as a combination of the PFR and MFR to a certain degree. First, let us show a representative pulse response curve for each of the ideal reactors in Figure 3.5. As seen in the figure, the response to a step input of tracer in a PFR is identical to the input function, whereas the response in a CSTR exhibits an exponential decay. The response curves as shown in Figure 3.5 are called *washout functions.* The input function of the inert tracer concentration $[I]$ can be mathematically expressed as

$$[I] = [I]_0 \quad \text{for } t < 0 \tag{3.17a}$$

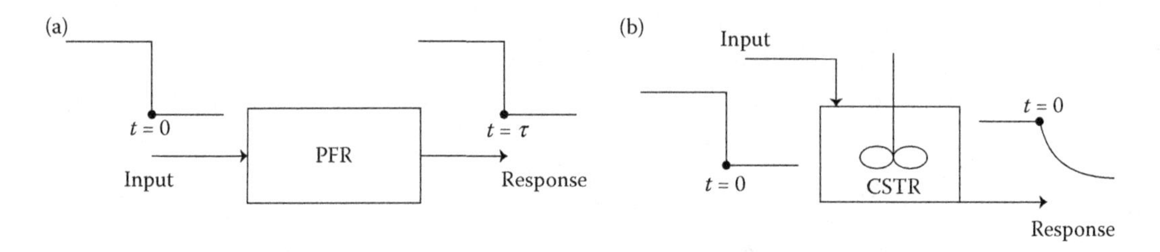

Figure 3.5 Response to a step input of a tracer from a perfect (a) PFR and (b) MFR.

$$[I] = 0 \quad \text{for } t \geq 0 \tag{3.17b}$$

The inert tracer experiment

while the differential equation governing the process for a constant-volume CSTR is

$$V\frac{d[I]}{dt} = -Q_{out}[I] \tag{3.18}$$

The solution to this equation is

$$\frac{[I]}{[I]_0} = \exp\left(-\frac{Q_{out}t}{V}\right) = \exp\left(-\frac{t}{\bar{t}}\right) \tag{3.19}$$

We specify the washout function as the ratio $[I]/[I]_0 = W(t)$, which represents the fraction of molecules that had a residence time of t or longer. As can be inferred from Figure 3.5a, the washout function for a PFR is

The washout function

$$W(t) = 1, \quad 0 < t < \bar{t} \tag{3.20a}$$

$$W(t) = 0, \quad t > \bar{t} \tag{3.20b}$$

We now define a *cumulative distribution function* $F(t)$ as the fraction of the molecules leaving the system with residence time t or less, or mathematically:

The cumulative distribution function

$$F(t) = 1 - W(t) \tag{3.21}$$

Finally, we define the *density function* $f(t)$ as

The density function

$$f(t) = \frac{dF}{dt} = -\frac{dW}{dt} \tag{3.22}$$

Note that

$$\int_0^{\infty} f(t)dt = 1 \tag{3.23}$$

The function $f(t)$ is called the *residence time distribution* (RTD). It is denoted by a curve that represents, at any given time, the amount of fluid with ages between t and $t + dt$ flowing out in the exit stream. The time

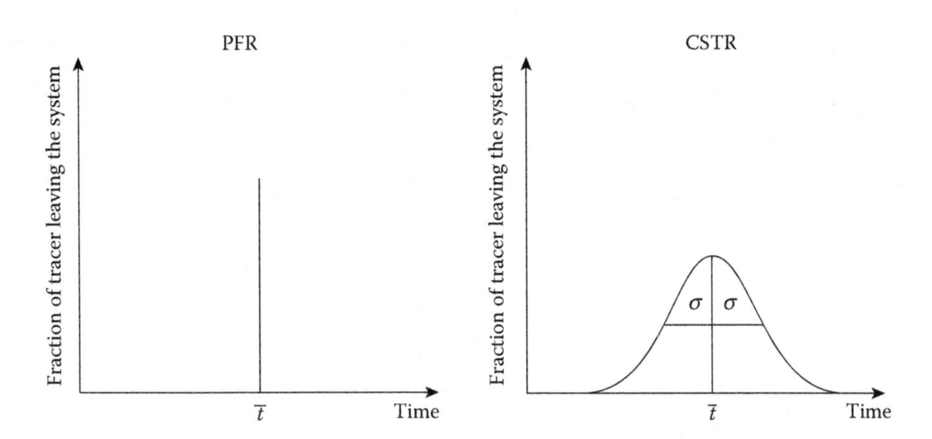

Figure 3.6 Normalized distribution functions for the two ideal reactors. The spread of the RTD (denoted by σ) is zero for the PFR.

spent by each element can vary from zero to infinity. This equation is displayed in Figure 3.6 again for the PFR and CSTR ideal reactor cases.

The two important characteristics of any distribution are the spread, which is characterized by its mean ($\bar{t}$ in our case), and the shape, which is characterized by its standard deviation σ. The mathematical tool most commonly used to determine these parameters is the analysis of moments, which is fully described in several books, for example, Nauman and Buffham (1983), Nauman (1987), and Levenspiel (1972, 1993). The expression for $\bar{t}$ is

$$\bar{t} = \int_0^\infty t f(t) dt \tag{3.24}$$

Mean residence time

and the standard deviation σ, or the spread of the distribution of t, is given by

$$\sigma^2 = \int_0^\infty [t - \bar{t}]^2 f(t) dt \tag{3.25}$$

Spread of the distribution

Types of distribution

The plug-flow limit is represented by the Dirac delta function

$$f(t) = \delta(\bar{t}) \tag{3.26}$$

which shows that $f(t) = 0$ at all times except at $t = \bar{t}$. The fully mixed limit is given by the exponential distribution

$$f(t) = \frac{1}{\bar{t}} \exp\left(-\frac{t}{\bar{t}}\right) \tag{3.27}$$

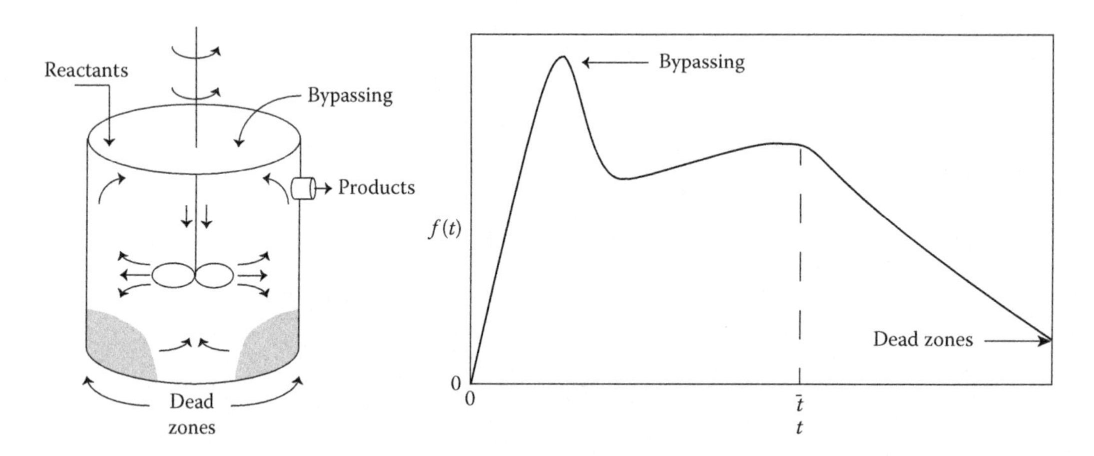

Figure 3.7 Dead zones and bypasses in a CSTR and the corresponding RTD plot. (From Fox, R.O. *Computational Models for Turbulent Reacting Flows*, Cambridge, UK, 2003.)

The standard deviation for the delta function is seen to be zero, a consequence of the fact that the pulse is perfectly sharp. On the other hand, it is equal to the residence time for exponential distribution, which denotes that this is the broadest distribution possible. Real (nonideal) RTDs lie between these extremes.

Clearly, then, the effect of RTD should be included in any reactor design, particularly when there is a strong reason to believe that the operation would be nonideal (i.e., neither PFR nor CSTR can be assumed). But before this can be done, the nature of the mixing problem must be identified (Figure 3.7).

Example 3.1: Limitations of the RTD model

The limitations of the RTD model

Here, we present a classic example from Danckwerts (1958) and Zwietering (1959): We can arrange a PFR and a CSTR in series in two different ways: either as the first reactor followed by the other, as shown in Figure 3.8.

As shown in Figure 3.8, the RTDs of the reactors are identical. On the one hand, the conversions at the exit of these reactors will be substantially different for reaction orders different from 1 especially when a complex reaction scheme is involved. On the other hand, RTD theory predicts that the reactor behavior should be identical whether the PFR or the CSTR is the first reactor in the combination. Therefore, with this example, the need for a more sophisticated theory than RTD is clearly demonstrated. In the next section, we will review some of the existing theories of mixing.

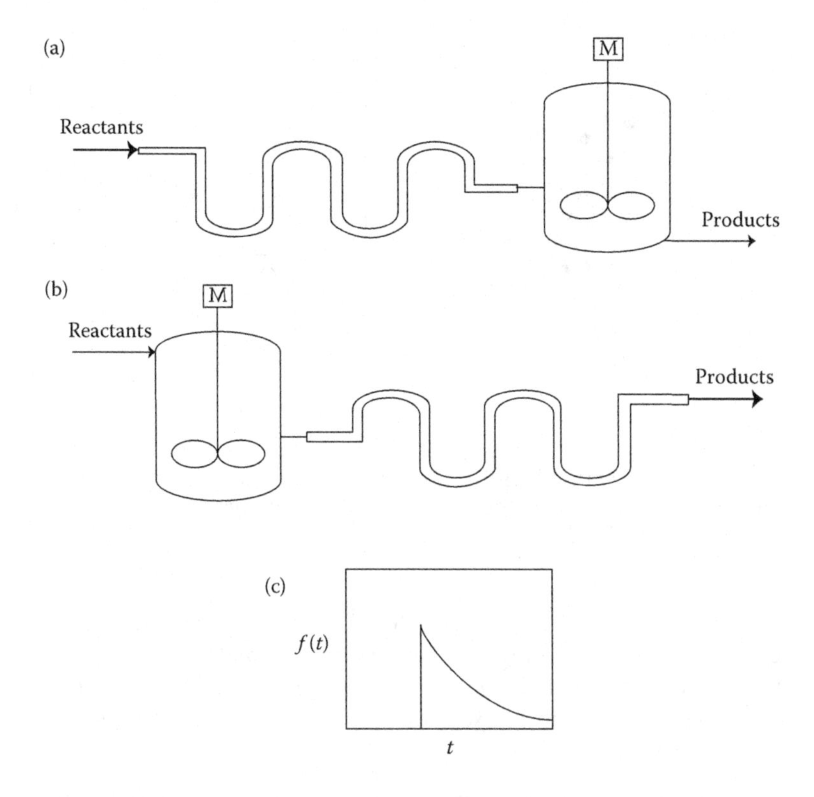

Figure 3.8 Schematics of the PFR–CSTR reactor arrangements. Both these arrangements, shown in parts (a) and (b), give rise to the same RTD, as shown in part (c).

Concept of mixing

Regions of mixing

The limits of mixing as we have understood them so far are plug flow (no mixing) and mixed flow (full mixing). However, there is a region in the vicinity of the fully mixed boundary where different "degrees of full mixing" can exist. In one limit of this region, clumps or aggregates of molecules enter the reactor and move through it without interacting with each other. Within each clump, however, there is complete mixing of the molecules at the molecular level. The residence time of each molecule within the clump is the same as that of the clump itself. Since the clumps are fully separated from one another, this kind of flow is known as *segregated flow*. On the other hand, in the other extreme, there are no clumps and mixing occurs at the molecular level. The kind of mixing considered in the earlier chapters refers to this perfectly mixed condition. It will be noted that we have now introduced a second "fully mixed" condition, segregated flow. Clearly, there can be degrees of mixing between molecules of the clumps, leading to various degrees of

Segregated flow

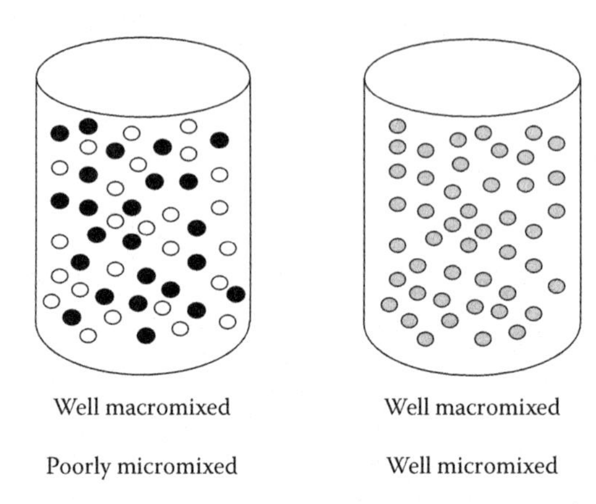

Figure 3.9 Poorly micromixed versus a well-micromixed system in an MFR. (From Fox, R.O. *Computational Models for Turbulent Reacting Flows*, Cambridge, UK, 2003.)

mixing or segregation at this level. Thus, one can discern two broad regions of mixing. The region between plug flow and fully segregated flow is referred to as the *macromixing* region and that between fully segregated and perfectly mixed flows is referred to as the region of *micromixing* (Figure 3.9).

Macromixing

Micromixing

One more level of mixing must be specified before we complete the bounds within which real reactors operate. This arises out of the fact that the only RTD that characterizes perfect mixing is exponential distribution. But each RTD (other than exponential) has its own limit of perfect mixing. This limit is referred to as *maximum mixedness* and represents the perfect mixing equivalent of *any* distribution other than exponential distribution.

RTD theory fails to describe reactor behavior for interacting fluid elements!

Briefly we treated the perfectly mixed reactor RTD in the mathematical analysis provided above. It is important to note from Example 3.1 that the RTD theory is not fully capable of explaining the behavior of the reactors, especially when the fluid elements are interacting. Thus, we give examples for a few other models here and refer the reader to an excellent text by Fox (2003) for a more in-depth analysis of these models in the turbulent flow regime. Four broad classes of micromixing models are sketched in Figure 3.10.

The schematic representation of maximum mixedness and segregated flow reactor conditions is shown in Figure 3.10. Whether fully segregated or perfectly mixed flow occurs depends on the nature of the fluid. The other representation of the segregated flow reactor is the MFR with ping-pong ball batch reactors filled with the reactive fluid, which will be analyzed below. The fluids that tend to largely macromix are referred to as *macrofluids* and those that tend to largely micromix are referred to as *microfluids*.

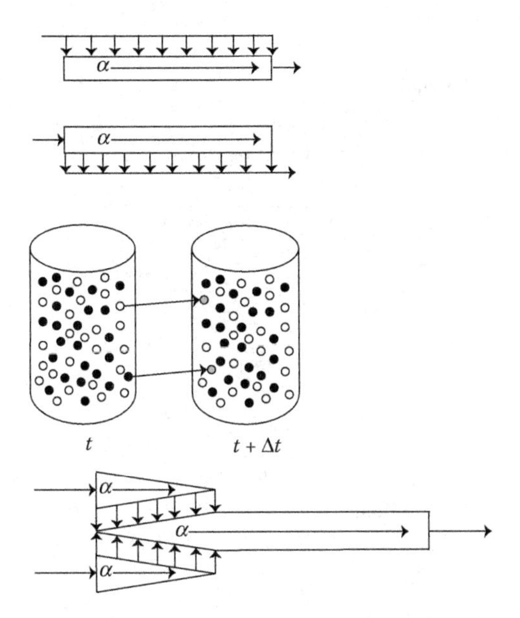

Figure 3.10 From top to bottom: maximum mixedness, segregated flow (minimum mixedness), coalescence–redispersion, and three-environment models. (From Fox, R.O. *Computational Models for Turbulent Reacting Flows*, Cambridge, UK, 2003.)

Fully segregated flow

In view of the various types of mixing explained above, the design equation for a reactor will depend on the region of mixing under consideration. Thus, we have: (1) a PFR corresponding to zero macromixing, (2) an MFR corresponding to zero micromixing (fully segregated flow), and (3) an MFR corresponding to perfect (molecular level) mixing or zero segregation. Cases (1) and (3) correspond, respectively, to the PFR and CSTR considered in Chapter 1. We develop the design equation for case (2), that is, fully segregated flow (Figure 3.11), and then consider situations where there can be partial macromixing or partial micromixing.

Mixing in relation to PFR and CSTR

We assume that each clump behaves as a batch reactor. The total reaction in the reactor is then given by the integral

$$\left(\frac{[\bar{A}]}{[A]_0}\right)_f = \int_0^\infty \left(\frac{[A]}{[A]_0}\right)_t f(t)dt \qquad (3.28)$$

where $([A]/[A]_0)_t$ represents the reaction in a batch of fluid of age t and $f(t)$ represents the time distribution of the little batches or clumps. The equation for $[A]/[A]_0$ in each clump depends on the order of the reaction and can readily be written from the batch reactor equations given in Chapter 1. Thus, for an nth-order reaction, we have

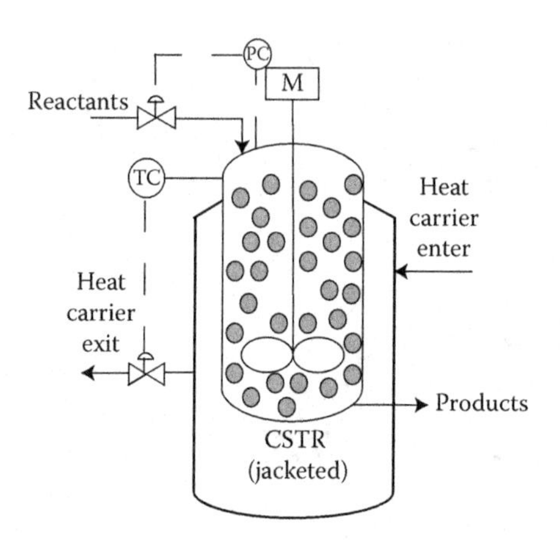

Figure 3.11 Fully segregated mixed-flow reactor. There is no exchange between the ping-pong balls.

$$\left(\frac{[A]}{[A]_0}\right)_i = \left(\frac{[A]}{[A]_0}\right)_{\text{batch}} = \{1-(1-n)[A]_0^{n-1}k\bar{t}\}^{1/(1-n)} \quad (3.29)$$

Substituting this equation into Equation 3.22 and assuming exponential distribution, we obtain

First-order reactions are not influenced by the degree of segregation

$$\left(\frac{[\bar{A}]}{[A]_0}\right)_f = \frac{1}{\bar{t}}\int_0^{\infty}\{1-(1-n)[A]_0^{n-1}k\bar{t}\}^{1/(1-n)}e^{-t/\bar{t}}\,dt \quad (3.30)$$

Note that when $n = 1$, micromixing has no effect.

Micromixing policy

Any micromixing policy to maximize conversion is irrelevant for a first-order reaction since it is unaffected by the degree of segregation. For nonfirst-order reactions, micromixing policies depend on two important considerations: whether the reaction order is greater or less than unity, and in the case of bimolecular reactions whether the feed is premixed or unpremixed.

The rate equations can be classified as concave up ($n > 1$), linear ($n = 1$), or concave down ($n < 1$). Since the second derivative of the rate equation is usually continuous, the following postulations can be made:

$n > 1$ segregation maximizes conversion

$n < 1$ maximum mixedness maximizes conversion

1. For $n > 1$, $d^2r/d[A]^2 > 0$, and segregation maximizes conversion, whereas maximum mixedness minimizes it.
2. For $n = 1$, $d^2r/d[A]^2 = 0$, and the extent of segregation has no effect on conversion.
3. For $n < 1$, $d^2r/d[A]^2 < 0$, and maximum mixedness maximizes conversion, whereas segregation minimizes it.

The above policy also holds for an irreversible bimolecular reaction provided the two species are premixed. For reversible reactions, again the same policy holds provided the rate constant of the forward reaction is higher than that of the reverse reaction. No such generalization seems possible for unpremixed feed.

Models for partial mixing

Clearly, the regions between the limits of fully segregated flow (minimum mixedness) and maximum mixedness are equally important, and procedures are necessary for designing reactors operating in these regions. From the discussion of macro- and micromixing presented above, it is important to note that two classes of partial mixing models are possible: for the macromixing region between plug flow and mixed flow, and for the micromixing region between full segregation and no segregation. Before we introduce new models, let us remember the axial dispersion model and tanks-in-series model presented for the TR and MFR configurations.

Axial dispersion model Here, let us work through the RTD for the axial dispersion model. The equation to solve is

Axial dispersion model

$$\frac{\partial[A]}{\partial t} + u\frac{\partial[A]}{\partial z} = D_{ez}\frac{\partial^2[A]}{\partial z^2} \tag{3.31}$$

For $Pe > 16$, an approximate solution for the washout function W is given by

$$W = \frac{[T]}{[T]_0} = 1 - \int_0^{\tau=t/\bar{t}} \frac{Pe}{4\pi t^3}\exp\left[\frac{-Pe(1-t)^2}{4t}\right]dt \tag{3.32}$$

The graphical representation of Equation 3.32 is given in Figure 3.12.

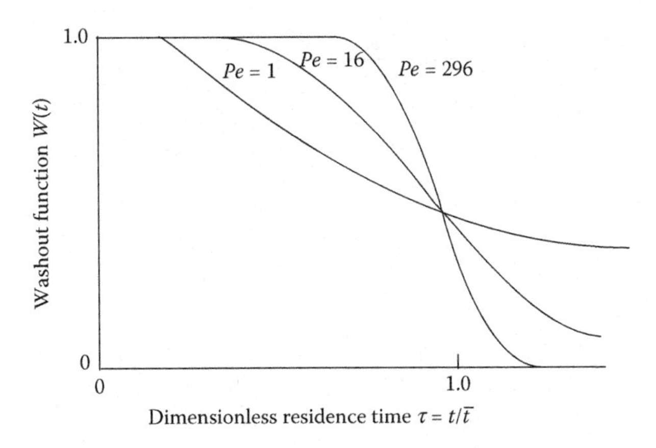

Figure 3.12 Effect of the Peclet number on the washout function. (From Nauman, E.B. *Chemical Reactor Design*, Wiley, NY, 1987.)

Tanks-in-series model The tanks-in-series model represents the mathematical situation where a sequence of CSTRs is used to simulate various degrees of partial mixing. The washout function and the density function for N tanks in series is given by

Tanks-in-series model

$$W(t) = e^{-Nt/\bar{t}} \sum_{i=0}^{N-1} \frac{N^i t^i}{i!(\bar{t})^i} \tag{3.33}$$

$$f(t) = \frac{N^N t^{N-1} e^{-Nt/\bar{t}}}{(\bar{t})^{N-1}(N-1)!} \tag{3.34}$$

The effect of the number of tanks on the washout function is shown in Figure 3.13. The striking similarity between this figure and Figure 3.12 is a clear evidence that both tanks-in-series and axial dispersion models can yield similar results. Both these models assume some sort of symmetry along the direction of flow and are hence unable to account for such common occurrences as short circuiting and channeling. A more comprehensive model that accounts for these features is the combined or compartment model in which plug flow, complete mixing, and short circuiting are treated as separate components that simultaneously contribute to the flow (Cholette and Cloutier, 1959).

Models for partial micromixing Since the introduction of the parameter J, defined by Equation 3.35, to describe the degree of segregation (Danckwerts, 1953; Zwietering, 1959), there has been an explosion of models to describe the degree of segregation (i.e., of partial micromixing)

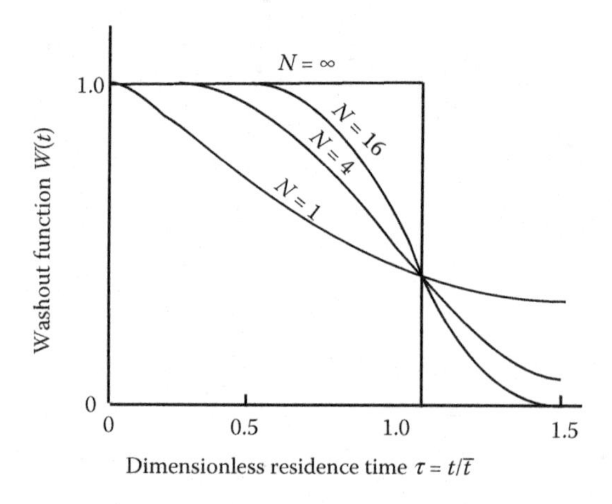

Figure 3.13 Effect of the number of tanks in series on the washout function. (From Nauman, E.B. *Chemical Reactor Design*, Wiley, NY, 1987.)

$$J = \frac{\overline{\left(\bar{t} - t\right)^2}}{\overline{\theta^2}} \tag{3.35}$$

Degree of segregation

where t is the age of the fluid element, $\bar{t}$ is the average of t, and

$$\overline{\theta^2} = \int_0^\infty \overline{\left(\bar{t} - t\right)^2} f(t)dt \tag{3.36}$$

with $J = 1$ for complete segregation and $J = 0$ for no segregation.

Degree of segregation defined by the age of the fluid at a point The basic assumption in all these models is that they give the same RTD and that differences in performance are attributable to different levels of micromixing as sought to be simulated by different physical postulations. The different situations are best analyzed using the general class of population balance models that treat the reaction fluid as a collection of discrete elements, each consisting of a number of molecules—that can be as low as one for the ideal mixer. These models can be divided into three main categories:

1. Two-environment models in which one environment is in a state of complete segregation and another is in a state of maximum mixedness.
2. Fluid element or particle models where the fluid is broken up into small elements, with mass transfer occurring by coalescence and redispersion or diffusion.
3. Fluid flow models, where a simple fluid mechanical model is constructed by dividing the reactor into different zones of macro- and micromixing; this is clearly an extension of the compartment models of macromixing but with zones of micromixedness added.

Turbulent mixing models

The basic assumption underlying the mixing models discussed so far is that the reaction is slow compared to mixing, so that the ultimate effect of mixing is manifested only through the prevailing state of mixing at the commencement of the reaction. Let us consider a fast bimolecular reaction such as precipitation, neutralization, azo coupling, some substitution, and many oxidation reactions, where reactant B is added to A present in the reactor. If the reaction is very fast, with a half-life of a few seconds, and the residence time is very high, conversion will be independent of residence time and RTD and the concentration of the limiting reactant would be close to zero. Mass transfer (i.e., mixing) and reaction no longer proceed consecutively but simultaneously, and the type of mixing involved is not macromixing characterized by RTD but turbulent-driven micromixing.

Characteristic timescales

It is important to have a sense of the timescales of mixing, reaction, and the rest of the processes. Some of the mixing models will depend strongly on these characteristic timescales. Therefore, in this section, we will provide a list of important timescales that we will also use when describing the multiphase reactions as well:

1. *Reaction timescale*, t_R: Since mixing does not influence the behavior of a first-order reaction, we will define the reaction timescale for a second-order reaction as the time required to decrease the reagent concentration to half its initial value. For a batch reactor, this is given by

Reaction timescale

$$t_R = \frac{1}{k[A]} \tag{3.37}$$

   where k is the second-order time constant.

2. *Time constant of micromixing by molecular diffusion*: Especially in two-phase reactions, when chemical species diffuse to the adjacent laminae as shown in Figure 3.14, the half-life of molecular diffusion with progressively thinning laminae is given by

Molecular diffusion time scale

$$t_{DS} = 2\left(\frac{\upsilon}{\epsilon}\right)^{1/2} \text{arcsin}\, h\left(\frac{0.05\upsilon}{D}\right) \tag{3.38}$$

   where D is the diffusivity, υ the kinematic viscosity, and ε the rate of energy dissipation per unit mass of solution in turbulent velocity fluctuations (Bourne, 2003).

3. *Time constant of micromixing by engulfment*: Transitory vortex tubes are characteristic of turbulent flow. Small energetic vortices acting near Kolmogorov scale* engulf the surround-

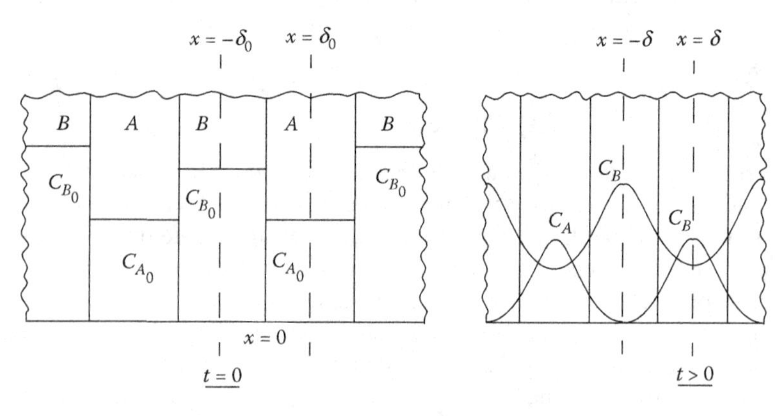

Figure 3.14 Concentration profiles due to molecular diffusion in thinning laminae. (From Bourne, J.R., *Org. Proc. Res. Dev.*, **7**, 471, 2003.)

* Kolmogorov scale is given by $\lambda_K = (v^3/\epsilon)^{1/4}$. The flow at sufficiently small scales are no longer turbulent.

ing fluid forming a short-lived laminated structure. Initially, when a small amount of B-rich solution engulfs A-rich surroundings (Figure 3.15), the rate of growth of the engulfed volume is given by

$$\frac{dV_E}{dt} = EV_E \tag{3.39}$$

Time constant for engulfment

where the engulfment-rate coefficient (E) and the time constant for engulfment, t_E, can be found from:

$$t_E = E^{-1} = 17\left(\frac{\nu}{\epsilon}\right)^{1/2} \tag{3.40}$$

4. *Reactor time constant*: Reactor time constant is the residence time defined by

$$\bar{t} = \frac{V}{Q} \tag{3.41}$$

Reactor time constant

5. *Turbulence integral timescale*: This is the ratio of the kinetic energy of the eddies divided by the rate of energy dissipation of the eddies given by

$$t_U = \frac{k_e}{\epsilon} \tag{3.42}$$

Turbulence integral timescale

For PFR, $t_U \propto d/u_z$, where d is the tube diameter and u_z is the linear velocity in the z-direction.

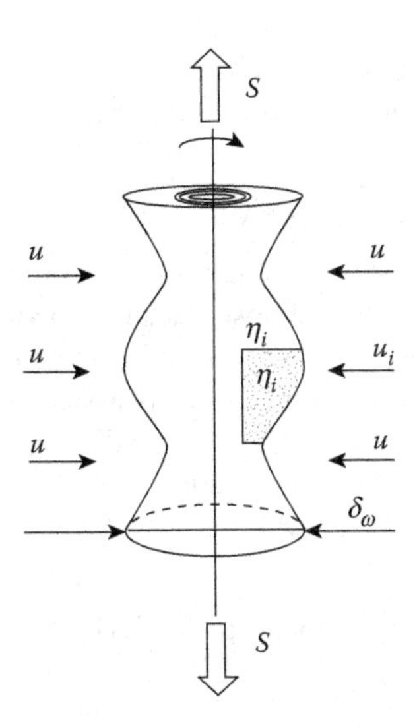

Figure 3.15 Vortex tube engulfing the surrounding fluid to form laminated structure. (From Bourne, J.R. *Org. Proc. Res. Dev.*, **7**, 471, 2003.)

Recirculation time

6. *Recirculation time*: Recirculation time can simply be taken as the time a fluid element spends away from the impeller. A ratio of the residence time to recirculation time of about 100 is an indication of well mixedness in MFRs.

Engulfment-deformation diffusion model

Here, we will briefly describe the salient characteristics of the engulfment-deformation diffusion (EDD) models of Bourne and his colleagues (see, e.g., Bourne et al., 1981; Bourne, 1984; Baldyga and Bourne, 1989). Basically the EDD models postulate the formation of deforming eddies (laminated structures) of one fluid in the other followed by reaction and diffusion in the eddies (Figure 3.15). The latest version of these models suggests that engulfment is the dominating process, so that they are now simply referred to as the E-model. When $t_E > t_{DS}$, deformation and diffusion rapidly homogenize the growing vortex at the molecular scale. A material balance on $[A]$ in the vortex is given by

Engulfment deformation diffusion model

$$\frac{d(V_E[A])}{dt} = EV_E\left(\langle[A]\rangle - [A]\right) + r_A V_E \tag{3.43}$$

when combined with

$$\frac{dV_E}{dt} = EV_E$$

simplifies to

$$\frac{d([A])}{dt} = E\left(\langle[A]\rangle - [A]\right) + r_A \tag{3.44}$$

These three equations and the engulfment time constant given in the previous section constitute the basis for the model.

Interaction by exchange with a mean

IEM model

This model was developed by Villermaux and his colleagues (see, e.g., Villermaux, 1985). The model postulates two environments with probabilities p_1 and $p_2 = 1 - p_1$, where p_1 is the volume fraction of stream 1 at the reactor inlet. In the IEM model, p_1 is assumed constant. If p_1 is far from 0.5, the IEM model yields poor predictions. For such situations, E-model that accounts for the evolution of p_1 should be employed. The concentration in environment n is

$$\frac{d[A]^{(n)}}{d\alpha} = \frac{1}{t_{\text{IEM}}}\left(\langle[A(\alpha)]\rangle - [A]^{(n)}\right) + (r_A V)^{(n)}$$

with

$$[A(0)]^{(n)} = [A]_{\text{in}}^{(n)} \tag{3.45}$$

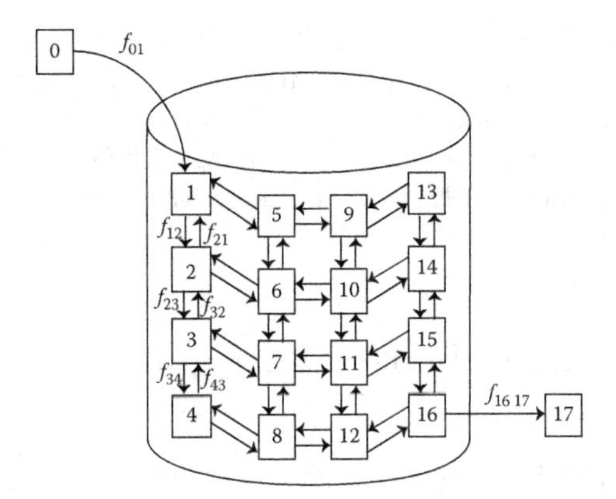

Figure 3.16 Sketch of a 16-zone model of a CSTR. (From Fox, R.O. *Computational Models for Turbulent Reacting Flows*, Cambridge, UK, 2003.)

where α represents the age of the fluid element. The average concentrations are obtained from

$$\langle [A(\alpha)] \rangle = \sum_{n=0}^{2} p_n \int_0^{\infty} [A(\beta)]^{(n)} f(\alpha,\beta) d\beta \qquad (3.46)$$

where

$$f(\alpha, \beta) = \delta(\beta - \alpha) \quad \text{for PFR and} \quad f(\alpha, \beta) = f_{\text{CSTR}}(\beta) \qquad (3.47)$$

Zone model Unlike RTD theory, zone models employ an Eulerian framework that ignores the age distribution of fluid elements inside each zone. This model ignores micromixing while providing a model for macromixing for large-scale inhomogeneity inside the reactor. The schematics of the reactor divided into zones are given in Figure 3.16. The transport between the zones is given by f_{ji}. Once the transport rates are identified, the zone model poses no particular mathematical difficulty; just the solution of coupled ordinary differential equations (ODEs) given in Equation 3.48.

A simple unsteady-state balance equation for the *i*th zone is given by

$$\frac{d[A]^{(i)}}{dt} = \sum_{j=0}^{N+1} (f_{ji}[A]^{(j)} - f_{ij}[A]^{(i)}) + (r_A V)^i \qquad (3.48)$$

Zone model

It is important to note that in the 16-zone model considered by Fox (2003) in Figure 3.16, there is no backflow in the 0th and 17th zones, that is

$$f_{10} = f_{17\,16} = 0$$

Joint PDF

Joint PDF can simply be explained in relation to the RTD. The RTD function is the PDF of the fluid-element ages as they leave the reactor. But unlike RTD, PDF also accounts for the spatial variations within the reactor itself. The power of the PDF over RTD lies in its capacity to account for the interaction between the fluid elements. Thus, for complex reactions in nonideal reactors, PDF is a powerful tool. We refer the reader to more advanced texts on mixing such as by Fox (2003), or Baldyga and Bourne (1999) in computational fluid dynamics (CFD) that provides the details of the theory.

Practical implications of mixing in chemical synthesis

General considerations

For a given reaction, one can choose the most appropriate type of reactor, such as plug flow, fully mixed, recycle, and so on, to maximize the conversion or yield, depending on whether the reaction is simple or complex. The main features of these reactors were considered earlier in Chapters 1 and 2. In general, where macromixing is the chief mixing phenomenon (as in the above cases), PFR is the reactor of choice for maximizing the yield of an intermediate in a complex reaction. However, for a reaction of the type

$$A \rightarrow R \rightarrow S \tag{R1.1}$$

$$A + A \rightarrow T \tag{R1.2}$$

the recycle reactor can be the preferred choice.

Consider another reaction system

$$A + B \rightarrow R \tag{R2.1}$$

$$R + B \rightarrow S \tag{R2.2}$$

The factors that control the chemical kinetics of a reaction are concentration and stoichiometric ratio. Thus, increasing the dilution often enhances the yield, and increasing the ratio $[A]/[B]$ in reaction R2 raises the yield of R. The factors that affect mixing are the type of mixer and stirrer speed. These two categories of factors are quantified in terms of t_R and t_D defined by Equations 3.37 and 3.38, respectively. Example 3.2, based on the results of Bourne et al. (1988), clearly brings out the importance of these equations in analyzing the role of micromixing. They also demonstrate, in a general way, the importance of an often-ignored fact: the role of *addition sequence* in reactions involving more than two reagents.

Example 3.2: Experiments to illustrate the role of micromixing in determining the yield of a fast reaction and the importance of addition sequence

A good example of reaction R2 is the coupling of 1-naphthol with diazotized sulfanilic acid producing two dyestuffs whose concentrations can be readily measured spectroscopically.* The reactions may be represented as

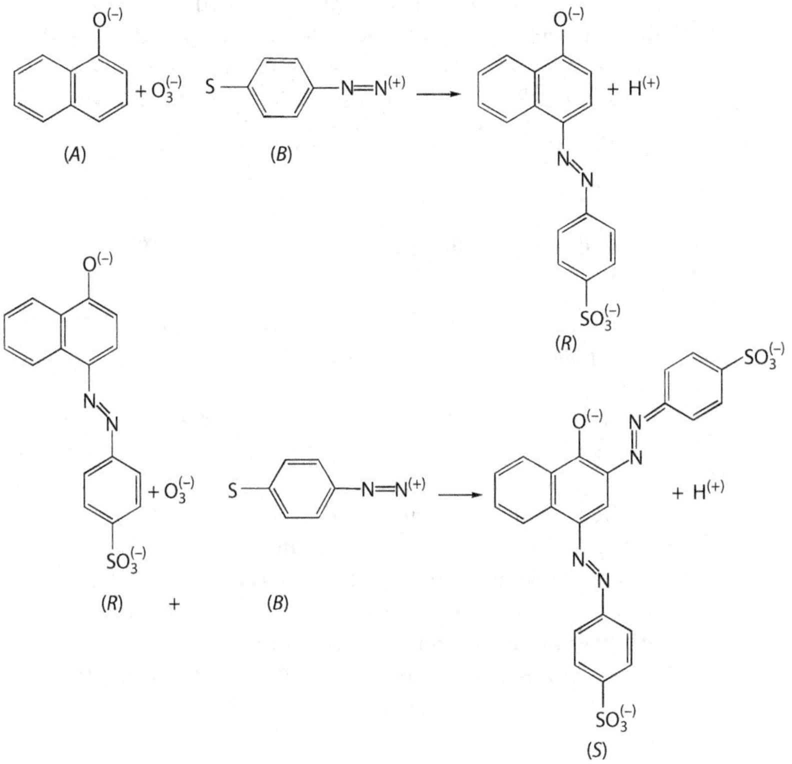

(R3)

In experiments carried out in a 1-L beaker stirred by a 5-cm diameter turbine at a speed of 300 rpm, Bourne et al. (1988) report the following details: $[A] = 1.1071$ mol/m^3, $\nu = 10^{-6}$ m^2/s, $D = 8.5 \times 10^{-10}$ m^2/s, $[A]/[B] = 1.05$, volumetric ratio of the two solutions = B added/A in beaker = 1/25, diazonium ion concentration = 2.754 mol/m^3 before mixing, buffers used to control pH: P1 = (Na_2CO_3 + $NaHCO_3$) for pH 10, P2 = (KH_2PO_4 + Na_2HPO_4) for pH 7. Three sets of conditions were studied, in which the rates were calculated from the ionic preequilibria and pK values of the reagents (Bourne et al., 1981).

1. *pH 10*: A was first buffered with P1 to pH 10, and 20 mL of B (pH ≅ 2) slowly added with stirring. The calculations gave $t_R \cong 8 \times 10^{-4}$ s, $t_D = 0.035$ s, and $e = 7.2 \times 10^{-2}$ W/kg, yield of R = 98.1%.

* This series–parallel reaction has been extensively used as a model reaction for studying the role of micromixing in complex reactions by Bourne et al. (1981).

2. *pH 7*: Same as above (with the same addition sequence) but with buffer P2. The calculations gave $t_R = 0.14$ s, $t_D = 0.035$ s (unchanged), and yield of $R \cong 99.9\%$.
3. *pH 2–10*: B (20 mL, pH 2) rapidly added to 500 mL of unbuffered A in the vessel, then 20 mL of buffer P1 added over 4 min so that coupling could proceed, and yield of $R \cong 99.9\%$. Note that the addition sequence was changed in this experiment.

CONCLUSIONS

The following conclusions can be drawn from the experiments reported above:

1. The first experiment, with the addition sequence A–P–B, corresponds to $t_D \gg t_R$. Therefore, it was controlled by mixing and was independent of kinetics.
2. By decreasing the pH to 7 in the second experiment (but with the same sequence A–P–B), the preequilibrium concentrations of the reactive species were changed. This resulted in a drastic reduction of the reaction rate, giving $t_D < t_R$, that is, the reaction was chemically controlled.
3. When the addition sequence was changed (A–P–B to A–B–P), it was found that $t_D = 0$, that is, $t_D < t_R$ and the reaction was again controlled by kinetics.
4. The controlling mechanism (and frequently the yield) can be affected by the addition sequence of the reagents.

Dramatic illustration of the role of addition sequence of reagents

The effect of addition sequence was only mildly apparent in the example given above. It has been more dramatically illustrated in the esterification reaction between maleianic acid (1) and thionyl chloride (2) (Kumar and Verma, 1984).

(R4)

When reactants (1) and (2) are mixed and the mixture is poured in any absolute alcohol (3), the same compound (4) results. On the other hand, when (1) is dissolved in absolute methanol or ethanol (3), and (2) is then added dropwise while shaking, compound (5) is formed. Thus, the change of sequence from 1–2–3 to 1–3–2 leads to a completely different product. This is ascribed to the mixing effect, but a firmer confirmation of this is needed.

Explore yourself

1. See if you can quickly generate answers to the following questions:
   a. What is backmixing?
   b. Rank the ideal reactors with respect to the degree of increasing backmixing.
   c. Can you use the axial dispersion model to measure the degree of backmixing?
   d. What are the limitations of the RTD theory?
   e. Describe the coalescence/redispersion model in less than three sentences.
   f. What is the relation between mixing and selectivity?
2. Brainstorm for the methods of ensuring well-mixed conditions without using an impeller for the following situations:
   a. Two gases
   b. Two miscible liquids
   c. Two immiscible liquids
   d. A gas stream and a liquid stream, gas is soluble in the liquid
   e. A gas stream and a liquid stream, gas is sparingly soluble in the liquid
   f. A liquid stream has to be contacted with a solid catalyst, you are free to choose the solid shape and size.
   g. A liquid and a gas reactant has to contact with a solid catalyst, you are free to choose the solid shape and size.
3. Do a literature and/or patent search for the existing solutions for the situations described in question 2. How well did you do during the brainstorming?
4. List as many strategies as possible to ensure a well-mixed condition in a liquid phase MFR if your reactor was 3 m in diameter and 8 m tall. Test your strategies against industrial solutions.
5. How would you induce backmixing through diffusion? List as many strategies as possible.
6. Describe the strategies to induce mixing in microfluidic reactors. Check yourself against the literature. How well did you do?

References

Baldyga, J. and Bourne J.R., *Turbulent Mixing and Chemical Reactions*, Wiley, NY, 1999.

Bourne, J.R., *Org. Proc. Res. Dev.*, **7**, 471, 2003.

Bourne, J.R., Ravindranath, K., and Thoma, S., *J. Org. Chem*, **53**, 5166, 1988.

Danckwerts, P.V., *Chem. Eng. Sci.*, **2**, 1, 1953.

Danckwerts, P.V., *Chem. Eng. Sci.*, **8**, 93, 1958.

Eldridge, J.W. and Piret, E.L., *Chem. Eng. Prog.*, **46**, 290, 1950.

Fox, R.O., *Computational Models for Turbulent Reacting Flows*, Cambridge, UK, 2003.

Kumar, B. and Verma, R.K., *Synth. Commun.*, **14**, 1359, 1984.

Nauman, E.B., *Chemical Reactor Design*, Wiley, NY, 1987.

Villermaux, J., in *Chemical Reactor Design and Technology, Proceedings of the NATO: Advanced Study Institute on Chemical Reactor Design and Technology*, London, Ontario, Canada, 1985.

Westerterp, K.R., Van Swaaij, W.P.M., and Beenackers, A.A.C.M., *Chemical Reactor Design and Operation*, 2nd ed., John Wiley and Sons, NY, 1983.

Zwietering, T.N., *Chem. Eng. Sci.*, **11**, 1, 1959.

Bibliography

On analysis of moments:

Levenspiel, O., *Chemical Reaction Engineering*, 2nd ed. Wiley, NY, 1972.

Levenspiel, O., *Chemical Reactor Omnibook*, Corvallis, OR, 1993.

Nauman, E.B. and Buffham, B.A., *Mixing in Continuous Flow Systems*, Wiley, NY, 1983.

On compartment model:

Cholette, A. and Cloutier, L., *Can. J. Chem. Eng.*, **37**, 105, 1959.

On EDD models:

Baldyga, J. and Bourne, J.R., *Chem. Eng. J. Biochem. Eng.*, **42**, 83, 1989.

Bourne, J.R., in *International Symposium on Chemical Reaction Engineering*, (ISCRE) 8, I. *Chem. Symp. Ser.*, 87, Edinburgh, 1984.

Bourne, J.R, Kozicki, F., and Rys, P., *Chem. Eng. Sci.*, **36**, 1643, 1981.

On predicting Peclet numbers for various reactors:

Wen, C.Y. and Fan, L.T., *Models for Flow Systems and Chemical Reactors*, Dekker, NY, 1975.

Interlude II

In the previous chapter, we focused our attention on how mixing can be accounted for in reactor analysis. In this part, we will give examples describing how mixing (or lack of it) can be used to our advantage, for improved conversions, yields and selectivity.

Limits of mean field theory

So far in this book, we assumed that the mean field theory of chemical kinetics hold true. In other words, the rate of the reactions is proportional to the average concentrations of the species in the reactor volume. This assumption can be valid for homogeneous systems, where the reaction rate is comparable to the rate of mixing provided by molecular collisions or through mechanical stirring. However, when dealing with catalytic reactions, we can no longer use the mean-field approximations for the reaction rates. The interactions between the reactants and the catalyst surface can induce aggregation of the adsorbed molecules on the surface. One very beautiful example for such systems is the CO oxidation reaction taking place on Pt surfaces. Before we can discuss these, we should briefly tackle the predator–prey problem.

The predator–prey problem or surface mixing

In a forest, the rabbit and wolf populations balance each other, including many other components of the overall ecosystem. For the sake of simplicity, we will focus only on the rabbits and foxes. The growth rate of the rabbit population is a function of two parameters: the birth rate of the rabbits and the rate that rabbits expire. The latter depends strongly on the wolf population, if the wolves are the only predators of the rabbits in this virtual environment. If the wolf population is large, they consume a large amount of rabbits, which brings the rabbit population to a level of extinction. Eventually, the wolf population declines due to the scarcity of the rabbits. Once the wolf population declines, without the predators, the rabbit population increases, which then improves the wolf population. In this particular example, the prey has a constant supply of food (open system); otherwise, the system rapidly reaches equilibrium.

In kinetic terms, this can be written as follows:

$$\frac{dR}{dt} = aR - bRW \tag{II.1}$$

$$\frac{dW}{dt} = -cW + dRW \tag{II.2}$$

where R and W represent the rabbit and wolf populations, respectively. Many textbooks addressed these problems. We will refer the curious reader to the recent book by Holmes (2009). The solution of these coupled differential equations is not in the scope of this book. It must be obvious that these pairs of equations can describe the oscillatory behavior of the predator and prey populations.

After studying Chapter 5, it will be more obvious that in order to estimate the reaction rate between CO and O_2, one needed to know the fraction of the catalyst surface populated by each of these reactants. Once a catalyst surface with its own chemistry enters the picture, some unexpected phenomena may take place. For example, attractive forces (such as van der Waals) necessary to initiate vapor–liquid transitions may enter the picture when the adsorbed molecules are in close proximity of one another. As such, a surface condensation may take place at temperatures much higher than the saturation temperatures at the given pressure. When such surface condensations occur, the surface of the catalyst is no longer a randomly populated matrix of a substrate with the adsorbates. On the contrary, the surface looks like a very ordered system (Figure II.1).

Gerhard Ertl, Nobel Prize in chemistry, 2007

In an earlier publication, Ertl and his coworkers have demonstrated the periodic fluctuations in CO oxidation reaction shown in Figure II.2. This and similar contributions have led to Ertl receiving the Nobel Prize in chemistry in 2007.

Figure II.1 The 2D patterns formed by adsorbed CO and O on Pt surfaces. (From Ertl, G., *Science*, **254**, 1750, 1986.)

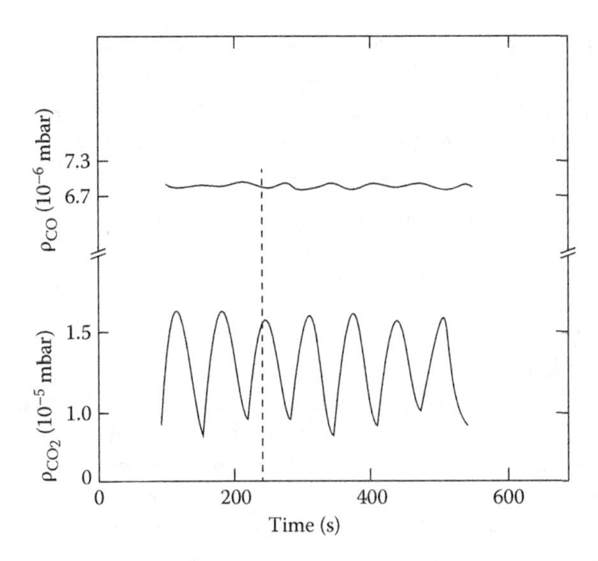

Figure II.2 The fluctuations in CO oxidation reaction over Pt(110) surfaces. (From Ertl, G., *Science*, **254**, 1750, 1986.)

The mathematical problem we encounter here is very similar to the temperature runaway problem we have seen when solving the energy balance across a CSTR in Chapter 1.

Mixing problem addressed

Short contact time reactors

Short contact time reactors become immediately popular and were adopted in the chemical industry eventually after the initial publication of Hickman and Schmidt (1993). The concept was very simple: If you have a series reaction $A \rightarrow B \rightarrow C$, you should keep the contact time of the reactor low in order to increase the yield of the intermediate compound B. The concept has replaced the endothermic steam-reforming reaction for syngas production, or the complicated autothermal reforming, where in part of the reactor you would burn part of your reactant and in the rest of the reactor, you would use the thermal energy released along with combustion products CO_2 and H_2O to produce syngas. The biggest virtue of the short contact time reactors is that you would use a very small reactor to ensure the microsecond timeframe of the contact time between the reactive gases and catalysts.

Questions to ponder

What is the major driving force behind producing syngas economically?

What are the downstream processes of syngas manufacture?

Can syngas be used as fuel in turbines?

What is IGCC?

$$CH_4 + \frac{1}{2}O_2 \rightarrow CO + H_2 \qquad \text{(II.1)}$$

Short contact times are needed to limit the mixing between the reactants and the surface intermediates in such a way that more favorable H_2

Figure II.3 Monoliths with different shapes and cell densities for different applications. (Courtesy of Kaleporselen.)

oxidation reaction would not take place, while CO would leave the surface partially oxidized. The detailed surface chemistry needed to fully understand this reaction system can be found in Hickmann and Schmidt (1993). The backmixing that can result from the pore diffusion is eliminated by using structured reactors.

Structured reactors used in this system are monoliths as shown in Figure II.3.

Microfluidic reactors

The microfluidic reactor technologies emerged recently as a result of collision of many needs and availability of many technologies. Volume optimization has always been the primary concern of the chemical reaction engineer. To carry out the same reaction in a smaller volume, the rates should be faster. In the upcoming chapters we will establish that the transport disguises, such as heat and mass transfer, inhibit reactions. We will learn how to assess the level of inhibition experimentally (Chapter 7) and account for these disguises through mathematical modeling (Chapter 6).

Silicon manufacturing technologies have enabled mass production of microscopic structures. Ironically, the mass production of these structures requires a substantial background in chemical engineering. This marriage of chemical engineering and the microelectronics manufacturing has lead to the birth of the new interdisciplinary field of the microfluidics. The involvement of chemical engineers in the microfabrication processes enabled fast diffusion of the technology back to the home field.

As seen in Figure II.4, it is possible to introduce well-ordered flow patterns in the microchannels in contrast to the large-scale units. Coalescence does not emerge as a problem as well. On the other hand, we will run into two major problems. First of all, due to the smaller sizes

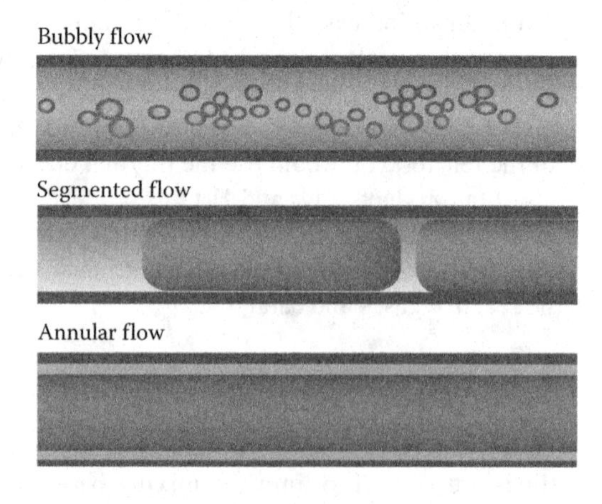

Figure II.4 The gas–liquid flow characteristics in a microchannel.

of the capillaries, flow fields will be dominated by the surface forces. Second, mixing will be the most important upstream issue.

Passive devices for mixing and pumping

At the scale where molecular forces start to dominate, using active devices for pumping or mixing fluids become impractical. In such a case, the fundamental understanding of the fluid flow is used to design systems.

Knudsen pump The thermal gradients in microfluidic devices are used to pump fluid in a certain direction. A gas at low pressures is transferred from a cold chamber to a hot one due to the thermal gradients which generates slip velocity in the flow opposite to the direction of the tangential heat flux (Gad-el-Hakk, 1999). Knudsen pumps are useful at especially high Knudsen numbers defined in the sidebar.

$$Kn = \frac{\text{mean free path}}{\text{dimensions of the conduit}}$$

Mixing

Slug flow as a mixer As we have seen in Chapter 3 in detail, mixing is a very important parameter in determining the selectivity of a reaction. In microfluidic systems, mixing becomes one of the most important components of the device manufacturing, since special caution must be taken for proper mixing of the fluids. The flow characteristics of a microfluidic reactor can be used to one's advantage as was presented by Gunther et al. (2005) that creating a slug flow system with alternating gas streams between liquid droplets can create enough drag such that two miscible fluids fed to the liquid droplets can perfectly mix under the conditions (Figure II.5).

Dean flow as a static mixer The use of the fluid inertia across a curved channel can create turbulence. The flow of fluids in the meandering channels undergoes mixing due to the inertia of the fluid in the flow direction. An example is shown in Figure II.6, while the Dean number is defined in the side bar.

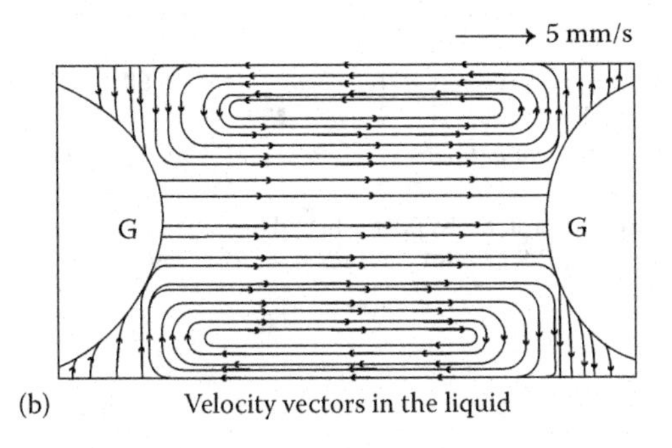

Figure II.5 The velocity vectors in a G–L slug flow reactor. (From Gunther, A. et al., *Langmuir*, **21**(4), 1547, 2005.)

$$\text{Dean number, } K = \mathrm{Re}\left(\frac{D}{2R}\right)^{1/2}$$

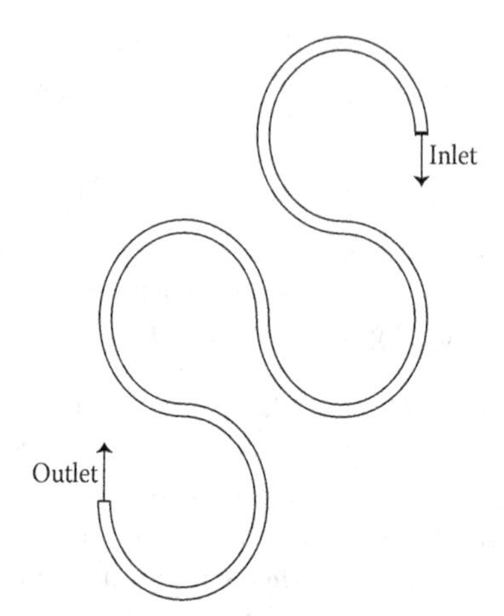

Figure II.6 A 4-element meandering mixer and its dimensions. (From Jiang, F. et al., *AIChE*, **50**, 2297, 2004.)

Among many published review articles and textbooks on static mixers, we particularly found Hessel et al. (2005) and de Mello (2006) easy to follow for the beginner.

Elastic turbulence One final topic we will mention here before closing the chapter is the elastic turbulence. The term describes the turbulence caused by the elastic forces arising in non-Newtonian fluids. In the situations where mixing is needed, the elastic forces in a non-Newtonian fluid can be used to create turbulence. The analysis by Joo and Shaqfeh (1992) is particularly useful for the flow of non-Newtonian fluids in microchannels but is beyond the scope of the text here.

References

de Mello, A.J., *Nature*, **442**, 394, 2006.

Ertl, G., *Science*, **254**, 1750, 1986.

Gad-el-Hakk, M., *J. Fluids Eng.*, **121**, 5, 1999.

Gunther, A., Jhunjhunwala, M., Thalmann, M., Schmidt, M.A., and Jensen K.A., *Langmuir*, **21**(4), 1547, 2005.

Hessel, V., Löwe, H., and Schönfeld, F., *Chem. Eng. Sci.*, **60**, 2479, 2005.

Hickmann, D.A. and Schmidt, L.D., *Science*, **259**, 343, 1993.

Holmes, M.H., *Introduction to the Foundation of Applied Mathematics*, Springer-Verlag, New York, 2009.

Jiang, F., Drese, K.S., Hardt, S., Küpper, M., and Schonfeld, F., *AIChE*, **50**, 2297, 2004.

Joo, Y.L. and Shaqfeh, E.S.G., *Phys. Fluids A*, **4**, 524, 1992.

Part II
Building on fundamentals

Originality does not consist in saying what no one else has said before, but in saying exactly what you think yourself.

James Stephens

Introduction

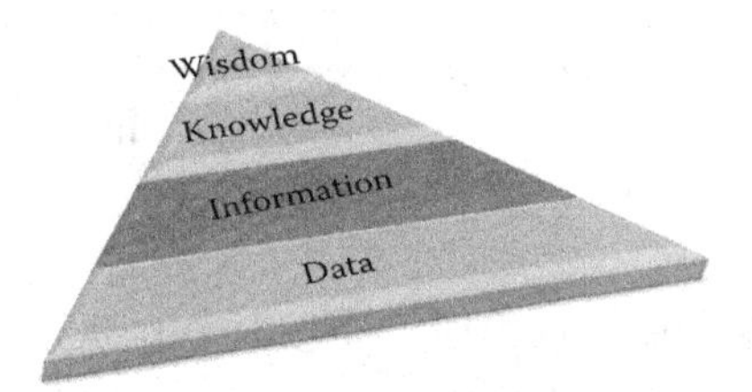

This part will serve at the information level of the pyramid of wisdom. In other words, we will collate information from different aspects of chemical engineering science and technology that will help a chemical engineer understand the operations of reactors or design better reactors. The basics given in Part I—Fundamentals Revisited was meant to connect with different aspects of the field assisting student's emergence as a full-blown chemical reaction engineer. Now, thermodynamics and transport phenomena will enter into the picture.

The different tools of the trade

The fundamental tools of trade of the profession of chemical engineering are (in addition to a solid basis in science) kinetics, thermodynamics, and transport phenomena. This part of the book will be concerned with these in the first three chapters. The final chapter will be the capstone one on experimental tools for obtaining good-quality data.

Relationship between thermodynamics and chemical reaction engineering

The connection between thermodynamics and chemical reaction engineering is very strong. First of all, we need to establish whether the conversions we desire at the temperatures and pressures involved are achievable—whether we reached the limits of thermodynamic equilibrium or are still left with room to maneuver. The second connection is the relation between the chemical and phase equilibria. If we are to design a reactor for a multiphase reaction, the phase equilibria become an immediate problem that we have to solve. The more novel connection comes later, when we intend to combine reaction with separation. A good *a priori* estimation in designing systems with a multitude of functionalities, such as a distillation column reactor or a membrane reactor, requires the solution of the chemical reaction problems along with the phase equilibria and other defining constraints that come into play.

Relationship between transport phenomena and chemical reaction engineering

The connection between chemical reaction engineering and transport phenomena also stems from multiphase reactions. For solid catalyzed gas or liquid reactions, mass transfer in the bulk or on the surface may become a problem. For gas–liquid reactions, the transport of the species to the reaction zone has to be considered. Similar problems arise for liquid–liquid reactions. Thus, we intend to give a brief introduction to these problems and, in the process, introduce dimensionless quantities such as the Thiele modulus, Damköhler number, Hatta modulus, effectiveness factor, and enhancement factor, and use them in designing reactors.

Relationship between chemical reaction engineering and kinetics

The accurate design of chemical reactors depends on an accurate perception of the chemical phenomena. In real-life situations, the heterogeneity of the chemical reactions and broad use of heterogeneous catalysts are common; therefore, we dedicated a full chapter to give a theory of chemical kinetics in the bulk (for homogeneous reactions) and on the surface (for heterogeneous reactions).

Chemical reaction engineering as an experimental and theoretical science

Chemical reaction engineering started out as an experimental science, but has increasingly become a combination of both experiment and mathematical theory. In this book, we will limit ourselves to describing good skills for collecting kinetic data in the so-called gradientless reactors (i.e., those in which gradients of mass and temperature are absent), after emphasizing the sources of gradients in the preceding chapters. Good-quality data free from transport disguises is, in itself, a very good catalyst characterization tool, and the best way of deriving a good reaction mechanism for further improvement of the catalyst or for designing a reactor. We believe that the students would have been exposed to experiment design and data analysis in the undergraduate courses in statistics as part of any undergraduate chemical engineering curriculum and hence will not cover this in this book. However, the curious student may like to consult the literature cited under bibliography for further depth.

Chapter 4 Rates and equilibria

The thermodynamic and extrathermodynamic approaches

Chapter objectives

Upon completion of this chapter, the successful student must be able to

- Perform chemical equilibria calculations for ideal and nonideal fluids in gas phase.
- Perform chemical equilibria calculations for ideal and nonideal fluids in liquid phase.
- Perform equilibrium calculations for systems involving adsorption–desorption equilibria.
- Apply extrathermodynamic approach to determining the equilibrium properties.

Introduction

In this chapter, we summarize the fundamental thermodynamic relationships relevant to chemical equilibria. Particular attention will be given to the thermodynamics of adsorption and the derivation of adsorption isotherms. Equations relating the effect of temperature on chemical equilibria are derived. The solution methods of chemical equilibria for complex reaction systems are presented.

In any reversible reaction such as

$$\upsilon_A A + \upsilon_B B \leftrightarrow \upsilon_R R + \upsilon_S S \tag{R1}$$

the system inevitably moves toward a state of equilibrium, that is, maximum probability. This equilibrium state is very important in analyzing chemical reactions because it defines the *limit to which any reaction can proceed.*

For molecules reacting in the liquid phase, the effects of reactant structure and of the solvent (medium) in which the reaction occurs (the so-called *solvation effects*) are not included in the conventional macroscopic approach to thermodynamics. The treatment of liquid phase reactions, therefore, tends to be less exact than of gas phase reactions involving simpler molecules without these influences.

A convenient way of approaching this problem is to start with the conventional macroscopic or thermodynamic approach and add enough microscopic detail to allow for the effects of solute (reactant) structure and the medium. This approach is called the *extrathermodynamic* approach and may be regarded as bridging the gap between the two rather disparate fields of rates and equilibria represented, respectively, by kinetics and thermodynamics.

Basic thermodynamic relationships and properties

Basic relationships

An important consideration in process calculations is the change that results in the basic thermodynamic properties, internal energy (U), enthalpy (H), Helmholtz free energy (A), and Gibbs free energy (G) when a closed system of constant mass moves from one macroscopic state to another. In the case of a homogeneous fluid, these change equations can be expressed in terms of four exact differential equations that can then be written in difference form by employing the operator Δ to represent the change from state 1 to state 2:

$$dU = T\,dS - P\,dV \text{ or in integral form } \Delta U = T\Delta S - P\Delta V \quad (4.1)$$

$$dH = dU + P\,dV \text{ or in integral form } \Delta H = \Delta U + P\Delta V \quad (4.2)$$

$$dA = dU - T\,dS \text{ or in integral form } \Delta A = \Delta U - T\Delta S \quad (4.3)$$

$$dG = dH - T\,dS \text{ or in integral form } \Delta G = \Delta H - T\Delta S \quad (4.4)$$

Of these, the enthalpy and free energy change equations are the most frequently used in the analysis of reactions.

Heats of reaction, formation, and combustion

Consider the reaction

$$A + B \rightarrow R + S \quad (R2)$$

The heat of reaction is given by

$$\Delta H_r = \Delta H_{f_R} + \Delta H_{f_S} - \Delta H_{f_A} - \Delta H_{f_B} \quad (4.5)$$

where the terms on the right refer to the enthalpies of formation of R, S, A, and B, respectively. The standard heat of reaction, denoted by ΔH_r^0, is defined as the difference between the enthalpies of the products in their standard states and of the reactants in their standard states, all at the same temperature.

The enthalpy of formation, usually known as the *heat of formation*, of a compound is the heat evolved during its formation from its constituent elements. The enthalpies of formation of elements are assumed to be zero. Thus, for the special case of heat of formation, each of the terms on the right in Equation 4.5 is zero, and the heat of reaction becomes equal to the heat of formation.

Usually, all the enthalpies in reaction R2 refer to the reactants and products in the ideal gaseous state, and appropriate corrections must be made for change of state if some of the components are in the liquid or solid state. To appreciate this fact fully, let us consider the combustion of a typical organic compound. Depending on the atoms present in the molecule, the final products of combustion would be H_2O, CO_2, SO_2, N_2, and HX (where X is a halogen). The heat evolved in such a reaction is called the *heat of combustion*. Standard heats of combustion are often listed with H_2O in the liquid state (i.e., as water); thus, a suitable correction must be made to get the values with all the products in the gaseous state. This is illustrated in Example 4.1.

Example 4.1: Calculating the heat of reaction from the heats of combustion

The heats of combustion of gaseous methyl alcohol and dimethyl ether (with H_2O as liquid water) are 182.6 and 347.6 kcal/mol, respectively. Calculate the heat of reaction for the dehydration of methyl alcohol to methyl ether when the reactants and products are all in the gaseous state and when they are all in the liquid state.

The combustion of methyl alcohol and methyl ether is represented by the reactions

$$2CH_3OH\ (g) + 3O_2\ (g) \rightarrow 2CO_2\ (g) + 4\ H_2O\ (l) + 365.2\ (\text{i.e., } 2\times 182.6)\ \text{kcal} \qquad (R3)$$

$$CH_3OCH_3\ (g) + 3O_2\ (g) \rightarrow 2CO_2\ (g) + 3H_2O\ (l) + 347.6\ \text{kcal} \qquad (R4)$$

Combining these, we obtain

$$2CH_3OH\ (g) \rightarrow CH_3OCH_3\ (g) + H_2O\ (l) + 17.6\ \text{kcal} \qquad (R5)$$

Thus, the heat of reaction is $17.6/2 = 8.8$ kcal/mol of methyl alcohol undergoing reaction, when all the reactants and products are in the gaseous state except water. Even if water is to be in the gaseous state, we must subtract the heat required to vaporize 1 mol of water (i.e., its heat of vaporization). We thus obtain

$$2CH_3OH\ (g) \rightarrow CH_3OCH_3\ (g) + H_2O\ (g) + 7.6\ \text{kcal} \quad (R6)$$

The heat of reaction with all the products in the gaseous state is, therefore, 7.6/2 or 3.8 kcal/mol of methyl alcohol reacting.

When the reactants and products are all in the liquid state, the heats of vaporization of methyl alcohol and methyl ether should also be considered. Thus, we should subtract the heat absorbed in vaporizing 2 mol of methyl alcohol and add the heats evolved in condensing 1 mol each of methyl ether and H_2O:

$$2(\Delta H_r^0) = 7.6 - 2(8.4) + 4.8 + 10 = 5.6$$

$$\Delta H_r^0 = 2.8\,\text{kcal/mol}$$

Implications of liquid phase reactions

The thermodynamic implications of reactions in the liquid state are important. Let us consider the case where a gas, liquid, or solid is dissolved in a solvent and the products also remain in solution (i.e., the reaction occurs in the liquid phase). The method illustrated in the above example is applicable to such cases. Since all these involve energy changes associated with condensation as well as dissolution and mixing with solvents, they are far more complicated than reactions in the gaseous state. As will be emphasized below, the formal thermodynamic approach fails to give predictive correlations for such cases, and resort to empirical combinations of the microscopic effects of solvents with the formal macroscopic approach becomes necessary.

Free energy change and equilibrium constant

Standard free energy change and equilibrium constant The concept of free energy is useful in defining the possibility of a reaction and in determining its limiting or equilibrium conversion. The formal definition of the equilibrium state of a chemical reaction is the state for which the total free energy is a minimum. Thus, the well-known rule: Reaction is spontaneous if ΔG is negative; it is not spontaneous if ΔG is positive. We shall now present the main features of the equilibrium state for ideal as well as nonideal gases.

Standard free energy change for ideal gases

Ideal gases Consider reaction R1. The free energy change accompanying this reaction is given by the well-known equation

$$\Delta G_r^\circ = -R_g T \ln K \quad (4.6)$$

where

$$K = \frac{k_{\text{forward}}}{k_{\text{reverse}}} = \frac{P_R^{\upsilon_R} P_S^{\upsilon_S}}{P_A^{\upsilon_A} P_B^{\upsilon_B}} \tag{4.7}$$

is the *thermodynamic equilibrium constant* of the reaction. It also represents a state where the two *rates* (not rate constants) are equal, so that the slightest parametric disturbance will drive it in any one direction.

Standard free energy change for nonideal gases

Nonideal gases In applying the concepts presented above to nonideal gases, it is necessary to introduce quantities that may be regarded as *nonideal gas equivalents* of ideal gases. These are *fugacity, fugacity coefficient, fugacity in a mixture,* and *fugacity coefficient in a mixture.* The first is defined by the equation

$$f_j = \phi_j P \tag{4.8}$$

where f_j is the *pure-component fugacity* (the nonideal gas equivalent of pressure) and ϕ_j is the *pure-component fugacity coefficient*, which is independent of composition and is a measure of nonideality. Pure-component fugacities can be estimated from generalized charts of f/P as a function of reduced pressure and temperature (see, e.g., Sandler, 2006).

When nonideal gases form an ideal mixture, the equation for K becomes

$$K = \frac{P_R^{\upsilon_R} P_S^{\upsilon_S}}{P_A^{\upsilon_A} P_B^{\upsilon_B}} \left(\frac{\phi_R \phi_S}{\phi_A \phi_B} \right) \tag{4.9}$$

On the other hand, when they form a nonideal mixture, the fugacity coefficient is replaced by the fugacity coefficient in the mixture $\bar{\phi}_j$ defined as

$$\bar{f}_j = \bar{\phi}_j y_i P \tag{4.10}$$

Notice that $\bar{\phi}_j$ is not independent of composition, and hence it is not possible to estimate it with any degree of certainty.

Temperature dependence of equilibrium constant

Temperature dependence of K The Gibbs–Helmholtz relationship

$$\frac{\partial}{\partial T}\left(\frac{G}{T} \right) = -\frac{H}{T^2} \tag{4.11}$$

can be used to derive the relationship between the temperature and the equilibrium constant, known as the van't Hoff relationship:

$$\frac{\partial \ln(K)}{\partial T} = \frac{\Delta H}{RT^2} \tag{4.12}$$

When all the effects, including the effect of temperature on the enthalpy change of the reaction, are taken into account, the following thermodynamic expression can be derived for K as a function of T:

$$\ln K = \int \frac{\left(\Delta H + \int_{298\,K}^{T} v_i \Delta C_p dT\right)}{RT^2} dT \quad (4.13)$$

and can be estimated from a known value of ΔH^{o} at any one temperature, usually 298 K.

Equilibrium compositions in gas phase reactions Of primary importance in conducting any reaction is knowledge of the equilibrium conversion and the composition of the reaction mixture at equilibrium. Noddings and Mullet (1965) have considered the most general reaction

$$\upsilon_A A + \upsilon_B B + \upsilon_C C + \upsilon_D D \leftrightarrow \upsilon_F F + \upsilon_G G + \upsilon_H H + \upsilon_I I + \upsilon_L L \quad (R7)$$

and given extensive tables of equilibrium composition versus equilibrium constant for several simplified forms of this general stoichiometry. Listed in Table 4.1 are the conversion–equilibrium constant relationships for five commonly encountered reaction types in chemical synthesis/technology.

Accounting for condensed phase(s) In the reactions considered in Table 4.1, all the components are in the ideal gaseous state. To treat

Table 4.1 Expressions for Equilibrium Conversion for a Few Common Types of Reactions

Reaction	Expression for K[a]	Pressure Dependence
1. $A \leftrightarrow R$	$\frac{X}{1-X}$	No (no volume change)
2. $A \leftrightarrow R + S$	$\frac{X^2 P}{(1-X)(1+X)}$	Direct (volume increase)
3. $A + B \leftrightarrow R$	$\frac{X(2-X)}{(1-X)^2 P}$	Inverse (volume decrease)
4. $A + B \leftrightarrow R + S$	$\frac{X^2}{(1-X)^2}$	No (no volume change)
5. $A \leftrightarrow 2R$	$\frac{4X^2 P}{(1-X)(1+X)}$	Direct (volume increase)

[a] Example: Reaction R2.

$$K = K_N \left(\frac{P}{N_t}\right)^{\Delta v}, \quad \text{where } K_N = \frac{N_R N_S}{N_A}$$

For 1 mol A and a conversion of X, $K = X^2 P/(1 - X)(1 + X)$.

reactions where all the components are in the liquid state, it should be remembered that vaporization at saturation pressure does not produce a change in free energy. Hence, (mole fraction × vapor pressure) may be used in place of partial pressure, and the equilibrium composition then calculated from equations similar to those in Table 4.1. However, where one of the components is a liquid, its activity with respect to a standard state of ideal gas can be taken as the vapor pressure at the temperature of the system. Thus, if R is in the liquid phase in reaction R3 of Table 4.1, the equilibrium constant is given by

$$K = \frac{P_{vR}}{P_A P_B} \tag{4.14}$$

Thermodynamics can also be usefully employed in analyzing the role of the catalyst. A typical example, illustrated below, is the Gattermann–Koch reaction:

Example 4.2: Thermodynamics of the Gattermann–Koch reaction

In this reaction, a –CHO group is introduced into a molecule such as benzene with the assistance of a catalyst of the type ($HCl + AlCl_3$) according to the reaction

$$C_6H_6\ (l) + CO\ (g) \rightarrow C_6H_5CHO\ (l) \tag{R8}$$

It has been postulated that formyl chloride is a transient intermediate in this reaction. The free energy change of the reaction leading to its formation

$$HCl\ (g) + CO\ (g) \leftrightarrow ClCHO\ (g) \tag{R9}$$

may be calculated as (–43.7) – (–22.78 – 32.82) = 12.43 kcal/mol (Dilke and Eley, 1949). In view of the positive value of ΔG^0_{298}, this compound clearly could not be isolated, but the fact that it is formed is indirectly substantiated by the negative value of ΔG^0_{298} for the following reaction:

$$ClCHO\ (g) + C_6H_6\ (g) \leftrightarrow C_6H_5CHO\ (g) + HCl\ (g) \tag{R10}$$

$$\Delta G^0_{f298} \quad -43.17 \quad 30.99 \quad -1.07 \quad -22.78$$

$$\Delta G^0_{r298} = -11.67 \text{ kcal/mol}$$

The calculations presented above are not conclusive enough. Hence, another line of investigation was pursued by Dilke and Eley (1949). The thermodynamics of formation of the catalytic complex with different chlorides such as those of Al, Sn, Fe, and Sb was studied to see whether complex formation does occur, and if so whether a consistent ranking of the catalysts might be discerned through increments in estimated values of enthalpy, entropy, and free energy or equilibrium constant. For this purpose, two quantities were measured: (1) the calorimetric enthalpy of mixing of the solid halide with liquid benzaldehyde, that is, the enthalpy of complex formation, and (2) the concentration of the species present at equilibrium. This yielded the equilibrium constant K in the mixing experiments in which complex formation occurs according to the reaction

$$C_6H_5CHO\,(l) + \tfrac{1}{2}M_2Cl_6(s) \leftrightarrow C_6H_5CHO \cdot MCl_3(s) \quad \text{(R11)}$$

where M = Al, Sn, Fe, Sb. The results (Stull et al., 1969) clearly show that complex formation does occur. Further, the increments in the enthalpy, free energy (or equilibrium constant), and entropy all point to the same order of performance of the different catalysts.

Chemical equilibria of complex reactions

Complex equilibria

Voluminous literature exists on the calculation of reaction equilibria in complex networks. The following two procedures are particularly useful: simultaneous solution of the equilibrium equations, and minimization of free energy.

Simultaneous solution of equilibrium equations A simple and direct method of determining the equilibrium composition of a complex reaction is to simultaneously solve all the equations comprising the complex network. The actual number of equations to be solved is equal to the number of independent reactions of the network. Chapter 2 deals formally with the treatment of complex reactions, and the method outlined therein can be applied to the present problem. Where the number of reactions is relatively small, say 2 or 3, simple, less formal methods can be used as illustrated in the example below.

Example 4.3: Equilibrium composition in a complex reaction

Consider the following reaction scheme reported by Stull et al. (1969):

$$CH_3Cl\,(g) + H_2O\,(g) \leftrightarrow CH_3OH\,(g) + HCl\,(g)$$

$$2CH_3OH\,(g) + (CH_3)_2O\,(g) + H_2O\,(g)$$

The equilibrium constant of the first reaction is given as 0.00154 and of the second as 10.6 at 600 K, and it is desired to calculate the equilibrium composition of the mixture produced by reacting methyl chloride with water.

Let us start with 1 mol each of methyl chloride and water. Assuming that X moles of HCl and Y moles of dimethyl ether are formed, the amounts of the different constituents at equilibrium would be: $CH_3Cl = (1 - X)$, $HCl = X$, $CH_3OH = X - 2Y$, $(CH_3)_2O = Y$, and $H_2O = (1 - X + Y)$. The equilibrium compositions of the two reactions can thus be expressed as

$$0.00154 = \frac{(X - 2Y)X}{(1 - X)(1 - X + Y)} \tag{4.15}$$

$$10.6 = \frac{Y(1 - X + Y)}{(X - 2Y)^2} \tag{4.16}$$

Solution of these equations gives

$$X = 0.048$$

$$Y = 0.009$$

The equilibrium composition can then be readily calculated from these values.

Extension to a nonideal system One can extend the treatment to a nonideal reaction by using Equation 4.9. Further, for any complex scheme such as

$$A + B \leftrightarrow 2R \tag{R12}$$

$$3R + B \leftrightarrow 3S \tag{R13}$$

it is more convenient to use the concept of *extent of reaction* (moles converted by a given reaction) than conversion (see Chapter 2). Thus, we define

ξ_1 = moles of A or B converted by reaction 1

ξ_2 = moles of B converted by reaction 2

based on which the number of moles of each component at a given extent of each reaction (starting with N_{A0} moles of A and N_{B0} moles of B) can be written and added up to give the total number of moles $N_t = N_{A0} + N_{B0} - \xi_2$. The equilibrium constants of the two reactions can then be written as

$$K_1 = \left(\frac{N_R^2}{N_A N_B}\right) K_{\phi 1}, \quad K_2 = \left(\frac{N_S^3}{N_B N_R^3}\right)\left(\frac{N_t}{P}\right) K_{\phi 2} \tag{4.17}$$

where

$$K_{\phi 1} = \frac{\phi_R^2}{\phi_A \phi_B}, \quad K_{\phi 2} = \frac{\phi_S^3}{\phi_B \phi_R^3} \tag{4.18}$$

and ϕ_S represent the pure-component fugacity coefficients. Equations 4.17 and 4.18 can be readily solved by using any good nonlinear equation solver.

Minimization of free energy

Minimization of free energy The total free energy G of a reacting system reaches a minimum at equilibrium. For an ideal gaseous system, the component partial pressures of any reaction, simple or complex, are related to this free energy by the equation

$$G = \sum_{j=1}^{N} N_j \bar{G}_j \tag{4.19}$$

where $P_j = y_j P$ and y_j is the mole fraction of j, that is, N_j/N_t. We now find the number of moles of each component *at equilibrium* by requiring that they produce a minimum in G in Equation 4.19 and simultaneously satisfy the elemental balance

$$\sum_{j=1}^{N} n_{ij} N_j - a_i = 0 = \varphi_i, \; i = 1,2,3,\ldots \tag{4.20}$$

where n_{ij} is the number of atoms of element i in component j, a_i the total number of atoms of element i in the system, and N_i the component I in the system. The minimization of the objective function (Equation 4.19) with the constraint (Equation 4.20) will be carried out with the help of *Lagrange multipliers*. We will carry out the development for a single-phase reaction.

Lagrange multipliers

The minimization will be carried out over the function

$$\psi = G - \lambda \phi_i \tag{4.21}$$

The minimum of function ψ with respect to the N_i in Equation 4.21 will be looked for by simply taking the derivative of the function with respect to each component such that

$$\left(\frac{\partial \psi}{\partial N_i} \right) = 0 = \bar{G}_i - \lambda n_{ij} \tag{4.22}$$

Equation 4.22 is solved along with the overall and component material balance equations (Equations 4.23 and 4.24) for the Lagrange multipliers for the global minimum of the Gibbs free energy change.

$$\sum_{j=1}^{N} \frac{n_{ij} N_j - a_i}{N_T} = 0, \quad i = 1,2,3,\ldots \tag{4.23}$$

$$\sum_{j=1}^{N} x_j = 1 \tag{4.24}$$

The system of equations derived from the previous section can be solved by the following procedure:

1. Chemical components that are predicted in the system are listed.
2. The Gibbs free energies of the chemical components are determined.

3. The feed composition and feed conditions are stated.
4. A meaningful initial guess is provided for each unknown in the system of equations.
5. By using an appropriate program or method, Equations 4.22 through 4.24 are solved.

Example 4.4: Determination of equilibrium conversions of the methanol synthesis reaction by Gibbs free energy minimization

Methanol is a very important chemical commodity. The methanol synthesis reaction is mildly exothermic and therefore reversible at high temperatures. On the other hand, high temperatures are needed such that the rates and therefore conversions are high. Combined together, this creates an optimization problem for low enough temperatures for high equilibrium conversions and high enough temperatures for higher reaction rates. As a result, design of methanol synthesis reactors depends on accurate information on the equilibrium conversion. By using the Gibbs free energy minimization method, we can only specify the species involved in the system, without having to specify the reactions. In this example, we will determine the equilibrium conversions of the methanol synthesis reaction at 50 atm. The feed gas composition and thermochemical data are given below:

Component (All in Gas Phase)	Mole Fraction	ΔG_f^0 (kJ/mol)	ΔH_f^0 (kJ/mol)
CH_3OH	0	−162.0	−200.7
H_2O	0.03	−228.6	−241.8
H_2	0.69	0.0	0.0
CO	0.25	−137.2	−110.5
CO_2	0.03	−394.4	−393.5

By using the above algorithm and a nonlinear equation solver (for this problem, we used MATHCAD), mole fractions of the species as a function of temperature were determined. The results are shown in Table 4.2.

Thermodynamics of reactions in solution

Most treatments of reactor design focus on the gaseous state. Many organic reactions are carried out in the liquid state, often in solvents, and hence we consider in this section the thermodynamics of reactions in solution.

Partial molar properties

Partial molar properties

A number of chemical reactions involve at least two chemical species, sometimes more. Any thermodynamic property M of a system of, say, N components (N_1, N_2, ..., N_N) can be defined as

Table 4.2 Equilibrium Compositions of the Methanol Synthesis Reaction as a Function of Temperature, Determined by the Method of Lagrange Multipliers

T (K)	y_{CH_3OH}	y_{CO}	y_{CO_2}	y_{H_2}	y_{H_2O}
300	0.631	1.30×10^{-7}	0.002	0.302	0.065
340	0.602	7.72×10^{-6}	0.015	0.332	0.051
380	0.556	1.78×10^{-4}	0.035	0.381	0.028
400	0.535	6.95×10^{-4}	0.044	0.402	0.018
450	0.485	0.013	0.054	0.443	0.005
480	0.410	0.048	0.051	0.487	0.003
500	0.330	0.088	0.047	0.532	0.003
530	0.194	0.156	0.039	0.609	0.003
560	0.090	0.209	0.032	0.667	0.003
600	0.026	0.242	0.027	0.701	0.005
700	0.001	0.261	0.018	0.707	0.012

$$M = f(T, p, N_1, N_2, \ldots, N_N) \tag{4.25}$$

We now formally define a *partial molar quantity* represented by $\bar{M}_i$ as

$$\bar{M}_i = \left[\frac{\partial M}{\partial N_i}\right]_{T,p,N\neq N_i} \tag{4.26}$$

where M is the partial molar thermodynamic property of a solution of constant composition. Thus, the mixture property can now be estimated:

$$M = N_1\bar{M}_1 + N_2\bar{M}_2 + \cdots + N_i\bar{M}_i + \cdots \tag{4.27}$$

Medium and substituent effects

Medium and substituent effects on standard free energy change, equilibrium constant, and activity coefficient

General considerations In this section, we will address dilute and concentrated solution cases separately. In the case of dilute solutions, the concentration can be used directly as an exact measure of its activity. This is often justified in organic synthesis since normally solvents are used in large excess. But where the reactant concentration is high, its activity in solution cannot be replaced by concentration without an appropriate correction factor.

For the case of dilute solutions, the dependence of the partial molar free energy of any component i on its concentration is expressed by the equation

$$\bar{G}_i = \bar{G}_i^{\circ} + RT \ln [i] \tag{4.28}$$

In analogy with Equation 4.6, the following relationship can be derived:

$$\Delta\bar{G}^\circ = -RT \ln K_C \tag{4.29}$$

where $\Delta\bar{G}^\circ$ is the partial molar standard free energy change for the reaction and K_C is the concentration-based equilibrium constant. For any reaction, such as R1, K_C is given by

$$K_C = \frac{[R]^{\nu_R}[S]^{\nu_S}}{[A]^{\nu_A}[B]^{\nu_B}} \tag{4.30}$$

It should be noted that the constant K_C is different from the K of Equation 4.7 based on partial pressures, and one can be converted to the other by using the gas law, $P = RT[i]$. Clearly, where all the reaction orders are unity, the RT terms cancel out, and the two equilibrium constants would be equal.

Equation 4.28 applied to concentrated solutions becomes

$$\Delta\bar{G}^\circ = -RT \ln K_a \tag{4.31}$$

with

$$K_a = K_\gamma K_c \tag{4.32}$$

where K_γ is a correction factor based on the activity coefficients γ of the different components, and for reaction R1, it is given by

$$K_\gamma = \frac{\gamma_R^{\nu_R}\gamma_S^{\nu_S}}{\gamma_A^{\nu_A}\gamma_B^{\nu_B}} \tag{4.33}$$

Solvent and solute operators

Solvent and solute operators It is useful to correlate any thermodynamic property M as the difference δ_M between the value for a given solute–solvent combination and that for a selected "standard" combination. The *operator* δ can represent the effect of changing the solvent structure on the reaction of a given solute (the *solvent operator*) or of changing the solute structure on reaction in a given solvent (the *solute* or *substituent operator*).

Thus, considering ΔG, equations for the solvent and solute operators would be as follows.

Solvent operator:

$$\delta_s \Delta\bar{G}_i^\circ = \Delta\bar{G}^\circ_{i,\text{solvent } S} - \Delta\bar{G}^\circ_{i,\text{standard solvent}} \tag{4.34}$$

Recall that the operator Δ refers to the effect of chemical reaction, that is, the change in free energy accompanying reaction, and should be distinguished from the newly defined solvent operator δ_s.

Substituent operator:

$$\delta_R \Delta \bar{G}^\circ = \Delta \bar{G}_R^\circ - \Delta \bar{G}_{R_0}^\circ \tag{4.35}$$

where R and R_0 represent the two substituent groups. R_0 is usually but not necessarily the hydrogen atom.

A striking example of the solvent effect is revealed in the keto-enol tautomerization of benzoyl camphor (Hammett, 1940). A similar substituent effect is seen in the acid dissociation constants in various solvents (Davis and Hetzer, 1958).

Comments

In spite of extensive theoretical and experimental studies on solvent and solute effects based on the operators defined above, there are no general theoretical models available as yet that can predict these effects with any certainty. From a practical point of view, therefore, one must look into methods other than those based on the application of formal thermodynamics. Thus, we turn to the extrathermodynamic approach.

Extrathermodynamic approach

Basic principles

The basis of the extrathermodynamic approach is simple. Consider any property such as the heat of formation. The effect, for instance, of an amino group in benzene is assumed to be the same as its effect in any aromatic compound (e.g., toluene or any of the xylenes). The principle is best illustrated for the ionization constant of carboxylic acids in a number of solvents:

$$\text{RCOOH} \overset{\text{water}}{\leftrightarrow} \text{RCOO}^- + \text{H}^+ \tag{R14.1}$$

$$\text{RCOOH} \overset{\text{alcohol}}{\leftrightarrow} \text{RCOO}^- + \text{H}^+ \tag{R14.2}$$

The extrathermodynamic relation would be

$$\log K_{a,w} = m \log K_{a,a} \tag{4.36}$$

where $K_{a,w}$ and $K_{a,a}$ are the ionization constants of a given acid in water and alcohol, respectively. The relationship can otherwise be

$$\log K_{a,R} = n \log K_{a,R'} \tag{4.37}$$

in a given solvent such as water, where R and R' represent different substituent groups.

In its most primitive form, the structure of the parent molecule is inconsequential. Thus, if the value of the substituent group is known, it can be used in any molecule regardless of its structure. Linear relationships based on such similarity of effects are referred to as extrathermodynamic relationships, since the approach does not call for information on the microscopic nature of the structures. To that extent, it retains the fundamental character of the thermodynamic approach.

Group contributions or additivity principle

As a result of the fortuitous simplifying circumstance mentioned earlier, a molecule can be divided into "action" and "neutral" zones. The latter zone usually occupies the bulk of the molecule's size. The word "action" is more appropriate than "reaction" as used by many authors since the concept is applicable even where no reaction occurs. Changes in properties are assumed to occur only as a result of changes in the action zone. Rules can be formulated for the effect of different substituent groups in the action zone. These are the so-called *additivity rules*, or the *rules of group contributions*. It must be noted, however, that in the interest of greater accuracy in properties estimation, it may often be necessary to introduce higher-order approximations that violate the neutrality of the neutral zone, but one pays a price for this: an increase in the number of empirical parameters.

Additivity rules
or
Rules of group contributions

Two general aspects of additivity methods are noteworthy:

1. Consider a compound with two functional groups, such as succinic acid ($HOOC(CH_2)_2COOH$). If we are interested in only one of the COOH groups acting as the reactive site and the other as substituent, we must divide the observed rate constant by a *statistical factor* of 2. For m reactive sites, the statistical factor is m, and if there are two reagents with m and n reactive sites, then the statistical factor is mn.
2. The same order of approximation should be used for all *part-structures* (generally referred to as *groups*). Use of even a single lower-order approximation would tend to subvert the accuracy of the higher-order approximations. Let us take, for instance, the value of N in NO_2 and NH_2. If different values are used for N in the two groups, then a similar higher-order approximation should be used in the case of other atoms also, such as O in COOH and CO.

A simple way to address the question of order of approximation is to assume that the contributions from any two groups are exactly additive if they are sufficiently apart within a given structure (Benson and Buss, 1958). Thus, for a molecule CH_3XCH_3, where X is the intervening structure, the zeroth-order approximation would be CH_3CH_3 (i.e., no intervening structure). For a first-order approximation, X could be a simple atom such as N or O or a CH_2 group. Higher-order approximations would involve correspondingly more CH_2 groups and even branched chains.

Extrathermodynamic relationships between rate and equilibrium parameters

Polanyi and Brønsted relationships

Polanyi and Brønsted relations

Consider any two reactions in a family of reactions such as isomerization of hydrocarbons or dehydration of alcohols. The difference in the activation energies of the two reactions is assumed to be directly proportional to the difference in their heats of reaction:

Polanyi relationship

$$\delta_p E = \alpha \delta_p (-\Delta H_r^\circ) \tag{4.38}$$

where δ_p may be regarded as a Polanyi operator. Using this relationship, the following important expression can be derived relating the rate constant of a reaction to its thermodynamic equilibrium constant:

Brønsted relation

$$k = (\text{constant})\, K^{\alpha} \tag{4.39}$$

This is referred to as the Brønsted relation. The proportionality constant and the constant α are characteristic of a given family of reactions.

When applied to reactions catalyzed by acids or bases, the Brønsted relation has a slightly different connotation. Examples of these are the base-catalyzed halogenation of ketones and esters and the acid-catalyzed dehydration of acetaldehyde hydrate. For an acid-catalyzed reaction, we have

$$k_a = (\text{constant}) K_a^{\alpha} \tag{4.40}$$

where K_a is the dissociation equilibrium constant of the acid (say HA) given by

$$HA \overset{K_a}{\leftrightarrow} A^- + H^+ \tag{R15}$$

Equation 4.40, which is the Brønsted relation for acid-catalyzed reactions, states that the rate constant of a reaction catalyzed by an acid is proportional to some power of the dissociation equilibrium constant of the acid used as catalyst.

Similarly, for reactions catalyzed by a base, we have

$$k_b = (\text{constant}) K_b^{\beta} \tag{4.41}$$

where k_b is the rate constant of the reaction and K_b is the dissociation constant of the base.

A striking example of the application of Equation 4.41 is the isomerization of substituted 5-aminitriazoles in ethylene glycol shown in reaction R16. Figure 4.1 is a log–log plot of the rate constant versus equilibrium constant for this reaction (Leffler and Grunwald, 1963).

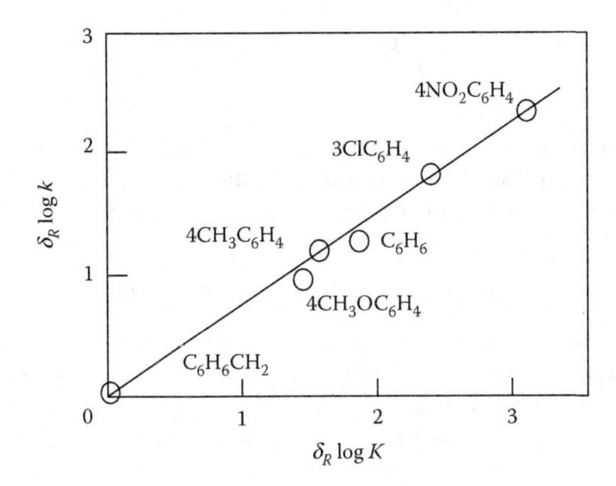

Figure 4.1 Rate equilibrium relationship in 5-aminotriazole rearrangement. (From Leffler, J.E. and Grunwald, E., *Rates and Equilibrium of Organic Reactions*, Dover, New York, 1963. With permission.)

$$\text{N(Ar)–N=N–C(}\phi\text{)=C–NH}_2 \xrightleftharpoons[\text{glycol}]{\text{in ethylene}} \text{N(H)–N=N–C(}\phi\text{)=C–N(H)–Ar} \qquad \text{(R16)}$$

Hammett relationship for dissociation constants

Hammett relationship

A widely used extrathermodynamic relationship for organic reactions is the Hammett linear free energy relationship. Although developed specifically for dissociation constants, the method in principle is applicable to any organic reaction.

Consider the family of reactions involving the ionization of benzoic acid in water followed by reaction with ethyl alcohol, and those of *para-* or *meta-*substituted benzoic acids in water also followed by reaction with ethyl alcohol.

$$C_6H_5COO^- + C_2H_5OH \overset{K_{a,0}}{\leftrightarrow} C_6H_5COOC_2H_5 + H_3O^+ \qquad \text{(R17)}$$

$$\text{m- or } p\text{-}XC_6H_4COO^- + C_2H_5OH \xleftrightarrow{K_a} \text{m- or } p\text{-}XC_6H_4COOC_2H_5 + H_3O^+ \qquad \text{(R18)}$$

where $K_{a,0}$ and K_a are the acid dissociation constants for the two reactions. The basic feature of the method is that the ratio σ of the dissociation constants of the two reactions is used as the correlating parameter for other members of the family, provided the reaction conditions are the same. Thus, we can write

$$\log \frac{k}{k_0} = \log \frac{K_a}{K_{a,0}} = \rho\sigma \tag{4.42}$$

where the constant ρ is a function of the reaction and the reaction conditions used. A detailed compilation of σ values for various reactions is given by McDaniels and Brown (1958) and Shorter (1982).

Selectivity

Extrathermodynamic approach to selectivity

The *selectivity* for a given product in a complex reaction is an important practical consideration in carrying out a reaction. Let us first define the *reactivity* of a reagent. Clearly, for any bimolecular reaction, it has any meaning only in relation to the second reactant, the substrate. However, as long as the substrates are reasonably similar, each reagent exhibits a characteristic substrate-independent reactivity that enables a broad ordering of reagents. We may, therefore, choose (or postulate) a "standard substrate" and define reactivity as

$$\begin{bmatrix}\text{Reactivity of} \\ \text{a reagent}\end{bmatrix} = \begin{bmatrix}\text{Rate constant } k_0 \text{ for reaction} \\ \text{with a standard substrate}\end{bmatrix} \tag{4.43}$$

Consider the reaction

$$A + B \xrightarrow{k_1} P \tag{R19.1}$$

$$A + C \xrightarrow{k_2} R \tag{R19.2}$$

From a thermodynamic point of view, it is convenient in a complex reaction of this type to view selectivity (say for P) as the preference of A to react with B over C.

Theoretical analysis We now seek to obtain a relationship between reactivity represented by the overall reaction of A with B and C (i.e., its conversion to P and R). Clearly, as the overall reactivity increases, every encounter between A and B and between A and C would be a successful one, leading to reaction with a selectivity determined by the randomness of the process. On the other hand, a continuous lowering of reactivity to zero can enhance selectivity to unity. Although not a universal law, this is the basis of the commonly observed increase in selectivity with decrease in conversion.

The reactivity–selectivity relationship can be analyzed by making the following postulations with respect to the reactants: Reactant A is designated as A_H if it is highly reactive (hot) and as A_C if it is less reactive (cold), and the second reactant (which is also referred to as the substrate) is designated as B_H if it is more reactive and as B_C if less. Thus, the selectivity will vary depending on different combinations of A and B.

Consider the reactions

$$A_H + B_H \rightarrow \text{Products} \quad \text{(R20.1)}$$

$$A_H + B_C \rightarrow \text{Products} \quad \text{(R20.2)}$$

and

$$A_C + B_H \rightarrow \text{Products} \quad \text{(R21.1)}$$

$$A_C + B_C \rightarrow \text{Products} \quad \text{(R21.2)}$$

where the less reactive substrate B_C can be, say, benzene, and the more reactive one B_H can be a substituted benzene. Reactions R20.1 and R20.2 represent the effect of changing the substrate structure from hot to cold for a highly reactive A, that is, hot A. Similarly, reactions R21.1 and R21.2 denote the effect of substrate structure for a less reactive or cold A.

Thermodynamics of adsorption

Adsorption

The thermodynamics of adsorption is particularly important for analyzing the catalytic reactions as well as the surface characterization of the catalysts. In this section, first, we will derive the Gibbs adsorption isotherm, and then we will use the Gibbs adsorption isotherm to derive the more useful isotherms such as Langmuir's and Fowler Guggenheim isotherms.

Gibbs isotherm

Reconsider Equation 4.3 for a system composed of an adsorbate, s, and a nonvolatile adsorbent, a:

$$dA = -SdT - PdV + \mu_a\, dN_a + \mu_s\, dN_s \quad (4.44)$$

In the absence of adsorbate, the Helmholtz free energy of a clean surface becomes

$$dA_{0a} = -S_{0a}\, dT - PdV_{0a} + \mu_{0a}\, dN_a \quad (4.45)$$

Upon subtraction, we obtain

$$d(A - A_{A0}) = -(S - S_{0a})\, dT - Pd(V - V_{0a}) + (\mu_a - \mu_{0a})\, dN_a + \mu_s\, dN_s \quad (4.46)$$

Adsorbate: the gas molecule

According to Gibbs adsorption isotherm, the adsorbent is considered inert. Thus, upon subtraction, the relevant components belonging to the adsorbent are eliminated and the remaining values $(A - A_{0a})$, $(S - S_{0a})$, $(V - V_{0a})$ are the corresponding values for the sorbate, while we define $-\Phi = \mu_a - \mu_{0a}$ such that

Adsorbent: the surface

$$dA_s = -S_s\, dT - PdV_s - \Phi dN_a + \mu_s\, dN_s \quad (4.47)$$

The same analysis can be applied to Equations 4.1 through 4.4 with similar consequences yielding the term $-\Phi\, dN_a$ in the definitions. Now, we will elaborate on this term. Given that Equations 4.1 through 4.4 are exact differentials, it is possible to write

$$-\Phi = \left(\frac{\partial U_s}{\partial N_a}\right)_{S_s,V_s,N_s} = \left(\frac{\partial H_s}{\partial N_a}\right)_{S_s,P,N_s} = \left(\frac{\partial A_s}{\partial N_a}\right)_{T,V_s,N_s}$$
$$= \left(\frac{\partial G_s}{\partial N_a}\right)_{T,P,N_s} \tag{4.48}$$

The surface area of the adsorbent, α, is directly proportional to N_a such that one can define a spreading pressure, π:

Spreading pressure

$$\Phi\, dN_a = \pi\, d\alpha \tag{4.49}$$

Furthermore

$$\pi = \left(\frac{\partial U_s}{\partial \alpha}\right)_{S_s,V_s,N_s} \tag{4.50}$$

corresponds to the difference between the surface tension of a clean surface and a surface covered with adsorbate. Now, we write the Gibbs free energy change for the adsorbate

$$dG_s = -S_s\, dT + V_s\, dP - \Phi dN_a + \mu_s\, dN_s \tag{4.51}$$

At constant temperature and pressure, the system reaches the equilibrium at $dG_s = 0$ such that

$$-\Phi\, dN_a + \mu_s\, dN_s = 0 \tag{4.52}$$

or

$$\pi\, d\alpha = N_s\, d\mu_s \tag{4.53}$$

Given that the adsorbate at the surface is in equilibrium with the gas phase (in this case, we will treat an ideal gas phase, which is true especially when the adsorption is done at low pressures), the chemical potential of the adsorbate will be equal to the gas phase chemical potential, which is given by

$$\mu_s = \mu_g = \mu_g^0 + RT \ln\left(\frac{P}{P^0}\right) \tag{4.54}$$

Thus

$$d\mu_s = \frac{RTdP}{P} \tag{4.55}$$

and

$$\alpha\left(\frac{\partial \pi}{\partial P}\right)_T = \frac{RT}{P} N_s \tag{4.56}$$

or if we define the surface coverage θ as $\theta = N_s/\alpha$, the final form of the Gibbs isotherm is

$$\left(\frac{\partial \pi}{\partial P}\right)_T = \frac{RT}{P}\theta \tag{4.57}$$

Gibbs isotherm

We have to remember that the isotherm equation was derived based on the assumption that the adsorbate is inert and is not modified upon adsorption. This assumption is not broadly valid and we know that adsorption modifies the surface to a significant extent. However, for all practical purposes, the Gibbs isotherm equation and the isotherms we will derive from it provide practical equations.

Henry's law

Henry's law

When the equation of state for the adsorbed phase corresponds to the ideal gas law

$$\pi\alpha = N_s RT \tag{4.58}$$

the Gibbs isotherm takes the form

$$\left(\frac{\partial \pi}{\partial P}\right)_T = \frac{\pi}{P} \tag{4.59}$$

leading to

$$\pi = KP \tag{4.60}$$

Substituting π from the ideal gas law presented above, we obtain

$$\frac{N_s}{\alpha} = \theta = \frac{KP}{RT} = K'C \tag{4.61}$$

N_s/α is the surface coverage of the adsorbate and we obtain a linear relationship between the surface coverage and the gas phase concentration of the sorbate. This is the famous Henry's law.

Langmuir isotherm

When the equation of state is of the $\pi(\alpha - b) = N_sRT$ form (we subtract the area occupied by the molecule, b, from the total area, α, for improved accuracy)

$$\left(\frac{\partial \pi}{\partial \alpha}\right)_T = -\frac{N_sRT}{(\alpha - b)^2} \tag{4.62}$$

Langmuir isotherm

When substituted in the Gibbs isotherm, we have

$$\frac{dP}{P} = -\frac{\alpha d\alpha}{(\alpha - b)^2} \tag{4.63}$$

If we assume $b \ll \alpha$ and neglect b^2 in the denominator, after integration, we obtain

$$KP = \frac{2b/\alpha}{1 - 2b/\alpha} = \frac{\theta}{1-\theta} \tag{4.64}$$

if we take $\theta = 2b/\alpha$.

The Langmuir adsorption isotherm equation can also be derived from the following postulates:

1. All of the sites on the surface are energetically equivalent, that is, there is no variation in the adsorption energies of the sites.
2. The surface is only capable of holding one monolayer of adsorbate.

The second postulate gives us a site balance as follows:

$\theta = \theta_v + \theta_A = 1$, where θ_A and θ_v indicate the fractional surface coverage of the adsorbate A and the vacant sites.

If the adsorption is considered as a reaction step

$$A + * \overset{K_{eq}}{\Leftrightarrow} A \tag{4.65}$$

The reaction at equilibrium with equal forward and reverse rates yields

$$K_{eq}\, P_A \theta_v = \theta_A \tag{4.66}$$

Using the site balance along with the equilibrium relationship yields

$$\theta_A = \frac{K_{eq} P_A}{1 + K_{eq} P_A} \tag{4.67}$$

which is called the Langmuir adsorption isotherm, since the equilibrium constant K_{eq} is determined at constant temperature.

For the derivation of Volmer's and Fowler Guggenheim isotherms from the gas phase equations of state and Gibbs adsorption isotherms, we will refer the reader to Ruthven (1984).

Inhomogeneities expressed in terms of a site-energy distribution

It is possible to extend the analysis for the surface nonidealities. For example, a surface can be composed of patches, each of which has a uniform site energy distribution such that the Langmuir adsorption isotherm explains the coverage over the patch. One can assume (or derive from the experimental data) a distribution function of the adsorption heats, Q, as $\delta(Q)$ such that the overall isotherm is

$$\theta_m = \int \frac{KP}{1+KP}\delta(Q)dQ \tag{4.68}$$

Various forms of the distribution functions for different adsorption isotherms can be found in Doraiswamy (1991).

Two-dimensional equations of state and their corresponding adsorption isotherms

The adsorption of gases (or liquids) on solid surfaces can also be represented as the two-dimensional equations of state, especially when the attractive–repulsive interactions of the adsorbed layers prevail. It is beyond the objectives of this book to go into details of the two-dimensional equations of state, but it is important to remind the reader that, as with their gas phase equivalents, such as the van der Waals equation of state, the two-dimensional equations of state with lateral interactions predict the separation of phases of the adsorbed layers since they give rise to the S-shaped adsorption isotherms. The key feature of the S-shaped isotherm equation is that the adsorbed layer splits into a high-density and a low-density region upon disturbance. The implications of these high-density and the low-density regions in catalysis is still being explored. One of the very interesting examples is the demonstration of the phase transitions occurring during CO oxidation over precious metal surfaces (von Oertzen et al. 1998), an investigation that eventually brought the Nobel Prize to the laboratories of Gerhard Ertl. Equally important is the demonstration of the surface reconstructions in the presence of adsorption, which challenges the assumptions in the derivations of Gibbs isotherm and beyond (Somorjai, 1994).

One of the well-known adsorption isotherms accounting for the adsorbate–adsorbate interactions is the Fowler Guggenheim isotherm. The paradox of constant heats of adsorption assumption and the variations and eventually constant heat prevailing has its roots in the fundamental Gibbs approach in such a way that the basic assumption is the inertness of the surface. In reality, upon adsorption, the surface is modified, electronically and structurally. The implications of these are not yet clearly in place of the isotherm equations for the development of better understanding and using simpler equations are no longer popular. Table 4.3 summarizes the two-dimensional equations of state and adsorption isotherms on homotattic surfaces.

Fowler Guggenheim isotherm

Table 4.3 Two-Dimensional Equations of State and Adsorption Isotherms on Homotattic Surfaces

Isotherm Description	Two-Dimensional Equation of State	Isotherm Equation	Commonly Called
Two-dimensional ideal gas	$\pi = \theta RT$	$KP = \theta$	Henry's law
Localized, no interaction	$\pi = RT\ln\left(\frac{1}{1-\theta}\right)$	$KP = \frac{\theta}{1-\theta}$	Langmuir's isotherm
Mobile, no interaction	$\pi = RT\left(\frac{\theta}{1-\theta}\right)$	$KP = \left(\frac{\theta}{1-\theta}\right)\exp\left(\frac{\theta}{1-\theta}\right)$	Volmer's isotherm
Localized, with interaction	$\pi = RT\left[\ln\left(\frac{1}{1-\theta}\right) - \frac{z\omega\theta^2}{2RT}\right]$	$KP = \left(\frac{\theta}{1-\theta}\right)\exp\left[\left(\frac{\theta}{1-\theta}\right) - \frac{z\omega\theta^2}{2RT}\right]$	Fowler Guggenheim

Source: Doraiswamy, L.K., *Prog. Surf. Sci.*, **37**, 1, 1991.

Appendix

Derivation of chemical equilibrium relationships for simple reactions

We will use the extent of the reaction, ξ, defined in Chapter 2, to write the amount of each species as

$$N_i = N_{i,0} + \upsilon_i\xi$$

The total Gibbs free energy of a mixture is

$$G = \Sigma N_i\bar{G}_i = \Sigma(N_{i,0} + \upsilon_i\xi)\bar{G}_i \tag{4.69}$$

Since at constant T and P, the only variable in a system is the reaction coordinate, ξ, such that the equilibrium condition is written as

$$\left(\frac{\partial G}{\partial \xi}\right)_{T,P} = 0 \tag{4.70}$$

Verbally stated, Equation 4.70 indicates that under constant temperature and pressure, the total Gibbs free energy is a minimum with respect to the changes in the reaction coordinate. When Equation 4.70 is imposed upon Equation 4.69, the result is

$$\left(\frac{\partial G}{\partial \xi}\right)_{T,P} = \Sigma N_i\left(\frac{\partial \bar{G}_i}{\partial \xi}\right)_{T,P} + \Sigma \bar{G}_i\left(\frac{\partial N_i}{\partial \xi}\right)_{T,P} \tag{4.71}$$

The first term on the right-hand side disappears via the Gibbs–Duhem relationship (for details, see Sandler, 2006). Finally, the fundamental relationship of the chemical equilibrium for single chemical reactions is obtained:

$$\Sigma \upsilon_i \bar{G}_i(T,P,y) = 0 \tag{4.72}$$

Determination of the partial molar Gibbs free energies *per se* is not a trivial task. On the other hand, fugacity (meaning tendency to escape in Greek) can be invoked, defining the Gibbs free energy along an isotherm

$$\begin{aligned} f &= P\exp\left\{\frac{\underline{G}(T,P) - \underline{G}^{\mathrm{IG}}(T,P)}{RT}\right\} \\ &= P\exp\left\{\frac{1}{RT}\int_0^P\left(\underline{V} - \frac{RT}{P}\right)dP\right\} \end{aligned} \tag{4.73}$$

The superscript IG indicates the ideal gas state. Note that once an equation of state is available, then it is easy to determine the integral on the right-hand side of Equation 4.73. Furthermore, the fugacity coefficient of a component i in a mixture can be defined as follows:

$$\bar{\phi}_i = \frac{\bar{f}_i}{y_i P} = \exp\left\{\frac{\bar{G}_i(T,P,y_i) - \bar{G}_i^{\mathrm{IGM}}(T,P,y_i)}{RT}\right\} \tag{4.74}$$

In the above equation, the fugacity and the fugacity coefficient allows us to determine the properties in a mixture where nonidealities are present in terms of the attractive–repulsive interactions between the molecules, resulting in a net change in total energy and/or total volume upon mixing. The superscript IGM indicates the corresponding values in an ideal gas mixture. The fugacity coefficient can be calculated when an equation of state in the form of a *PVT* relationship is available. Reiterating the definition of the partial molar Gibbs free energy, $\bar{G}$, in relation to the molar Gibbs free energy, G, through the fugacity of the component in the mixture, $\bar{f}_i$, and the pure-component fugacity, f_i:

$$\bar{G}_i(T,P,y) = \underline{G}_i(T,P) + RT\ln\frac{\bar{f}_i(T,P,y)}{f_i(T,P)} \tag{4.75}$$

And substitution of Equation 4.75 into Equation 4.72 results in

$$\Sigma \upsilon_i \bar{G}_i(T,P,y) = \Sigma \upsilon_i \underline{G}_i(T,P) + RT\,\Sigma \upsilon_i \ln\frac{\bar{f}_i(T,P,y)}{f_i(T,P)} = 0 \tag{4.76}$$

Rearranging yields the chemical equilibrium condition for a single reaction in its most general sense:

$$\frac{-\Sigma \upsilon_i \underline{G}_i(T,P)}{RT} = -\frac{\Delta \underline{G}^0}{RT} = \Sigma \upsilon_i \ln \frac{\bar{f}_i(T,P,y)}{f_i(T,P)}$$

$$= \ln \prod \left(\frac{\bar{f}_i(T,P,y)}{f_i(T,P)} \right)^{\upsilon_i} \tag{4.77}$$

Reactions in gas phase Equation 4.77 can be rewritten in terms of the fugacity coefficients as

$$\frac{-\Sigma \upsilon_i \underline{G}_i(T,P)}{RT} = -\frac{\Delta \underline{G}^0}{RT} = \Sigma \upsilon_i \ln \frac{\bar{f}_i(T,P,y) y_i P / y_i P}{f_i(T,P) P / P}$$

$$= \ln \prod \left(y_i \frac{\bar{\phi}_i(T,P,y)}{\phi_i(T,P)} \right)^{\upsilon_i} \tag{4.78}$$

When reactions are taking place under the ideal gas conditions, then all the fugacity coefficients become 1 to yield

$$-\frac{\Delta \underline{G}^0}{RT} = \ln \prod y_i^{\upsilon_i} \tag{4.79}$$

Reactions in liquid phase On the other hand, when the reactions are taking place in the liquid state, the fugacity coefficients are expressed in terms of the activity coefficients as

$$\bar{f}_i(T,P,x) = x_i \gamma_i f_i(T,P) \text{ such that}$$

$$-\frac{\Delta \underline{G}^0}{RT} = \ln \prod (x_i \gamma_i)^{\upsilon_i} \tag{4.80}$$

However, when the liquid mixture is ideal, that is, $\gamma_i = 1$, then

$$-\frac{\Delta \underline{G}^0}{RT} = \ln \prod x_i^{\upsilon_i} \tag{4.81}$$

The groups on the right-hand side of Equations 4.79 through 4.81 is the equilibrium constant, K, of the reaction.

Now, we will arrive at this point in mathematics from another direction. At equilibrium, the net rate of the reaction has to be zero, that is

$$r_{A_f} = r_{A_r} \quad \text{or} \quad k_f f(C_i) = k_r g(C_i)$$

where $f(C_i)$ or $g(C_i)$ are the mathematical functions of concentrations of the relevant species in the reaction. The ratio of the rate constants also yield the equilibrium constant:

$$K_{eq} = \frac{k_f}{k_r} \quad (4.82)$$

Equation 4.82 is identical to one of Equations 4.79 through 4.81. With them, we establish the fundamental link between kinetics and thermodynamics.

Explore yourself

1. Why do we use four different forms of energy (U, H, A, and G) when analyzing systems and solving problems?
2. Why do we use Gibbs free energy to specify the equilibrium of a chemical reaction?
3. What is the fundamental basis of the Brønsted–Polanyi correlation?
4. What are the critical assumptions of the Gibbs isotherm?
5. What is the difference between Langmuir and Fowler–Guggenheim isotherms?
6. Discuss the assumption of inert substrate in the derivation of the Gibbs isotherm. How realistic is such an assumption? How would you remove this assumption? How much improvement do you anticipate if you lifted this assumption?
7. List as many effects as possible of a solvent on the chemical kinetics of a liquid phase reaction. How can you identify whether the solvent effects are important or not?

References

Benson, S.W. and Buss, J.H., *J. Chem. Phys.*, **29**, 550, 1958.
Davis, M.M. and Hetzer, H.B., *J. Res. Natl. Bur. Stand.*, **60**, 569, 1958.
Dilke, M.H. and Eley, D.D., *J. Chem Soc.*, 2601, 1949.
Doraiswamy, L.K., *Prog. Surf. Sci.*, **37**, 1, 1991.
Hammett, L.P., *Physical Organic Chemistry*, McGraw Hill, NY, 1940.
Leffler, J.E. and Grunwald, E., *Rates and Equilibrium of Organic Reactions*, Dover, NY, 1963.
McDaniels, D.H. and Brown, H.C., *J. Org. Chem.*, **23**, 420, 1958.
Noddings, C.R. and Mullet G.M., *Handbook of Compositions at Thermodynamic Equilibrium*, Wiley Interscience, NY, 1965.
Ruthven, D.M., *Principles of Adsorption and Adsorption Processes*, John Wiley and Sons, NY, 1984.
Sandler, S.I., *Chemical, Biochemical and Engineering Thermodynamics*, 4th edition, Wiley, NY, 2006.
Shorter, J.A., *Correlation Analysis of Organic Reactivity*, Research Studies Press, Wiley, NY, 1982.
Somorjai, G.A., *Annu. Rev. Phys. Chem.*, **45**, 721, 1994.

Stull, D.R., Westrum, E.F. Jr., and Sinke, G.C., *The Chemical Thermodynamics of Organic Compounds*, Wiley, NY, 1969.
von Oertzen, A., Mikhailov, A.S., Rotermund, H.H., and Ertl, G., *J. Phys. Chem.*, **B102**, 4966, 1998.

Bibliography

Adsorption thermodynamics:
Fowler, R.H. and Gugenheim, E.A., *Statistical Thermodynamics*, The Macmillan Company, NY, 1940.
Ruthven, D.M., *Principles of Adsorption and Adsorption Processes*, John Wiley and Sons, NY, 1984.

On minimization of free energy:
Prausnitz, J.M., *Molecular Thermodynamics of Fluid Phase Equilibria*, Prentice-Hall, Englewood Cliffs, NJ, 1969.
Walas, S.M., *Phase Equilibria in Chemical Engineering*, Butterworth, Boston, 1985.

Interlude III

Reactor design for thermodynamically limited reactions

We will continue the discussion of ammonia synthesis we briefly covered in the Overview.

$$N_2 + 3H_2 \rightarrow 2NH_3 \quad \text{(R1)}$$

Understanding ammonia synthesis technology is important for several reasons:

1. High pressures are necessary to drive a reaction of the net negative stoichiometry.
2. The net exothermic reaction needs high temperatures for reasonable reaction rates.
3. The reaction is mildly exothermic. Therefore, at temperatures the reaction rates are reasonable, the equilibrium conversions are quite low, imposing a thermodynamic limitation.

Once these restrictions are imposed on the reaction, the design of the reactor has to be optimized by taking all these effects into account. First, let us remember the effect of temperature on equilibrium conversion of an exothermic reaction. We will start with the well-known identity

$$\Delta G = \Delta H - T\Delta S = -RT \ln K \quad \text{(III.1)}$$

As one can easily deduce from Equation III.2, for an exothermic reaction ($\Delta H < 0$), at elevated temperatures, the reaction equilibrium is attained at conversions lower than 1.

Kinetics

The most widely known rate expression for ammonia synthesis was provided by Temkin in the early 1940s (Temkin and Pyzhev, 1940):

$$r = k_1^{'} P_{N_2} \left(\frac{P_{H_2}^3}{P_{NH_3}^2} \right)^{\alpha} - k_2^{'} \left(\frac{P_{NH_3}^2}{P_{H_2}^3} \right)^{1-\alpha} \quad \text{(III.2)}$$

where

$$\alpha = 0.4 - 0.5$$

Various modifications have been reported in the literature. In this part, we will use the one reported by Rossetti et al. (2006). The details of the analysis and calculations can be found elsewhere (Aslan, 2012).

$$\frac{d\eta}{d\tau} = k_f \lambda(q) \frac{(a_{N_2})^{0.5}[(a_{H_2})^{0.375}/(a_{NH_3})^{0.25}] - (1/Ka)[(a_{NH_3})^{0.75}/(a_{H_2})^{1.125}]}{1 + K_{H_2}(a_{H_2})^{0.3} + K_{H_2}(a_{NH_3})^{0.2}}$$

where η is the extent of the reaction, τ the space time, λ the conversion factor, and a_i the activity of species i.

Optimization of temperatures and pressures Setting the value of the rate to zero, it is possible to determine the relationship between equilibrium

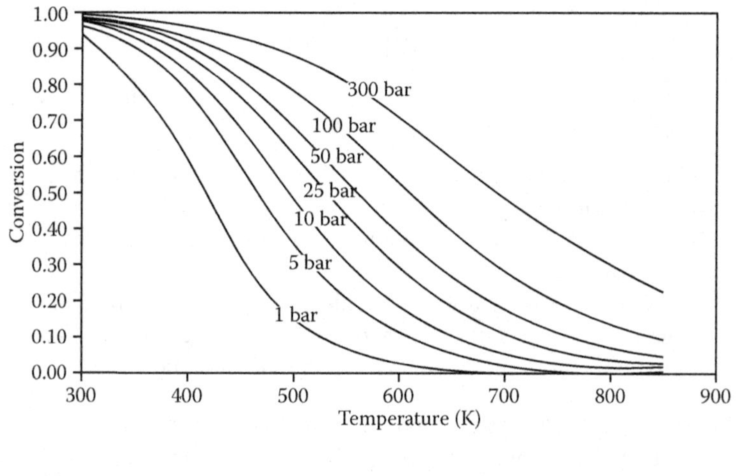

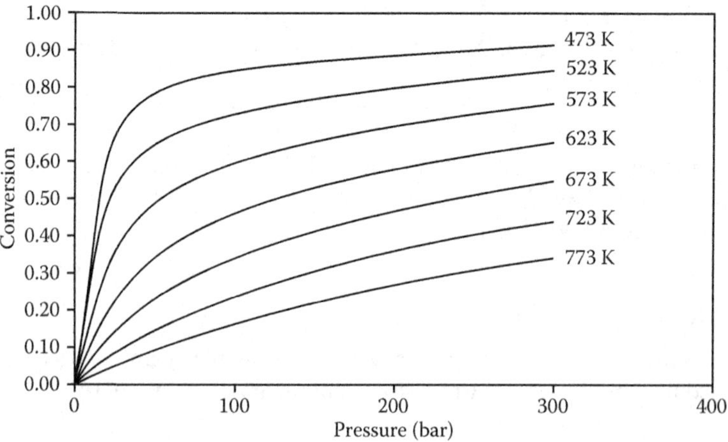

Figure III.1 Relationship between the equilibrium conversion temperature and pressure. The data were calculated using the modified Temkin kinetics.

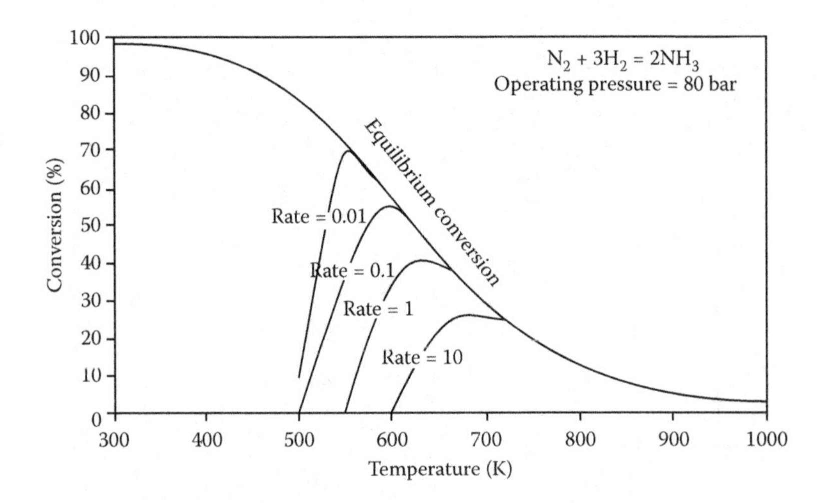

Figure III.2 Constant rate curves at a total pressure of 80 bar. Note that $r_a = 0$ is the equilibrium curve.

conversion and pressure and the relation between equilibrium conversion and temperature are given in Figure III.1. Being mildly exothermic, ammonia synthesis equilibrium conversions decrease as the temperatures increase where most industrial reactions are carried out (ca. 600 K). The favorable effect of increased pressure is clearly seen in Figure III.1.

The relationship between kinetics and thermodynamics should also be taken into account. Since this reaction is a reversible one, there is a trade-off between the temperature and pressure of the reaction. At low temperatures, the equilibrium conversions, or maximum attainable conversions are high, while the rates are low as dictated by Arrhenius relationship between the temperature and the reaction rate. Thus, an optimum temperature has to be found.

In Figure III.2, constant rate curves based on the modified Temkin equation are given. We will use these data to size our reactor. High temperatures are needed for faster kinetics. But as we go higher in temperature of the mildly exothermic reactions, we lose to the thermodynamics, that is, the constant rate curve starts to bend down, and the reverse component of the reaction rate starts to dominate.

If we are to design an adiabatic reactor, we should add the energy balance equation in Figure III.3. The energy balance for an adiabatic reactor is given by the line aa. Once the system heats up close to the equilibrium temperature, the rates slow down. The system must be moved away from the equilibrium by either providing heat exchange, which means a horizontal line at the exit conversion of the reactor (bb in Figure III.3a), or by cold feed injection which means an operating line with a positive slope (bb in Figure III.3b) bringing the reactor conversion after the cold feed injection to a somewhat lower value due to the increase in the amount

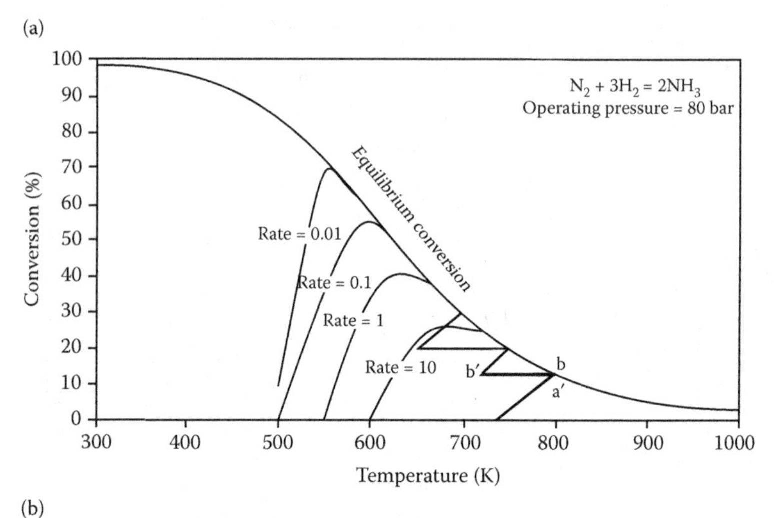

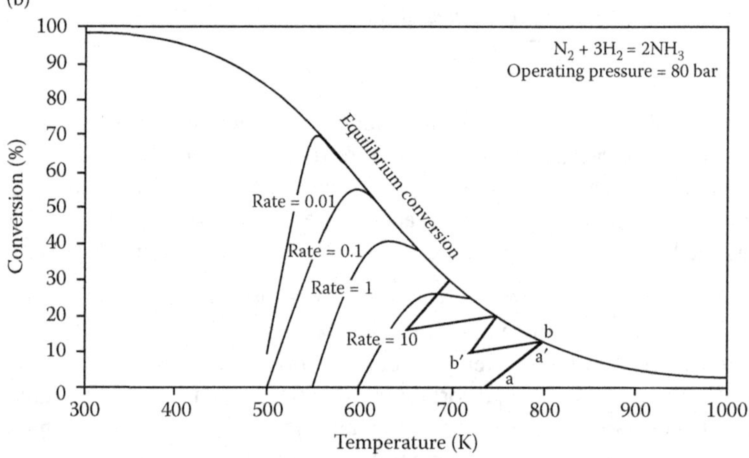

Figure III.3 Equilibrium conversion, constant rate, and energy balance curves for an adiabatic ammonia synthesis reactor: (a) intermediate cooling and (b) cold feed injection.

of the reactants. The final design is determined after the optimization of parameters such as the mode of heat exchange, the number of catalytic beds, the exit conversion, separation costs, and the compressor duty of the recycle stream.

References

Aslan M.Y. Ru based ammonia synthesis catalysts, MS thesis, Middle East Technical University Ankara, 2012.

Rossetti, I., Pernicone, N., Ferrero, F., and Forni, L., *Ind. Eng. Chem. Res.*, **45**, 4150, 2006.

Temkin, M.I., and Pyzhev, V. *Acta Physicochim.*, **12**, 327, 1940.

Chapter 5 Theory of chemical kinetics in bulk and on the surface

Chapter objectives

After successful completion of this chapter, the students are expected to

- Differentiate the collision theory and the transition state theory in terms of their postulates, assumptions, and limitations.
- Determine the kinetic parameters of elementary reactions from collision theory.
- Evaluate vibrational, rotational, and translational partition functions from the physical data and relate the numerical results to the kinetic parameters.
- Interpret the meanings of the order and the activation energy of a reaction.
- Propose and justify a mechanism for a given reaction with experimentally determined orders and activation energies.
- Apply pseudo-steady-state and pseudoequilibrium approximations for the derivation of rate expressions from the postulated reaction mechanisms.
- Derive the rate for a heterogeneously catalyzed reaction following Langmuir–Hinshelwood–Hougen–Watson (LHHW) kinetics.
- Perform microkinetic analysis for heterogeneously catalyzed reactions.
- Critically evaluate and justify the mechanism steps and their kinetic parameters.

Chemical kinetics

In the earlier chapters, we presented the reaction rates as a function of temperature and the concentration of the species involved. In this chapter, we will elaborate on the theories behind the concentration dependencies, temperature dependencies, and how the overall reaction mechanisms are proposed and elucidated. In such a context, it is reasonable to start from

Power law rate expression

a general description of the reaction rate:

$$\text{Rate} = k e^{-E_A/RT}[A]^{\alpha}[B]^{\beta} \tag{5.1}$$

We have already defined the rate constant and its relation to thermodynamics in Chapter 4. Here, we will delve into the two fundamental theories behind the overall rate expression given by Equation 5.1, the so-called collision theory and transition state theory. There are some excellent textbooks on chemical kinetics; the reader is directed to the list in Bibliography for further details.

It is not possible to cover all of the history or the theory of the chemical kinetics in the context of this chapter. However, the authors' intention is to give the student an essential minimum in the theory of chemical kinetics to be able to follow the literature and to incorporate in the design of the chemical reaction units. This chapter is divided into two sections: in the first part, the homogeneous kinetics will be covered in detail, covering the collision theory and the transition state theory for the determination of the rate constants and reaction rate expressions. Old but still valid approximations of pseudo-steady-state and pseudo-equilibrium concepts will be given with examples. In the second part, the heterogeneous reaction kinetics will be discussed from a mechanistic point of view.

Collision theory

Collision theory

The collision theory gives the fundamental perspective of how the colliding molecules give rise to the chemical conversion. The kinetic theory of gases gives us an estimate on the collision frequency or collision number as a function of mean molecular diameter, the temperature, and the reduced mass of the colliding particles as

$$Z = \sigma_{A,B}^2\left[8\pi k_B T\left(\frac{m_A + m_A}{m_A m_B}\right)\right]^{1/2} \tag{5.2}$$

where $\sigma_{A,B}$ is the mean molecular (collision) diameter of A and B and k_B is the Boltzmann constant (the ideal gas constant R divided by the Avagadro's number). The derivation of the above expression is based on the fact that the collision frequency between A and B molecules is proportional to the volume swept by molecule A in a given time multiplied by the concentration of B molecules. The volume swept by molecule A is estimated as the volume of a cylinder with a diameter equal to the mean molecular diameter, $\sigma_{A,B}$ multiplied by the Maxwell–Boltzmann velocity distribution given by

Maxwell–Boltzmann velocity distribution

$$v_{A,B} = \sqrt{\frac{8k_B T}{\pi\mu_{AB}}} \tag{5.3}$$

where

$$\mu_{AB} = \left(\frac{m_A m_B}{m_A + m_B}\right) \tag{5.4}$$

According to the collision theory, the preexponential factor k of Equation 5.1 is equal to Z given in Equation 5.2.

The shortcomings of the collision theory have been realized very early through experimentation, such that the preexponentials predicted from experimental data did not agree with the predictions of the theory. In most of the situations, the reaction rates were found to take place much more slowly; thus, a correction factor to the collision frequency, called the probability factor or steric factor, was added to the prediction. Thus, the final form of the rate constant k was as follows:

Preexponentials predicted by the collision theory are larger than the experimental values.

$$k = PZ\, e^{-E_A/RT} \tag{5.5}$$

Steric factor

The second and more important shortcoming of the collision theory was the thermodynamic consistency in the case of reversible reactions. When the equilibrium constant of the reaction $A_2 + B_2 \Leftrightarrow 2AB$ is predicted from the rate constants of the forward and reverse reactions, the equilibrium constant is evaluated as

Thermodynamic consistency problem of the reversible reactions in collision theory

$$K = \frac{k_1}{k_2} = e^{-E_f - E_r/RT} = e^{-\Delta H/RT} \tag{5.6}$$

When evaluating Equation 5.6, it must be noticed that due to the nature of the reaction stoichiometry, the collision numbers evaluated from Equation 5.2 should be identical for the forward and the reverse reactions. This can only be true at absolute zero or if the reaction does not involve any entropy change. But if the steric factor, P, is introduced in the definition of the preexponential, the consistency with the thermodynamics is regained, but the fundamental understanding is somewhat lost. The derivation of the relationship between the ratio of P_f/P_r and the entropy change is left to the reader and will be clear in the next section.

Transition state theory

The transition state theory requires the conception of an "activated complex" named by H. Eyring (1935) or "transition state" by M.G. Evans and M. Polanyi (1935). According to the transition state theory, the rate of a reaction is defined as

Transition state theory

$$\text{Rate} = c^{\ddagger\prime}\,\frac{\bar{v}}{\delta} \tag{5.7}$$

Here, $c^{\ddagger\prime}$ represents the number of activated complexes in unit volume lying in a length of δ representing the activated state on top of the barrier, and $\bar{v}$ is the mean velocity of crossing, such that $\bar{v}/\delta$ is the frequency at which the activated complex crosses the barrier. The activated complex molecules are considered having lost one vibrational degree of freedom and gained one translational degree of freedom in the reaction coordinate. Thus, their concentration is estimated as

Concentration of the activated complex

$$c^{\ddagger\prime} = c^{\ddagger} \frac{(2\pi \mu k_B T)^{1/2} \delta}{h} \tag{5.8}$$

The factor $(2\pi\mu k_B T)^{1/2}\ \delta/h$ is the partition function for the translation in the reaction path. So, combining Equations 5.3 and 5.8 in Equation 5.7 yields

$$\text{Rate} = c^{\ddagger} \frac{k_B T}{h} \tag{5.9}$$

where h is the Planck's constant. In Equation 5.9, a universal frequency was defined as $k_B T/h$, which is independent of the reactants and the type of the reaction but only depends on the temperature.

Now, we have to define the rate constant in terms of the transition state thermodynamic properties:

$$k = \frac{k_B T}{h} e^{-\Delta G^{\ddagger}/RT} \tag{5.10}$$

or

$$k = \frac{k_B T}{h} e^{-\Delta H^{\ddagger}/RT} e^{\Delta S^{\ddagger}/R} \tag{5.11}$$

Careful readers will remember that we have derived similar expressions for the overall thermodynamics of the reaction in Chapter 4. It is not possible to obtain the enthalpy, entropy, and Gibbs free energy change of the transition state molecule with respect to a standard reference state. On the other hand, it is possible to evaluate the equilibrium constant $K^{\ddagger}$ in terms of the partition functions of the species involved. Their fundamental derivation involves the methods of statistical mechanics that we do not intend to cover in detail. The partition function of a molecule/volume measures the probability of finding that molecule in a given volume and is equal to the sum of the $e^{-\varepsilon/k_B T}$ terms for all forms of the energy. The forms of the energy mentioned here include translational, vibrational, and rotational forms of energy of the molecular motion as well as nuclear and electronic energies possessed by the molecule. Each of these terms must be appropriately weighed according to the degeneracy of the particular energy level.

Partition function

timescales

At this point, it might be useful to remember the timescales of the processes during a catalytic reaction. The electronic processes of the potential energy surface of the reaction have characteristic times of 10^{-15} s, while the vibrational motions of the atoms are in the order of 10^{-12} s. The timescales of the bond formation and breaking of the catalytic processes are reported to be in the order of 10^{-4} to 10^{2} s (van Santen and Neurock, 2006). Under these circumstances, it is fair to assume that all vibrational motions are equilibrated with the exception of those on the reaction coordinate (this condition is satisfied when $E_A > 5k_BT$ (Kramers, 1940). Under these conditions, the rate at the transition state is expressed as

$$\text{Rate} = \Gamma \frac{k_B T}{h} \frac{q^{\ddagger}}{q_0} e^{-(E_b - E_0)/RT} \tag{5.12}$$

where $E_b - E_0$ refers to the barrier height of the reaction. Γ is the transmission factor, and it can safely be taken as one unless the tunneling effects are important. The other condition is that if the solution is viscous or diffusion limitations are present, Γ can be less than 1. $q^{\ddagger}$ and q_0 are the partition functions of the transition state and the initial state, respectively.

The partition function can be represented as the multiple of vibrational, rotational, and translational vibrational partition functions as

$$q = q_{\text{trans}} q_{\text{rot}} q_{\text{vib}} \tag{5.13}$$

with

$$q_{\text{trans}} = \left(\frac{2\pi\mu k_B T}{h^2} \right)^n \tag{5.14}$$

where n is the degree of freedom of the translational motion (for a gas phase species, n is 3/2). For a AB-type heterodiatomic molecule, the rotational partition function is

$$q_{\text{rot}} = \frac{8\pi^2 I k_B T}{h^2} \tag{5.15}$$

where I is the moment of inertia μR_{eq}^2, μ the reduced mass, and R_{eq} the atomic distance. For harmonic frequencies, the vibrational partition functions are

$$q_{\text{vib}} = \prod_i \left(\frac{1}{1 - e^{-h\nu/k_B T}} \right) \tag{5.16}$$

These equations will allow us to predict the rate constants of the individual reaction steps in a complex mechanism, such as in combustion or in a catalytic reaction. We refer the curious reader to texts in physical chemistry or kinetics (e.g., Laidler, 1987) for detailed examples on how to apply the transition state theory (TST) for the kinetic parameter estimation.

Proposing a kinetic model

NIST databases

Proposing a kinetic model is important in successful design of a reactor. The kinetic models of various gas phase and combustion reactions are available in databases. One of the most frequently used and widely covered databases is NIST (http://kinetics.nist.gov/CKMech/). It is possible to reach a vast amount of data at the NIST website provided that the kinetic mechanism is obtained in the temperature and pressure range of interest. The models available in these databases can be utilized as they are available, or they can be modified according to the needs of the engineer. When modifying an existing reaction model, one must be careful about satisfying the thermodynamic consistency tests.

For the surface reactions, a database is not yet available. However, one can easily start from the gas phase mechanisms, if available, and propose a surface reaction mechanism analogous to the gas phase mechanism in fundamental respects. One can take a gas phase mechanism and, carefully keeping thermodynamic consistency in mind, modify a mechanism to represent a surface reaction. A more rigorous approach is to use density functional theory (DFT) when looking for a plausible mechanism. Although the method offers fundamental rigor in estimating the transition states and discriminating the mechanism steps, the CPU demands and challenges faced with defining the catalyst surface are the major drawbacks. Interested reader can refer to van Santen and Neurock (2006), which gives a nice overview of the DFT as it is applied to the catalytic reactions.

Thermodynamic consistency tests

Once the model is proposed, it has to be tested against the experimental data. The experimental data can be in the form of conversion versus temperature or concentration versus temperature or the empirical rate expression may have been obtained. But here, we will first describe the classical analytical tools for obtaining the rate from a mechanism. The thermodynamic consistency tests involve the determination of the overall enthalpy change and overall entropy change of the mechanism obtained from the enthalpy and entropy changes of the individual steps. Once these changes are calculated, they are compared against the tabulated thermodynamic data. A nice recent example of the thermodynamic consistency test was given by Mhadeswar and Vlachos (2005a) for CO oxidation reaction over Pt surfaces.

Example 5.1: Bodenstein reaction

We will focus our attention to the Bodenstein reaction, a historically important reaction due to the fact that for the first time a nonelementary kinetic mechanism was postulated (Laidler, 1987):

$$H_2 + I_2 \rightarrow 2HI \qquad \text{(R1)}$$

Bodenstein approximations

Obtain a rate expression for the following mechanism:

1. $I_2 \rightleftharpoons 2I$	fast	(R2)
2. $I_2 + H_2 \rightleftharpoons HI + H$	fast	(R3)
3. $H + I_2 \rightarrow HI + I$	slow	(R4)

SOLUTION

For the fast steps, we can invoke pseudoequilibrium such that

$$[I] = (K_1[I_2])^{1/2} \quad (E5.1.1)$$

$$[H] = \frac{K_2[I][H_2]}{[HI]} = K_2(K_1[I_2])^{1/2}\frac{[H_2]}{[HI]} \quad (E5.1.2)$$

The rate of the reaction is determined from the slow step as

$$\text{Rate} = k_3[H][I_2] = k_3K_1^{1/2}K_2[I_2]^{3/2}\frac{[H_2]}{[HI]} \quad (E5.1.3)$$

It is clear that the rate is not proportional to $k[H_2][I_2]$ due to the nonelementary kinetics.

Exercise: By choosing different steps as slow or fast, derive alternative rate expressions.

Brief excursion for the classification of surface reaction mechanisms

The construction of a model based on one of the adsorption–reaction–desorption steps being the limiting step constitutes the core of the semiempirical approach considered in this section. In this approach, the microscopic origins of the observed macroscopic effects of catalysts (as described by many authors, e.g. Plath, 1989) are ignored. The models thus developed are commonly known as Langmuir–Hinshelwood models among chemists and as Hougen–Watson models among chemical engineers. We choose to call them *Langmuir–Hinshelwood–Hougen–Watson models.*

Langmuir–Hinshelwood–Hougen–Watson models

Langmuir–Hinshelwood–Hougen–Watson models

In the interest of generality, we consider hypothetical reactions and derive rate equations for a few typical *LHHW models.* As the Langmuir isotherm is the basis of all LHHW models, we begin by a simple derivation of this isotherm.

Langmuir isotherm Unlike in homogeneous reactions where the rate is proportional to the reactant concentration (say, $[A]$), in catalytic reactions, it is proportional to the surface concentration $[A]_s$. Since $[A]_s$ is not usually known, it is convenient to express it in terms of $[A]$ by equating the rates of adsorption and desorption for the reaction:

$$A + s \rightleftharpoons A_s \quad (R5)$$

$$k_{Aa}[A][s]_v = k_{Ad}[A]_s \quad (5.17)$$

Giving

Fractional coverage

$$\theta_A = K_A[A]\theta_v = K_A[A](1 - \theta_A) \tag{5.18}$$

where θ_A and θ_v are, respectively, and the fractions of adsorbed A and vacant site (s) on the surface are given by

$$\theta_A = \frac{[A]_s}{[s]_t}, \quad \theta_v = \frac{[s]_v}{[s]_t} \tag{5.19}$$

thus

$$\theta_A = \frac{K_A[A]}{1 + K_A[A]}, \quad \theta_v = \frac{1}{1 + K_A[A]} \tag{5.20}$$

where other components besides A are also adsorbed, such as B, R, I (inert), Equation 5.20 becomes

$$\theta_A = \frac{K_A[A]}{1 + K_A[A] + K_B[B] + K_R[R] + K_I[I]} \tag{5.21a}$$

$$\theta_v = \frac{1}{1 + K_A[A] + K_B[B] + K_R[R] + K_I[I]} \tag{5.21b}$$

To use these expressions in formulating LHHW rate models, an understanding of the "slowest" or the rate-determining step is necessary.

Rate-determining step

Rate-determining step The basic assumption of the LHHW models is that the slowest of many possible steps involving adsorption, surface reaction, and desorption is rate-controlling. It is helpful to regard the rate-limiting step as the one that consumes most of the available driving force. A clarifying analogy is that of a current passing through a set of resistances in series: Although the current (corresponding to the rate) is the same, the conductivity of any one of the resistors can be lower than that of the others, making it the rate-limiting resistor.

We now develop rate equations by assuming any one of several steps involved in a given reaction as the rate-determining step.

Basic procedure Consider the reaction

$$A + B \rightleftharpoons P + R \tag{R6}$$

The various steps comprising the reaction are as follows.

Adsorption of A and B:

$$A + s \rightarrow A_s, \quad B + s \rightarrow B_s \tag{R7.1}$$

Surface reaction:

$$A_s + B_s \rightarrow P_s + R_s \tag{R7.2}$$

Desorption of P and R:

$$P_s \rightarrow P + s, \quad R_s \rightarrow R + s \tag{R7.3}$$

According to the concept of the rate-determining step, any one of these steps can be controlling.

Focusing now on the case where surface reaction is controlling, the basic procedure in developing an LHHW model is to write the rate equation in terms of the surface coverage θ_A of reactant A rather than its concentration $[A]$. Sometimes, as in reactions requiring a second (vacant) site for adsorbing a product (e.g., $A \rightarrow P + R$), the rate will also directly depend on the fraction of surface covered by vacant sites θ_v; and when there is dissociation of a reactant, a pair of adjacent vacant sites should be available, so that the rate of adsorption would now be proportional to θ_v^2 rather than θ_v. One of the characteristics of the surface reactions is that in most of the situations the adsorption is much faster than the rest of the steps. In such a situation, we can easily assume that the adsorption step is at pseudoequilibrium, indicating that the rate of adsorption is equal to the rate of desorption:

$$A + s \rightleftharpoons A_s$$

$$r_{\text{ads}} = k_f[A]\theta_v \tag{5.22}$$

$$r_{\text{des}} = k_r\theta_A \tag{5.23}$$

Pseudoequilibrium hypothesis: For that particular step of the reaction, forward and reverse rates are so fast that the reaction step is taken to be at equilibrium.

For this simple case, the coverage of A is simply what can be predicted from Langmuir adsorption isotherm (Equation 5.20). Another very frequently encountered situation is the dissociative adsorption, or adsorption requiring more than one site.

$$A_2 \rightleftharpoons 2A$$

$$r_{\text{ads}} = k\theta_A\theta_v^2 \tag{5.24}$$

$$r_{\text{des}} = k_r\theta_A^2 \tag{5.25}$$

such that the coverage relates to $[A_2]$ as follows:

$$\theta_A = \frac{(K_A[A_2])^{1/2}}{1 + (K_A[A_2])^{1/2}} \tag{5.26}$$

Let us now turn to the case where adsorption of one of the components is controlling. Thus, consider the reaction

$$A \rightleftharpoons R \tag{R8}$$

with adsorption of A controlling. Clearly, adsorption equilibrium does not exist, and we cannot use Equation 5.20 to get θ_A. Instead, since now the reaction would be in equilibrium, we use

$$-r_{wA} = k_{Aa}\left(\theta_A - \frac{\theta_R}{K'}\right) \tag{5.27}$$

where K' is the equilibrium constant for the surface reaction. This constant must be distinguished from the true thermodynamic equilibrium constant K given by $[R]/[A]$. Simple algebraic manipulations then lead to expressions for θ_A, θ_B, θ_R, θ_S, and θ_v. Using these, the final expression for the rate can be developed.

Bodenstein approximation: At some point during the reaction, the rate of formation and the rate of disappearance of an intermediate are identical causing a steady-state concentration, or surface coverage, of the reactive intermediate.

There may be situations where the rates of the individual steps are all of the same order of magnitude and we could not invoke the pseudoequilibrium hypothesis. In such a case, we can use the pseudo-steady-state hypothesis, or Bodenstein (1927) approximation: At some point during the reaction, the rate of formation and the rate of disappearance of an intermediate are identical causing a steady-state concentration, or surface coverage, of the reactive intermediate.

$$\frac{d\theta_A}{dt} = 0 \tag{5.28}$$

All LHHW models can be consolidated into a single general form

General form of LHHW

$$\text{Rate} = \frac{(\text{kinetic term})\,(\text{potential term})}{(\text{adsorption term})^n} \tag{5.29}$$

in which the exponent n in the adsorption term denotes the number of sites participating in the catalysis. Yang and Hougen (1950) list the various terms for several classes of reactions. It is a relatively simple matter to construct a full LHHW model from this table. A more elaborate method of accounting for all possible models (often over a hundred) for a given reaction has been proposed by Barnard and Mitchell (1968).

Example 5.2: N_2O + CO mechanism

The kinetics of the reaction between N_2O and CO was studied by McCabe and Wong (1990) between 550 and 700 K and reactant partial pressures of 0.6 and 7 Torr. They measured the apparent reaction orders of -1 ± 0.15 in CO partial pressure and 0.65 ± 0.1 in N_2O partial pressures under differential reaction conditions at temperatures between 564 and 583 K. Derive a mechanism such that orders are consistent with the experimental values.

SOLUTION

Uner (1998) has postulated the following mechanism:

1. $N_2O + * \rightleftharpoons N_2O*$
2. $N_2O* + * \rightleftharpoons N* + NO*$
3. $NO* + * \rightleftharpoons N* + O*$
4. $CO + 2* \rightleftharpoons *CO*$ (*the bridge-bonded species*)

or alternatively

4a. $CO + * \rightleftharpoons CO*$
4b. $CO* + * \rightleftharpoons *CO*$
5. $2N* \rightarrow N_2 + 2*$
6. $*CO* + O* \rightarrow CO_2 + 3*$

The following assumptions were made:

1. All the steps that are shown to be reversible are in equilibrium.
2. Bridge-bonded CO is the reactive one.
3. N_2O dissociatively interacts with CO and this dissociation is a two-step process.
4. Surface is mostly covered by CO.
5. The rate-determining step is the dissociation of surface NO into surface N and O.

These assumptions along with the elementary reactions lead to relationships such as

$$k_1 P_{N_2O}\theta_v = k_{-1}\theta_{N_2O} \quad \text{(E5.2.1)}$$

$$k_2\theta_{N_2O}\theta_v = k_{-2}\theta_{NO}\theta_N \quad \text{(E5.2.2)}$$

$$\text{Rate} = k_3\theta_{NO}\theta_v \quad \text{(E5.2.3)}$$

From Equations E5.2.1 and E5.2.2, we obtain the surface coverage of NO in terms of the partial pressure of N_2O:

$$\theta_{NO} = \frac{K_1 K_2 P_{N_2O}\theta_v^2}{\theta_N} \quad \text{(E5.2.4)}$$

where $K = k_i/k_{-i}$.

To determine the coverage of surface N, the steady-state condition between the two products must be used. In other words, the rate of CO_2 production must be equal to the rate of N_2 production, such that

$$k_5\theta_N^2 = k_6\theta_{CO}\theta_O \quad \text{(E5.2.5)}$$

As a result of assumption 4, the surface coverage of carbon monoxide is taken as 1, and the resulting expression for the surface coverage of atomic nitrogen is obtained as

$$\theta_N = \left(\frac{k_6\theta_O}{k_5}\right)^{1/2} \tag{E5.2.6}$$

A steady-state balance for the surface oxygen species results in the following expression:

$$\theta_O = \frac{k_5\theta_{NO}\theta_v}{k_6\theta_{CO}} \cong \frac{k_3\theta_{NO}\theta_v}{k_6} \tag{E5.2.7}$$

for surface oxygen coverage. Substituting Equation E5.2.7 into Equation E5.2.6, we obtain

$$\theta_N = \left(\frac{k_3\theta_{NO}\theta_v}{k_5}\right)^{1/2} \tag{E5.2.8}$$

To determine the vacant size concentration, we will use assumptions 2 and 4 (i.e., the surface is nearly saturated with CO and the reactive form of CO is bridge bonded). If we assume that the adsorption of CO takes place via step 4 in the mechanism, the CO coverages can be predicted from

$$K_4 P_{CO}\theta_v^2 = \theta_{CO} \tag{E5.2.9}$$

On the other hand, we may choose steps (R4a) and (R4b) as CO adsorption pathways to bridge-bonded species. In such a case, the coverage of CO can be determined from

$$K_{4a}K_{4b}P_{CO}\theta_v^2 = \theta_{CO} \tag{E5.2.10}$$

Equations E5.2.9 and E5.2.10 are equivalent in terms of the coverage and partial pressure functionality. Therefore, the simpler form (i.e., Equation E5.2.9) will be used in the analysis. The vacant site concentration will be determined from Equation E5.2.9 and the site balance equation

$$\theta_v + \theta_{CO} = 1 \tag{E5.2.11}$$

The following second-order polynomial can be obtained in terms of the vacant site concentration

$$\theta_v^2 + (K_4P_{CO})^{-1}\theta_v - (K_4P_{CO})^{-1} = 0 \tag{E5.2.12}$$

The physically meaningful root of Equation E5.2.12 is

$$\theta_v = \frac{-1 + (1 + 4K_pP_{CO})^{1/2}}{2K_4P_{CO}} \tag{E5.2.13}$$

The value of K_4 was estimated in the orders of 10^{-2}–10^{-3} Torr^{-1} from the adsorption rate data (McCabe and Wong, 1990) and

Table 5.1 Activation Energy Data for the Proposed Model

Step	E_{af} (kcal/mol)	Reference	E_{ar} (kcal/mol)	Reference
1	0	McCabe and Wong (1990)	5	McCabe and Wong (1990)
2	0	Lombardo and Bell (1991)	21	Lombardo and Bell (1991)
3	19	Lombardo and Bell (1991)	—	—
4	0	McCabe and Wong (1990)	18–31	Thiel et al. (1979)
				Batteas et al. (1993)
	Bridge bonded		41.5 ± 1.0	Broadbelt and Rekoske (1997)
5	31	Belton et al. (1993)	—	—
6	24–27	Shoustorowich (1986)	—	—

desorption activation energy of CO given in Table 5.1. Therefore, $(4K_4P_{CO})^{1/2} \gg 1$, which simplifies Equation E5.2.13 to

$$\theta_v = (K_pP_{CO})^{-1/2} \quad \text{(E5.2.14)}$$

Substituting Equation E5.2.14 into Equation E5.2.8 and then both in Equation E5.2.4 yields the coverage of NO as

$$\theta_{NO} = \frac{[K_1K_2P_{N_2O}/(k_3/k_5)^{1/2}]^{2/3}}{(K_4P_{CO})^{1/2}} \quad \text{(E5.2.15)}$$

and the rate of nitrous oxide decomposition is obtained as

$$\text{Rate} = \frac{(k_3k_5{}^{1/2}K_1K_2)^{2/3}K_4^{-1}P_{N_2O}^{2/3}}{P_{CO}} \quad \text{(E5.2.16)}$$

Eley–Rideal mechanism

Eley–Rideal mechanism

Consider reaction R1 again:

$$A + B \rightleftharpoons P + R \quad \text{(R9)}$$

It is also possible that one of the reactants, say B in the above reaction, is not adsorbed. In such a mechanism (known as the Eley–Rideal mechanism), we simple use p_B or $[B]$ for B (and not θ_B). While the LHHW mechanism requires the adsorption of all reactants on the surface, the Eley–Rideal mechanism proceeds with one adsorbed reactant and one gas phase species. Depending on the interaction between the adsorbate and the adsorbent, one of the species may be so weakly bound to the surface that it is essentially not adsorbed. Furthermore, some of the reactions may proceed via a nonadsorbed intermediate. In addition to catalytic reaction kinetics, the Eley–Rideal mechanism is frequently encountered during the crystal growth processes.

Mars–van Krevelen mechanism

Mars–van Krevelen mechanism

This particular mechanism involves the red-ox modification of the catalyst. Usually the reactant is oxidized by the lattice oxygen of the catalyst, which is regenerated at the end of the catalytic cycle. Similar mechanisms also prevail with the compounds of sulfur and nitrogen such as during hydrodesulfurization or hydrodenitrogenation. The principle is that the lattice is depleted in one of the components, such as an oxygen atom, and it is replaced during the reaction. The control of selectivity in partial oxidation reactions is usually easier when the lattice oxygen is used.

Example 5.3: Use of CeO_2 as an oxygen pump

The emission control catalysts used in the exhaust mufflers have precious metals supported over oxides. The support oxides serve several functions, including providing mechanical stability and necessary surface acidity. But one of the oxides, CeO_2, is placed in the catalytic converters to serve as oxygen pump between the gas phase oxygen and surface reaction taking place over the precious metal.

Oran and Uner (2004) have carried out extensive reaction tests and determined the reaction orders of CO oxidation as –2 and 1 with respect to CO and oxygen over Pt/γ-Al_2O_3, while the corresponding values were reported as –1 and 0 over Pt/CeO_2 and Pt/CeO_2/γ-Al_2O_3. Propose and justify a mechanism for this system.

SOLUTION

Over Pt/γ-Al_2O_3, the assumptions are listed below:

i. The surface is poisoned by the adsorbed CO atoms, that is, CO coverage is nearly 1.
ii. The rate-limiting step is the dissociative adsorption of oxygen.

The surface reactions are depicted as follows:

1. $CO + * \rightarrow CO*$
2. $O_2 + 2* \rightarrow 2O*$
3. $CO* + O* \rightarrow CO_2 + 2*$

As a result of the first assumption, the vacant site coverage could be written as

$$\theta_v = 1 - \theta_{CO} \quad \text{(E5.3.1)}$$

and the second assumption leads to

$$\text{Rate} = k_2 P_{O_2} \theta_v^2 \quad \text{(E5.3.2)}$$

Given that the first step proceeds under pseudoequilibrium conditions

$$\theta_{CO} = \frac{K_1 P_{CO}}{1 + K_1 P_{CO}} \quad \text{(E5.3.3)}$$

such that the reaction rate takes the following form:

$$\text{Rate} = \frac{k_2 P_{O_2}}{(1 + K_1 P_{CO})^2} \quad \text{(E5.3.4)}$$

For Pt/CeO_2 or $Pt/CeO_2/\gamma\text{-}Al_2O_3$ catalysts, the assumptions are

i. The Pt surface is poisoned by the adsorbed CO atoms, that is, CO coverage over Pt is nearly 1.
ii. The rate-limiting step is the oxygen exchange at the ceria and Pt interface.
iii. Even in the absence of gas phase oxygen, ceria surface is oxidized; therefore, the surface coverage of oxygen over ceria can be taken as independent of the gas phase oxygen partial pressure.

The surface reactions are depicted as follows:

4. $CO + * \rightarrow CO*$
5. $O_2 + 2\otimes \rightarrow 2O\otimes$
6. $O\otimes + * \rightarrow O* + \otimes$
7. $CO* + O* \rightarrow CO_2 + 2*$

where * denotes a catalytic site on the Pt surface and ⊗ denotes a site over ceria surface. As a result of the first assumption, the vacant site coverage over Pt could be written as

$$\theta_v = 1 - \theta_{CO} \quad \text{(E5.3.5)}$$

And the second assumption leads to

$$\text{Rate} = k_2 \theta_{O\otimes} \theta_v \quad \text{(E5.3.6)}$$

The third assumption renders the surface coverage of oxygen over ceria almost constant such that

$$\text{Rate} = k_2' \theta_v \quad \text{(E5.3.7)}$$

where

$$k_2' = k_2 \theta_{O\otimes} \quad \text{(E5.3.8)}$$

Given that the first step proceeds under pseudoequilibrium conditions

$$\theta_{CO} = \frac{K_1 P_{CO}}{1 + K_1 P_{CO}} \quad \text{(E5.3.9)}$$

such that the reaction rate takes the following form:

$$\text{Rate} = \frac{k_2'}{1 + K_1 P_{CO}} \quad \text{(E5.3.10)}$$

Michelis–Menten mechanism

Enzyme-catalyzed biochemical reactions generally follow the Michelis–Menten mechanism. The mechanism is conceptually similar to the mechanisms discussed so far, with the exception that the solid catalyst is replaced by the enzyme. Here we will see the reaction between an enzyme, E, performing the catalytic action, a substrate, S being converted onto the product, P. The reaction steps are

$$E + S \Leftrightarrow ES \tag{R10}$$

$$ES \rightarrow E + P \tag{R11}$$

Similar to the pseudoequilibrium and rate-determining step hypotheses done so far, we assume that reaction R10 is fast and can be easily treated at equilibrium, while reaction R11 is the rate-determining step. Given that the total enzyme concentration is constant, we can write

Michelis–Menten mechanism

$$[E]_0 = [E] + [ES] \tag{5.30}$$

The concentration of $[ES]$ complex can be determined by assuming pseudo-steady state for the species, that is, the rate of formation and the rate of depletion of $[ES]$ must be equal

$$k_1[E][S] = k_{-1}[ES] + k_{\text{cat}}[ES] \tag{5.31}$$

or

$$\frac{[E][S]}{[ES]} = \frac{k_{-1} + k_{\text{cat}}}{k_1} = K_m \tag{5.32}$$

Combining Equations 5.30 and 5.32, we obtain for the substrate concentration

$$[ES] = \frac{[E]_0[S]}{K_m + [S]} \tag{5.33}$$

The rate of formation of P can thus be written as

$$r_P = k_{\text{cat}}[ES] = k_{\text{cat}}\frac{[E]_0[S]}{K_m + [S]} \tag{5.34}$$

This general expression is widely applicable to most enzyme-catalyzed reactions that take place in a single-step reaction.

Influence of surface nonideality

When proposing a surface reaction model, several precautions must be taken. The models themselves may suffer from a few genetic defects and have therefore been the subject of some criticism and much commentary. The chief limitations are

1. All sites are equally active, with equal heats of adsorption; this is not true since there is usually a distribution of activity on the surface, which is ignored in the LHHW models.
2. Interaction between adsorbed molecules is negligible; again, not true.
3. Molecules are always adsorbed at random on the surface; also, not true. Similar molecules may tend to adsorb in contiguity forming their own islands; this leads to a completely different mechanism of surface reaction.

A number of isotherms that dispense with assumptions (1) and (2) have been proposed (see Doraiswamy, 1991). Some studies (e.g., Kiperman et al., 1989) indicate that it may not be possible to model certain reactions without invoking the role of surface nonideality. Fortuitously, the use of more rigorous isotherms does not materially affect the companion problem of the diffusion–reaction behavior of systems (Shendye et al., 1993)—a topic considered in the next section.

Paradox of heterogeneous kinetics An interesting feature of LHHW kinetics is worth noting. Many reactions on surfaces known to be nonideal surprisingly follow the ideal LHHW models, a situation that can only be described as the "paradox of heterogeneous kinetics" (see Boudart, 1986). In the same vein but with less justification, it has also been argued for over four decades—for example, from Weller (1956) to Bouzek and Roušar (1996)—that rate data for a given reaction can be correlated equally well by simple power law kinetics (thus dispensing with the surface science approach altogether). In general, LHHW models supplemented by rigorous methods of parameter estimation do represent a valid mechanistic approach that can be accepted as a reasonably sound basis for reactor design. A more puristic approach would, however, require a firmer anchoring to the methods of surface science.

Microkinetic analysis

The microkinetic analysis is a relatively new tool in identifying the structure–activity relationships. The seminal textbook of Dumesic et al. (1993) and several review articles (e.g., Stoltze, 2000; Waugh, 1999) describe the merits and the methods in great detail. The method is a powerful tool in catalyst design, requiring *a priori* knowledge or a postulate of the surface intermediates and the capability of the estimation of the individual rate parameters, the activation energy, and the preexponential in each surface step. Once the mechanism is constructed based on the spectroscopic evidence for the surface intermediates, and experimentally measured or theoretically quantified rate constants, the surface reaction mechanism can be solved without any choice of the rate-determining step, or a postulate of the pseudoequilibrium.

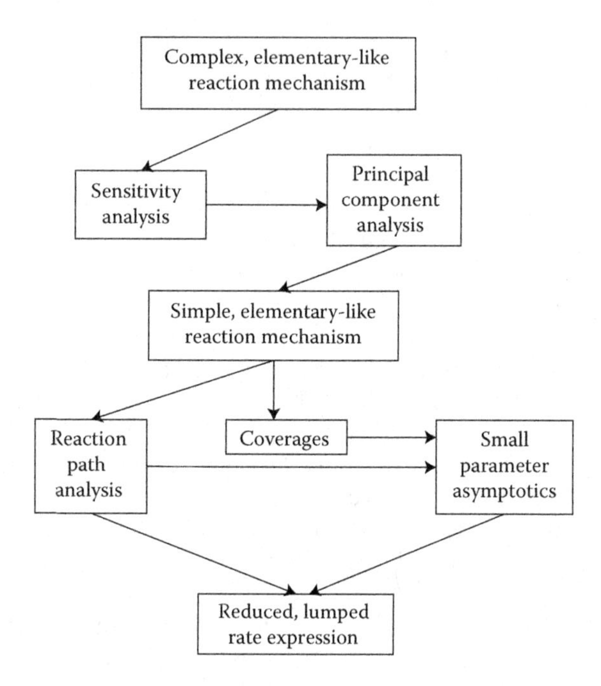

Figure 5.1 Flowchart of the mechanism reduction methodology. (Adapted from Mhadeswar, A.B. and Vlachos, D.G., *Comb. Flame,* **142**, 289, 2005a.)

The rate constants of the individual steps are strong functions of temperature. With changing temperatures, the rate constants change and the same mechanism step can be rate determining at one temperature, while it can approach equilibrium at a different temperature.

The detailed understanding at the molecular level can provide tremendous insights into designing the reactors in terms of the operational parameters, such as temperature, pressure, concentration of the reactants, and the flow characteristics. But it must be carefully kept in mind that the time- and lengthscales of the reactions at the molecular level and reactors operating in industrial scale could be different by several orders of magnitude (Maestri, 2012). The concept of "seamless chemical engineering" mentioned briefly in the Overview chapter becomes very important, especially when transferring kinetic information from the molecular level to the design problem to be carried out at the metric scale. Since the mechanism postulation and reduction is a formidable task, one of the heuristics is presented here for reference (Figure 5.1).

The difficulty of combining molecular rigor with macroscopic accuracy makes it rather difficult to combine a design protocol with microkinetic analysis. However, with better understanding of the surface phenomena, better catalyst formulations and better choices of operational parameters will be possible. As the methodology is already clearly laid out by an earlier book by Dumesic and coworkers (1993), we will summarize here some characteristics of the microkinetic analysis.

Postulate a mechanism

The first step of the microkinetic analysis is to postulate an accurate reaction network, composed of elementary reactions. To establish the independence of the reaction network is important. One of the rules of thumb is to determine the rank of the stoichiometric coefficient matrix as we have covered in Chapter 2. The second rule of thumb is to have as many surface reactions as the surface intermediates and making sure that for each reaction step, there is one new chemical species. Then, it is possible to construct the relationship with the gas phase by invoking the adsorption–reaction–desorption steps as mentioned earlier in this chapter.

Determine the kinetic parameters

The second step is to find good estimates of the rate parameters. The rate parameters can be obtained from collision theory, transition state theory, as well as first principles calculations such as DFT. Calorimetric measurements of heats of adsorption is possible for the surface intermediates with a gas phase precursor. Otherwise, the surface energetic must be estimated. As we have mentioned earlier, the computational cost of DFT is overriding its utility and accuracy in the present-day capabilities. Eventually, the parameter space must be constructed with two major constraints. The first constraint requires the consistency with the thermodynamics and the second constraint requires that the macroscopic rate data can be reproduced. Unity bond index-quadratic exponential potential (UBI-QEP) method of Shustorovich (1986, 1998) offers a relatively accurate and affordable estimation of the surface energetics.

Simplify the mechanism

The detailed mechanism in most cases will be too complicated to be handled effectively, especially when macroscopic phenomena is under scrutiny. Therefore, the mechanism should be reduced carefully to a manageable size, systematically. Sensitivity analysis based on a constraint and a choice of parameters will render some of the mechanism steps ineffective in the overall analysis, similar to the pseudoequilibrium hypothesis done in earlier kinetic analysis work. But this time, elimination is based on some rigorous analysis with substantial information on the kinetics, and not on a simplifying assumption to be validated against data fitting. For sensitivity analysis, one has to select model responses, such as conversion, selectivity, and rate. Then, the sensitivity of the model response to the parameters is analyzed. For example, the sensitivity analysis of reaction rate r_i with respect to the Arrhenius preexponentials can be done by constructing a sensitivity matrix with the elements of

Table 5.2 Principal Component Analysis (PCA) Results

Eigenvalues	First Eigenvector	Second Eigenvector	Third Eigenvector	Reaction Pair Numbers	Reaction Pair
2.1×10^{0}	$\mathbf{-2.6 \times 10^{-4}}$	$\mathbf{2.7 \times 10^{-2}}$	$\mathbf{7.4 \times 10^{-2}}$	$\boldsymbol{R_1}$–$\boldsymbol{R_2}$	$H_2 + 2^* \leftrightarrow 2H^*$
3.9×10^{-4}	1.5×10^{-7}	-6.2×10^{-6}	4.4×10^{-6}	R_3–R_4	$O_2 + 2^* \leftrightarrow 2O^*$
3.1×10^{-5}	6.3×10^{-8}	2.3×10^{-6}	-1.6×10^{-6}	R_5–R_6	$OH +^* \leftrightarrow H^* + O^*$
1.7×10^{-10}	$\mathbf{-1.6 \times 10^{-2}}$	$\mathbf{5.4 \times 10^{-1}}$	$\mathbf{-8.3 \times 10^{-1}}$	$\boldsymbol{R_7}$–$\boldsymbol{R_8}$	$H_2O^* + {}^* \leftrightarrow H^* + OH^*$
1.2×10^{-13}	7.0×10^{-8}	2.8×10^{-6}	-2.5×10^{-6}	R_9–R_{10}	$H_2O^* + O^* \leftrightarrow 2OH^*$
4.8×10^{-16}	3.2×10^{-8}	1.3×10^{-6}	-1.3×10^{-6}	R_{11}–R_{12}	$OH + {}^* \leftrightarrow OH^*$
Very small	$\mathbf{-3.2 \times 10^{-5}}$	$\mathbf{1.8 \times 10^{-4}}$	$\mathbf{2.8 \times 10^{-4}}$	$\boldsymbol{R_{13}}$–$\boldsymbol{R_{14}}$	$H_2O + {}^* \leftrightarrow H_2O^*$
eigenvalues	3.2×10^{-9}	1.5×10^{-7}	-4.6×10^{-7}	R_{15}–R_{16}	$H +^* \leftrightarrow H^*$
	6.1×10^{-8}	2.2×10^{-6}	-1.5×10^{-6}	R_{17}–R_{18}	$O + {}^* \leftrightarrow O^*$
	-5.2×10^{-8}	$\mathbf{-1.6 \times 10^{-4}}$	$\mathbf{-4.7 \times 10^{-5}}$	$\boldsymbol{R_{19}}$–$\boldsymbol{R_{20}}$	$CO + {}^* \leftrightarrow CO^*$
	-1.6×10^{-6}	$\mathbf{-3.8 \times 10^{-5}}$	$\mathbf{1.2 \times 10^{-5}}$	$\boldsymbol{R_{21}}$–$\boldsymbol{R_{22}}$	$CO_2 + {}^* \leftrightarrow CO_2^*$
	5.6×10^{-8}	2.5×10^{-6}	-2.1×10^{-6}	R_{23}–R_{24}	$CO_2^* + {}^* \leftrightarrow CO^* + O^*$
	$\mathbf{-4.7 \times 10^{-2}}$	$\mathbf{-6.4 \times 10^{-1}}$	$\mathbf{-4.8 \times 10^{-1}}$	R_{25}–$\boldsymbol{R_{26}}$	$CO_2^* + H^* \leftrightarrow CO^* + OH^*$
	-1.8×10^{-7}	-7.1×10^{-6}	5.4×10^{-6}	R_{27}–R_{28}	$COOH + {}^* \leftrightarrow COOH^*$
	$\mathbf{-3.5 \times 10^{-2}}$	$\mathbf{-5.5 \times 10^{-1}}$	$\mathbf{-2.6 \times 10^{-1}}$	$\boldsymbol{R_{29}}$–$\boldsymbol{R_{30}}$	$COOH^* + {}^* \leftrightarrow CO_2^* + H^*$
	$\mathbf{-2.2 \times 10^{-4}}$	$\mathbf{-3.7 \times 10^{-3}}$	$\mathbf{9.6 \times 10^{-4}}$	$\boldsymbol{R_{31}}$–$\boldsymbol{R_{32}}$	$COOH^* + {}^* \leftrightarrow CO_2^* + H^*$
	$\mathbf{-1.0 \times 10^{0}}$	$\mathbf{4.0 \times 10^{-2}}$	$\mathbf{4.5 \times 10^{-2}}$	$\boldsymbol{R_{33}}$–$\boldsymbol{R_{34}}$	$CO^* + H_2O^* \leftrightarrow COOH^* + H^*$
	4.3×10^{-8}	1.7×10^{-6}	-1.6×10^{-6}	R_{35}–R_{36}	$CO_2^* + OH^* \leftrightarrow COOH^* + O^*$
	-1.8×10^{-7}	-7.0×10^{-6}	5.4×10^{-6}	R_{37}–R_{38}	$CO_2^* + H_2O^* \leftrightarrow COOH^* + OH^*$
	2.7×10^{-8}	2.4×10^{-7}	2.5×10^{-7}	R_{39}–R_{40}	$HCOO + 2^* \leftrightarrow HCOO^{**}$
	-1.5×10^{-7}	-5.8×10^{-6}	4.0×10^{-6}	R_{41}–R_{42}	$CO_2^* + H^* \leftrightarrow COOH^*$
	5.2×10^{-8}	2.0×10^{-6}	-1.8×10^{-6}	R_{43}–R_{44}	$CO_2^* + OH^* + {}^* \leftrightarrow HCOO^{**} + O^*$
	1.2×10^{-8}	-2.0×10^{-7}	4.2×10^{-7}	R_{45}–R_{46}	$CO_2^* + H_2O^* + {}^* \leftrightarrow HCOO^{**} + OH^*$

Source: Adapted from Mhadeswar, A.B. and Vlachos, D.G., *Catal. Today*, **105**, 162, 2005b.

Note: The dominant eigenvalues are shown in the first column. The eigenvectors corresponding to the three largest eigenvalues of the matrix $\mathbf{S}^T\mathbf{S}$ are shown in columns 2–4, respectively. The corresponding reaction pair number and the reaction are shown in the last two columns. Elements indicated in bold meet the cutoff threshold of 10^{-5}.

$$S_{ij} = \frac{\partial(\ln R_i)}{\partial(\ln A_j)} \tag{5.35}$$

The sensitivity analysis establishes the basis for the principle component analysis. The principle component analysis yields a number of important reaction intermediates that are significant in predicting the reaction behavior. The rest of the steps are too fast to influence the global kinetics and therefore can be eliminated from the overall analysis.

Compare the model predictions with the kinetic data

The final step is to compare the predictions of the microkinetic model with the global kinetic measurements. The ultimate goal of the sophisticated calculations and methodologies is to be able to predict the behavior with improved accuracy and rigor while maintaining the prediction power.

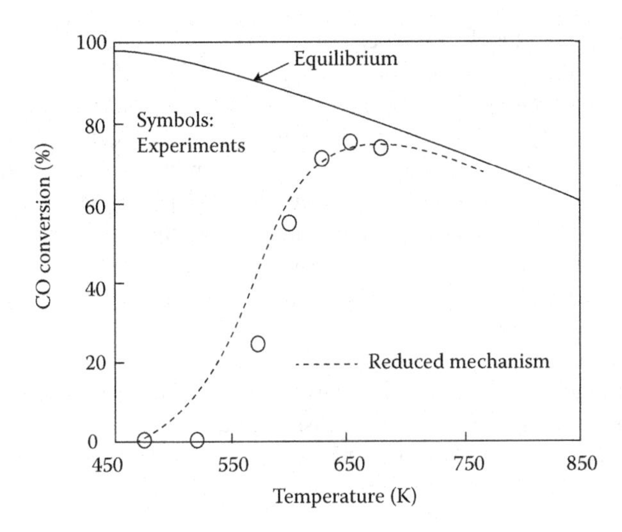

Figure 5.2 Comparison of the model predictions. (Adapted from Mhadeswar, A.B. and Vlachos, D.G., *Catal. Today,* **105**, 162, 2005b.)

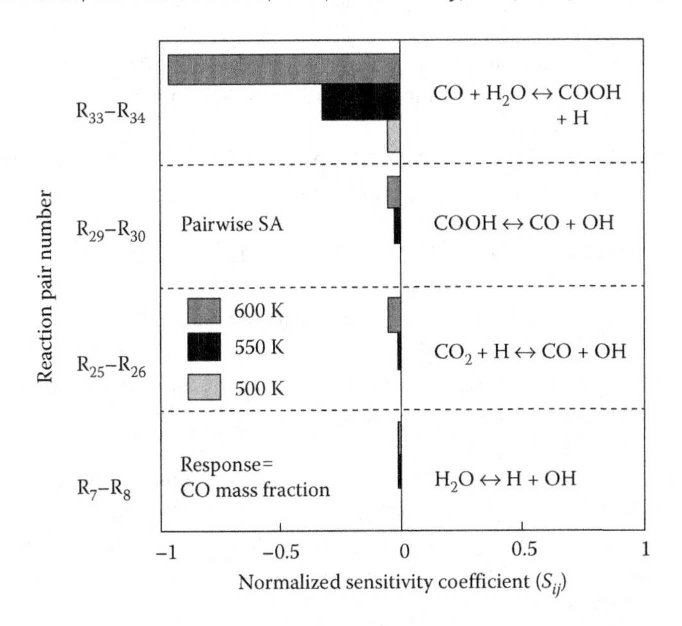

Figure 5.3 Sensitivity of CO mass fraction with respect to preexponentials. (Adapted from Mhadeswar, A.B. and Vlachos, D.G., *Catal. Today,* **105**, 162, 2005b.)

Example 5.4: Simplification of the mechanism through the principal component analysis

In this example, we will discuss the reduction of a thermodynamically consistent mechanism to a smaller but manageable size through principle component analysis as reported by Mhadeswar and Vlachos (2005b). They performed a principle component analysis with the outcomes as reported in Table 5.2. Through the elimination of the reaction steps with very small eigenvalues, and maintaining the thermodynamically and stoichiometrically consistent steps, they were able to reduce the reaction network from 46 steps to 18 steps. It is important to note that the reduced reaction network still follows the full mechanism and is also thermodynamically consistent.

As seen in Figure 5.2, the 18-step mechanism can capture the essentials of the kinetics as much as the 46-step mechanism without any loss of accuracy. Simplification improves computational cost and improves the ease of analysis.

In Figure 5.3, the results of the sensitivity analysis toward CO mass fraction is shown. The message to take from this plot is to identify $CO^* + H_2O^* \Leftrightarrow COOH^* + H^*$ as the most important one.

Explore yourself

1. See if you can quickly generate answers to the following questions:
   a. What are the shortcomings of the collision theory?
   b. Define the transition state and activated complex. How are they related?
   c. Why is pseudo-steady state pseudo?
   d. Explain Bodenstein approximation.
   e. List the assumptions in the LHHW mechanism.
   f. What is the difference between the LHHW and Eley–Riedal mechanisms?
2. Discuss the advantage of using density functional theory (DFT) for mechanism elucidation against the empirical approach where you use statistical methods to fit a mechanism to a given reaction. List at least five advantages and five disadvantages of each of the methods.
3. List at least five advantages and five disadvantages of using UHV studies for elucidating the surface reaction mechanisms of the catalytic reactions. Suggest methods for extrapolating the data obtained under UHV conditions to real-life situations at high pressures.
4. How can you extrapolate a surface reaction mechanism obtained for a gas phase reaction for the same reaction carried out in the liquid phase?
5. How can you extrapolate a gas phase radical reaction mechanism to a catalytic surface reaction?
6. Photosynthetic reactions can be broadly classified as C3, C4, and CAM mechanisms. Find out about these mechanisms and comment on the differences and similarities between them. How would you extrapolate what you learned from photosynthesis for designing chemical reactors?

References

Barnard, J.A. and Mitchell, D.S., *J. Catal.*, **12**, 376, 1968.

Batteas, J.D., Gardin, D.E., Van Hove, M.A., and Somorjai, G.A., *Surf. Sci.*, **297**, 11, 1993.

Belton, D.N., DiMaggio, C.L., and Ng, K.Y.S., *J. Catal.*, **144**, 273, 1993.

Bodenstein, M., *Ann. Phys.*, **82**, 836, 1927.

Boudart, M., *Ind. Eng. Chem. Fundam.*, **25**, 656, 1986.

Boudart, M., Mears, D.E., and Vannice, M.A., *Ind. Chim. Belge*, **32**, 281, 1967.

Bouzek, K. and Roušar, I., *J. Chem. Technol. Biotechnol.*, **66**, 131, 1996.

Broadbelt, L.J., and Rekoske, J.E., in *Dynamics of Surface and Reaction Kinetics in Heterogeneous Catalysis* (Eds. Froment, G.F. and Waugh, K.C.) p. 341. Elsevier, Amsterdam, 1997.

Doraiswamy, L.K., *Prog. Surf. Sci.*, **37**, 1, 1991.

Dumesic, J.A., Rudd, D., Aparicio, L.M., Rekoske, J.E., and Trevino, A.A., *The Microkinetics of Heterogeneous Catalysis*, ACS, Washington, DC, 1993.

Eyring, H., *J. Chem. Phys.*, **3**, 107, 1935.

Kiperman, S.L., Kumbilieva, K.E., and Petrov, L.A., *Ind. Eng. Chem. Res.*, **28**, 376, 1989.

Kramers, H.A., *Physica*, **7**, 284, 1940.

Laidler, K.J., *Chemical Kinetics*, 3rd edition, Harper & Row, New York, 1987.

Lombardo, S. and Bell, A.T., *Surf. Sci. Rep.*, **13**, 1, 1991.

Maestri, M., Microkinetic analysis of complex chemical processes at surfaces, in *New Strategies in Chemical Synthesis and Catalysis* (Ed. Pignataro, B.) Wiley-VCH, Weinheim, Germany, 2012.

McCabe, R.W. and Wong, C.J., *J. Catal.*, **121**, 422, 1990.

Mhadeswar, A.B. and Vlachos, D.G., *Comb. Flame*, **142**, 289, 2005a.

Mhadeswar, A.B. and Vlachos, D.G., *Catal. Today*, **105**, 162, 2005b.

Oran, U. and Uner, D., *Appl. Catal. B Environ.*, **54**, 183, 2004.

Plath, P., Ed., *Optimal Structures in Heterogeneous Reaction Systems*, Springer-Verlag, Berlin, 1989.

Polanyi, M. and Evans, M.G., *Trans. Faraday Soc.*, **31**, 875, 1935.

Shendye, R.V., Dowd, M.K., and Doraiswamy, L.K., *Chem. Eng. Sci.*, **48**, 1995 1993.

Shustorovich, E., *Surf. Sci.*, **176**, L863, 1986.

Shustorovich, E. and Sellers, H., *Surf. Sci. Rep.*, **31**, 5–119, 1998.

Stoltze, P., *Prog. Surf. Sci.*, **65**, 65, 2000.

Thiel, P.A., Williams, E.D., Yates, J.T., Jr., and Weinberg, W.H., *Surf. Sci.*, **84**, 54, 1979.

Uner, D.O., *J. Catal.*, **178**, 382, 1998.

Van Santen, R.A. and Neurock, M., *Molecular Heterogeneous Catalysis: A Conceptual and Computational Approach*, Wiley VCH, Weinheim, 2006.

Waugh, K., *Catal. Today*, **53**, 161, 1999.

Weller, S., *AIChE J.*, **2**, 59, 1956.

Yang, K.H. and Hougen, O.A., *Chem. Eng. Prog.*, **46**, 14, 1950.

Bibliography

On general chemical kinetics:

Asperger, S., *Chemical Kinetics and Inorganic Reaction Mechanisms*, Kluwer Academic/Plenum Press, NY, 2003.

Chorkendorff, I. and Niemantsverdriet, J.W., *Concepts of Modern Catalysis and Kinetics*, Wiley-VCH, Weinheim, 2007.

Glasstone, S., Laidler, K.J., and Eyring, H., *The Theory of Rate Processes*, McGraw Hill, Kogakusha, 1941.

Masel, R.I., *Principles of Adsorption and Reaction on Solid Surfaces*, John Wiley and Sons, NY, 1996.

On use of adsorption models:

Boudart, M. and Djega-Mariadassou, G., *Kinetics of Heterogeneous Catalytic Reactions*, Princeton University Press, Princeton, NJ, 1984.

Butt, J.B., *Reaction Kinetics and Reactor Design*, Prentice-Hall, Englewood Cliffs, NJ, 1980.

Doraiswamy, L.K. and Sharma, M.M., *Heterogeneous Reactions: Analysis, Examples and Reactor Design, Vol. 1*, John Wiley, NY, 1984.

Hougen, O.A. and Watson, K.M., *Chemical Process Principles, Vol. III*, Wiley, NY, 1947.

Satterfield, C.N., *Heterogeneous Catalysis in Practice*, McGraw Hill, NY, 1980.

Yang, K.H. and Hougen, O.A., *Chem. Eng. Prog.,* **46**, 14, 1950.

Chapter 6 Reactions with an interface

Mass and heat transfer effects

Chapter objectives

After successful completion of this chapter, the students are expected to

- Differentiate the parallel and series resistances of mass transfer in heterogeneous reactions.
- Compare and contrast fluid–fluid and fluid–solid reaction systems for the governing differential equations.
- Derive equations describing simultaneous mass transfer with chemical reaction and identify the boundary conditions in fluid–fluid and fluid–solid systems.
- Differentiate two-film theory, penetration theory, and surface renewal theory of mass transfer across interfaces.
- Explain the role of mass transfer on the reaction rate and selectivity.
- Identify the modes of diffusion and apply to derive models of the mass transfer across a pellet.
- Define, evaluate, and use the concepts of effectiveness factor, Thiele modulus, and Weisz modulus.
- Define, evaluate, and use the concepts of enhancement factor and Hatta modulus.

Introduction

When more than one phase is involved in a reaction, inevitably an interface formed, creating its own resistance to transport. This situation has to be understood clearly and discussed carefully when designing reactors involving more than one phase. The concept of resistance is mostly a disguise. However, when complex reaction schemes with selectivity issues are present, resistance to unwanted products or reaction pathways can offer very good solutions. To choose the best alternative, the design engineer should understand the chemical reactions from the perspective of resistances across the interfaces. Before we begin the detailed treatment of the mass and heat transfer effects in chemical reactors, we will define the terminology.

Diffusivity

The process can be broadly classified as bulk diffusion, Knudsen diffusion, and surface diffusion. The molecular driving force of diffusion is the chemical potential difference created by a local population of a chemical species. Molecules tend to distribute uniformly across the space while migrating among like or unlike molecules. Bulk diffusion is the predominant mechanism when the pressures are high and pore sizes are large. On the other hand, at lower pressures, Knudsen diffusion prevails, when the mean free path of the molecules are larger than the pore size. When the molecules are adsorbed strongly on the pores or the pore sizes are too small, the mechanism of diffusion becomes surface diffusivity.

Diffusivities in gases Binary gaseous diffusion coefficients are important parameters in the design of reactors for two-phase reactions involving a gas and a liquid or solid (either as catalyst or as reactant). The recommended equation for low pressures is a modified form of the theoretical Chapman–Enskog equation, but a more readily usable equation is (Gilliland, 1934)

$$D_{AB} = 0.0043\frac{T^{3/2}}{P\left(V_A^{1/3} + V_B^{1/3}\right)^2}\sqrt{\frac{1}{M_A} + \frac{1}{M_B}} \tag{6.1}$$

where D is in cm²/s, P is in atmospheres, T is the absolute temperature in K, and V_i are the molar volumes at their normal boiling points in cm³/mol. No reliable method is available for estimating D_{AB} at elevated pressures. A rule of thumb is to assume that ρD_{AB} is constant provided that $\rho_r < 1$.

In the case of a multicomponent system, the diffusivity of A is estimated by assuming that it diffuses through a stagnant film of the other gases. The overall diffusivity, say, of component 1 can then be calculated from the various constituent binary diffusivities using one of the following two equations (the second being slightly more accurate):

$$\frac{1}{\bar{D}_j} = \frac{1}{1 - y_i}\sum_{i \neq j}^{N}\frac{y_i}{D_{ji}} \tag{6.2}$$

$$\frac{1}{\bar{D}_j} = \frac{\sum_{i \neq j}^{N}(1/D_{ji})[y_i - y_j(\upsilon_i/\upsilon_j)]}{1 - y_j\sum_{i \neq j}^{N}\upsilon_i/\upsilon_j} \tag{6.3}$$

where $\bar{D}_j$ is the diffusion coefficient of j in a mixture of $1 + 2 + \ldots i \ldots + N$ components, D_{ji} are the binary diffusion coefficients of j and i $(i \neq j)$, υ_i is the stoichiometric coefficient of i in the reaction, and y_i its mole fraction.

Diffusivities in liquids We restrict the treatment here to binary mixtures of, say, A and B in which A diffuses in B at infinite dilution, that is, at very low concentrations of A in B. Reference may be made to Reid et al. (1987) for correlations at high concentrations.

All correlations proposed for estimating the diffusivity at infinite dilution are modifications of the original Stokes–Einstein equation

$$D_{AB} = \frac{RT}{6\pi\mu r_a} \tag{6.4}$$

Stokes-Einstein equation for gas diffusivity

where μ is the solvent viscosity and r_a is the radius of the solute assumed to be spherical. We refer the curious reader to the bibliography on the diffusivities at the end of this chapter. The diffusivities reported in these articles relate the diffusivity to viscosity (as in the Stokes–Einstein equation), but impart greater generality by including the molar volume of A or those of A and B. As an example, we give the following simplified form of the Tyn–Calus correlation (1975):

$$D_{AB} = 8.93 \times 10^{-8} \frac{V_B^{0.267}}{V_A^{0.433}} \frac{T}{\eta_B} \tag{6.5}$$

with the molar volume being obtained from

$$V_i = 0.285(V_c)_i^{1.048}, \quad i = A, B \tag{6.6}$$

where V_i and $(V_c)_i$ are the molar and critical volumes, respectively, of species i.

Effective diffusivity A definition for the effective diffusivity is necessary due to the nonuniformities in the pores and channels of a solid:

Effective diffusivity

$$D_{eA} = \frac{D_A \phi_p \delta}{\bar{t}} \tag{6.7}$$

where

$$\bar{t} = \text{Tortuosity} = \frac{\text{Actual distance between the two points}}{\text{Shortest distance between these two points}} \tag{6.8}$$

Tortuosity

$$\phi_p = \text{Porosity} = \frac{\text{Volume of void space}}{\text{Total volume}} \tag{6.9}$$

Porosity

and

$$\delta = \text{Constrictivity} \propto \frac{\text{Diameter of the diffusing particle}}{\text{Pore diameter}} \tag{6.10}$$

Constrictivity

Transport between phases

The ideal reactors that we discussed so far are called ideal because of their uniformity of concentration and temperature either throughout the reactor (in CSTRs and in well-mixed batch reactors) or in radial coordinates (in PFRs). We assumed homogeneity ad hoc, when deriving the equations and making the analyses. In reality, the systems lose their ideality due to the different states of aggregation of matter that needs to be contacted to accomplish a specific chemical conversion. Operational parameters may also lead to the loss of ideality. For example, in a CSTR, the mixing may not be perfect; it has already been shown that for a reactor to be called as well mixed, the residence time in the reactor should be 100 times or greater than the circulation times of the impeller (Westerterp et al., 1984). When the mixing is not perfect, the uniform concentration across the reactor assumption immediately fails, giving rise to the concentration gradients. In a tubular flow reactor, laminar flow conditions can reflect themselves in parabolic velocity profiles, which may give rise to radial concentration gradients. Although we have given greater emphasis to the treatment of such imperfections in Chapter 3, we will offer a brief introduction of how to take the mass transfer effects into account in this chapter.

General remarks

Before fully appreciating any concept of mass transport, an understanding of the frequently used term *flux* is necessary. It is simply the amount of fluid transported per unit time per unit area (perpendicular to the direction of transport). According to the well-known Fick's law, the flux is proportional to the concentration difference of a "moving" gas between two points divided by the distance separating the points, and the proportionality constant is known as the *diffusion coefficient* or *diffusivity* (with units of distance²/time).

Fick's first law:

Fick's first law

$$J_A = aj_A = -aD\frac{\partial[A]}{\partial z} \tag{6.11}$$

where a is the interfacial area and D is the diffusion coefficient. Similarly, Fick's second law in one dimension is expressed as

Fick's second law

$$\frac{\partial[A]}{\partial t} = D\frac{\partial^2[A]}{\partial z^2} \tag{6.12}$$

When transport occurs between phases, we usually express the flux as a function only of the concentration difference of the "diffusing gas." The proportionality constant here is known as the *mass transfer coefficient* (with units of distance/time). More puristically, it is referred to as the *phenomenological mass transfer coefficient*, to distinguish it from other, often more useful, definitions. This is usually the chemical engineer's way of looking at mass transport and is essentially a lumped parameter model in which the transport is "localized" and is not a function of distance. Simply expressed in the following equation is the rate of a mass transfer process across an interface of area a_L and a concentration gradient of $[A]^* - [A]_b$, where $[A]^*$ is the interface concentration and $[A]_b$ is the bulk concentration:

$$r'_A a_L = r_A = k'_K a_L([A]^* - [A]_b) \tag{6.13}$$

Rate of mass transfer across an interface

The concept of diffusion is used whenever one is dealing with transport within a phase as a function of position. For example, when a chemical reaction occurs in a catalyst pellet, the reactant has to diffuse through the catalyst and react while it is still diffusing. Thus, in any rational analysis of such a situation, we are concerned with diffusion. On the other hand, when one has often to deal with immiscible or sparingly miscible fluids, (s)he has to contend with the problem of transport of desired species across an interface such as gas–liquid, liquid–liquid, and fluid–solid. In such a situation, one has to fall back on the concept of mass transfer coefficient by defining a hypothetical *film* across which transport occurs. Although the two models (one based on diffusion and the other on mass transfer coefficient) are related, we shall largely be concerned with the latter in dealing with interfacial phenomena.

In cases where the reactants are present in two different phases, one of the reactants must diffuse from its phase into the other for reaction to occur there. If the distribution coefficients of the two reactants do not favor any particular phase, the reaction can occur in both the phases (particularly if both are liquid). Clearly, therefore, the rates of *mass transfer* of reactants between phases become an important consideration in heterogeneous systems in general.

Three major theories of mass transfer have been in vogue to explain interphase transport: Lewis and Whitman film theory (1924), Higbie (1935) and Danckwerts (1950, 1951a) versions of the penetration theory, and Danckwerts (1951b, 1953) surface renewal theory. The *film theory* in essence asserts that adjacent to any interface there is a stagnant *film* of thickness δ through which transport of any species occurs by molecular conduction (there is no convection). Conditions in the rest of the fluid, called the *bulk*, are assumed to remain constant, so that the driving force for transport is consumed entirely by the film, as depicted in Figure 6.1. If the diffusing molecule participates in a reaction in this phase after or during transport, the reaction can occur in the film, the bulk, or both.

Another theory of mass transfer is based on the postulation that elements of the fluid impinge on the interface where they remain for a specified period of time during which they shed their load of reactant and then return to the body of the fluid. The contact time of an element with the interface can be constant for all elements (Higbie, 1935) or vary from element to element (Danckwerts, 1953). Such a postulation, sketched in

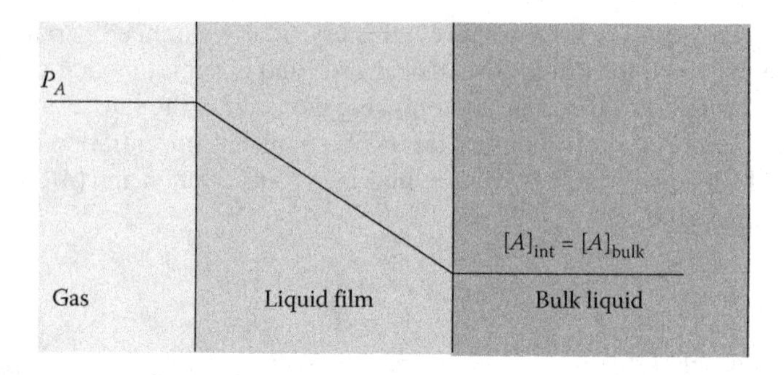

Figure 6.1 The film theory for mass transfer.

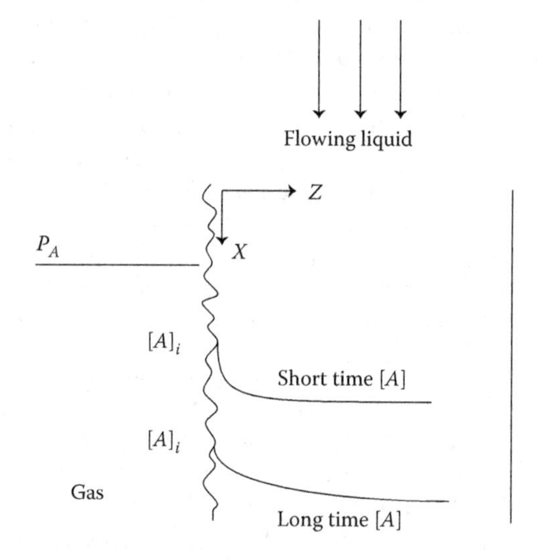

Figure 6.2 The penetration theory for mass transfer.

Figure 6.2, is the basis of the so-called *penetration theory*. The film and penetration theories differ from each other in the definition of k_L'.

Film theory

Film theory

The advantage of using Equation 6.13 for the mass transfer coefficient lies in the assumption that under similar hydrodynamic conditions, the film thickness δ is constant, so that the flux across the film, on the one hand, is equal to the convective flux, with a mass transfer coefficient of k'_L, and, on the other hand, is equal to a diffusive flux around a linear concentration gradient given by

$$N_A = \frac{D}{\delta}([A_i] - [A]) = k_L'([A_i] - [A]) \tag{6.14}$$

such that the relationship between the diffusivities and the mass transfer coefficients is

$$k'_L = \frac{D_A}{\delta} \tag{6.15}$$

Penetration theory

Astarita (1967) has combined the Higbie and Danckwerts approaches to give the following general equation for the mass transfer coefficient:

$$k'_L = \sqrt{\frac{D_A}{t_D}} \tag{6.16}$$

where t_D is an *equivalent diffusion time* that replaces t^* (the time each element spends at the surface in the Higbie model), and $1/s$ (the average life of surface elements, s being the rate of surface renewal) of the Danckwerts model.

Thus, we have the film and penetration theories to choose from. In making a choice, it should be noted that one is often concerned with the *ratio* of diffusivities of the two reactants of a two-phase system and not the diffusivity of just one diffusing component. Since the diffusion coefficients of organic compounds in many of the solvents normally used do not greatly differ from one another, the difference between D_B/D_A and $\sqrt{D_B/D_A}$ tends to be negligibly small. Hence, one is often justified in using the film model except in situations where unsteady-state behavior is clearly indicated. As a result, our treatment of interphase transport will largely be based on the film theory, the penetration theory being invoked only in some special cases.

Our discussion about the penetration theory will, first of all, be based on the concept of diffusion into a falling film. The problem is well known and solved in many texts of transport phenomena and diffusion (e.g., Cussler, 2001). We will not go into the details of the problem and the solution; we will rather give the final result and the equation for the flux of the species at the interface for the coordinates shown in Figure 6.2 when the bulk concentration of $[A]$ is negligible:

$$j_A|_{z=0} = \sqrt{\frac{Dv_{\max}}{\pi x}}\,[A]_i \tag{6.17}$$

Here, $v_{\max}$ is the maximum velocity of the laminar velocity profile of the falling liquid film. This equation still prevails, for the situation depicted in Figure 6.2; when we have appreciable bulk concentration of $[A]$, the solution becomes

$$j_A|_{z=0} = \sqrt{\frac{Dv_{\max}}{\pi x}}\,([A]_i - [A]) \tag{6.18}$$

$k' = \frac{D}{\delta}$

Film theory

The total flux at the gas–liquid interface, indicated by N_A then is equal to $j_{A|z} = 0$. The total flux at the interface given by Equation 6.13 is a point value, when averaged over x:

$$N_A = \frac{1}{L}\int_0^{\delta} \sqrt{\frac{Dv_{max}}{\pi x}}([A]_i - [A])dx \tag{6.19}$$

will yield

$$N_A = 2\sqrt{\frac{Dv_{max}}{\pi\delta}}\,([A]_i - [A]) \tag{6.20}$$

where δ is the thickness of the liquid film. The convective mass transfer equation across this film is

$k' = \sqrt{D/\tau}$

Surface renewal theory

$$N_A = k\,([A]_i - [A]) \tag{6.21}$$

A comparison of Equations 6.20 and 6.21 yields

$$k = 2\sqrt{\frac{Dv_{max}}{\pi\delta}} \tag{6.22}$$

The quantity v_{max}/δ is called the contact time, t.

Surface renewal theory

We will continue with the flux into an infinite slab, replacing v_{max}/δ with the contact time, t.

$$n\,|_{z=0} = j_A\,|_{z=0} = \sqrt{\frac{D}{\pi t}}([A]_i - [A]) \tag{6.23}$$

$k' = 2\sqrt{Dv_{max}/\pi\delta}$

Penetration theory

Now, we have small surface elements remaining at the interface at a limited period of time. If we assume the interface behaving like a CSTR and recalling the residence time distribution of the CSTR from Chapter 3 as

$$E(t) = \frac{e^{-t/\tau}}{\tau} \tag{6.24}$$

For a quickly renewing surface for small τ, the surface is infinitely large such that

$$N_1 = \int_0^{\infty} E(t)n|_{z=0}\,dt = \sqrt{\frac{D}{\tau}}\,([A]_i - [A]) \tag{6.25}$$

When we compare Equations 6.14 and 6.25, we see that

$$k' = \sqrt{\frac{D}{\tau}} \tag{6.26}$$

a very similar result to the penetration theory is indeed obtained. The derivation of the relationship between the mass transfer coefficient and the diffusivity of the film theory is left to the reader.

Characteristic times for diffusion, reaction, and mass transfer

From Equations 6.12 and 6.16, it is evident that the characteristic time for diffusion is $t_D = D/k_L'^2$. On the other hand, for a first-order reaction, the characteristic time for reaction is $t_R = k_R^{-1}$, where k_R is the reaction rate constant. Finally, the characteristic time for the mass transfer is $t_M = 1/k_L'a$.

$t_D = \frac{D}{k_L'^2}$

$t_R = k_R^{-1}$

$t_M = \frac{1}{k_L'a}$

We will use these characteristic times to define some dimensionless numbers, such as Damköhler number, Thiele modulus, and Hatta modulus. Here, we will give the broad definitions, and the true meanings will be clear as we move along the chapter:

$$\text{Damköhler number, } Da = \frac{\text{Characteristic fluid transport time}}{\text{Characteristic reaction time}}$$

Damköhler number

$$\text{Thiele modulus, } \phi = \left(\frac{\text{Characteristic diffusion time}}{\text{Characteristic reaction time}}\right)^{1/2} \text{ for solids}$$

Thiele modulus

$$\text{Hatta modulus, } M_H = \left(\frac{\text{Characteristic diffusion time}}{\text{Characteristic reaction time}}\right)^{1/2} \text{ for liquids}$$

Hatta modulus

Two-film theory of mass and heat transfer for fluid–fluid reactions in general

Mass transfer To provide a basis to account for the influence of phase heterogeneity in reaction analysis, the theories presented above were based on a single film—associated with a single phase. In applying them to real systems, two films must be considered, one on either side of the interface, as shown in Figure 6.3.

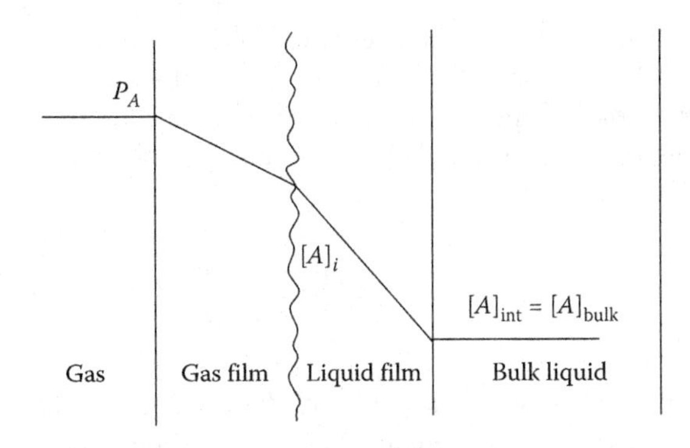

Figure 6.3 Two-film theory of mass transfer.

From the schematics depicted in Figure 6.3, it is easy to conclude that the rate of mass transfer between the gas and the gas–liquid interface is

$$r'_A = k'_G([A]^* - [A]_{\text{int}}) \tag{6.27}$$

while that between the gas–liquid interface and the liquid phase is

$$r'_A = k'_L([A]_{\text{int}} - [A]) \tag{6.28}$$

Eliminating $[A]_{\text{int}}$ between the equations above will give us an overall rate for the mass transfer defined as

$$r'_A = k'_{GL}([A]^* - [A]) \tag{6.29}$$

where the overall mass transfer coefficient, with the units of m/s, is defined as

$$\frac{1}{k'_{GL}} = \frac{1}{k'_G} + \frac{1}{k'_L} \text{ (gas–liquid)} \tag{6.30}$$

$$\frac{1}{k'_{L_1L_2}} = \frac{1}{k'_{L_1}} + \frac{1}{k'_{L_2}} \text{(liquid–liquid)} \tag{6.31}$$

Heat transfer The basic concepts of mass transfer can be readily extended to heat transfer by writing the rate of heat transfer (from phase 1 to phase 2) across a surface as

$$Q = UA_h(T_1 - T_2) \tag{6.32}$$

where U is an overall heat transfer coefficient defined as

$$\frac{1}{U} = \frac{1}{h_1} + \frac{1}{h_2} \tag{6.33}$$

h_1 and h_2 being the individual coefficients for the two films (see Figure 6.4). It must be noted, however, that, unlike in mass transfer, a third mode of transport is also involved in heat transfer: conduction through the wall (of thickness d_w). Thus, Equation 6.33 should be modified as

$$\frac{1}{U} = \frac{1}{h_1} + \frac{d_w}{\lambda}\frac{A_1}{A_m} + \frac{A_1}{A_2}\frac{1}{h_2} \tag{6.34}$$

where A_1 and A_2 are the surface areas of the two sides of the wall (one containing the reaction mixture and the other the control fluid), and A_m is the logarithmic mean of A_1 and a solid boundary.

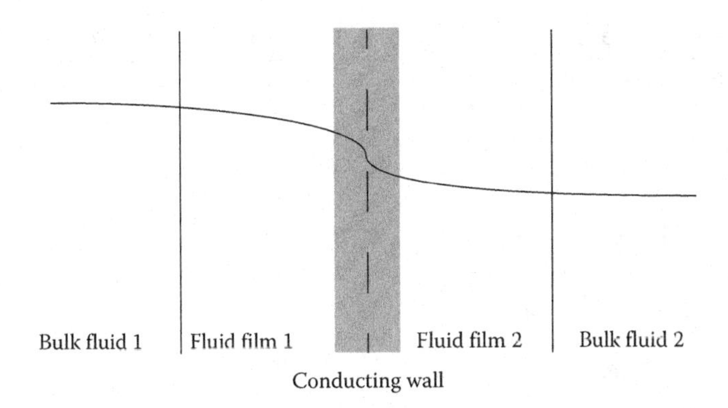

Figure 6.4 Two-film representation of heat transfer across a curved solid boundary.

Mass transfer across interfaces: Fundamentals

Mass transfer across the interfaces is the broad topic under discussion in this chapter. Before we start discussing the specifics of the matter, we would like to present the general equations of change for the processes that involve mass transfer and reactions. Our intention here is not to give a comprehensive analysis of mass transfer, in general, but to demonstrate the key principles of the problem and some commonalities between the processes that we will discuss in this and subsequent chapters.

Before we proceed further, we need to differentiate a homogeneous and a heterogeneous reaction system. We will do so based on the relative rates of mass transfer and reaction kinetics. A fast reaction can only take place at the interface; therefore, the rate expression is expressed in the boundary condition. On the contrary, a slow reaction at rates comparable to that of diffusion resides mathematically in the continuity equation. In vector form, we express the continuity equations as follows:

Fast reactions (also called heterogeneous):

$$\frac{\partial[A]}{\partial t} = D\nabla^2[A] - \underline{\nabla} \cdot [A]\underline{v} \tag{6.35}$$

Fast (heterogeneous) reactions

Slow reactions (also called homogeneous):

$$\frac{\partial[A]}{\partial t} = D\nabla^2[A] - \underline{\nabla} \cdot [A]\underline{v} + r_A \tag{6.36}$$

Slow (homogeneous) reactions

where ∇ indicates the gradient operator.

In Tables 6.1 and 6.2, the equation of continuity are given. In Table 6.1, the general form of the equation of continuity was given. In Table 6.2, the flux terms were given in open form for systems with constant density and diffusivity.

Table 6.1 Equation of Continuity for A in Various Coordinate Systems

Rectangular Coordinates

$$\frac{\partial[A]}{\partial t}+\frac{\partial N_{Ax}}{\partial x}+\frac{\partial N_{Ay}}{\partial y}+\frac{\partial N_{Az}}{\partial z}=r_A$$

Cylindrical Coordinates

$$\frac{\partial[A]}{\partial t}+\frac{1}{r}\frac{\partial}{\partial r}(rN_{Ar})+\frac{1}{r}\frac{\partial N_{A\theta}}{\partial \theta}+\frac{\partial N_{Az}}{\partial z}=r_A$$

Spherical Coordinates

$$\frac{\partial[A]}{\partial t}+\frac{1}{r^2}\frac{\partial}{\partial r}\left(r^2N_{Ar}\right)+\frac{1}{r\sin\theta}\frac{\partial}{\partial \theta}\left(N_{A\theta}\sin\theta\right)+\frac{1}{r\sin\theta}\frac{\partial N_{A}\phi}{\partial \phi}=r_A$$

Table 6.2 Equation of Continuity for Constant Density and Diffusivity

Rectangular Coordinates

$$\frac{\partial[A]}{\partial t}+v_x\frac{\partial[A]}{\partial x}+v_y\frac{\partial[A]}{\partial y}+v_z\frac{\partial[A]}{\partial z}=D_A\left(\frac{\partial^2[A]}{\partial x^2}+\frac{\partial^2[A]}{\partial y^2}+\frac{\partial^2[A]}{\partial z^2}\right)+r_A$$

Cylindrical Coordinates

$$\frac{\partial[A]}{\partial t}+v_r\frac{\partial[A]}{\partial r}+v_\theta\frac{1}{r}\frac{\partial[A]}{\partial \theta}+v_z\frac{\partial[A]}{\partial z}=D_A\left(\frac{1}{r}\frac{\partial}{\partial r}\left(r\frac{\partial[A]}{\partial r}\right)+\frac{1}{r^2}\frac{\partial^2[A]}{\partial \theta^2}+\frac{\partial^2[A]}{\partial z^2}\right)+r_A$$

Spherical Coordinates

$$\frac{\partial[A]}{\partial t}+v_r\frac{\partial[A]}{\partial r}+v_\theta\frac{1}{r}\frac{\partial[A]}{\partial \theta}+v_\phi\frac{1}{r\sin\theta}\frac{\partial^2[A]}{\partial \phi}=D_A\left(\begin{array}{l}\frac{1}{r^2}\frac{\partial}{\partial r}\left(r^2\frac{\partial[A]}{\partial r}\right)+\frac{1}{r^2\sin^2\theta}\frac{\partial}{\partial \theta}\\ \left(\sin\theta\frac{\partial[A]}{\partial \theta}\right)+\frac{1}{r^2\sin^2\theta}\frac{\partial^2[A]}{\partial \phi^2}\end{array}\right)+r_A$$

The continuity equation given in the most general form in Tables 6.1 and 6.2 can and will be simplified to the geometries we will use, and the relevant boundary conditions will be selected to solve problems that involve chemical reactions with mass transfer. In this chapter, we will analyze three broad classes of reactions as follows:

Solid catalyzed fluid-phase reactions

1. *Solid catalyzed fluid-phase reactions:* This is a very common class of reactions involving mass transfer and chemical reactions. The reaction system involves film mass transfer, diffusion through the pores, and surface reactions. We will solve the problem for a spherical pellet and give the results for a cylindrical pellet and for a slab.

Noncatalytic gas–solid reactions

2. *Noncatalytic gas–solid reactions:* In principle, these types of reactions have very similar characteristics to the solid catalyzed fluid-phase reactions, with one major difference: the

reactive interface changes position with respect to time. Thus, this system becomes an unsteady-state problem with a change in the radius of the spherical particle.

3. *Gas–liquid reactions:* We will solve this problem across a slab representing the liquid film across a gas–liquid interface. We will use the film theory to derive how the concentration changes across the interface and how the reaction rates are related to the overall transport rate.

Gas–liquid reactions

Of these fundamental classes of reactions, it is imperative to define the degree of heterogeneity with respect to the rates of diffusion and the reaction. For example, if a reaction is extremely fast, it would occur at the interface between the phases, and the mathematical description of the reaction rate should only appear in the relevant boundary condition. Such situations are commonly referred to as *heterogeneous* in the transport phenomena literature. On the other hand, if the reaction rates are comparable to the rates of diffusion or convective transport, then the rate expression is included in the continuity equation, as given in Tables 6.1 and 6.2, and the overall differential equations are solved accordingly.

Solid catalyzed fluid reactions

Solid catalysts by their very nature involve diffusion of reactant fluids within their matrix. These fluids react even as they diffuse. Thus, the problem of *internal diffusion* accompanied by reaction becomes important. Another problem of equal importance is the transport of reactants from fluid bulk to catalyst surface—often referred to as *external diffusion*.

Internal diffusion versus external diffusion

Overall scheme

For a solid catalyzed fluid-phase hypothetical reaction, let the pellets be placed in a flowing stream of reactants inside a tubular reactor. Restricting our attention now to a single pellet and its immediate environment, the various steps involved in the overall process are shown in Figure 6.5.

This physical–chemical circuit is built in analogy with the electrical circuit shown at the bottom of the figure. Clearly, the overall process is a complex combination of chemical and physical steps. Note, however, that the mathematical analysis of the parallel pathways (diffusion and reaction) is not based on the addition of reciprocal resistances as in parallel electrical circuits, but on the fact that the two occur simultaneously on a single pathway, that is, the molecule reacts even as it diffuses.

Role of diffusion in pellets: Catalyst effectiveness Catalysts are normally used in the form of pellets, except in fluidized bed reactors where powders are used. Thus, problems of resistance to diffusion within the pellets are common. These have been quite extensively studied and many

Catalyst effectiveness

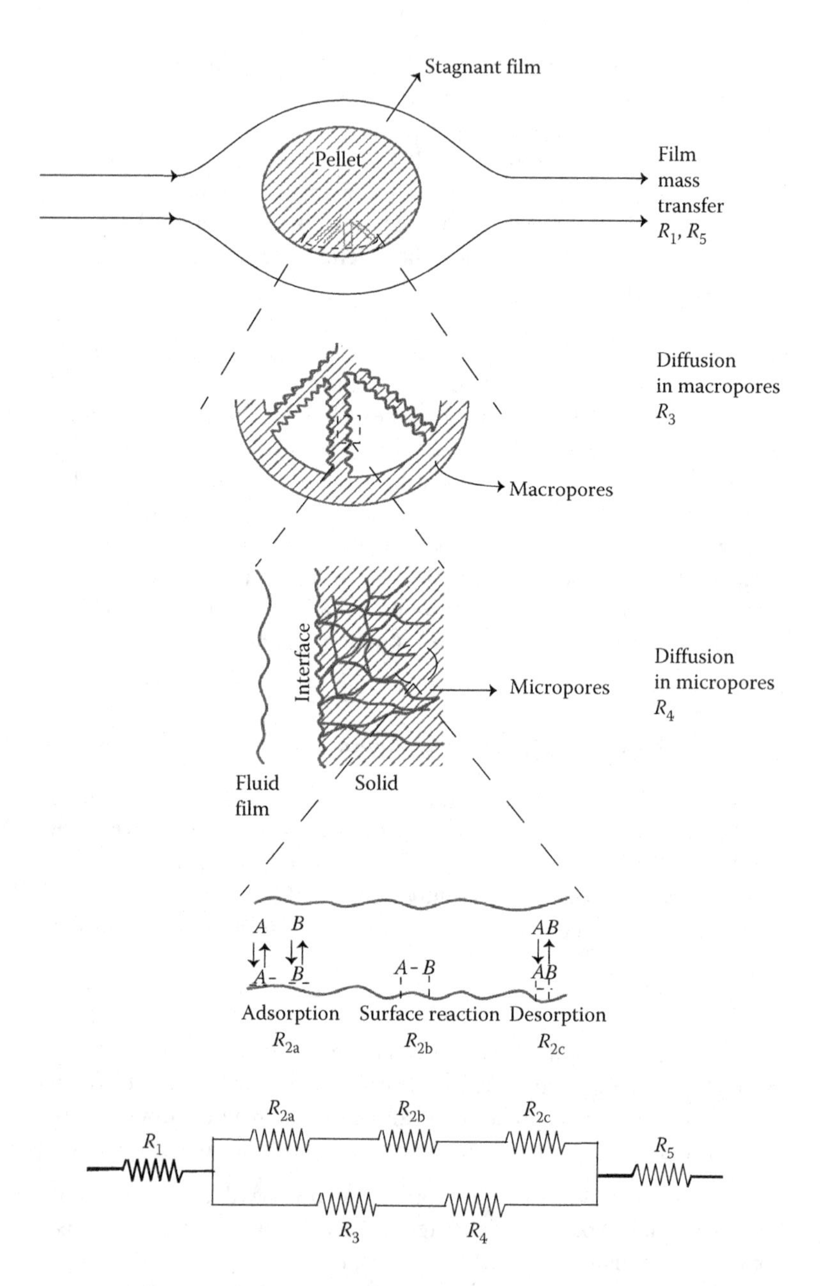

Figure 6.5 Major steps in the solid catalyzed reaction $A + B \rightarrow R + S$. The electrical resistance analogy is also shown.

Modes of diffusion

Bulk
Surface
Configurational
Knudsen

texts written (see, in particular, Aris, 1975). We present some basic equations here and outline methods for a quick evaluation of these effects.

The basic requirement in any study of internal diffusion is an understanding of the various modes of transport in a straight capillary: bulk, Knudsen, configurational, and surface. This knowledge is then extended to diffusion in the porous matrix of a pellet to formulate expressions for an *effective diffusion coefficient.* We confine our treatment in this book to listing in Table 6.3 the more important equations for direct use in estimation. In this table are given the equations for bulk (for macropores with $r_p > 200$ Å), Knudsen (for micropores with $r_p < 50$ Å), and combined diffusion (for all pore sizes) in a straight capillary. The last is also referred to as diffusion in the *transition regime.* These are followed by equations for effective diffusion coefficient in a pellet in these regimes. Two models are considered for the transition regime: (1) the parallel path model (Johnson and Stewart, 1965; Feng and Stewart, 1973) that accounts for a single overall pore size distribution; and (2) the *micro–macro* or *random pore model* (Wakao and Smith, 1962, 1964) that assumes a *bimodal distribution* of the pore structure, one for the space between particles in a pellet (usually the macropores), and the other for the pores within a particle (usually the micropores). *Multimodal structures* involving more than one macropore distribution are also possible (Cunningham and Geankoplis, 1968), but such complexity is almost never consistent with the quality of basic data that can be generated. Other pore structures have also been proposed, for example, the *pore network model* of Beekman and Froment (1982). Extensive testing of these models has shown the parallel path model to be slightly superior (Satterfield and Cadle, 1968a,b; Brown et al., 1969; Patel and Butt, 1974).

Effective diffusion coefficient

The resistance to diffusion (expressed in terms of an effective diffusion coefficient as defined above) has the effect of progressively reducing the concentration of the reactant molecule from the catalyst surface to the center. This leads to a lower reaction rate and to a lower value of the rate constant, that is,

$$k_a = \varepsilon k \tag{6.37}$$

Effectiveness factor

where k_a is the *actual rate constant,* k the *true* or *intrinsic rate constant,* and ε commonly referred to as the catalyst *effectiveness factor* (or *utilization factor*). It must be noted that being a codeterminant of the apparent rate, it is as important a factor as the true rate constant itself in the analysis and design of the catalytic reactors. Several detailed treatments of the subject are available, for example, Petersen (1965), Aris (1975), Carberry (1976), Luss (1977), Doraiswamy and Sharma (1984), and Froment and Bischoff (1990), and we restrict the treatment to a brief outline of the approaches and equations.

First-order isothermal reaction in a spherical catalyst Consider a simple first-order reaction:

$$A \rightarrow \text{Products}$$

With a first-order rate expression given by

$$-r_A = k_v[A] \tag{6.38}$$

Table 6.3 Equations for Estimating Diffusivities of A in Pellets in Various Regimes

Serial Number	Flow Regime/Pore Structure	Parameter to Be Estimated	Equation	Main Input Data
1	Bulk diffusion (independent of pore structure)	Bulk diffusivity D_{bA} (cm^2/s)	$D_{bA} = \frac{3.0 \times 10^{-3} T^{1.75}}{P\left(V_A^{1/3} + V_B^{1/3}\right)} \left(\frac{1}{M_A} + \frac{1}{M_B}\right)^{1/2}$	Molecular weights M_i, temperature, and molar volumes V_i
2	Bulk diffusion in a porous pellet	Effective bulk diffusivity D_{ebA}	$D_{ebA} = \frac{f_c}{\tau} D_{bA}$	Tortuosity factor τ, pellet porosity f_c
3	Knudsen diffusion in a straight capillary	Knudsen diffusivity D_{KA} (cm^2/s), $\bar{r}_p$ in Angstroms	$D_{KA} = 9.7 \times 10^{-5} \bar{r}_p \left(\frac{T}{M_A}\right)^{1/2}$	Capillary radius, temperature, and molecular weight
4	Knudsen diffusion in a porous pellet	Effective Knudsen diffusivity D_{eKA}	$D_{eKA} = \frac{f_c}{\tau} D_{KA}$	Tortuosity factor τ, pellet porosity f_c
5	Transition regime in a straight capillary	Combined diffusivity D_{cA}	$D_{cA} = D_{bA} \frac{\ln((1 - \alpha y_{AL} + R_D)/(1 - \alpha y_{A0} + R_D))}{\alpha(y_{A0} - y_{AL})}$	Bulk and Knudsen diffusivities from 1 and 3, and α from experiment or the Hoogschagen relation
6	Transition regime in a porous pellet	Effective diffusivity D_{eA}	$D_{eA} = D_{ebA} \frac{\ln((1 - \alpha y_{AL} + R_{D_e})/(1 - \alpha y_{A0} + R_{D_e}))}{\alpha(y_{A0} - y_{AL})}$	
7	Computational models a. Macropore–micropore model	Effective diffusivity D_{eA}	See Wakao and Smith (1962)	Bulk diffusivity, Knudsen diffusivities in micro- and macropores, molecular weights, macro- and microporosities, and mean micro- and macropore radii
	b. Rigorous equation accounting for pore size distribution	Effective diffusivity D_{eA}	$D_{eA} = \frac{k^p D_{bA}}{\alpha(y_{A0} - y_{AL})} \int_0^\infty \ln\left(\frac{1 - \alpha y_{AL} + R_D}{1 - \alpha y_{A0} + R_D}\right)$	Bulk and Knudsen diffusivities and pore size distribution; assume $k^p \approx 1/3$ (see Johnson and Stewart, 1965)

Note: $R_D = D_{bA}/D_{KA}$, $R_{D_e} = D_{ebA}/D_{eKA}$, $f(r_p)$ is volume of pores between r_p and $(r_p + dr_p)$ per unit volume of pellet; $\bar{r}_p$ = average pore radius; $\alpha = [1 + (N_{bA}/N_{bB})] = [1 - (\sqrt{M_B/M_A})]$, Hoogschagen's relation; y_{A0}, y_{AL} = mole fractions of A at $\ell = 0$ and $\ell = L$, respectively; D_{cA} and D_{eA} are combined diffusivities in a straight capillary and pellet, respectively, but since various other factors are also involved in diffusion in a pellet, the diffusivity in this case is called effective diffusivity; V_i may be estimated from the tables of Fuller et al. (1966).

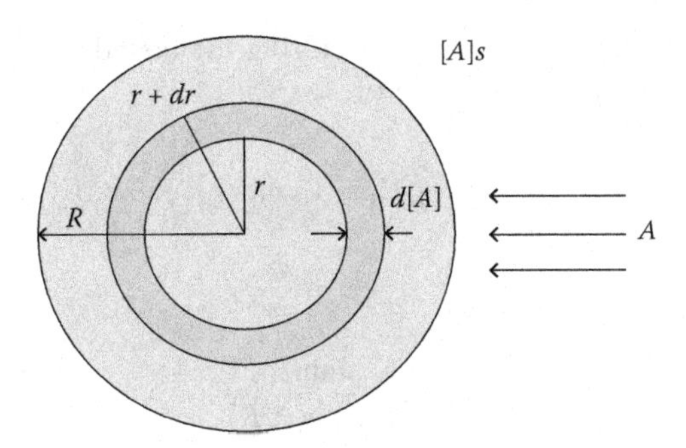

Figure 6.6 A differential element of a spherical pellet.

occurring in a spherical catalyst pellet, as shown in Figure 6.6. Note that, for convenience, we have switched from rates based on catalyst weight to rates based on catalyst volume. Selecting the appropriate continuity equation from Table 6.2 for spherical coordinates, and eliminating the irrelevant terms for a system where

- the process is at steady state
- there are no convective flows in r, θ, and ϕ directions
- variations in θ and ϕ directions are negligible

the resultant equation becomes

$$0 = D_A\left(\frac{1}{r^2}\frac{\partial}{\partial r}\left(r^2\frac{\partial[A]}{\partial r}\right)\right) + r_A \tag{6.39}$$

For a first-order reaction, we obtain, after rearrangement

$$\frac{d^2[A]}{dr^2} + \frac{2}{r}\frac{d[A]}{dr} = \frac{k_v[A]}{D_{eA}} \tag{6.40}$$

This can be recast in dimensionless form as

$$\frac{d^2[\hat{A}]}{d\hat{R}^2} + \frac{2}{\hat{R}}\frac{d[\hat{A}]}{d\hat{R}} = \phi_{s1}[\hat{A}] \tag{6.41}$$

where $[\hat{A}] = [A]/[A]_S$, $\hat{R} = r/R$ and

$$\phi_{s1} = L\left(\frac{k_v}{D_e}\right)^{1/2} \tag{6.42}$$

Thiele modulus

is known as the Thiele modulus for a first-order reaction in a sphere. The symbol ϕ represents the modulus in general and the subscripts specify pellet shape and reaction order. It is a measure of the relative rates of reaction and diffusion: low values denote chemical control, and high values diffusion control.

Equation 6.40 can be solved by specifying the boundary conditions at the surface and center. These are

$$\text{Surface}: [A] = [A]_S, \quad \text{Center}: \frac{d[A]}{dr} = 0 \tag{6.43}$$

and the solution is

$$[A] = \frac{R}{r} \frac{\sinh(\phi_{s1} r)}{\sinh(\phi_{s1} R)} \tag{6.44}$$

from which the concentration can be computed as a function of radial position for various values of the Thiele modulus.

We define an effectiveness factor as

Effectiveness factor

$$\varepsilon = \frac{\text{Actual rate within the pellet based on average concentration}}{\text{Rate based on surface conditions throughout the pellet}}$$

$$\varepsilon = \frac{(1/R)k_V \int_0^R [A]dr}{k_V [A]_S} \tag{6.45}$$

Thus, we merely combine Equations 6.42 and 6.43 and integrate between the limits $r = 0$ and $r = R$ to give

$$\varepsilon = \frac{3}{\phi_{s1}^2}(\phi_{s1}\coth(\phi_{s1}) - 1) \tag{6.46}$$

By procedures similar to that presented above for a sphere, we can derive expressions for other shapes as well. We consider two other shapes: flat plate (or slab) and cylinder. The equations for all the three shapes are given in Table 6.4 and the corresponding plots are included in Figure 6.7.

Equations for a single pore (item 4 in the table) are similar to those for the flat plate.

Since the equations for the three shapes produce three different curves, it is desirable to formulate a single Thiele modulus that will not only be applicable to these three shapes but also to any shape. It is also desirable to generalize the modulus to include reactions of any order n. The final equation obtained is

$$\frac{d^2[\hat{A}]}{d\hat{\Lambda}^2} + \frac{s}{\hat{\Lambda}} \frac{d[\hat{A}]}{d\hat{\Lambda}} = \phi^2[\hat{A}] \tag{6.47}$$

where $\hat{\Lambda} = \Lambda/\Lambda_0$ is the *normalized length coordinate*, Λ is the *generalized length coordinate* characteristic of any shape (r for the sphere or

Table 6.4 Intraphase Diffusion Parameters and Equations for a First-Order Reaction ($A \rightarrow$ Products) in Different Catalyst Shapes

Shape and Definition	Thiele Modulus	Normalized Thiele Modulus Based on Λ_0	Modified Distance Parameter	Equation for ε	Equation for ε Based on Generalized Modulus
1. Infinite slab a. One end open	$\phi_{p1} = L\left(\frac{k_v}{D_e}\right)^{1/2}$	$\Lambda_0\left(\frac{k_v}{D_e}\right)^{1/2}$	L	$\varepsilon = \frac{1}{\phi_{p1}}(\tanh \phi_{p1})$	$\varepsilon = \frac{1}{\phi}(\tanh \phi)$
b. Both ends open			$L/2$		
2. Sphere	$\phi_{s1} = L\left(\frac{k_v}{D_e}\right)^{1/2}$	$\Lambda_0\left(\frac{k_v}{D_e}\right)^{1/2}$	$\frac{R}{3}$	$\varepsilon \frac{3}{\phi_{s1}}\left(\frac{1}{\tanh \phi} - \frac{1}{\phi}\right)$	$\varepsilon = \frac{3}{\phi}\left(\frac{1}{\tanh(3\phi)} - \frac{1}{3\phi}\right)$
3. Infinite cylindrical rod (sealed ends)	$\phi_{c1} = R\left(\frac{k_v}{D_e}\right)^{1/2}$	$\Lambda = \left(\frac{k_v}{D_e}\right)^{1/2}$	$\frac{R}{2}$	$\varepsilon = \frac{3}{\phi_{c1}}\left(\frac{I_1(\phi_{c1})}{I_{c1}(\phi_{c1})}\right)$ where I_0 and I_1 are modified Bessel functions of the first kind of order 0 and 1, respectively	$\varepsilon = \frac{1}{\phi}\left(\frac{I_1(2\phi)}{I_0(2\phi)}\right)$
4. Single pore (open ends)	$\phi_{p1} = L_p\left(\frac{k_v}{D_e}\right)^{1/2}$ $\left[= L_p\left(\frac{2k_s}{r_p D_e}\right)^{1/2}\right]$	Same as for infinite slab with L replaced by L_p			

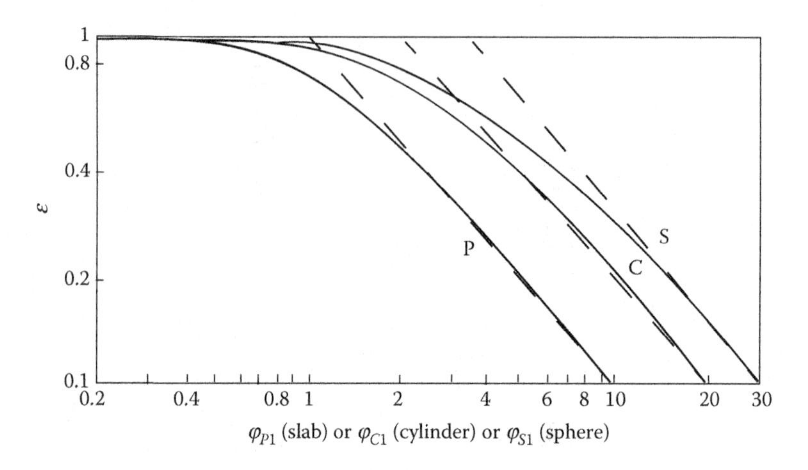

Figure 6.7 Effectiveness factors for a first-order reaction in a slab (P), cylinder (C), and sphere (S) as functions of Thiele moduli for the three shapes. (Adapted from Aris, R., *Elementary Chemical Reactor Analysis*, McGraw-Hill, New York, 1969.)

cylinder, and x for the flat plate), Λ_0 is the *generalized length parameter* characteristic of any shape

$$\Lambda_0 = \frac{\text{Volume of shape}}{\text{Surface area of shape}} \tag{6.48}$$

with $\Lambda_0 = L/2$ for the plate, $R/2$ for the cylinder, and $R/3$ for the sphere; s is a shape constant, with values of 2 for the sphere, 1 for the cylinder, and 0 for the flat plate; and ϕ is the *generalized Thiele modulus* applicable to any shape and any reaction order

Thiele modulus
Applicable to any shape
Any reaction

$$\phi = \Lambda_0 \sqrt{\frac{(n+1)k_v[A]_S^{n-1}}{2D_{eA}}} \tag{6.49}$$

The effectiveness factor equations obtained by solving the shape-generalized Equation 6.36 for a first-order reaction ($n = 1$) are included in Table 6.4.

Weisz modulus: Practical useful quantity

Weisz modulus

Recall that for calculating the Thiele modulus, knowledge of the rate constant is needed, which requires elaborate kinetic studies under conditions free of diffusional effects. A practically more useful modulus based on *observable quantities* can be obtained by recasting Thiele modulus in the form

$$\phi^2 = R^2 \frac{(\text{true rate})}{D_{eA}[A]_S} \tag{6.50}$$

and defining the new modulus as

$$\phi_a^2 = R^2 \frac{(\text{actual rate})}{D_{eA}[A]_S} = R^2 \frac{[\varepsilon(\text{true rate})]}{D_{eA}[A]_S} = \phi^2 \varepsilon \qquad (6.51)$$

Weisz modulus

This modulus is named after Weisz who first proposed it along with Prater in 1954 and can easily be prepared from the more common ε–ϕ plot.

Delineation of regimes Figure 6.8 shows a plot of the effectiveness factor against the Weisz modulus using the generalized length parameter Λ_0 and is hence valid for all shapes. The ε versus Thiele modulus plot is also shown in the figure. Notice that the two curves coincide with each other except for a small range in the shaded region. Three regions can be identified: chemical control, diffusion control, and combined control (shaded). These are clearly marked on the figure with corresponding values of ϕ_1 and ϕ_a.

Nonisothermal effectiveness factors

Generation of heat inside a pellet due to reaction and its transport through the pellet can greatly affect the reaction rate. For endothermic reactions there is a fall in temperature within the pellet. As a result, the rate falls, thus augmenting the retarding effect of mass diffusion. On the other hand, for exothermic reactions, there is a rise in temperature within the pellet. This leads to an increase in rate that can more than offset the decrease due to lowered concentration. Thus, the effectiveness factor can actually be greater than 1.

In analyzing the effect of thermal diffusion within the pellet, the methodological solution involves the solution of equation of continuity and

The energy balance within the pellet

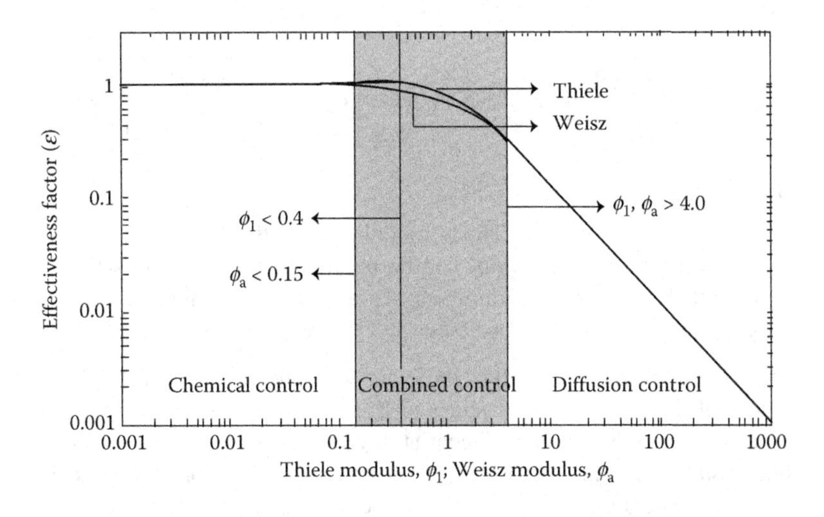

Figure 6.8 Effectiveness factor as a function of Weisz (observable) modulus ϕ_a and also Thiele modulus ϕ_1. The regimes of control are also shown.

equation of energy simultaneously. But for the sake of simplicity, we will lump the lateral variations in temperature by making use of the following fundamental heat balance:

$$D_{eA}(-\Delta H_r)([A]_S - [A]) = \lambda(T - T_s) \tag{6.52}$$

where $-\Delta H_r$ is the heat of reaction and λ the thermal conductivity of the pellet. This can be expressed in dimensionless form as

$$\frac{\Delta T}{T_S} = \beta_m(1 - [\hat{A}]) \tag{6.53}$$

where

$$\beta_m = \frac{(-\Delta H_r)D_{eA}[A]_S}{\lambda T_s} \tag{6.54}$$

represents the maximum temperature rise, that is, the rise when the inside concentration $[\hat{A}](=[A]/[A]_s)$ is zero. Another commonly used group is the Arrhenius parameter expressed at the surface temperature:

$$\alpha_S = \frac{E}{R_g T_s} \tag{6.55}$$

We now consider a differential section of a pellet of any shape and write the following continuity equation:

$$D_A\left(\frac{d^2[\hat{A}]}{d\hat{\Lambda}^2} + \frac{s}{\hat{\Lambda}}\frac{d[\hat{A}]}{d\hat{\Lambda}}\right) = -r_{vA}([A], T_S) \tag{6.56}$$

In dimensionless form, this becomes

$$\frac{d^2[\hat{A}]}{d\hat{\Lambda}^2} + \frac{s}{\hat{\Lambda}}\frac{d[\hat{A}]}{d\hat{\Lambda}} = \phi_1^2[\hat{A}] = (\phi_1)_S^2 \exp\left(\alpha_s\beta_m\frac{1-[\hat{A}]}{1+\beta_m(1-[\hat{A}])}\right)[\hat{A}] \tag{6.57}$$

where $(\phi_1)_S$ represents the Thiele modulus for a first-order reaction at surface temperature. Solutions can be obtained as effectiveness factor plots with β_m and α_s as parameters. An astounding number of studies have been reported on various aspects of the solutions (see Aris, 1975).

Nonisothermal effectiveness factors

The plots of Weisz and Hicks (1962) are reproduced in Figure 6.9. The nature of the curves at high values of β_m suggests multiple solutions. In other words, the reaction can occur at three steady states, two stable and one unstable. It is instructive to note that ε given by one of the solutions in the multiple steady-state region can be orders of magnitude higher than unity. Instabilities of this kind are essentially local in nature. The stable multiple solutions belong to the interior and exterior surfaces, the one

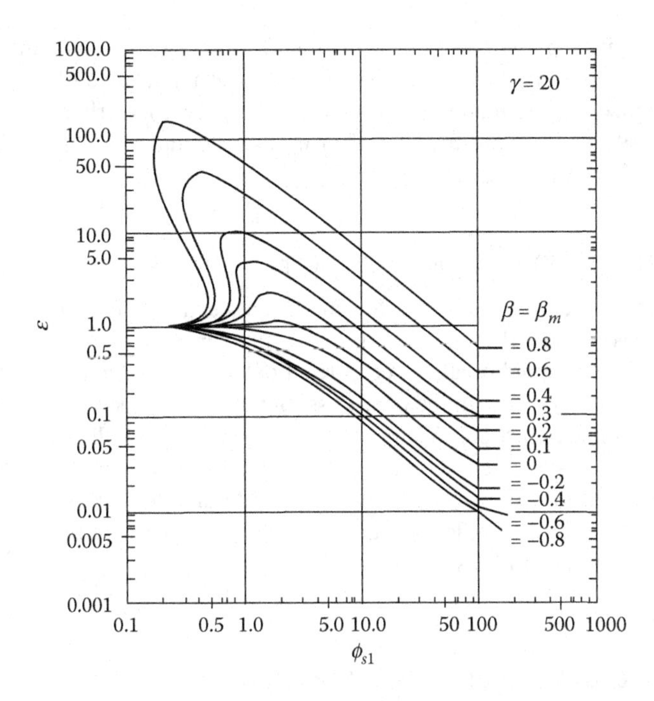

Figure 6.9 Effectiveness factor with first-order reaction in a spherical nonisothermal catalyst pellet. (Adapted from Weisz, P.B. and Hicks, J.S., *Chem. Eng. Sci.*, **17**, 265, 1962.)

with higher temperature depend on whether the reaction is endothermic or exothermic. It is important to note that the reactor as a whole can also exhibit multiple steady states, and a brief treatment of such instabilities is given in Chapter 1.

Multicomponent diffusion The mathematical description of diffusion of more than one component is a complex problem and not germane to our subject. However, with the present trend toward increasing use of solid catalysts in chemical synthesis, many situations do arise in which multicomponent diffusion is involved. We left the prediction of the multicomponent diffusivity outside the scope of this book, which can be found elsewhere (Doraiswamy, 2001). Once the multicomponent diffusivity is determined, the effective diffusivity can then be found from Equation 6.7.

Miscellaneous effects A number of factors can influence the effectiveness factor, some of which are particle size distribution in a mixture of particles/pellets, change in volume upon reaction, pore shape and constriction (such as ink-bottle-type pores), radial and length dispersion of pores, micro–macro pore structure, flow regime (such as bulk or Knudsen), surface diffusion, nonuniform environment around a pellet, dilution of catalyst bed or pellet, distribution of catalyst activity in a pellet (see, in particular, the review by Gavrilidis and Varma, 1993), transverse diffusion, and external surface of catalyst. Reference may be made to the books by Aris (1975), Carberry (1976), Butt (1980), Doraiswamy and Sharma (1984), and Lee (1985) for a discussion of these effects.

Extension to complex reactions

We considered in Chapter 2 the mathematical treatment of complex reactions. It was also shown how some simpler reaction schemes like parallel, series, and parallel–series reactions are amenable to analytical solution. We consider in the present section the role of pore diffusion in these complex reactions. We omit the mathematical details and present in Table 6.5 the salient features of the effect of pore structure, that is, monomodal or bimodal distribution, on yield and conversion in a few selected types of complex reactions. Product R (bolded in the table) is considered to be the desired product.

Question to ponder: What is the reason for high selectivity in enzymatic reactions?

Noncatalytic gas–solid reactions

We will continue with our excursion to the reaction systems with an interface but this time we will deal with noncatalytic gas–solid reactions. These types of reactions are quite common in industry, and even in everyday life, burning of coal being the most common example. The difference between the treatment of the gas–solid catalytic and gas–solid noncatalytic reactions are several fold. We will list the most important ones that will differentiate the analysis here:

1. The catalytic gas–solid reactions take place across an interface that is invariant. During the noncatalytic gas–solid reactions, the interface moves.
2. Owing to the time dependency of the size and the shape of the interface, the noncatalytic gas–solid reactions are of unsteady state in nature, which should be reflected in the mathematics.
3. The porosity of the reactive solid determines whether the reaction is taking place across just the interface or the whole volume of the solid is participating in the reaction.

With these we enlist the two fundamental approaches to the noncatalytic gas–solid reaction systems: The shrinking core model and volume reaction model. In the volume reaction model, the solid is porous, the fluid easily diffuses in or out of the solid, such that the reaction can take place homogeneously everywhere in the solid. On the other hand, with the shrinking core model (SCM), also called the sharp interface model (SIM), there is a sharp interface between the unreacted core and reacted shell of the particles.

Table 6.5 Effect of Pore Diffusion on Selectivities/Yields in the More Common Classes of Reactions

Reaction R = Desired Product	Pore Structure	Intrinsic Selectivity/ Yield	Actual Selectivity/ Yield	Main Features
1. Independent $A \xrightarrow{1} R$, $B \xrightarrow{2} S$	Monodispersed	$s = \dfrac{k_{v1}}{k_{v2}}$	$s_a = \left(\dfrac{s}{\alpha_D}\right)^{1/2}$	Greater diffusional resistance for reaction 2 enhances y_{Ra}
2. Parallel $A \xrightarrow{1,m} R$, $A \xrightarrow{2,n} S$	Monodispersed	$s = \dfrac{k_{v1}}{k_{v2}}$, $y_R = \dfrac{1}{1+p_{nm}}$ where $p_{nm} = \dfrac{k_{v2}}{k_{v1}[A]_s^{m-n}}$ $(m > n)$	$s_a = s^{1/2}$ $\dfrac{y_{Ra}}{y_R} = \dfrac{n+1}{2m-n+1}$ (Pawlawski, 1961; Roberts, 1972)	Plot of $\dfrac{y_{Ra}}{y_R}$ (0.3–1.0) vs. ϕ (0–10); curves p_{nm} = 0.10, 1.0, 10, 100; $m = 2$, $n = 1$; Max. decrease = 1/2
3. Consecutive $A \xrightarrow{1} R \xrightarrow{2} S$		$Y_R = \dfrac{s}{1-s}\left[(1-x_A)^{\tau} - (1-x_A)\right]$		
	Monodispersed	$s = \dfrac{k_{v1}}{k_{v2}}$, $\tau = \dfrac{1}{s}$	$\tau = \dfrac{1}{s^{1/2}}$	Plot of y_{Ra} (0–1.0) vs. X_A (0–1.0); s, α_D fixed; curves $\tau = 1/s$, $1/s^{1/2}$, $1/s^{1/4}$
	Bidispersed	$s = \dfrac{k_{v1}}{k_{v2}}$, $\tau = \dfrac{1}{s}$	$\tau = \dfrac{1}{s^{1/4}}$ (Carberry, 1962; Doraiswamy and Sharma, 1984)	

continued

Table 6.5 (continued) Effect of Pore Diffusion on Selectivities/Yields in the More Common Classes of Reactions

Reaction R = Desired Product	Pore Structure	Intristic Selectivity/Yield	Actual Selectivity/Yield	Main Features
4. Parallel–consecutive a. $A \xrightarrow{1} \mathbf{R} \xrightarrow{2} S$, $A \xrightarrow{3} S$	Monodispersed	$s = \frac{k_{v1}}{k_{v2}}$	$y_{Ra} = f(\phi, s_n)$	y_{Ra} vs. X_A (0 to 1.0); s_n fixed; ϕ; 0.2; 0.1
b. $A \xrightarrow{1} \mathbf{R} \xrightarrow{2} T$, $A \xrightarrow{3} S$	Monodispersed	$s_{13} = \frac{k_{v1} + k_{v3}}{k_{v2}}$	See Wirges and Raehse (1975) for full equation $s_n = s_1$ for $k_{v3} = 0$ $s_n = s_{13}$ for $k_{v3} \neq 0$	
c. $A + B \xrightarrow{1} \mathbf{R}$ $R + B \xrightarrow{2} S$	Monodispersed	$y_{RB} = \frac{[R]}{[B]_0}$ $y_{RA} = \frac{[R]}{[A]_0}$ $= y_{RB}\left(\frac{[B]_0}{[A]_0}\right)$	No analytical solution for $(y_{RB})_a$ or $(y_{RB})_a$, numerical solution by Wirges and Raehse (1975)	$(y_{RB})_a$ vs. X_A (0 to 1.0); $\frac{[B]_0}{[A]_0}$; 0.1; 10

d. Same as (a)	Monodispersed with the external surface constituting a finite fraction (f) of the total surface area of the catalyst	Same as for (a) or (b)	No analytical solution, numerical solution by Varghese et al. (1978)	y_{Ra} is a strong function of f. It increases with Bi_m. 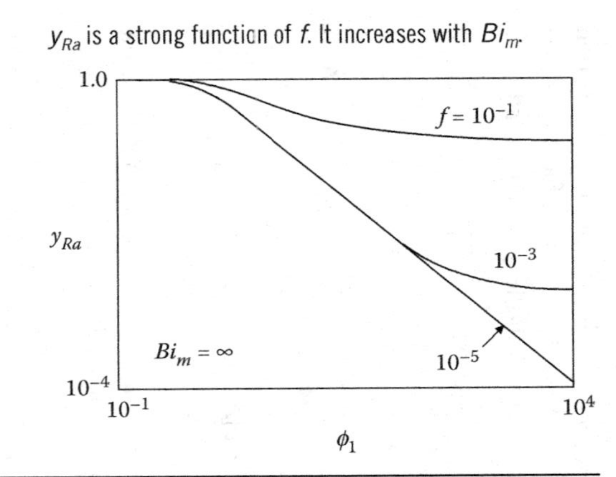

The shrinking core model

For the shrinking core model, we start with the continuity equation. Since the reaction is taking place at a fast rate at the interface, we do not have the rate in the continuity equation but in the boundary condition:

$$\frac{1}{r^2}\frac{\partial}{\partial r}\left(D_{eA}r^2\frac{\partial[A]_s}{\partial r}\right)=0 \tag{6.58}$$

$$\text{BC1:}\quad \text{at } r=r_c,\quad D_{eA}\left(\frac{\partial[A]_s}{\partial r}\right)=ak_s[A]_s[S]_0 \tag{6.59}$$

$$\text{BC2:}\quad \text{at } r=R,\quad D_{eA}\left(\frac{\partial[A]_s}{\partial r}\right)=k_g([A]-[A]_s^*) \tag{6.60}$$

The volume reaction model

On the other hand, for the volume reaction model, the set of differential equation and the boundary conditions are as follows:

$$\frac{1}{r^2}\frac{\partial}{\partial r}\left(D_{eA}r^2\frac{\partial[A]_s}{\partial r}\right)=r_A \tag{6.61}$$

$$\text{BC1:}\quad \text{at } r=0,\quad \left(\frac{\partial[A]_s}{\partial r}\right)=0 \tag{6.62}$$

$$\text{BC2:}\quad \text{at } r=R,\quad D_{eA}\left(\frac{\partial[A]_s}{\partial r}\right)=k_g([A]-[A]_s^*) \tag{6.63}$$

In Chapter 10, we will demonstrate the solution strategies and analyze different cases for both situations.

Gas–liquid reactions in a slab

Gas–liquid reactions constitute a very important class of reactions in chemical process industry. The mathematical treatment of the mass transfer between a gas film and a liquid film is rather similar to the developments of this chapter. Thus, in the spirit of keeping the analogy, we will briefly treat the fundamentals of the gas–liquid reactions here, leaving the rest of the detail to the Chapter 11 dedicated to gas–liquid reactions.

We will begin by deriving a general equation for the gas–liquid reaction

$$A(g)+B(l)\rightarrow P(l)$$

with a simple rate expression given by

$$r=k[A]\,[B] \tag{6.64}$$

We will take the reaction taking place in the bulk after the mutual diffusion of species $[A]$ and $[B]$ to the reaction locus via diffusion on the geometry shown in Figure 6.10.

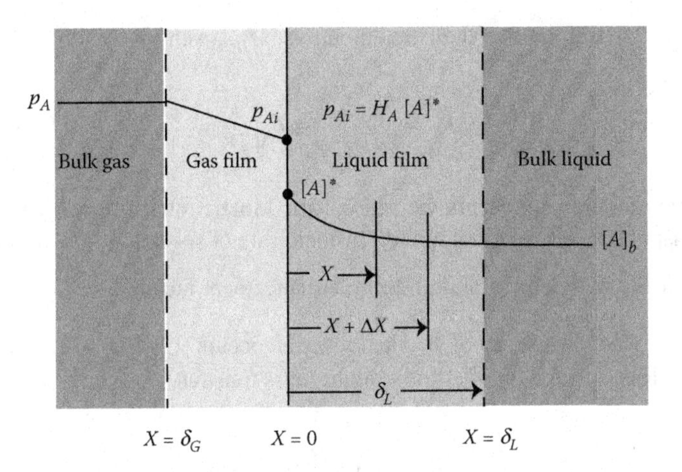

Figure 6.10 The geometry of reaction diffusion planes.

Gas–liquid reactions are classified according to the relative rate of the reaction in comparison to the rate of mass transfer as follows:

- Very slow reactions taking place only in the bulk of the liquid
- Slow reactions taking place in the liquid film as well as bulk
- Fast reactions taking place in the liquid film
- Instantaneous reactions taking place at a plane within the film
- Instantaneous reactions taking place at the gas–liquid interface

Two-film theory

Slow reactions For slow reactions, the rate of diffusion is comparable to the rate of reaction and therefore the differential equation describing the process contains both pieces of information. The differential equations for the distribution of [*A*] and [*B*] across the gas and liquid interfaces can be written, with the boundary conditions, as

$$D_A \frac{d^2[A]}{dx^2} = r \quad \text{BC1: at } x = 0, \quad [A] = [A]^* \qquad \text{BC2: at } x = \delta, \quad [A] = [A]_b \tag{6.65}$$

$$D_B \frac{d^2[B]}{dx^2} = br \quad \text{BC1: at } x = 0, \quad \frac{d[B]}{dx} = 0 \qquad \text{BC2: at } x = \delta, \quad [B] = [B]_b \tag{6.66}$$

Solutions of these differential equations are relatively straightforward, and the concentration of *A* is given by

$$[A] = \frac{[A]^* \sinh[M_H(1 - (x/\delta))] + [A]_L \sinh(M_H(x/\delta))}{\sinh(M_H)} \tag{6.67}$$

where M_H is known as the Hatta modulus, M_H, given by

Hatta modulus:

Compare to the Thiele modulus!

$$M_H = \left(\frac{t_D}{t_R}\right)^{1/2} = \delta\sqrt{\frac{k}{D_A}} \tag{6.68}$$

Hatta modulus represents the ratio of the kinetic rate in the absence of transport effects to maximum diffusional rate of species A into a liquid.

Then, we will define the liquid film enhancement factor.

The enhancement factor:

Compare to the effectiveness factor

$$\eta = \frac{\text{Rate of take-up of } A \text{ when reaction occurs}}{\text{Rate of take-up of } A \text{ for straight mass transfer}} = \frac{j_A}{k_L([A]^* - [A]_L)} \tag{6.69}$$

If we ignore the concentration of A in bulk and define

$$j_A = -D_A \left.\frac{d[A]}{dx}\right|_{x=0} = \frac{M_H}{\tanh M_H}\left[1 - \frac{[A]_L}{[A]^*}\frac{1}{\cosh M_H}\right] k_L[A]^* \tag{6.70}$$

We obtain

$$\eta = \frac{M_H}{\tanh M_H}\left(1 - \frac{[A]_L}{[A]^*}\frac{1}{\cosh M_H}\right) \tag{6.71}$$

For negligible A in the bulk

$$\eta = \frac{M_H}{\tanh M_H} \tag{6.72}$$

This definition is analogous to the effectiveness factor, where mass transfer was *inhibiting*. In this case, mass transfer is being enhanced in the presence of chemical reaction, and the enhancement factor is always greater than 1.

Note that unlike the definition of the effectiveness factor for the catalytic reactions where the normalizing rate was the rate of the reaction, here, the normalizing rate is the rate of mass transfer. Thus, the reaction is considered as the *intruder* (albeit benevolent, or *enhancing*), whereas for catalytic reactions, diffusion was the intruder (often, but not always, *retarding*).

Instantaneous reactions In the case of fast reactions, the rate is so fast that it does not appear in the diffusion volume. The kinetic information is only a part of the boundary condition. The geometry is shown in Figure 6.11.

$$D_{eA}\frac{d^2[A]}{dx^2} = 0 \quad BC1\text{: } at\ x = 0,\quad [A] = [A]^* \qquad BC2\text{: } at\ x = \lambda,\quad [A] = 0 \tag{6.73}$$

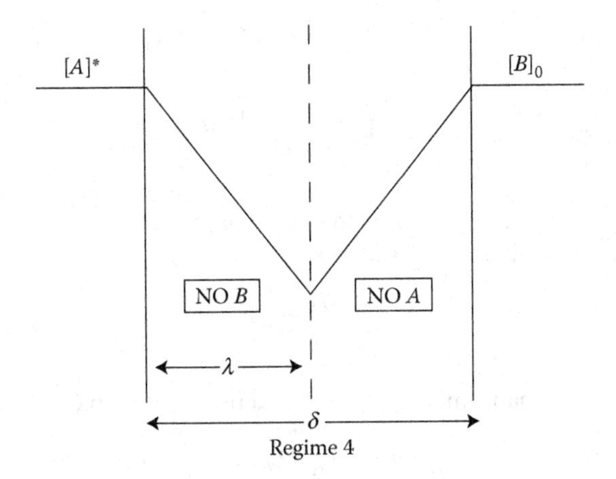

Figure 6.11 The geometry and the concentrations for instantaneous reaction systems.

Giving

$$[A] = [A]^*\left(1 - \frac{x}{\lambda}\right)$$

$$D_{eB}\frac{d^2[B]}{dx^2} = 0 \quad BC1: at\ x = \delta,\ \ [B] = [B]_b \qquad BC2: at\ x = \lambda,\ \ [B] = 0 \tag{6.74}$$

Giving

$$[B] = [B]_b\left(\frac{x - \lambda}{\delta - \lambda}\right)$$

Such that the fluxes of A and B across the interface as follows:

$$j_A = \frac{D_{eA}}{\lambda}[A]^* \tag{6.75}$$

$$-j_B = \frac{D_{eB}}{\delta - \lambda}[B]_b \tag{6.76}$$

The stoichiometry of the reaction $\nu_B A + \nu_B B \rightarrow products$ leads to

$$\frac{j_A}{\nu_A} = -\frac{j_B}{\nu_B} \tag{6.77}$$

Combination of Equations 6.75, 6.76, and 6.77 gives us

$$\frac{\delta}{\lambda} = \left(1 + \frac{\nu_A}{\nu_B}\frac{D_{eB}}{D_{eA}}\frac{[B]_b}{[A]^*}\right) \tag{6.78}$$

Noting that the flux of A with the reaction is given by Equation 6.75 while the flux without the reaction is

$$j_A = \frac{D_{eA}}{\delta}[A]^* \tag{6.79}$$

Such that the enhancement factor defined by Equation 6.69 is simply

$$\eta = \left(1 + \frac{\nu_A}{\nu_B}\frac{D_{eB}}{D_{eA}}\frac{[B]_b}{[A]^*}\right) \tag{6.80}$$

Now, if we look at the whole picture of the gas–liquid reactions, we have the following situations.

If the rate is slow, Hatta modulus is small. If the rate is fast, the Hatta modulus is large. For very large Hatta modulus, the enhancement factor is equal to Hatta modulus, since tanh $\infty = 1$. Finally, for instantaneous reactions, the enhancement factor has a limiting value given by Equation 6.80. A qualitative plot of enhancement factor versus Hatta modulus is given in Figure 6.12. The merits of the figure and how it can be used for equipment selection will be discussed in Chapter 11.

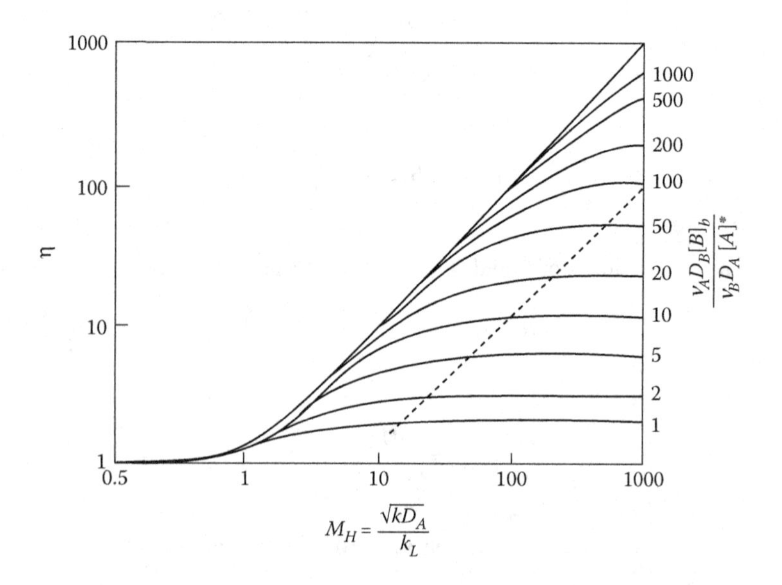

Figure 6.12 Enhancement factor as a function of the Hatta modulus.

Effect of external mass and heat transfer

External effectiveness factor

It is useful to define an *external effectiveness factor* along the lines of the effectiveness factor for diffusion within the solid ε, which may now be more appropriately called the "internal effectiveness factor."

External effectiveness factor

$$\text{External effectiveness factor } \varepsilon_{\text{ext}} = \frac{\text{Actual rate in the presence of external diffusional resistance}}{\text{Rate under conditions where the surface and bulk concentrations are the same}}$$

The mathematical representation for a first-order reaction becomes

$$\varepsilon_{\text{ext}} = \varepsilon \frac{[A]_s}{[A]} \tag{6.81}$$

Given that

$$\frac{[A]_s}{[A]} = \frac{1}{1 + \phi^2(\varepsilon / Bi_M)} \tag{6.82}$$

One obtains

Biot number for mass transfer

$$\frac{1}{\varepsilon_{\text{ext}}} = \frac{1}{\varepsilon} + \frac{\phi^2}{Bi_M} \tag{6.83}$$

where Bi_M is the Biot number for mass transfer given by $k_m R/D_e$.

As will be shown later, heat transfer is considerably more important than mass transfer in the external film. Detailed theoretical analyses of the external film problem are available (Carberry and Kulkarni, 1973; Carberry, 1975) and will not be considered here.

Combined effects of internal and external diffusion

We have thus far considered the two effects separately. The combined effects of internal and external diffusion can be accounted for as follows.

In such a case, we need to solve Equation 6.36 with a modified surface boundary condition, which accounts for the diffusional resistance across the fluid film on the surface and hence gives the true surface concentration $[A]_s$. Thus, when the boundary condition will not be $[A] = [A]_s$ but must be modified to

$$\frac{D_{eA}}{L} = \left(\frac{d[A]}{dr}\right) = k_G'\{[A]_b - [A]\} \tag{6.84}$$

or

$$[\hat{A}] = [\hat{A}]_b - \frac{1}{Bi_m}\frac{d[\hat{A}]}{dr} \tag{6.85}$$

The solution for a flat plate ($S = 0$) is then

$$\varepsilon = \frac{\tanh \phi_{p1}}{\phi_{p1}(1 + (\phi_{p1} \tanh \phi_{p1}/Bi_m))} \tag{6.86}$$

A practical way of plotting this equation is to use the observable quantity $\phi_a = \varepsilon\phi_1^2$ instead of ϕ_1, where ϕ_1 is independent of shape. Such a plot is shown in Figure 6.13.

Relative roles of mass and heat transfer in internal and external diffusion

Gas phase reactants An appreciation of the relative magnitudes of the heat and mass transfer effects in internal and external diffusion is useful. A measure of the relative magnitudes is the ratio of the mass to heat Biot numbers:

$$B = \frac{Bi_m}{Bi_k} = \frac{k'_G L/D_e}{hL/\lambda_e} = \left[\frac{(\Delta C_{\text{int}}/\Delta C_{\text{ext}})}{(\Delta T_{\text{int}}/\Delta T_{\text{ext}})}\right] = \left[\frac{(\Delta C/\Delta T)_{\text{int}}}{(\Delta C/\Delta T)_{\text{ext}}}\right] \tag{6.87}$$

As this ratio usually has values in the range 10–500 for gas–solid systems, it may be concluded that ΔC_{int} and ΔT_{ext} must be very high. In other words, heat transfer would be the controlling resistance externally and mass transfer internally. This can be understood by considering a

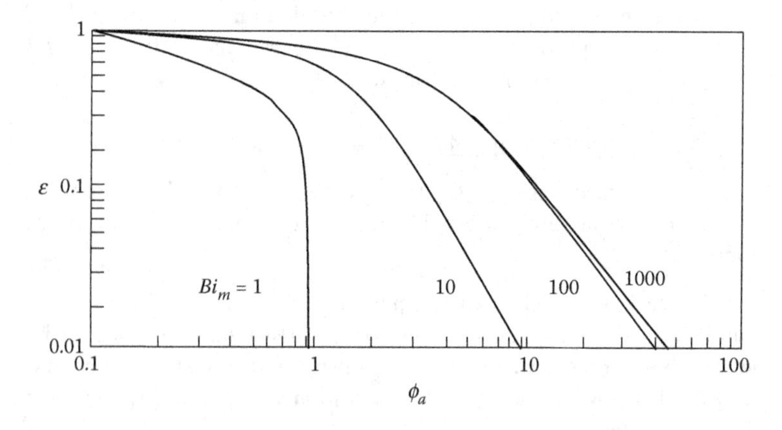

Figure 6.13 Effectiveness factor versus Weisz modulus for different Biot numbers. (Adapted from Carberry, J.J., *Chemical and Catalytic Reaction Engineering*, McGraw-Hill, New York, 1976.)

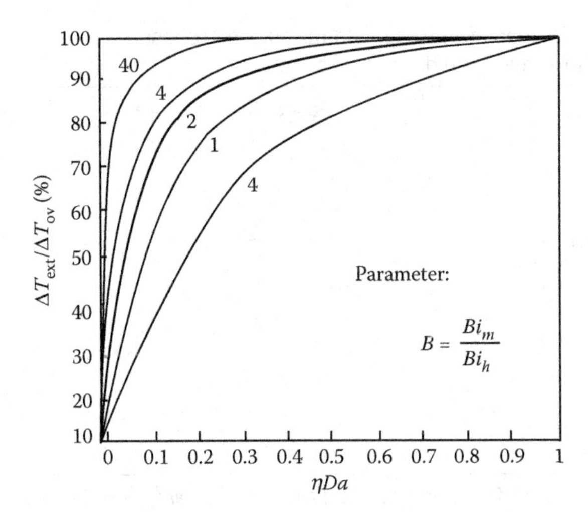

Figure 6.14 Ratio of external to total ΔT versus the observable quantity ηDa for solid catalyzed reactions, with the ratio of mass to heat transfer Biot numbers B as parameter. (Adapted from Carberry, J.J., *Chemical and Catalytic Reaction Engineering*, McGraw-Hill, New York, 1976.)

highly exothermic reaction for which the heat generated within the pellet is transported quickly enough to the surface, but further dissipation across the film is slow, leading to increased surface temperatures and to enhanced reaction rates.

A quantitative analysis of the relative magnitudes of the temperature gradients across the external film and within the pellet leads to a very useful relationship based only on observable quantities (Carberry, 1975), that is

$$\frac{\Delta T_{ext}}{\Delta T_{ov}} = \frac{B\varepsilon_{ext} Da}{1 + \varepsilon_{ext} Da(B - 1)} \tag{6.88}$$

where $\varepsilon_{ext} Da$ for a first-order reaction is $k_{va}/k_G'a = r_a/k_G'a[A]_b$ (an observable quantity). Figure 6.14, which is a graphical representation of this equation, clearly shows a marked rise in the external gradient with increase in the Biot numbers ratio.

Liquid phase reactants Where the reactants are liquids (as in many organic reactions), the values of B as defined in Equation 6.87 are much less than 1, and hence the conclusions would be quite the converse of those for gas phase reactants: the major fraction of the temperature gradient resides within the solid, while the concentration gradient is largely confined to the external film. Indicative value ranges for gas–solid and liquid–solid systems are given in Table 6.6.

Table 6.6 Ranges of Important Intra- and Interphase Parameter Values for Gas–Solid and Liquid–Solid Reactions

Parameter	Gas–Solid	Liquid–Solid
$\alpha_b = \dfrac{E}{R_g T_b}$	5–40	5–40
$\beta_{m,\text{int}}\left[= \dfrac{\Delta T_{\text{int}}}{T_0}\right]$	0.001–0.250	0.001–0.100
$\beta_{m,\text{ext}}\left[= \dfrac{\Delta T_{\text{ext}}}{T_0}\right]$	0.01–2.00	0.001–0.050
$B = \dfrac{Bi_m}{Bi_h}$	$10–10^4$	$10^{-4}–10^{-1}$

Source: Adapted from Carberry, J.J., *Chemical and Catalytic Reaction Engineering*, McGraw-Hill, New York, 1976.

Regimes of control

As the controlling regime changes, the values of the kinetic parameters (E and n) also change. The limiting values of E for different regimes are indicated in Figure 6.15, which is a representation of these changes on an Arrhenius plot for reactions with nonnegligible heat effects.

It will be noted that E can change from a high positive value (for chemical control) to a negligible value for external diffusion control. There can also be a region of negative activation energy corresponding to surface diffusion control, but this is almost never observed and is not considered

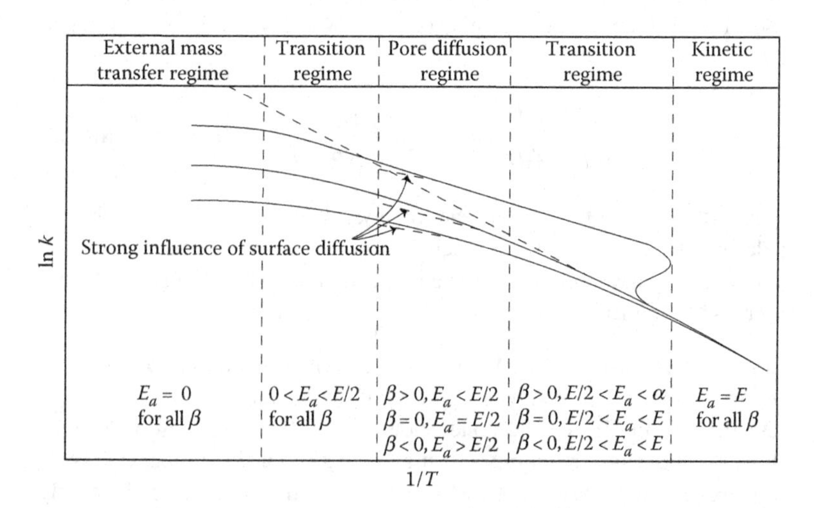

Figure 6.15 Schematic representation of regimes of operation on an Arrhenius diagram. (Adapted from Rajadhyaksha, R.A. and Doraiswamy, L.K., *Catal. Rev. Sci. Eng.*, **13**, 209, 1976.)

here. For highly exothermic reactions, the activation energy can rise to almost infinity as the temperature is raised. Also, the value in the limit of pore diffusion control depends on the thermal nature of the reaction and is equal to $E_{true}/2$ for reactions with no heat effect. The limiting values of E and n for different conditions are discussed by Languasco et al. (1972), Rajadhyaksha and Doraiswamy (1976), Doraiswamy and Sharma (1984), and Dogu (1989), and are included in Figure 6.15.

Explore yourself

1. See if you can quickly generate answers to the following questions:
   a. What is the difference between film mass transfer and bulk diffusion?
   b. What is the difference between bulk diffusion and surface diffusion?
   c. What is the difference between Thiele modulus and Weisz modulus?
   d. The mathematical description of Biot number for heat transfer is very similar to that of Nusselt number. What is their physical significance?
   e. The mathematical description of Biot number for mass transfer is very similar to that of Schmidt number. What is their physical significance?
   f. In Figure 6.15, the measured activation energy is $E/2$ in the presence of mass transfer. Elaborate on the reason.
   g. What is the physical significance of the Hatta modulus? How does it compare to the Thiele modulus?
   h. What is the physical significance of the effectiveness factor? Where and how do we use the effectiveness factors?
2. What is the physical significance of the enhancement factor? Where and how do we use the enhancement factors? How does the enhancement factor compare to the effectiveness factor, conceptually and mathematically?
3. Develop analogies and outline the differences between gas–solid catalytic and gas–solid noncatalytic reactions? In addition to a conceptual description, also provide your answer mathematically.
4. Describe the mass transfer processes around enzymes. List the typical values of the rates and diffusivities. Would you model an enzyme as a solid catalyst? What would be a typical value for the Thiele modulus? What would be the effectiveness factor? Comment on the calculated value of the effectiveness factor and its implications on the backmixing and selectivity.
5. Coal gasification requires contacting fine micrometer to millimeter range in diameter) particles with oxygen and steam in a

reactor. The reaction proceeds through pyrolysis, that is, high-temperature decomposition of the organic structure to methane and small organic molecules, followed by the oxidation of the char. Derive balanced equations to describe each of these processes. Compare your differential equations and the boundary conditions to the model equations reported in the literature. Which one has higher porosity, coal or char? How would your choice of boundary conditions depend on the porosity? How would your model equation depend on the porosity?

6. What is the reason for the multiple values for the nonisothermal effectiveness factors? What is the reason for the values of effectiveness factors > 1? What physical significance does this bear?

References

Astarita, G., *Mass Transfer with Chemical Reactions*, Elsevier, Amsterdam, 1967.

Aris, R., *Elementary Chemical Reactor Analysis*, McGraw-Hill, NY, 1969.

Aris, R., *The Mathematical Theory of Diffusion and Reaction in Permeable Catalysts*, Vols. 1 and 2, Clarendon Press, Oxford, UK, 1975.

Beekman, J.W. and Froment, G.F., *Ind. Eng. Chem. Fundam.*, **21**, 243, 1982.

Brown, L.F., Hanyes, H.W., and Manogue, W.H., *J. Catal.*, **14**, 220, 1969.

Butt, J.B., *Reaction Kinetics and Reactor Design*, Prentice-Hall, Englewood Cliffs, NJ, 1980.

Carberry, J.J., *Chem. Eng. Sci.*, **17**, 675, 1962.

Carberry, J.J., *Ind. Eng. Chem. Fundam.*, **14**, 129, 1975.

Carberry, J.J., *Chemical and Catalytic Reaction Engineering*, McGraw-Hill, NY, 1976.

Carberry, J.J. and Kulkarni, A.A., *J. Catal.*, **31**, 41, 1973.

Cunningham, R.S. and Geankoplis, J., *Ind. Eng. Chem. Fundam.*, **7**, 535, 1968.

Cussler, E.L., *Diffusion: Mass Transfer in Fluid Systems*, Cambridge University Press, London, UK, 2001.

Danckwerts, P.V., *Trans. Faraday Soc.*, **46**, 300, 1950.

Danckwerts, P.V., *Trans. Faraday Soc.*, **47**, 1014, 1951a.

Danckwerts, P.V., *Chem. Res.*, **43**, 1460, 1951b.

Danckwerts, P.V., *Chem. Eng. Sci.*, **2**, 1, 1953.

Dogu, G. Diffusion limitations for reactions in porous catalysts, in *Handbook of Heat and Mass Transfer, Volume 2: Mass Transfer and Reactor Design* (Ed. Cheremisinoff, N.P.) pp. 433–489, Gulf Publishing Co, Houston, TX, 1989.

Doraiswamy, L.K. *Organic Synthesis Engineering*, Oxford, NY, 2001.

Doraiswamy, L.K. and Sharma, M.M., *Heterogeneous Reactions: Analysis, Examples and Reactor Design*, Vol. 1, John Wiley, NY, 1984.

Feng, C.F. and Stewart, W.E., *Ind. Eng. Chem. Fundam.*, **12**, 143, 1973.

Froment, G.F. and Bischoff, K.B., *Chemical Reactor Analysis and Design*, Wiley, NY, 1990.

Fuller, E.N., Schettle, P.D., and Giddings, J.C., *Ind. Eng. Chem.*, **58**,19, 1966.

Gavrilidis, A. and Varma, A., *Catal. Rev. Sci. Eng.*, **35**, 399, 1993.

Gilliland, E.R., *Ind. Eng. Chem.*, **26**, 681, 1934.

Higbie, R., *Trans. Am. Inst. Chem. Eng.*, **31**, 365, 1935.

Johnson, M.L.F. and Stewart, W.E., *J. Catal.*, **4**, 248, 1965.

Languasco, J.M., Cunningham, R.E., and Calvelo, A., *Chem. Eng. Sci.*, **27**, 1459, 1972.

Lee, H.H., *Heterogeneous Reactor Design*, Butterworth, Boston, MA, 1985.
Lewis, W.K. and Whitman, W.G., *Ind. Eng. Chem. Res.*, **16**, 1215, 1924.
Luss, D., In *Chemical Reactor Theory: A Review* (Eds. Lapidus, L. and Amundson, N.R.) Prentice-Hall, Englewood Cliffs, NJ, 1977.
Patel, P.V. and Butt, J.B., *Ind. Eng. Chem. Proc. Des. Dev.*, **14**, 298, 1974.
Pawalski, J., *Chem. Eng. Tech.*, **33**, 492, 1961.
Petersen, E.E., *Chemical Reaction Analysis*, Prentice-Hall, Englewood Cliffs, NJ, 1965.
Rajadhyaksha, R.A. and Doraiswamy, L.K., *Catal. Rev. Sci. Eng.*, **13**, 209, 1976.
Reid, R.C., Prausnitz, J.M., and Poling, B.E., *The Properties of Gases and Liquids*, McGraw-Hill, NY, 1987.
Roberts, G.W., *Chem. Eng. Sci.*, **27**, 1409, 1972.
Satterficld, C.N. and Cadle, P.J., *Ind. Eng. Chem. Fundam.*, **7**, 202, 1968a.
Satterfield, C.N. and Cadle, P.J., *Ind. Eng. Chem. Fundam.*, 256, 1968b.
Tyn, M.T. and Calus, W.F., *Processing,* **21**(4), 16, 1975.
Varghese, P., Varma, A., and Carberry, J.J., *Ind. Eng. Chem. Fundam.,* **17**, 195, 1978.
Wakao, N. and Smith, J.M., *Chem. Eng. Sci.*, **17**, 825, 1962.
Wakao, N. and Smith, J.M., *Ind. Eng. Chem. Fundam.*, **3**, 123, 1964.
Weisz, P.B. and Hicks, J.S., *Chem. Eng. Sci.*, **17**, 265, 1962.
Weisz, P.B. and Prater, C.D., *Adv. Catal.,* **6**, 143, 1954.
Westerterp, K.R., Van Swaaij, W.P.M., and Beenackers, A.A., *Chemical Reactor Design and Operation*, Wiley, NY, 1984.
Wirges, H.P. and Rahse, W., *Chem. Eng. Sci.,* **30**, 647, 1975.

Chapter 7 Laboratory reactors

Collection and analysis of the data

Chapter objectives

After successful conclusion of this chapter, students should be able to

- Design experiments to measure reaction rates.
- Decide the sequence of data measurement for multivariate situations.
- Choose the most appropriate reactor configuration for the measurement.
- Eliminate the effects of heat and mass transfer during the measurement.
- Extract the rate parameters such as reaction order and activation energy, from kinetically relevant data.
- Apply linear and nonlinear regression to the experimental data to extract kinetic parameters.
- Apply graphical and numerical integration/differentiation techniques to analyze data.
- Evaluate the quality in terms of the errors associated with the data.
- Evaluate the quality of the fitted parameters based on the statistical information.
- Discriminate different kinetic models based on the experimental evidence and tools of statistics.

Chemical reaction tests in a laboratory

Chemical reactions can cover a broad range of phases from homogeneous to multiple heterogeneous states. It is very important to select the correct reactor type and method to collect kinetically meaningful data. The simplest of known systems are batch reactors for liquid phase homogeneous reaction tests. We will begin this chapter by giving broad details of chemical reaction tests in a laboratory starting with a simple homogeneous chemical reaction test and progressively modify analysis for complex tests, including monitoring of reaction intermediates over the catalyst surface. It has to be borne in mind that laboratory tests are just the beginning stage of chemical reaction tests. As mentioned in the earlier chapters, industrial reactors possess a number of complexities and most of which can be classified in the transport limitations category. Thus, measuring good kinetic data in a laboratory is just the beginning of an appropriate design of a chemical reactor on the large scale. On the other hand, the experimenter must be careful when performing the reaction tests. Inaccurate measurements of kinetic data mostly results from transport disguises such as heat and/or mass transfer which could not be eliminated due to poor design of the measurement system.

A perspective on statistical experimental design

For an engineer, the most important factor is optimization. In this section, we will concentrate on the optimization with respect to the experimentation time. It is possible to perform as many experiments as possible with the standard methodology of varying one parameter while keeping all others constant if the variables are independent, not interrelated, and known, a rarely met condition in reality. Under situations when ambiguities prevail, it might be wise to factor out some of the variables and choose a statistical experimental design methodology to identify correlations and cross-correlations, and to obtain a maximum amount of information with minimum number of tests. As with any other experimental science, the reader is advised to acquire basic skills in statistics such that the forthcoming material is meaningful. Many texts are available; we will follow Box et al. (2005).

The utility of statistical experiment designs are many fold. First of all, *a priori* design of experiments requires that the researcher carefully consider the dependent and independent parameters. Second, cross-correlations between the independent parameters can be explored. Last but most important of all, the designed experiments minimize experimental effort while maximizing the obtained information. Several experimental design protocols and procedures exist in the literature, and the interested reader is directed to the textbooks by Montgomery and Runger (1994), Box et al. (2005), and Lazic (2004).

Classical experimental designs require the investigation of one parameter at a time. In contrast, full factorial designs include all design points and when combined with the statistical methods of data analysis provide maximum amount of information with the minimum amount of experimentation.

Example 7.1: Factoring out the parameters in a test

Suppose that you are looking for the composition of a catalyst in terms of support, active material, and promoter. A catalyst screening test for a combination of all parameters may require a large number of expensive experimentation and a large number of samples. Instead, factoring out these three independent parameters in 2^3 full factorial design (Table E7.1.1) will enable you to determine the focal point of the optimum composition. The measured variable is the reaction rate. Careful measures should be taken to determine the reaction rate free from artifacts which will be explained in the later sections of this chapter.

In this scheme, the experimentalist has to choose two levels of each variable labeled as (–) and (+). The first experiment, which is conducted with all (–) levels of all independent variables, is called the reference trial. The basic effects and mutual interactions are determined from simple algebraic relationships given below:

$$E_{X1} = \frac{Y_2 + Y_4 + Y_6 + Y_8}{4} - \frac{Y_1 + Y_3 + Y_5 + Y_7}{4}$$

$$E_{X2} = \frac{Y_3 + Y_4 + Y_7 + Y_8}{4} - \frac{Y_1 + Y_2 + Y_5 + Y_6}{4}$$

$$E_{X3} = \frac{Y_5 + Y_6 + Y_7 + Y_8}{4} - \frac{Y_1 + Y_2 + Y_3 + Y_4}{4}$$

$$E_{X1X2} = \frac{Y_1 + Y_4 + Y_5 + Y_8}{4} - \frac{Y_2 + Y_3 + Y_6 + Y_7}{4}$$

$$E_{X2X3} = \frac{Y_1 + Y_2 + Y_7 + Y_8}{4} - \frac{Y_3 + Y_4 + Y_5 + Y_6}{4}$$

$$E_{X1X3} = \frac{Y_1 + Y_3 + Y_6 + Y_8}{4} - \frac{Y_2 + Y_4 + Y_5 + Y_7}{4}$$

$$E_{X1X2X3} = \frac{Y_2 + Y_3 + Y_5 + Y_8}{4} - \frac{Y_1 + Y_4 + Y_6 + Y_7}{4}$$

Table E7.1.1 Full Factorial Design, 2^3

Experiment Number	Variable X_1	Variable X_2	Variable X_3	Response
1	–	–	–	Y_1
2	+	–	–	Y_2
3	–	+	–	Y_3
4	+	+	–	Y_4
5	–	–	+	Y_5
6	+	–	+	Y_6
7	–	+	+	Y_7
8	+	+	+	Y_8

These individual effects and mutual interactions will help the experimenter design further experiments where interactions are either emphasized or eliminated.

Batch laboratory reactors

The type of reactor to be used in laboratory experiments to obtain data for process evaluation or kinetic modeling depends on the nature of the reaction, that is, homogeneous gas phase or liquid phase reaction, gas–liquid or liquid–liquid reaction, gas phase or liquid phase reaction on a solid catalyst, or three-phase slurry reaction. Solid phase reactions are also possible but they are quite rare in organic synthesis/technology. Laboratory reactors can roughly be divided into two categories: reactors for gathering data in a chemist's laboratory with the object of developing a feasible synthetic route for a chemical, and those used to obtain precise kinetic data under isothermal conditions which also take into account the mass and heat transfer features of the reaction. Figure 7.1 lists the main laboratory reactors used for different reaction systems along with an indication of the chapters in which they are considered in this book. In these reactors, one either fully eliminates mass transfer resistances or accounts for them wherever they are inevitably

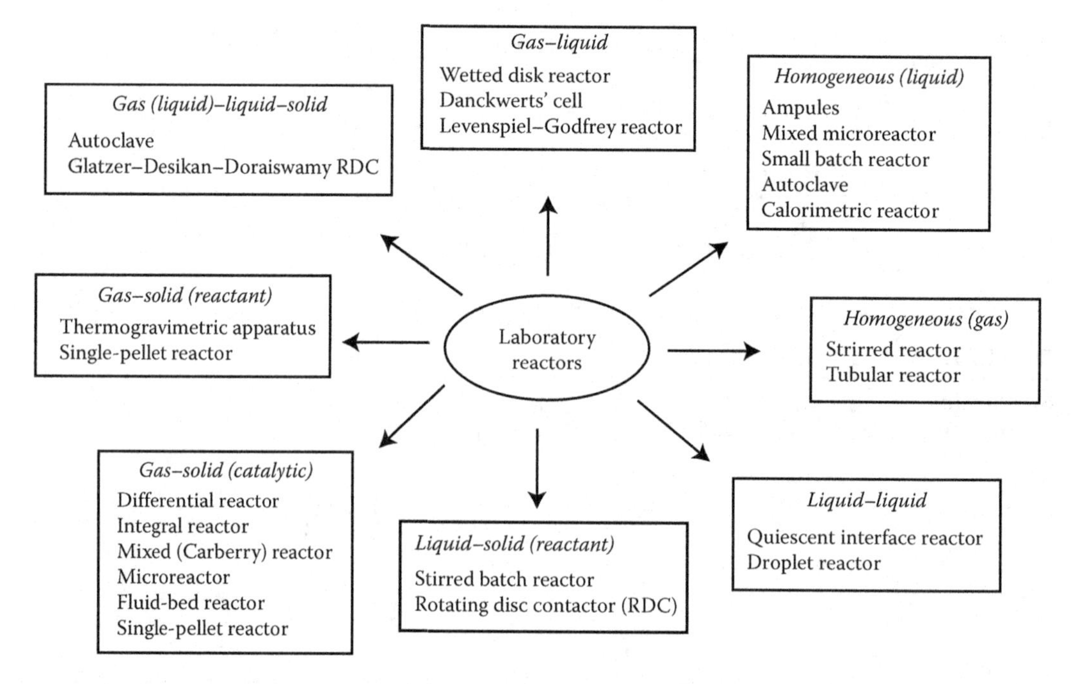

Figure 7.1 Main types of laboratory reactors for studying various homogeneous and heterogeneous reactions.

present. Designs that eliminate mass and heat transfer effects are called *gradientless reactors*. They may be regarded as heterogeneous reactor equivalents of homogeneous reactors which, by their very nature, are gradientless (except for very fast reactions involving two miscible liquids, where local gradients in the vicinity of reaction spots may exist).

For homogeneous reactions, a number of laboratory designs are in use. Several designs are sketched in Figure 7.2. When high pressures are needed, autoclave configurations may be preferred. The heat flow calorimetric reactor (Figure 7.2g) is a particularly interesting design for it eliminates chemical analysis, depending only on the rate of heat evolution as a function of time to determine the conversion as a function of time.

Rate parameters from batch reactor data

In this section we will briefly go over the situation where we will use batch reactor data for determining the rate constant and reaction order for a given reaction. First, we begin by collecting the concentration-time data. In gas phase reactions, pressure can be advantageously used in place of concentration. Also, the rate equations can be used in their differential or integrated forms.

From concentration data

Let us take the nth-order irreversible reaction

$$A \rightarrow \text{Products} \tag{R1}$$

with rate expression

$$-r_A = k[A]^n \tag{7.1}$$

The rate expression can also be written as

$$\ln(-r_A) = \ln k + n \ln[A] \tag{7.2}$$

In the differential method of estimating the rate parameters, we first plot $[A]$ as a function of time and differentiate it either graphically or by curve-fitting. The slope thus obtained gives the rate directly for a reaction with no volume change, based on which the kinetic parameters n and k can be determined as shown in Figure 7.3.

But for a reaction with volume change, the rate is obtained from the following modified form:

$$-\frac{d[A]}{dt} = (-r_A)(1 + X_A\varepsilon_A) \tag{7.3}$$

Let us now consider a bimolecular reaction

$$A + B \rightarrow R + S \tag{R2}$$

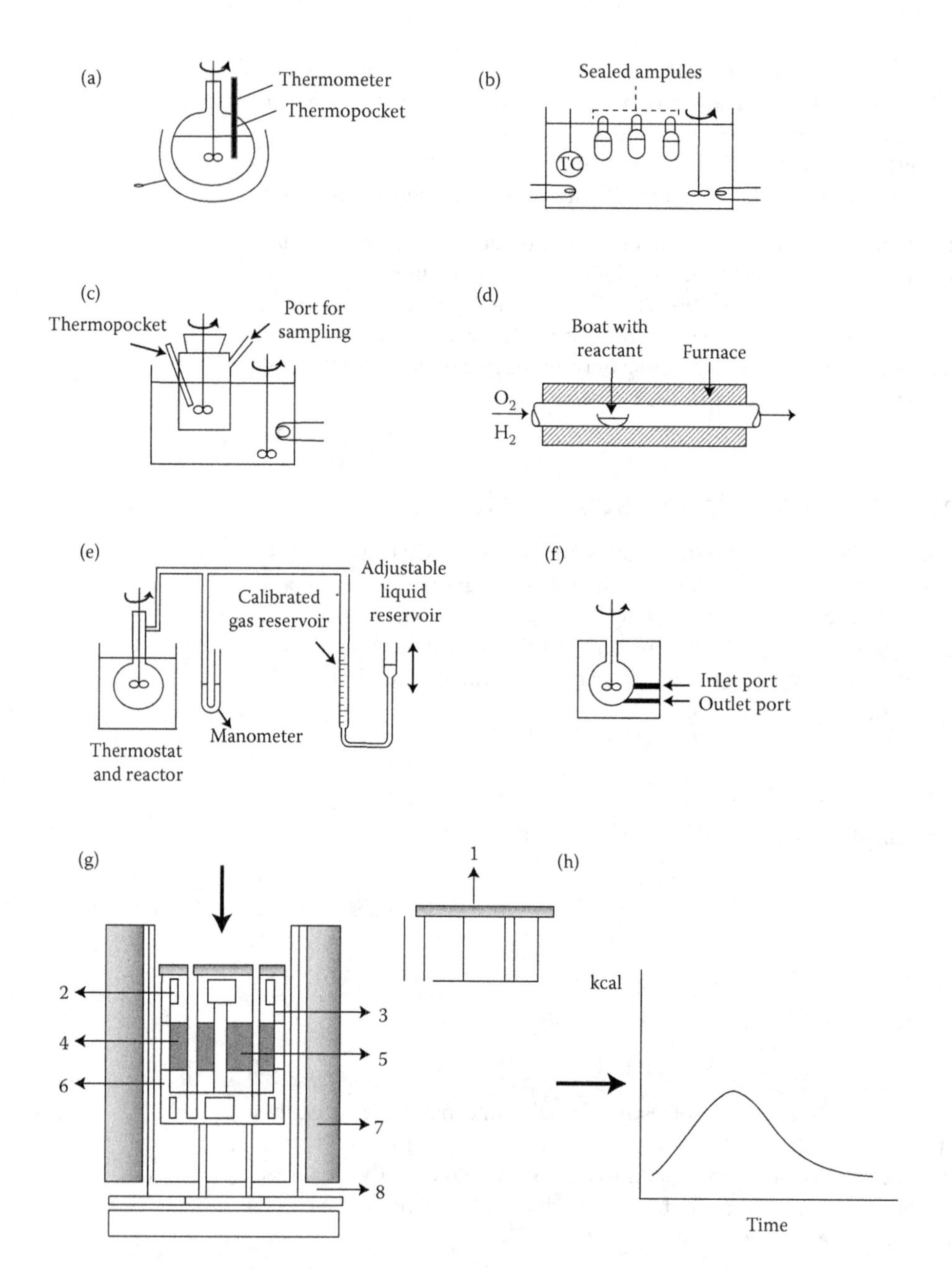

Figure 7.2 Alternative configurations of batch laboratory reactors to obtain kinetic data, mainly from homogeneous mixtures. (a) Round-bottomed flask in a heating mantle, (b) ampules in a thermostat, (c) small bench-scale reactor in a thermostat, (d) boat containing liquid reactant in a furnace with or without a flowing gaseous reactant, (e) reactor with provision for measuring evolving gas, (f) mixed microreactor, (g) calorimetric reactor, and (h) output from calorimetric reactor. 1, Removable lid; 2, thermal buffer zone; 3, heating elements; 4, thermopiles; 5, experimental area; 6, calorimetric block; 7, insulation layers; and 8, cooling circuit.

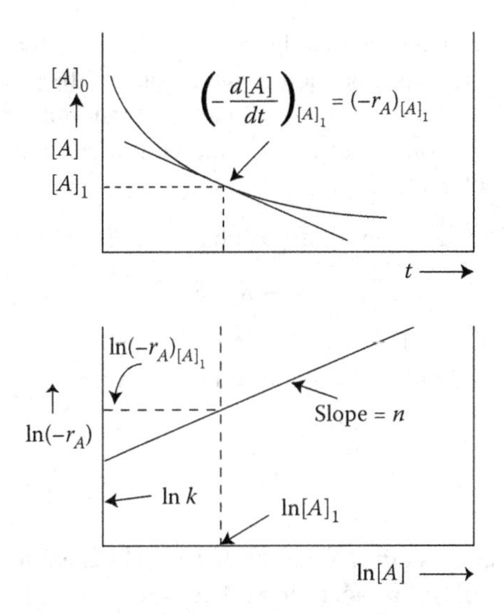

Figure 7.3 Rate constant from concentration–time data for $[-r]_A = k[A]^n$.

The rate in its most general form may be expressed as

$$-r_A = k[A]^m\,[B]^n \tag{7.4}$$

and may be measured as described above for a unimolecular reaction. Equation 7.4 can now be recast as

$$\ln(-r_A) = \ln k + m\ln[A] + n\ln[B] \tag{7.5}$$

and the parameters k, m, and n determined by regression analysis. To determine m, n, and k, careful parameterization of the measurement should be performed. In other words, the experimental data should be collected under conditions where the changes in the parameters are accounted for. Two most commonly used strategies are mentioned below:

1. In determining m, variation in A is not influenced by variation in B. For this, a large excess of B should be employed. The converse should be applied when determining n. When a large excess of one of the reactants was employed, the kinetics reduces from $(m \times n)$th order to pseudo nth order and Equation 7.5 reduces to Equation 7.2.
2. A factorized design can be performed to minimize the number of experiments as briefly explained in Example 7.1 and in the preceding section.

From pressure data

Typically, a liquid phase reaction is carried out in a closed vessel and the progress of reaction monitored by changes in concentration (by analysis of samples withdrawn periodically). Where gas phase reactions are concerned, or a liquid phase reaction in which one of the products is a gas, the progress of reaction can be more easily monitored by recording the pressure as a function of time. This method is particularly useful for reactions with volume change.

Thus for a typical (and common) gas phase reaction with volume change

$$A \rightarrow R + S \tag{R3}$$

the batch reactor mole balance $dN_A/dt = r_A V$ can be recast using the ideal gas law as the following expression can be derived:

$$\frac{dP}{dt} = \frac{k}{(RT)^{n-1}} P^n \tag{7.6}$$

Example 7.2: Decomposition of di-*t*-butyl peroxide in the vapor phase to acetone and ethane

The reaction

$$(CH_3)_2\, COOC(CH_3)_3 \rightarrow 2(CH_3)_2\, CO + C_2H_4 \tag{R4}$$

was carried out in a constant-volume batch reactor (Peters and Skorpinski, 1965) and its progress monitored by the increase in pressure. However, as the *P*–*t* data were not recorded in the original paper, they have been generated from the data given, and the values at two temperatures are given in Table E7.2.1.

Table E7.2.1 Pressure–Temperature Data for the Decomposition of Di-*t*-Butyl Peroxide

T = 154.6°C		*T* = 147.2°C	
Time (min)	*P* (mmHg)	Time (min)	*P* (mmHg)
0	173.5	0	182.6
2	187.3	2	190.5
3	193.4	6	201.7
5	205.3	10	213.6
6	211.3	14	224.3
8	222.9	18	235.0
9	228.6	20	240.4
11	239.8	22	245.4
12	244.4	26	255.6
14	254.5	30	265.2
15	259.2	34	274.4
17	268.7	38	283.3
18	273.9	40	288.0
20	282.0	42	292.0
21	286.8	46	300.2

Nitrogen was used as the diluent in the reaction. Its partial pressure remained approximately constant throughout the reaction (P_{N_2} at 154.6°C = 8.1 mm and at 147.2°C = 4.5 mmHg). The initial partial pressures of the peroxide (A) were 168 mm and 179 mmHg at 154.6°C and 147.2°C, respectively. Obtain a suitable rate equation for the reaction, assuming ideal gas.

SOLUTION

Material balance on the reaction gives

$$-\frac{1}{V_t}\frac{dN_A}{dt} = -\frac{d[A]}{dt} = k_v\frac{N_A}{V_t} \quad \text{or} \quad -\frac{d[A]}{dt} = k_v[A] \quad (7.7)$$

For constant-volume conditions, we have

$$\begin{aligned} N_t &= N_A + N_B + N_C + N_{N_2} \\ N_B &= 2N_C = 2(N_{A_0} - N_A) \\ N_t &= N_A + 2(N_{A_0} - N_A) + (N_{A_0} - N_A) + N_{N_2} \end{aligned}$$

giving

$$N_A = \frac{3N_{A_0} + N_{N_2} - N_t}{2} \quad (7.8)$$

Assuming ideal gas law, $[A] = p_A/RT$,

$$-\ln\left(\frac{[A]}{[A]_0}\right) = k_v t \quad \text{or} \quad \ln\left[\frac{p_{A_0}}{p_A}\right] = \ln\left[\frac{p_{A_0}}{3p_{A_0} + p_{N_2} - P_t}\right] = k_v t$$

$$(7.9)$$

Thus, a plot of $\ln[3p_{A_0} + p_{N_2} - p_{t_0}]$ versus t gives a slope of k_v. Such a plot is shown in Figure 7.4, from which

$$k_v = 0.0086\ \text{min}^{-1} \text{ at } 147.2°\text{C}$$

$$= 0.0193\ \text{min}^{-1} \text{ at } 154.6°\text{C}$$

Thus, the rate model is

$$-r_A = k_v[A] \quad (7.10)$$

where A represents butylperoxide and k_v is the first-order rate constant with the values given above.

Flow reactors for testing gas–solid catalytic reactions

Flow reactors are very convenient especially for catalyst testing. However, one must be very careful when collecting and interpreting the flow reactor

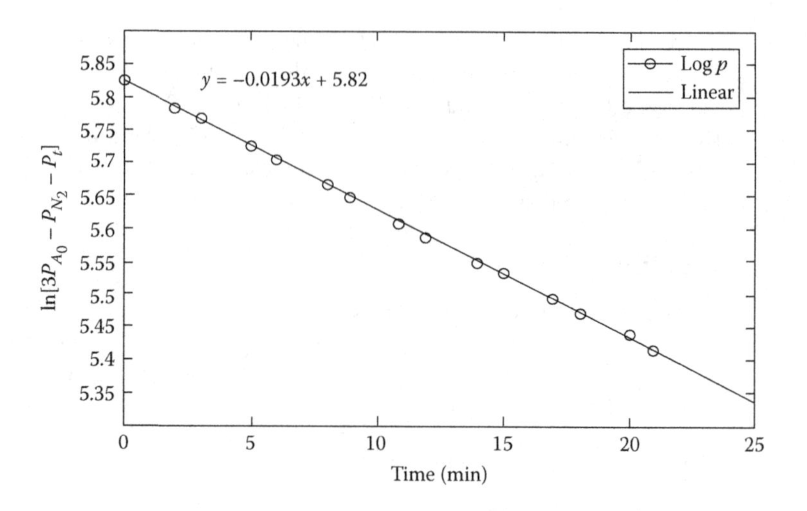

Figure 7.4 Time versus $\ln[3p_{A_0} + p_{N_2} - p_{t_0}]$ plot generated in MATLAB basic fitting option to determine the rate constant k at 154.6°C.

data. To demonstrate the true situation, we will start by comparing the design equations of a PFR, CSTR, a differential flow reactor (DFR), and a recycle reactor (RFR).

Differential versus integral reactors

There are two fundamental types of experimental reactors for measuring solid catalyzed reaction rates: integral and differential. The integral reactor consists essentially of a tube of diameter less than 3 cm filled with, say, W grams of catalyst. Each run comprises steady-state operation at a given feed rate, and based on several such runs a plot of the conversion X_A versus W/F_{A0} is prepared. Differentiation of this curve gives the rate at any given X_A (i.e., concentration) as

$$-r_{wA} = \frac{dX_A}{d(W/F_{A0})}, \quad \text{mol/gcat s} \tag{7.11}$$

This is illustrated in the top half of Figure 7.5. A differential reactor, on the other hand, uses a differential amount of catalyst (usually <1 g) in which a differential conversion (<1–2%) occurs, so that the rate may be directly obtained as

$$-r_{wA} = \frac{\Delta X_A}{\Delta(W/F_{A0})} \tag{7.12}$$

at the average concentration in the bed. This is illustrated in the bottom portion of the figure.

A convenient way of operating a differential reactor at integral conversions is to use a fully mixed reactor in which a constant concentration

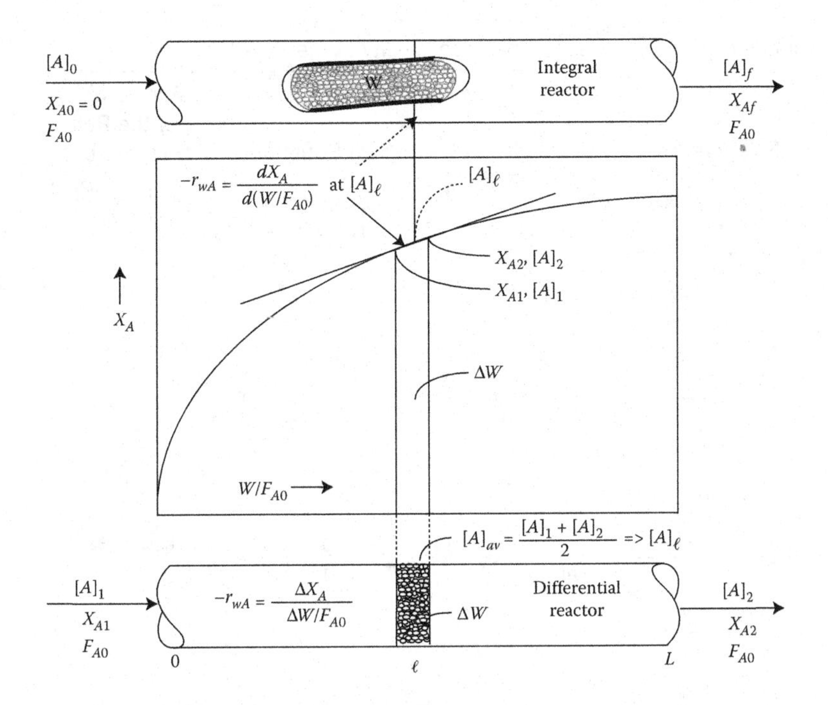

Figure 7.5 Integral and differential reactors for experimental rate determination.

within the reactor is imposed by efficient mixing. Several innovative designs to achieve mixing in a fixed bed of pellets have been proposed, some of which are: the spinning basket reactor (Carberry, 1976), the internal circulation reactor (Berty et al., 1969), the rotating pot reactor (Choudhary and Doraiswamy, 1972), and the recycle reactor (see, e.g., Satterfield and Roberts, 1968; Carberry, 1976). A few important designs are sketched in Figure 7.8 (shown later in the chapter). Finally it should be noted here that the material covered here is not comprehensive. Novel developments in the literature must always be monitored.

It is always best to operate an experimental reactor under conditions where all diffusional disguises are lifted (by using the criteria listed in the next section). A less acceptable alternative is to account for them through appropriate effectiveness factors and external transport coefficients. A number of highly sophisticated computer-controlled reactor systems are commercially available, such as the Berty recycle reactor. Many of them are available with software and appropriate interfacing that can set and implement the experiments for each of a series of sequential runs (see, e.g., Mandler et al., 1983), resulting in the emergence of the most acceptable model at the end of the exercise.

As shown in Table 7.1, the rate is not accessible from the PFR data as well as the RFR data for small R values. When the rate data are not accessible, it is not possible to perform generalized analyses for the kinetic parameters

Table 7.1 Comparison of Various Flow Geometries for the Accessibility of the Rate Information

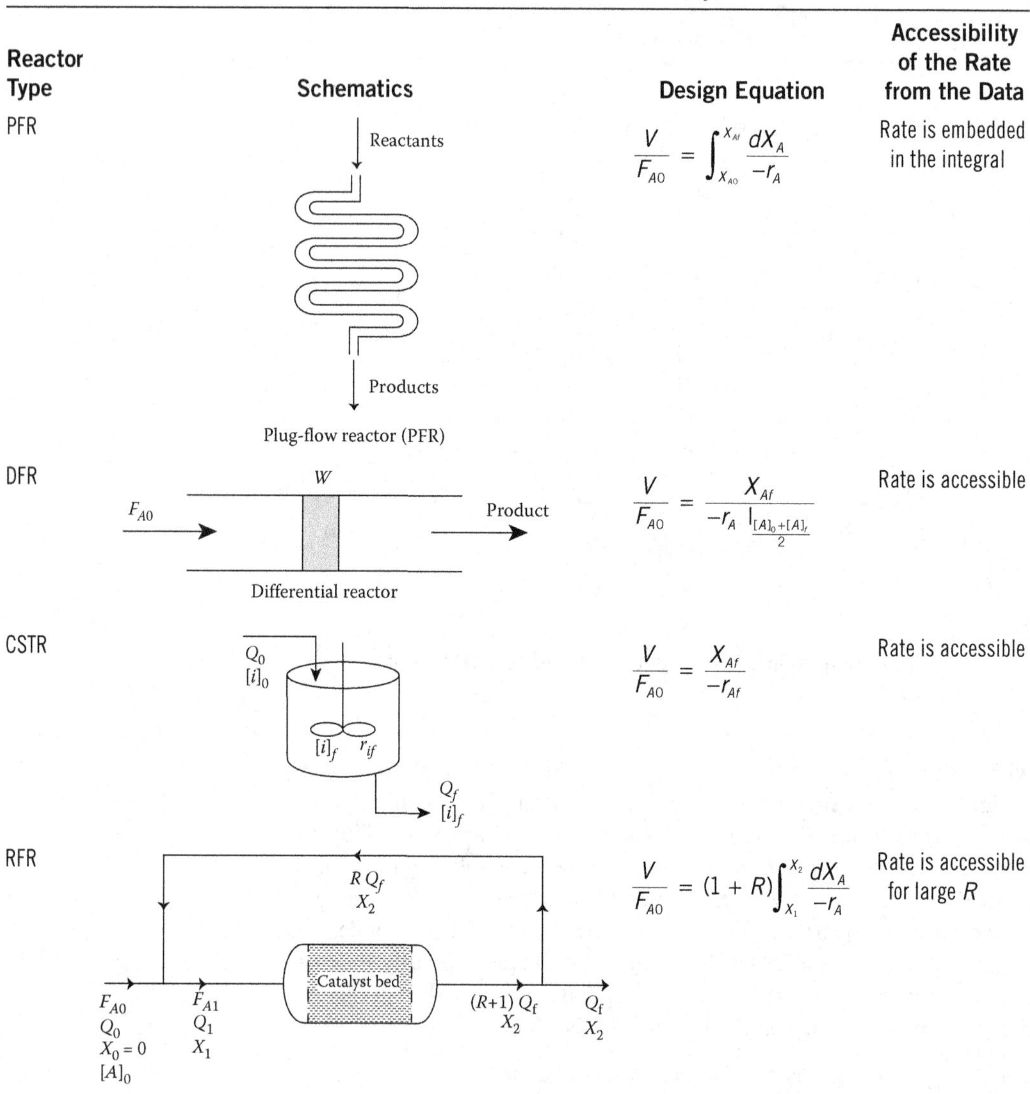

Reactor Type	Schematics	Design Equation	Accessibility of the Rate from the Data
PFR	Plug-flow reactor (PFR)	$\frac{V}{F_{A0}} = \int_{X_{A0}}^{X_{Af}} \frac{dX_A}{-r_A}$	Rate is embedded in the integral
DFR	Differential reactor	$\frac{V}{F_{A0}} = \frac{X_{Af}}{-r_A \mid_{\frac{[A]_0+[A]_f}{2}}}$	Rate is accessible
CSTR		$\frac{V}{F_{A0}} = \frac{X_{Af}}{-r_{Af}}$	Rate is accessible
RFR		$\frac{V}{F_{A0}} = (1+R)\int_{X_1}^{X_2} \frac{dX_A}{-r_A}$	Rate is accessible for large R

such as order or activation energy. Only if an assumption about the reaction order is made then the integral can be taken and an algebraic equation is available for data analysis—hence the title of integral reactor. As we have seen in Chapter 3, for small values of R the RFR behaved like a PFR and thus can be considered as an integral reactor as well.

A CSTR, on the other hand, for the well-mixedness condition, has a design equation for which the rate is algebraically accessible once the volume, the molar flow, and the conversion data are available. Thus, the rate can be easily extracted numerically so that it can be further processed for the extraction of rate parameters such as the orders and activation energies.

There are two more ways of accessing rate information under the integral of PFR design equation:

1. If the reactor size is kept very small such that the conversion across the reactor remains below 5%, then the integral sign can be safely dropped and the rate is evaluated at the average concentration of the species. Note that in this case we use the average concentration of the species, while we used the exit concentration of the species in the CSTR.
2. The small size of the reactor can only give us the initial rates or rates at very low conversions. To obtain rate information at higher conversions, a RFR is recommended. In such a configuration, by keeping the recycle ratio large, we provide large enough backmixing that the conversion between the inlet and exit of the reactor again becomes very close to each other. We can use the design equation for the DFR to determine the kinetic parameters.

One of the more important problems is the accurate measurement of the chemical kinetics especially when heterogeneous catalysts are involved. In such systems, the transport disguises may shadow the true kinetic data and the experimenter must be overly cautious to eliminate them.

Eliminating or accounting for transport disguises

The transport disguises we deal with can be heat transfer problems or mass transfer problems. We will begin by discussing the mass transfer problems and methods of elimination.

When testing catalysts (or gas–solid noncatalytic reactions), we primarily have two mass transfer resistances. One of them is due to pore diffusion and other is due to the film mass transfer resistance. If our aim is to measure true kinetics, we should be operating our reactor in a regime that such effects are eliminated.

Eliminating film mass transfer resistance

Eliminating the film mass transfer resistance The film mass transfer resistance can be eliminated by increasing the film mass transfer coefficient. The film mass transfer coefficient is directly proportional to the Reynolds number, thus increasing the linear velocity while keeping the space–time constant would increase the mass transfer rates—hence decrease the mass transfer resistance. If our conversions have a tendency to increase, then we should increase the linear velocity until the conversions are independent of the linear velocity as shown in Figure 7.6 where a plot of conversion versus linear velocity can give us the region of the linear velocities to be used to be free from film mass transfer resistances.

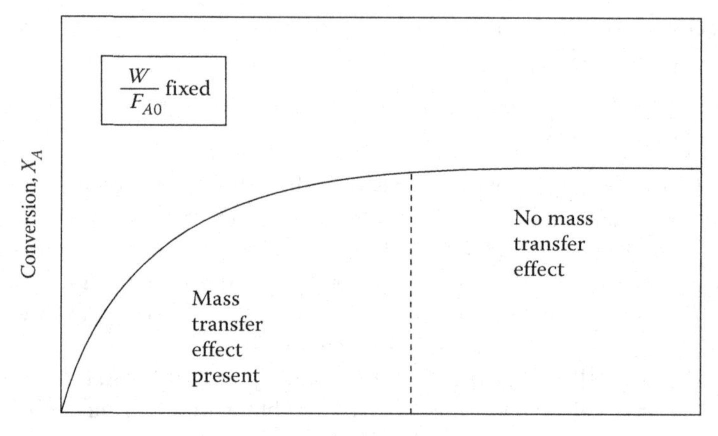

Figure 7.6 Experimental evaluation of the effect of external mass transfer. Velocity is varied by varying F_{A0} and W simultaneously so that W/F_{A0} (contact time) remains constant.

Eliminating pore diffusion resistance

Eliminating the pore diffusion resistances Pore diffusion resistances can be identified by the effect of the particle size on conversion. If the conversion is a function of the particle size, then the rate data are veiled by pore diffusion effects. To eliminate pore diffusion effects, the careful experimenter is advised to measure the conversion as a function of the particle diameter. If the conversions increase with decreasing particle diameters at constant space–time, then the data are veiled by pore diffusion limitations. The use of Arrhenius plots to elucidate the particle size temperature optimums is shown in Figure 7.7.

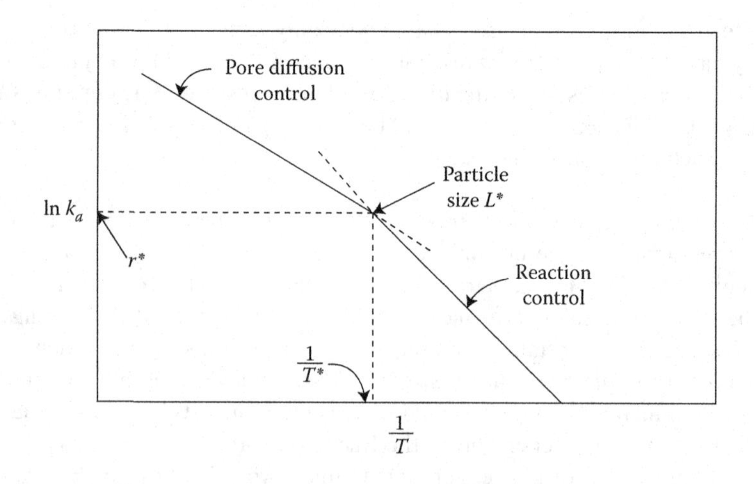

Figure 7.7 Arrhenius plot for formulating rate equations in chemical/pore diffusion regimes. * in the figure indicates the critical values.

Eliminating axial dispersion effects

Eliminating axial dispersion effects To eliminate axial dispersion effects a reactor that is sufficiently long should be used. For this purpose, a length of at least 40–50 pellet diameters is recommended.

Koros–Nowak criterion

Koros–Nowak criterion Ultimately, the only positive *diagnostic* test for the absence of these effects is to *verify directly that the reaction is in the kinetic regime*. For this purpose, the catalyst powder is diluted with an inert powder of the same size and characteristics (such as the unimpregnated support). If it is found that

$$\frac{\text{Rate (diluted pellet)}}{\text{Rate (undiluted pellet)}} = \frac{\text{Weight of catalyst in the diluted pellet}}{\text{Weight of catalyst in undiluted pellet}} \tag{7.13}$$

then it would be clear evidence that the reaction is in the kinetic regime (Koros and Nowak, 1967).

If experimental tests for heat or mass transfer effects cannot be carried out as described above, the criteria assembled in Table 7.2 can be used to confirm the absence of the transport disguises.

Catalyst dilution for temperature uniformity A basic requirement for obtaining precise kinetic data is the temperature uniformity of the catalyst pellet. Catalyst dilution with a highly conducting material such as SiC is very helpful in ensuring temperature uniformities in catalytic tests (see, e.g., Rihani et al., 1965). However, the exit product composition can be influenced (to the extent of 3–5%) by the manner of inert solids distribution in the bed and can become important in experiments of high precision. Van den Bleek et al. (1969) proposed the following criterion for neglecting this so-called *dilution effect*:

$$\frac{bd_p}{l\delta} < 4 \times 10^{-3} \tag{7.14}$$

where b is the dilution ratio (weight of total solid particles/weight of catalyst particles), l the undiluted height, and δ the experimental error. Sofekun et al. (1994) further examined this problem and proposed a more rigorous statistical criterion for the absence of the dilution effect.

Gradientless reactors

A convenient way of operating a differential reactor at integral conversions is to use a fully mixed reactor in which a constant concentration within the reactor is imposed by efficient mixing. A few important designs are sketched in Figure 7.8.

Transport disguises in perspective

It should be noted that, as the controlling regime changes, the values of the kinetic parameters (E and n) also change as already discussed in

Table 7.2 Mears (1971) Criteria to Operate under the Conditions That the Effectiveness Factor, $\varepsilon \cong 1 \pm 0.05$

	Intraparticle	Interphase	Interparticle
Concentration perturbations only	$\frac{R_p^2(-r_A)}{D_{eA}[A]_s} < \frac{1}{\lvert n \rvert}$	$\frac{R_p(-r_A)}{k'_G[A]_b} < \frac{0.15}{n}$	
Temperature perturbations only	$\frac{\lvert \Delta H \rvert R_p^2(-r_A)}{\lambda T_s} < \frac{T_s R_g}{E}$	$\frac{(-\Delta H)(-r_A)R_p}{hT_b} < 0.15\frac{T_b R_g}{E}$	$\frac{\lvert \Delta H \rvert R_{\text{reactor}}^2(-r_A)}{\lambda_{\text{bed}} T_s} < 0.4\frac{T_s R_g}{E}$
Both concentration and temperature perturbations are present	$\frac{R_p^2(-r_A)}{D_{eA}[A]_s} < \frac{1}{\left\lvert n - \left(\frac{E}{R_g T_s}\right)\frac{\lvert \Delta H \rvert D_{eA}[A]_s}{T_s} \right\rvert}$		

Note: λ, thermal conductivity of the particle; k'_G, film mass transfer coefficient; R_p, particle radius; R_{reactor}, reactor radius; h, heat transfer coefficient; T_b, bulk temperature; T_s, surface temperature; the rest of the symbols follow from the text.

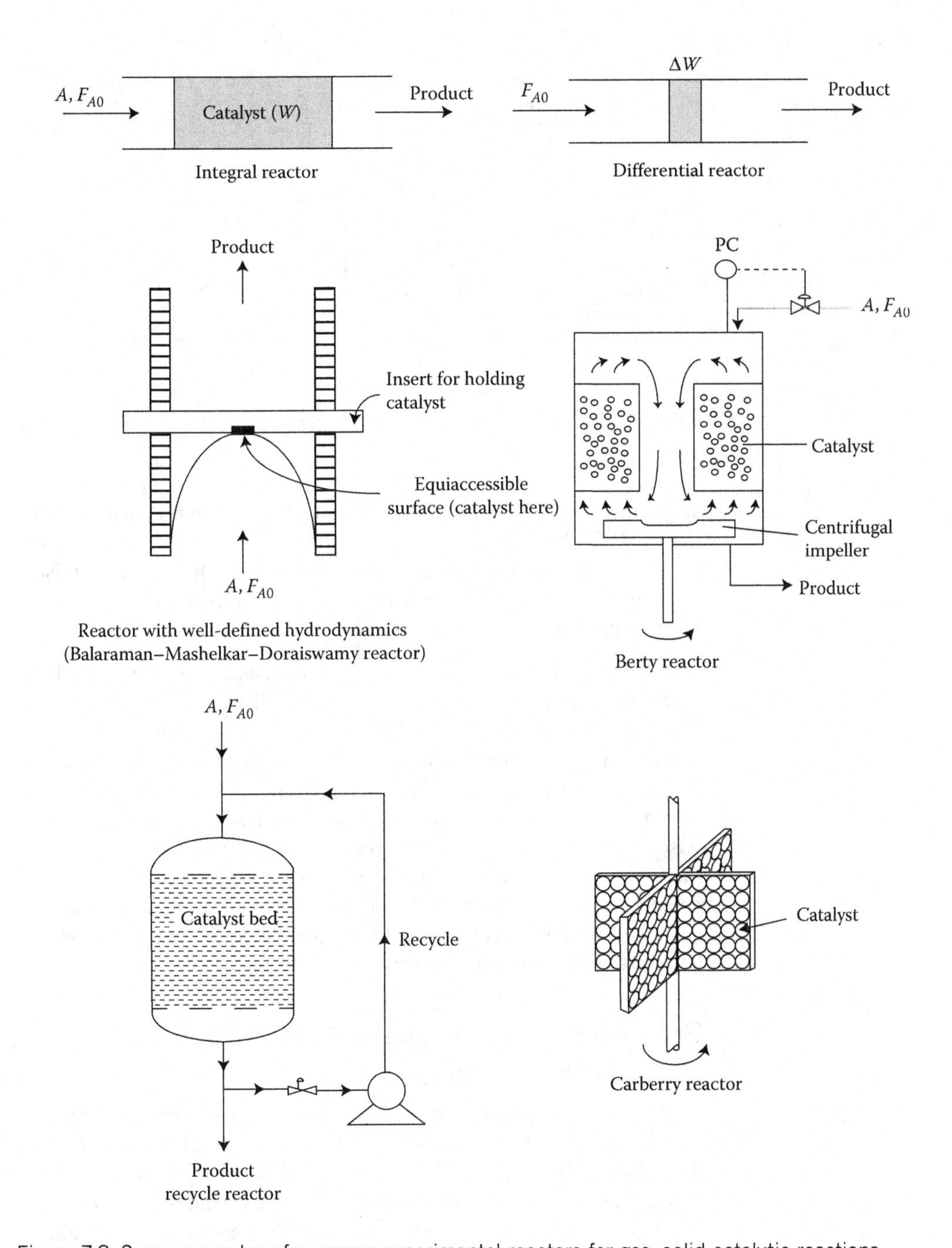

Figure 7.8 Some examples of common experimental reactors for gas–solid catalytic reactions.

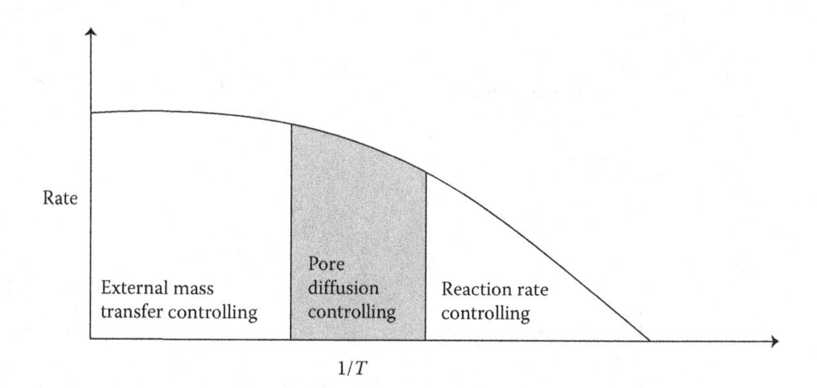

Figure 7.9 Schematic representation of regimes of operation on Arrhenius diagram.

detail in Chapter 6. A schematic simplified plot is reproduced here in Figure 7.9. A better understanding of the controlling regime is imperative in the choice of a laboratory reactor free from transport disguises. A comprehensive table comparing various features of laboratory reactors can be found in Joshi and Doraiswamy (2010).

It will be noted that activation energy can change from a high positive value (for chemical control) to a negligible value for external diffusion control. There can also be a region of negative activation energy corresponding to surface diffusion control, but this is almost never observed and is not considered here. For highly exothermic reactions, the activation energy can rise to almost infinity as the temperature is raised. Also, the value in the limit of pore diffusion control depends on the thermal nature of the reaction and is equal to $E_{true}/2$ for reactions with no heat effect. The limiting values of E and n for different conditions are discussed by Languasco et al. (1972), Rajadhyaksha and Doraiswamy (1976), and Doraiswamy and Sharma (1984).

Guidelines for eliminating or accounting for transport disguises

It is necessary to eliminate transport disguises in order to obtain precise kinetic data. The methods of accomplishing this have been discussed in detail by Doraiswamy and Sharma (1984). Where they cannot be eliminated, they should be accounted for, and the true values extracted from the "contaminated data." The following procedures/comments should be useful:

1. To eliminate external transport effects, run the reaction at a constant value of W/F_{A0} but changing the velocity. In other words, change the hydrodynamic conditions keeping the kinetic factor (W/F_{A0}) constant. The conversion increases and then levels off at a velocity beyond which external mass

transfer would have no effect (Figure 7.6). This can be done by changing the reactor diameter, or changing both W and F (higher F corresponding to higher velocity). One should be cautious in the very low-velocity region where the curve might show a deceptive peak before resuming the normal course (Doraiswamy and Sharma, 1984).

2. Carry out runs at different particle sizes. The rate will increase with decreasing particle size, reaching a constant value that would indicate the absence of pore diffusional effect. On an Arrhenius plot, the effect of pore diffusion for a given particle size will be as shown in Figure 7.7. This can be divided into two asymptotic regions, one corresponding to chemical control and the other to diffusion control. We are only concerned with the chemical control line.
3. Use a reactor that is sufficiently long to ensure the absence of axial diffusion. A length of at least 40–50 pellet diameters is recommended.
4. In case experimental tests for diffusional effects as described under 1–4 cannot be carried out, use the criteria assembled in Table 7.2 to confirm the absence of these effects in the runs that were carried out. Ultimately, the only positive diagnostic test for the absence of these effects is to verify directly that the reaction is in the kinetic regime. For this purpose, the catalyst powder is diluted with an inert powder of the same size and characteristics (such as the unimpregnated support). If it is found that

$$\frac{\text{Rate (diluted pellet)}}{\text{Rate (undiluted pellet)}} = \frac{\text{Weight of catalyst in the diluted pellet}}{\text{Weight of catalyst in undiluted pellet}}$$

then it would be clear evidence that the reaction is in the kinetic regime (Koros and Nowak, 1967).

Analyzing the data

Modeling of solid catalyzed reactions

The overall scheme Consider the catalytic reaction $A \rightarrow B$ taking place over catalyst pellets. Let the pellets be placed in a flowing stream of reactants inside a tubular reactor. The construction of a model based on one of the adsorption–reaction–desorption steps being the limiting step constitutes the core of the semiempirical approach considered in this chapter. In this approach, the microscopic origins of the observed macroscopic effects of catalysts (as described by many authors, e.g., Plath, 1989) are ignored. The models thus developed are commonly known as Langmuir–Hinshelwood models among chemists and as Hougen–Watson models among chemical engineers.

Table 7.3 LHHW Models for a Few Selected Reaction Types

Reaction	Controlling Mechanism	LHHW Equation for $-r_{wA}$
$A \leftrightarrow R$	Surface reaction	$\dfrac{k_w K_A([A] - ([R]/K))}{1 + K_A[A] + K_R[R]}$
$A + B \leftrightarrow R + S$	Surface reaction	$\dfrac{k_w K_A K_B([A][B] - ([R][S]/K))}{(1 + K_A[A] + K_B[B] + K_R[R] + K_S[S])^2}$
$A + B \leftrightarrow R + S$	Adsorption of B	$\dfrac{k_{Ba}([B] - ([R][S]/K[A]))}{1 + K_A[A] + K_B([R][S]/K[A]) + K_R[R] + K_S[S]}$
$A + B \leftrightarrow R + S$	Desorption of R	$\dfrac{k_{Rd}K(([A][B]/[S]) - ([R]/K))}{(1 + K_A[A] + K_B[B] + KK_R([A][B]/[S]) + K_S[S])}$
$A + B \leftrightarrow R + S$	Adsorption of A with dissociation	$\dfrac{k_{Aa}([A] - ([R][S]/K[B]))}{(1 + \sqrt{K_A[R][S]/K[B]} + K_B[B] + K_R[R] + K_S[S])^2}$
$A + B \leftrightarrow R$	Surface reaction with dissociation of A, B not adsorbed	$\dfrac{k_w K_A([A][B] - ([R]/K))}{(1 + \sqrt{K_A[A]} + K_R[R])^2}$

We choose to call them *Langmuir–Hinshelwood–Hougen–Watson (LHHW) models.*

LHHW models The Langmuir isotherm was treated in detail in Chapter 5. In the interest of generality, we consider hypothetical reactions and derive rate equations for a few typical *LHHW models*. The results of the derivations are presented in Table 7.3.

Selection of the most plausible model

General considerations Several methods are available for model selection. We shall only be concerned with statistical methods, which in their simplest form are based on selection of *the most plausible model* from among a number of candidates. It must be emphasized that these methods do not uniquely select a candidate as the only acceptable model, but converge to one that is statistically the most acceptable.

Preliminary short listing of models The first step is to write down the rate equations for all possible controlling steps. A preliminary short listing can then be done by reducing the models to initial conditions when no product would have formed. Under these conditions, each model exhibits a specific behavior with variation in total pressure. Table 7.4 lists the complete rate equations, initial rate equations (expressed in terms of the total pressure P), and the corresponding r_{WA0}–P plots for three representative models. A more complete treatment can be found in Yang and Hougen (1950). Similar effects can be produced by changing the ratio of reactants in a bimolecular reaction.

Table 7.4 Initial Rate-Pressure Relationships for Different Rate Forms

Reaction	Model Equation for $-r_{wA}$	Equation for the Initial Rate, $-r_{wA0}$	Behavior of $-r_{wA0}$ with Variation in Total Pressure P
1. $A \leftrightarrow R + S$ (surface reaction controlling)	$\frac{k_{wp}K_A(p_A - (p_R p_S/K))}{(1 + K_A p_A + K_R p_R + K_S p_S)^2}$	$\frac{aP}{(1 + bP)^2}$	Exhibits a maximum
2. $A + B \leftrightarrow R$ (surface reaction controlling, B not adsorbed)	$\frac{k_{wp}K_A(p_A p_B - (p_R/K))}{1 + K_A p_A + K_R p_R}$	$\frac{aP^2}{1 + bP}$	Initial rapid increase followed by a slow increase
3. $A \leftrightarrow R + S$ (desorption of R controlling)	$\frac{k_{Rd}K}{1 + K_A p_A + KK_R(p_A/p_S) + K_S p_S}$	$\frac{k_R}{K_R}$ = const.	No effect of P on rate

The basic steps in model selection Having narrowed down the field by the initial rate method, the following procedure can then be used to select the most probable model from the surviving list of contenders.

1. Write down all possible mechanisms and the corresponding rate equations.
2. Linearize the equations as illustrated below for a typical case.

$$-r_{wA} = \frac{kK_A}{1 + K_A[A] + K_R[R]}\left([A] - \frac{[R]}{K}\right) \tag{7.15}$$

   Recast this as

$$\bar{r} = a + b[A] + c[R] \tag{7.16}$$

   where

$$\bar{r} = \frac{[A] - [R]/K}{-r_{wA}}, \quad k = \frac{1}{b}, \quad K_A = \frac{b}{a}, \quad K_R = \frac{c}{a} \tag{7.17}$$

   Then minimize the sum of squares of the residuals

$$s^2 = \frac{\sum_{i=1}^{n}(y_i - \bar{y})^2}{n - 1} \tag{7.18}$$

   where $\bar{y}$ is the arithmetic mean of n measurements and y_i an individual measurement, and determine the rate and adsorption equilibrium constants at each temperature. This is the well-known linear least-squares adaptation of the regression theory.
3. However, regression theory requires that the errors be normally distributed around $-r_{wA}$, and not around $\bar{r}$ as in the linearized version just described. Hence, use the values determined above as initial estimates to obtain more accurate values of the constants by minimizing the sum of squares of the residuals of rates directly from the "raw rate equation" by nonlinear least-squares analysis.

Statistical *t*-tests

4. The analysis described under (3) can be carried out by the differential method in which the rates to be used in the equations are obtained directly in a differential reactor or by appropriate manipulation of integral data obtained in an integral reactor.
5. From standard statistical *t*-tests, make sure that all constants are significantly different from zero and discard models for which even one of these constants is significantly negative. This is because none of these constants can reasonably be negative.
6. From the models surviving step (5), reject those for which the rate constant k decreases with temperature, or the adsorption equilibrium constant K_A or K_R increases with temperature. This is because the rate constant for a single reaction step must always increase with temperature, and adsorption being exothermic K_A and K_R must decrease with temperature.
7. From the models surviving step (6), choose the one which best fits the experimental data.

Refinements in nonlinear estimation of parameters Himmelblau (1970), Kittrell (1970), and Huet et al. (1996) give very useful treatments of nonlinear estimation of kinetic parameters. A simple procedure is illustrated in detail for the isomerization of *n*-butene to *i*-butene by Raghavan and Doraiswamy (1977) (see also Huet et al., 1996). Among other useful illustrations are those of Franckaerts and Froment (1964), Dumez and Froment (1976), and Dumez et al. (1977). [Sequential design can also be used for updating the parameter values of the simpler power-law model with a high correlation between preexponential factor and activation energy, often a vexing problem (Dovi et al., 1994).] More recently, some good strategies have been formulated (Watts, 1994) in which the original LHHW models are reformulated and the parameters transformed to produce "well-behaved" estimates.

Comments Some of the criteria just outlined for model selection may not always be valid. For instance

1. More than one step can be rate controlling, as in the case of dehydrogenation of *sec*-butyl alcohol (Thaller and Thodos, 1960; Bischoff and Froment, 1962; Shah, 1965; Choudhary and Doraiswamy, 1972).
2. There are a few instances of endothermic adsorption, and in such cases the adsorption equilibrium constants will increase with temperature (see, e.g., Doraiswamy and Sharma, 1984).
3. Even if all the criteria are satisfied, there is still the possibility that the reaction may proceed according to a different mechanism. This is the limit of mechanism discrimination through statistics. From this point on, careful attention must be paid for validating the mechanism with information provided from surface science.

Influence of surface nonideality

In addition to the limitations specifically applicable to the procedure outlined above, the models themselves suffer from a few intrinsic defects, and have therefore been the subject of some criticism and much commentary. The chief limitations are

1. All sites are equally active, with equal heats of adsorption; this is not true since there is usually a distribution of activity on the surface which is ignored in the LHHW models.
2. Interaction between adsorbed molecules is negligible, again not true.
3. Molecules are always adsorbed at random on the surface, also not true; like molecules may tend to adsorb in contiguity forming their own islands; this leads to a completely different mechanism of surface reaction.

A number of isotherms which dispense with assumptions (1) and (2) have been proposed (see Doraiswamy, 1991). Some studies (e.g., Kiperman et al., 1989) indicate that it may not be possible to model certain reactions without invoking the role of surface nonideality. Fortuitously, the use of more rigorous isotherms does not materially affect the companion problem of the diffusion-reaction behavior of systems (Shendye et al., 1993)—a topic considered in the next section.

The paradox of heterogeneous kinetics An interesting feature of LHHW kinetics is worth noting. Many reactions on surfaces known to be nonideal surprisingly follow the ideal LHHW models, a situation that can only be described as the "paradox of heterogeneous kinetics" (see Boudart et al., 1967; Boudart, 1986). In the same vein but with less justification, it has also been argued for over four decades—for example, from Weller (1956) to Bouzek and Roušăr (1996)—that rate data for a given reaction can be correlated equally well by simple power law kinetics (thus dispensing with the surface science approach altogether). In general, LHHW models supplemented by rigorous methods of parameter estimation do represent a valid mechanistic approach that can be accepted as a reasonably sound basis for reactor design. A more puristic approach would, however, require a firmer anchoring to the methods of surface science.

Explore yourself

1. If you have not yet already acquired the skills of parameter estimation, teach yourself the least squares (LS) method of parameter optimization. The method, like the rest of the methods that will be covered in this part, is based on the minimization of the sum of the squares between the measured and estimated values. The estimation function variables are optimized as such.

2. Explore the Internet for the public software for parameter estimation. What are the newest developments? Can you write your own code for parameter estimation?
3. What are the latest and the most fashionable in the laboratory reactors market? What do they promise? Also, follow from the publications what do they deliver?
4. Why do most laboratories report data collected in home-built systems?
5. Yuceer et al. (2008) propose a new program called PARES for estimating the parameters for nonlinear functions. Download the program for further use. Make yourself familiar with the following optimization methods:
   a. Levenberg–Marquardt
   b. Nelder–Mead simplex
   c. Quasi Newton
   d. Sequential quadratic programming

References

Berty, J.M., Hanibrick, J.O., Malone, T.R., and Ullock, D.S., *64th National Meeting of the AIChE*, New Orleans, Paper no. 42E, 1969.

Bischoff, K.B. and Froment, G.F., *Ind. Eng. Chem. Fundam.*, **1**, 195, 1962.

Boudart, M., *Ind. Eng. Chem. Fundam.*, **25**, 658, 1986.

Boudart, M., Mears, D.E., and Vannice, M.A., *Ind. Chim. Beige*, **32**, 281, 1967.

Bouzek, K. and Rousăr, I., *J. Chem. Technol. Biotechnol*, **66**, 131, 1996.

Box, G.E.P., Hunter, J.S., and Hunter, W.G., *Statistics for Experimenters: Design, Innovation, Discovery*, Wiley-Interscience, New York, 2005.

Carberry, J.J., *Chemical and Catalytic Reaction Engineering*, McGraw-Hill, New York, NY, 1976.

Choudhary, V.R. and Doraiswamy, L.K., *Ind. Eng. Chem. Proc. Des. Dev.*, **11**, 420, 1972.

Doraiswamy, L.K., *Prog. Sur. Sci.*, **37**, 1, 1991.

Doraiswamy, L.K. and Sharma, M.M., *Heterogeneous Reactions: Analysis, Examples and Reactor Design*, Vol. 1, John Wiley, New York, NY, 1984.

Dovi, V.G., Reverberi, A.P., and Acevedo, D.L., *Ind. Eng. Chem. Res.*, **33**, 62, 1994.

Dumez, F.J., Hosten, L.H., and Froment, G.F., *Ind. Eng. Chem. Fundam.*, **16**, 298, 1977.

Franckaerts, J. and Froment, G.F., *Chem. Eng. Sci.*, **19**, 807, 1964.

Himmelblau, D.M., *Process Analysis by Statistical Methods*, Wiley, New York, NY, 1970.

Huet, S., Bouvier, A., Gruet, M.A., and Jolivet, E., *Statistical Tools for Numerical Regression*, Springer, New York, 1996.

Joshi, J.B. and Doraiswamy, L.K., Chemical reaction engineering. In *Albright's Chemical Engineering Handbook* (Ed. Lyle, F.), CRC Press, Boca Raton, FL, 2010.

Kiperman, S.L., Kumbilieva, K.E., and Petrov, L.A., *Ind. Eng. Chem. Res.*, **28**, 376, 1989.

Kittrell, J.R., In *Advances in Chemical Engineering*, Vol. 8 (Eds. Drew, T.B., and Hoops, J.W.), Academic Press, New York, 1970.

Koros, R.M. and Nowak, E.J., *Chem. Eng. Sci.*, **22**, 470, 1967.

Kubota, H., Yamanaka, Y., and Lana, I.G.D., *J. Chem. Eng. Jpn.*, **2**, 71, 1969.

Languasco, J.M., Cunningham, R.E., and Calvelo, A., *Chem. Eng. Sci.*, **27**, 1459, 1972.

Lazic, Z.R., *Design of Experiments in Chemical Engineering: A Practical Guide*, Wiley-VCH, Weinheim, 2004.

Mandler, J., Lavie, R., and Scheintuch, M., *Chem. Eng. Sci.*, **38**, 979, 1983.

Mears, D.E., *Ind. Eng. Chem. Proc. Des. Dev.*, **10**, 541, 1971.

Montgomery, D.C. and Runger, G.C., *Applied Statistics and Probability for Engineers*, John Wiley and Sons Inc., New York, 1994.
Peters, M.S. and Skorpinski, E.J., *J. Chem. Educ.*, **42**, 329, 1965.
Plath, P., Ed. *Optimal Structures in Heterogeneous Reaction Systems*, Springer-Verlag, Berlin, 1989.
Raghavan, N.S. and Doraiswamy, L.K., *J. Catal.*, **48**, 21, 1977.
Rajadhyaksha, R.A. and Doraiswamy, L.K., *Catal. Rev. Sci. Eng.*, **13**, 209, 1976.
Rihani, D.N., Narayanan, T.K., and Doraiswamy, L.K., *Ind. Eng. Chem. Proc. Des. Dev.*, **4**, 403, 1965.
Satterfield, C.N. and Roberts, G.W., *AIChE J.*, **14**, 159, 1968.
Shah, M.J., *Ind. Eng. Chem.*, **57**, 18, 1965.
Shendye, R.V., Dowd, M.K., and Doraiswamy, L.K., *Chem. Eng. Sci.*, **48**, 1995, 1993.
Sofekun, O.A., Rollins, D.K., and Doraiswamy, L.K., *Chem. Eng. Sci.*, **49**, 2611, 1994.
Thaller, L.H. and Thodos, G., *AIChE J.*, **6**, 369, 1960.
Van den Bleek, C.M., Van der Wiele, K., and Van den Berg, P.J., *Chem. Eng. Sci.*, **24**, 681, 1969.
Watts, D., *Can. J. Chem. Eng.*, **72**, 701, 1994.
Weller, S., *AIChE J.*, **2**, 59, 1956.
Yang, K.H. and Hougen, O.A., *Chem. Eng. Prog.*, **46**, 14, 1950.
Yuceer, M., Atasoy, I., and Berber, R., *Braz. J. Chem. Eng.*, **25**, 813, 2008.

Part III
Beyond the Fundamentals

Numquam ponenda est pluralitas sine necessitate.
Plurality must never be posited without necessity.

William of Occam, 1285–1348

Objectives

This part is dedicated to exploring the advanced reactor design territory. After the conclusion of this part, students must be able to

- Design and analyze the reactors for
  - Gas–solid noncatalytic reactions
  - Gas–solid catalytic reactions
  - Gas–liquid and liquid–liquid reactions
  - Multiphase reactions
- Explore the emerging reaction/reactor fields such as
  - Combo reactors
  - Homogeneous catalysis
  - Membrane reactors
  - Phase transfer catalysis

To accomplish these, you will learn how to

- Write down the momentum, material, and energy balance equations in more than one phase.
- Incorporate the kinetics, transport, and thermodynamics for systems comprising more than one phase.
- Use approximations whenever needed, with a clear understanding of the limitations of these approximations.

Introduction

The chapters in this part will combine the information brought in from different aspects of chemical engineering science to generate a knowledge basis to design better reactors. Nowadays, better reactors provide a means of sustainable production, with better material and energy economy and with higher selectivity. Unfortunately, not all the engineering problems can be solved through sheer analytical reasoning and solution of the differential equations derived from the fundamental balance equations. A practicing engineer has to frequently resort to the empirical knowledge base. What a chemical engineer understands from a pilot-scale unit is exactly this component of the design procedure: both testing the theory and also collecting empirical information at the same time.

The different tools of the trade

In this part, we address primarily the design of reactors for heterogeneous reactions. A difficulty in designing reactors for the reaction with an interface lies at the interface itself. Two (or three) different states of aggregation communicate with each other at the interface by thermodynamic mechanisms such as vapor–liquid equilibrium or adsorption–desorption equilibrium. The processes beyond that interface require a solid understanding of transport phenomena. Across the interfaces, resistances exist for momentum, heat, and mass transfer. These resistances sometimes act as a disguise, and sometimes serve as a means for improving selectivity. Whether the transport process is a friend or foe truly depends on the relative rates of the individual processes and the ultimate goal of the synthesis protocol.

The next important step in designing a reactor is the choice of the reactor type. Among many different alternatives for chemical reactors, in this part of the book, we will focus on three major systems due to the breadth of the processes that these systems are used: The packed-bed reactor, the fluidized-bed reactor, and gas–liquid(–solid) contact equipment. We have dedicated a separate chapter for each of these systems and elaborated on their design procedure through extensive examples.

Dutta and Gualy (2000) have very succinctly summarized all the necessary steps for robust chemical reactor design and scale up. In the quotation window, you see their definition of robustness. After the quotation, we give a brief summary of the protocol they proposed:

> "**What is a robust model?** Such a model is a practical, reliable, and useful package for analyzing, scaling up, designing and optimizing a given reaction and reactor system. It provides the best design for

a new system, revamp, or modernization, and the optimum operating conditions for an existing reactor. The model predicts the performance for a wide variety of designs and operating conditions, including those used in the commercial reactor. It also covers conditions beyond normal operations, to predict upset, specification, turndown and unsafe situations. Such a model should be based primarily on the fundamentals principles of reaction engineering and reactor hydrodynamics. It should use the minimum number of adjustable/experimental parameters and be solved by standard mathematical routines requiting minimum execution time. And it should be easy to integrate with other in-house or commercial simulation packages."

Dutta and Gualy (2000)

1. Define the reaction type: it is important if you are designing a reactor for a homogeneous or a heterogeneous system.
2. Design, build, and operate a test unit. This unit is also called a process development unit (PDU).
3. Collect and analyze data.
4. Establish the preliminary reaction mechanism and kinetics.
5. Study the safety aspects.
6. Define the reactor type and hydrodynamics.
7. Determine the details to be included in the models.
8. Choose the right balance equations.
9. Select evaluation procedures for all nonkinetic process parameters.
10. Determine the model structure and solution procedures.
11. Develop a preliminary model.
12. Include additional details for the final model.
13. Develop the final model by tuning the parameters.
14. Make scale-up projections and establish the optimum design.

In the subsequent chapters, we will cover the influence of the transport processes on the design heuristics. In Table PIII.1, a brief summary is presented.

Figure 7.1 gave us an idea of the number of forms a reactor can take depending on the physical state of the reactants/products involved (gas, liquid, solid, or combinations thereof) and the state of motion of the solids (fixed, fluidized, or moving). In the subsequent chapters, we consider fixed-bed reactors involving catalytic (nonreactive) and noncatalytic solids and a fluid (generally gaseous). Of these different types, we consider two in some detail in this part: fixed- and fluidized-bed reactors. In Figure PIII.1, most common types of fixed-bed reactors are shown schematically while in Figure PIII.2, the most common types of the fluidized-bed reactors are presented.

Table PIII.1 Relative Importance of Major Phenomena That May Influence Reactor Models

Phenomenon	Where It Usually Is More Important	Where It Usually Is Less Important	Where It Must be Considered
Pore-diffusion resistance	a. Reactions involving solid particle size >1/16 in. b. All fast, noncatalytic gas/solid (G/S) reactions like combustion and gasification	Catalytic bubbling fluidized-bed (BFB) and circulating fluidized-bed (CFB) reactors with particle size <100 μm	Fixed- and moving-bed G/S reactor models and fast reaction systems
Film diffusion (interphase mass transfer) resistance	a. All bubbling reactors like BFB and gas/liquid (G/L) and three phase (3-P) reactors b. All fast, noncatalytic gas/solid reactions like combustion and gasification	Catalytic fixed bed G/S reactors	All bubbling systems like BFB, G/L and 3-P reactors, and fast reaction systems
Pressure drop	Fixed-, moving-bed and deep BFB G/S reactors, and liquid phase reactors	CFB and entrained-bed reactors	G/S fixed- and moving bed reactor models and deep beds
Heat transfer resistance	a. Across two-phase interface in fast reactions b. Gas side of tube wall in liquid-cooled gas phase or G/S reactors	Within solid particles in solid/fluid reactions	Gas side of tube wall in liquid-cooled gas phase or G/S reactors
Heat loss to atmosphere	Small-diameter laboratory and pilot-plant units	Commercial reactors	
Axial dispersion	a. Low L/D and low Reynolds number (*Re*) flow conditions b. Vessel with baffles or internal obstructing flows	High L/D and high Re flow in open pipes	
Radial dispersion	Large-diameter reactors with low flow rates, and CFB reactors		[Usually ignored in preliminary models]
Wall effect	Small-diameter reactors with low *Re* flow condition and CFB reactors		[Usually ignored in preliminary models]
Temperature profile	Fixed- and moving-bed G/S reactors	Dense phase of BFB reactors	Fixed- and moving-bed G/S reactor models
Volume change	Gas phase G/S and G/L reactions particularly with no gas dilution (e.g. with N_2)	Reactions not involving gas phase	
Phase holdups	All 2-P and 3-P reactors involving liquid phase		All 2-P and 3-P reactors involving liquid phase
Bed/line voidage/ voidage profile	CFB reactors and solids' circulation-systems design		CFB reactors and solids' circulation-systems design

Source: Dutta, S. and Gualy, R., *Chemical Engineering Progress,* (October), 37–51, 2000.

Process intensification

Chemical reaction engineering at the heart of the chemical process plant is going through a transformation in the era of sustainability. Raw material and energy economy becomes inherent in any process design where

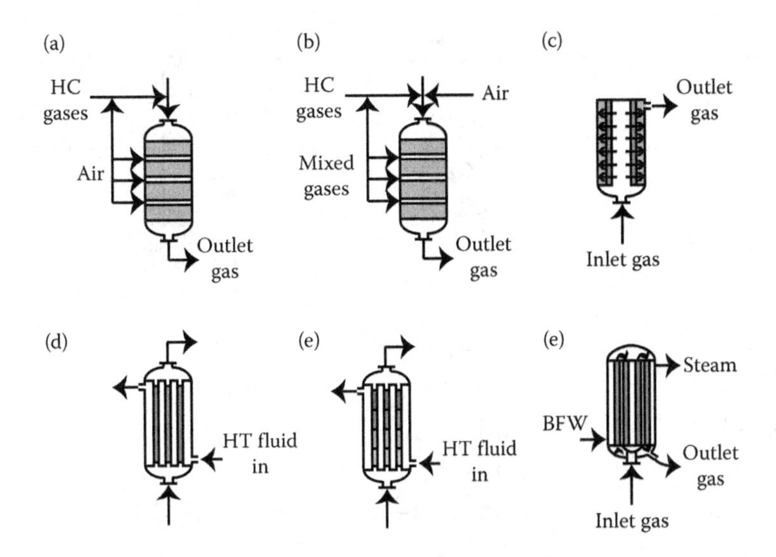

FIGURE PIII.1 Alternative fixed-bed reactor designs. (a) Adiabatic quench reactor with split air flow, (b) adiabatic quench reactor, (c) double-wall heat exchanger reactor, (d) multi-tubular fixed bed, (e) multi-tubular fixed bed with varying catalyst concentration, and (f) radial flow reactor. (Adapted from Dutta, S. and Gualy, R., *Chemical Engineering Progress,* (October), 37–51, 2000.)

process intensification is a necessity. The design of these systems is now based on the pressing need for improved selectivity and/or activity.

Microfluidics

These reactors are literally the new kids in town. The heat transfer problem that is so inhibitory in packed bed systems becomes a nuisance in the small sizes of the microfluidic reactors. The mass transfer problem can also be relinquished in the microdomains. The microdomains provide excellent flow control, as well as bring in the surface tension as an additional force field to inertial and viscous forces that we are used to dealing with so far. The surface and interface forces at this level of miniaturization become driving forces for better mixing domains.

Microfluidics is emerging as a multidisciplinary subject and has already started contributing immensely to reactor design. The manufacture of microfluidic reactors has its roots in microelectronics device manufacturing technologies. Larger (millimetric) scale systems are also considered as microfluidics, the smaller size systems enable utmost control of heat and mass transfer. Furthermore, the intricate design schemes allow for unique mixing protocols. The detailed exploration of the microfluidic reactors for improved transport characteristics is presently highly empirical while the benefits are being demonstrated on various reaction systems. We have limited our coverage of microfluidic reactors in

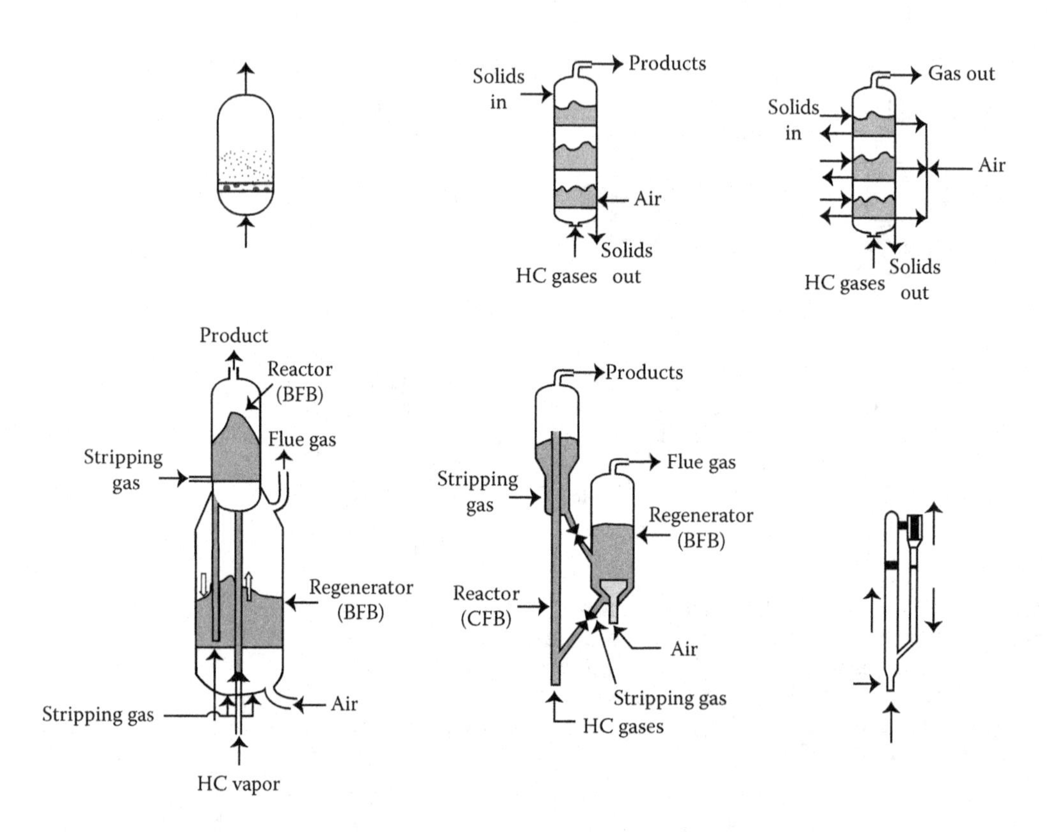

FIGURE PIII.2 Alternative fluidized-bed reactor designs. (a) Bubbling fluid bed (BFB), (b) multistage BFB, (c) multistage BFB with split air flow and temperature programming, (d) BFB/BFB combo with oxygen carrier as catalyst, (e) CFB/BFB combo with oxygen carrier as catalyst, and (f) circulating fluid bed (CFB). (Adapted from Dutta, S. and Gualy, R., *Chemical Engineering Progress,* (October), 37–51, 2000.)

this book to the utility of the technique in designing alternative mixing schemes in Part I and measurement of mass transfer in Part II, leaving the exploration of the most recent developments to the reader through specialized texts appearing in the literature (Ehrfeld et al. 2000; Hessel et al. 2004).

Membrane reactors

The need for the membrane reactors also primarily stems from the equilibrium conversion limitation of a reversible reaction. These reactors are used when the boiling point differences are not sufficient to use distillation column (combo) reactors. The thermal sensitivity of the reactive domain may inhibit the use of boiling point differences for the product separation, hence the use of distillation column reactors. A perm selective membrane can be used within the reactor providing product separation. The other advantage of the membrane reactors is the possibility they offer to run the reaction either in the gas or in the liquid phase, thus adding to the diversity of the reaction domain. We have covered the different types of membrane reactors in Chapter 13.

Combo reactors

These types of reactors are becoming increasingly common in chemical process industries. They are used to drive away one of the products that inhibit the progress of the reaction via an equilibrium limitation imposed by the Le Chatelier principle. Once the product is removed, the equilibrium shifts to the right and 100% conversions at low temperatures are possible. The key requirement is that there should be sufficient boiling point difference between the products and the reactants such that one of the products can be selectively removed by distillation. The design principles and selected examples are presented in Chapter 14.

Homogeneous catalysis

We have given a separate treatment to homogeneous catalysis in Chapter 15 from the point of view of chemical reaction engineering and the special subtopic of gas–liquid reactions with the realization that the selective manufacture of many fine and specialty chemicals is carried out using homogeneous catalysts. Use of homogeneous catalysts for biphasic systems is emerging as new methods in fine and specialty chemical synthesis (Duque et al. 2011). The continuous system analysis requires the use of pertinent tools provided by chemical reaction engineering, especially of the liquid–liquid reactions and the interface analysis required thereof.

Phase-transfer catalysis

The technology requires a quad in this case that enables the transfer of ions between two phases. The idea requires a phase-transfer agent in flux. Needless to say, these systems are highly controlled/inhibited by the mass transfer rate of the phase-transfer agents. A detailed coverage of the phase-transfer catalysts and their use in specialty chemical manufacture is also covered in Chapter 16.

References

Duque, R., Ochsner, E., Clavier, H., Caijo, F., Nolan, S.P., Mauduit, M., and Cole-Hamilton, D.J., *Green Chemistry*, **13**, 1187–1195, 2011.

Dutta, S. and Gualy, R., *Chemical Engineering Progress*, (October) 37–51, 2000.

Ehrfeld, W., Hessel, V., and Löwe, H., *Microreactors: New Technology for Modern Chemistry*, Wiley-VCH, Weinheim, 2000.

Hessel, V., Hardt, S., and Löwe, H., *Chemical Micro Process Engineering: Fundamentals, Modelling, and Reactions*,Wiley-VCH, Weinheim, 2004.

Chapter 8 Fixed-bed reactor design for solid catalyzed fluid-phase reactions

Chapter objectives

After successful conclusion of the chapter, the students must be able to

- Select the objective function to optimize for the best design configuration for a fixed-bed reactor.
- Collect relevant kinetic, thermodynamic, and transport parameters.
- Choose a reactor model given the reaction parameters.
- Choose a flow modality.
- Choose a heat-exchange protocol.
- Provide an optimum design for a given reaction with one or more of the following objective functions: yield, selectivity, material economy, and energy economy.

Introduction

Catalytic reactions are carried out in reactors with a fixed, fluidized, or moving bed of the catalyst. While the chemical kinetics of the reaction obviously remains the same for all these reactors, the hydrodynamic features vary considerably. Since no complete description of these features is possible, it is convenient to postulate different situations and develop mathematical models to represent these situations for each type of reactor. It is also important to note that wherever solid catalysts are used, the question of catalyst deactivation cannot be ignored. Several books and reviews covering a variety of situations have been collected in the Bibliography for further reference.

Effect of catalyst packing in a tubular reactor

Recall from Chapter 1 that the catalytic rate was expressed in terms of unit weight of the catalyst and not of unit volume as in the case of an unpacked tubular reactor (the PFR as described in Chapter 1). This distinction, obvious as it is in defining the rate of a catalytic reaction, becomes particularly relevant in the design of a packed tubular reactor. Volume, pressure drop, flow distribution, and overall catalyst effectiveness are all affected. Let us first consider the volume, which is entirely a function of the bulk density of the bed, and all terms, including the rate and the rate constant, must be suitably modified. Table 1.1 in Chapter 1 lists the appropriate corrections.

Second, the effect of packing is greatly influenced by the size of the particle/pellet comprising the bed. This effect is explicitly seen in pressure-drop behavior that defines the limiting flow as well as in the overall catalyst effectiveness. In Chapter 6, we saw how the size of the pellet affects the catalyst effectiveness; the larger the size, the lower the effectiveness factor. Thus, one has to contend with a trade-off between two opposite effects: decrease in limiting flow with decrease in pellet size but an increase in catalyst effectiveness. Clearly, then, the treatment of fixed-bed catalytic reactors is far more complex than nonpacked homogeneous reactors. These effects are qualitatively illustrated in Figure 8.1.

Figure 7.1 gave us an idea of the number of forms a reactor can take depending on the physical state of the reactants/products involved (gas, liquid, solid, or combinations thereof), and the state of motion of the solids (fixed, fluidized, or moving). In this chapter, we consider fixed-bed reactors involving catalytic (nonreactive) solids and a fluid (generally gaseous). Of these different types, we consider two types in some detail in this chapter: fixed- and fluidized-bed reactors. We also briefly look at the basic features of two other types of reactors: radial and wire-gauze reactors.

Fixed-bed reactor

The fixed-bed reactor is the most widely used reactor for solid catalyzed reactions. A number of design and operational variations are available. In principle, the design of a fixed-bed reactor is very similar to the design of a plug-flow reactor. But it is important to carefully take the transport resistances created by the solid packing into account. In Chapter 6, we saw that the particle size of a solid determines the magnitude of the diffusion resistance. For smaller particles, the diffusion resistances are low; therefore, their contribution to the rates was minimal. When the small particles are packed in a bundle of tubes, however, there is a significant limitation to the length of the packing that you are allowed due to the large pressure drops as a result of the fine packing. Therefore, there is a compromise between the rates and the pressure drop. Among the many alternative techniques developed, we pay particular attention to the radial-flow reactor initially developed by Haldor Topsøe Company for ammonia synthesis reaction, later in this chapter.

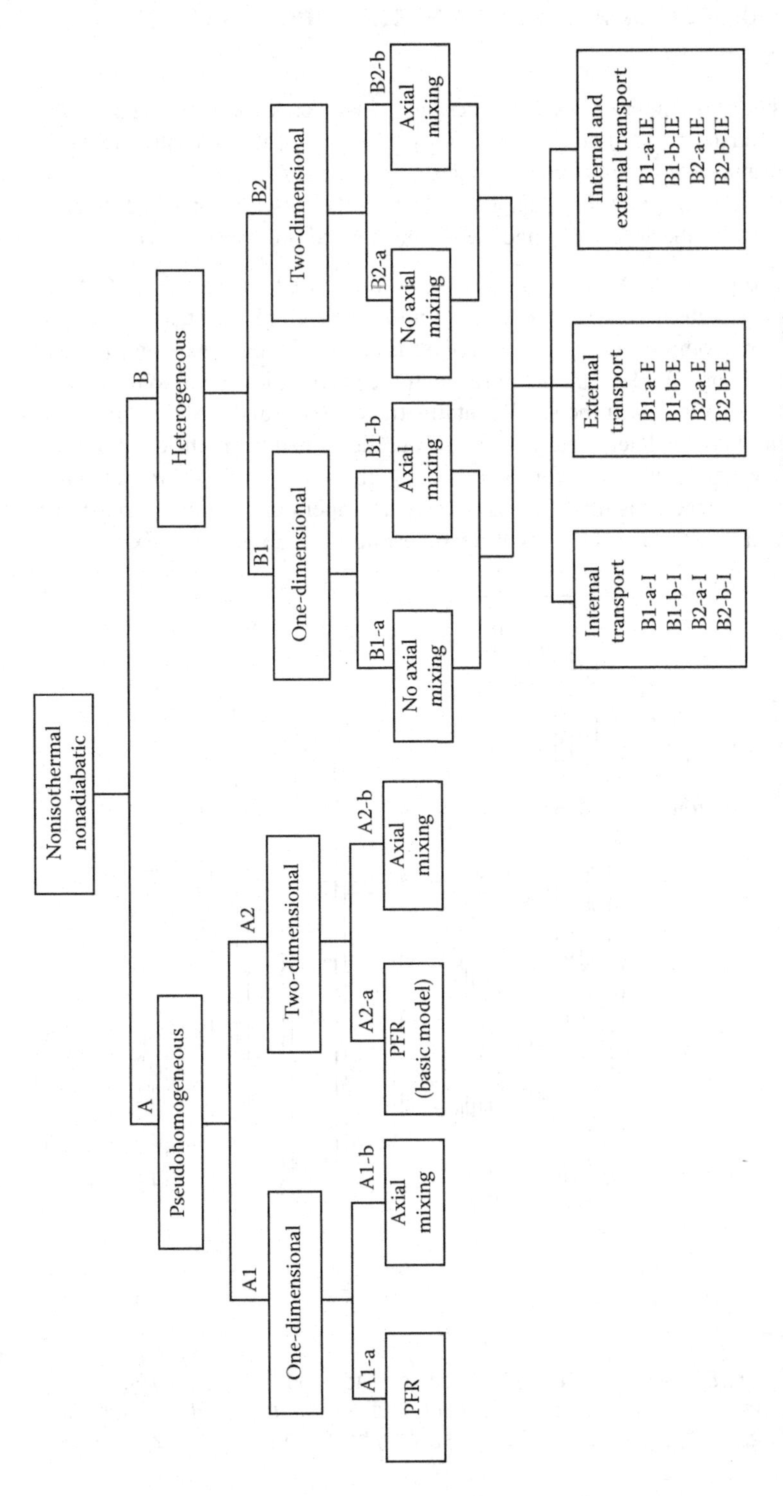

Figure 8.1 Comparison of the homogeneous and nonhomogeneous reactors for nonisothermal nonadiabatic conditions.

Nonisothermal, nonadiabatic, and adiabatic reactors

Multitubular reactor
Heat-exchanger reactor

The most important fixed-bed designs are the nonisothermal, nonadiabatic, fixed- (or packed)-bed reactor (NINA-PBR)* (also called the multitubular or heat-exchanger-type reactor), and the single or multistage adiabatic fixed-bed reactor (A-PBR), and it is important at the outset to note the difference between the approaches and the design of these two operational categories.

A typical NINA-PBR design is shown in Figure 8.2. It consists of a bundle of tubes, usually around 3–5 cm in diameter, placed in a shell. In the most common design, the catalyst is placed in the tubes and a heat-exchange fluid is circulated through the shell. A less common design is the one in which the shell contains the catalyst, and the heat-exchange fluid is circulated through the tube bundle embedded in the catalyst bed. The height of the reactor can be usually in the range of 3–10 m. A feature of the reactor is that there is usually a temperature profile for reactions with even a low heat effect (exothermic or endothermic). Exothermic

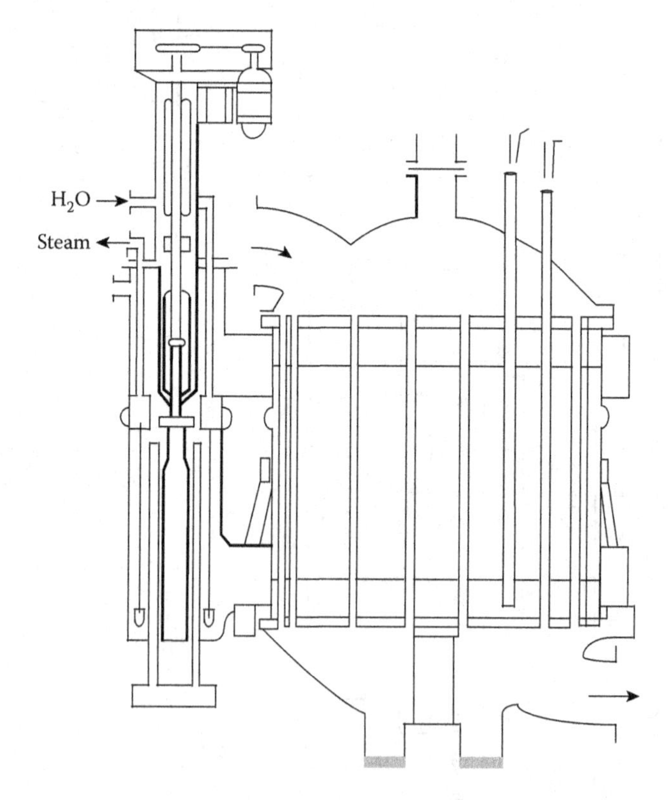

Figure 8.2 A typical NINA-PBR reactor. (Adapted from Joshi, J.B. and Doraiswamy, L.K. *Chemical Reaction Engineering, in Albright's Chemical Engineering Handbook*, CRC Press, Albright, Boca Raton, FL, 2009.)

* PBR was chosen to avoid confusion with the fluidized-bed reactor (FBR).

reactions are more difficult to handle because hot spots can develop within the reactor, leading to catalyst deactivation and adverse changes in conversion/selectivity. The objective of the design here is to find the height for a required conversion for tubes of a fixed diameter. Then, depending on selectivity, stability, radial and axial concentration, and temperature profile considerations, the diameter and height are recalculated for optimum performance.

Adiabatic packed-bed reactor

On the other hand, the approach to the design of an A-PBR is quite different. Since there is no abstraction or addition of heat in this reactor, radial transport of heat is absent and that of mass can usually be neglected. An important consequence of this is that the reactor diameter can be quite large, thus dispensing with the need for using small-diameter tubes as in the case of NINA-PBR. Hence, a single large-diameter reactor should, in principle, be all that is needed. In practice, however, if the heat effect is large, the temperature may become too high or too low beyond some axial position in the reactor, depending on whether the reaction is exothermic or endothermic. It would then be necessary to use more than one stage and cool or heat the process fluid between stages (see Figure 8.3). The design here calls for optimizing the inlet and outlet conditions (and therefore the stage height for a given diameter) for each stage to obtain the desired conversion/selectivity at the end of the final stage. The recommended method for optimizing a staged system like this is dynamic programming.

In this section, we consider the design procedures for both NINA-PBR and A-PBR and illustrate the use of the more important methods in the examples provided in this chapter.

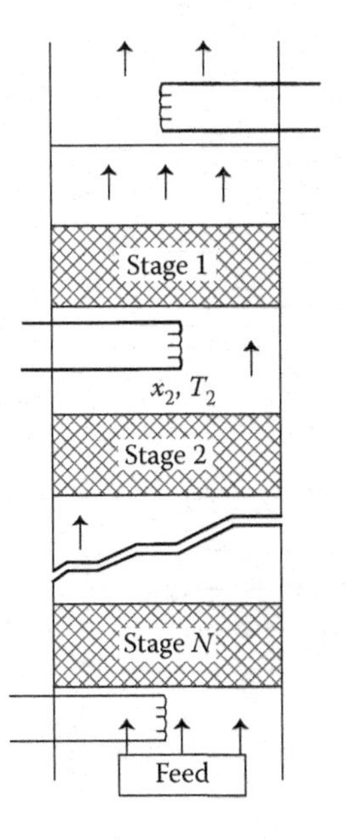

Figure 8.3 Multiple-stage reactor with interstage cooling.

For $L/D \gg 50$, plug-flow assumption is valid. Check if the system is free from

- -Axial dispersion problems!
- -Radial concentration and temperature gradients!

Design methodologies for NINA-PBR

The number of tubes to be used is obtained from the desired production rate and the allowable velocity in each tube. Velocities in the order of 1 m/s are considered normal. Then the objective of the design is to calculate the tube length for accomplishing the required conversion. For tubes with $L/D \gg 50$, plug flow can usually be assumed. However, this assumption may not always be valid because radial gradients can exist. Axial diffusion can also exist, thus modifying the conversion gradient that would normally exist due to flow as the reaction progresses from the inlet of the reactor to the outlet. These gradients can cause severe deviations from ideal reactor conditions. To account for all these gradients and other nonidealities that might exist, several broad classes of models have been proposed. Four such classes are sketched in Figure 8.4.

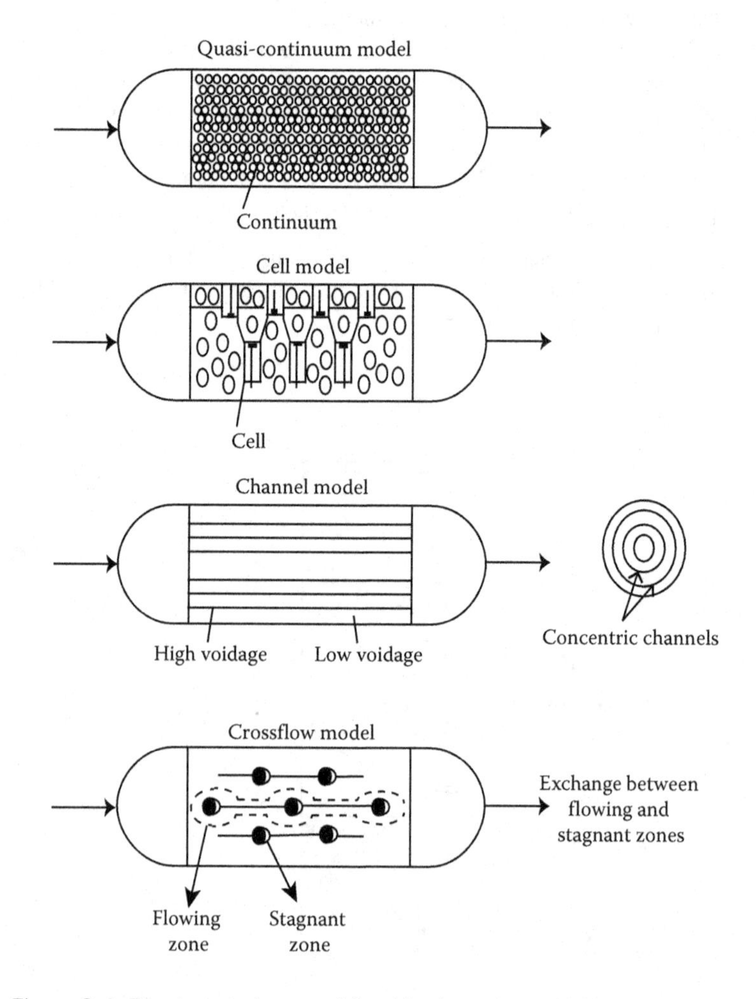

Figure 8.4 Four broad classes of fixed-bed reactor models.

Quasi-continuum models Of these, the quasi-continuum model is the most common. Here, the solid–fluid system is considered as a single pseudo-homogeneous phase with properties of its own. These properties, for example, diffusivity, thermal conductivity, and heat transfer coefficient, are not true thermodynamic properties but are termed as effective properties that depend on the properties of the gas and solid components of the pseudo-phase. Unlike in simple homogeneous systems, these properties are anisotropic, that is, they have different values in the radial and axial directions. Kulkarni and Doraiswamy (1980) have compiled all the equations for predicting these effective properties. Both radial and axial gradients can be accounted for in this model, as well as the fact that the system is really heterogeneous and hence involves transport effects both within the particles and between the particles and the flowing fluid.

Effective properties

Cell model Another important class of models is the cell model, which breaks up the reactor into a large number of cells, where each cell (or microreactor) corresponding to a single pellet and its immediate neighborhood. By allowing for flow between cells in both the radial and axial directions, one- (1-D) and two-dimensional (2-D) models as well as various degrees of mixing can be simulated. The equations involved in these models are algebraic and not differential as in the continuum models.

Models based on the pseudo-homogeneous assumption

For the simple isothermal situation where no axial or radial gradients exist, the design equations for a simple reaction derived in Chapter 1 are still valid. For nonisothermal and nonadiabatic reactions with axial and radial variations in the temperature and concentration, the simultaneous, numerical solutions of the most general forms of the equations are necessary. Depending on the absence of axial and/or radial diffusion, the diffusivity terms are set to zero, and the material and energy balance equations collapse to simpler forms. The material and energy balance equations for four models of varying degrees of simplification are listed in Table 8.1.

Homogeneous, pseudo-homogeneous, and heterogeneous models

A homogeneous reactor design protocol was described in Chapter 1. The design involved only one state of aggregation, and the fluid properties were also effective properties.

A pseudo-homogeneous model requires the determination of effective properties of a heterogeneous system in such a way that the homogeneous

Table 8.1 Some Models for Interparticle Transport: The Reactor Field Equations

	Model	Equation[a]	Boundary Conditions
A1-a	i. 1-D isothermal plug-flow reactor	$u\frac{d[A]}{dz}+(-r_A)=0$	$[A]=[A]_0$ at $z=0$
	ii. 1-D nonisothermal plug-flow reactor	$u\frac{d[A]}{dz}+(-r_A)=0$	$[A]=[A]_0$ at $z=0$
		$u\rho C_p\frac{d[T]}{dz}+\frac{4U}{d_T}(T-T_w)-(-\Delta H_r)(-r_A)=0$	$T=T_0$ at $z=0$
A1-b	i. 1-D isothermal reactor with axial dispersion	$D_{ez}\frac{d^2[A]}{dz^2}-u\frac{d[A]}{dz}-(-r_A)=0$	$-D_{ez}\frac{d[A]}{dz}-u([A]_0-[A])=0$ at $z=0$ $\frac{d[A]}{dz}=0$ at $z=L$
	ii. 1-D nonisothermal reactor with axial dispersion	$D_{ez}\frac{d^2[A]}{dz^2}-u\frac{d[A]}{dz}-(-r_A)=0$	$-D_{ez}\frac{d[A]}{dz}=u([A]_0-[A])$ at $z=0$ $\frac{d[A]}{dz}=0$ at $z=L$
		$\lambda_{ez}\frac{d^2T}{dz^2}-u\rho C_p\frac{dT}{dz}-\frac{4U}{d_T}(T-T_w)+(-\Delta H_r)(-r_A)=0$	$-\lambda_{ez}\frac{dT}{dz}=u\rho C_p(T_0-T)$ at $z=0$ $\frac{dT}{dz}=0$ at $z=L$
A2-a	2-D nonisothermal reactor with radial gradients and no axial dispersion (basic model)	$D_{er}\left(\frac{\partial^2[A]}{\partial r^2}+\frac{1}{r}\frac{\partial[A]}{\partial r}\right)-u\frac{\partial[A]}{\partial z}-(-r_A)=0$	$\frac{\partial[A]}{\partial r}=0$ at $z>0, r=0$ $\frac{\partial[A]}{\partial r}=0$ at $z>0, r=R$ $[A]=[A]_0$ at $z=0, r>0$

		$\lambda_{er}\left(\frac{\partial^2 T}{\partial r^2}+\frac{1}{r}\frac{\partial T}{\partial r}\right)-u\rho C_p\frac{\partial T}{\partial z}+(-\Delta H_r)(-r_A)=0$	$\frac{\partial T}{\partial r}=0$ at $z>0, r=0$ $-\lambda_{er}\frac{\partial T}{\partial r}=h_w(T-T_w)$ at $z>0, r=R$ $T=T_0$ at $z=0, r>0$
A2-b	2-D nonisothermal reactor with radial gradients and axial dispersion	$D_{er}\left(\frac{\partial^2[A]}{\partial r^2}+\frac{1}{r}\frac{\partial[A]}{\partial r}\right)+D_{ez}\frac{\partial^2[A]}{\partial z^2}-u\frac{\partial[A]}{\partial z}-(-zr_A)=0$ $\lambda_{er}\left(\frac{\partial^2 T}{\partial r^2}+\frac{1}{r}\frac{\partial T}{\partial r}\right)+\lambda_{ez}\frac{\partial^2 T}{\partial z^2}-u\rho C_p\frac{\partial T}{\partial z}+(-\Delta H_r)(-r_A)=0$	$-D_{ez}\frac{\partial[A]}{\partial z}=u([A]_0-[A])$ at $z=0, r>0$ $\frac{\partial[A]}{\partial z}=0$ at $z=L, r>0$ $\frac{\partial[A]}{\partial r}=0$ at $z>0, r=0$ $\frac{\partial[A]}{\partial r}=0$ at $z>0, r=R$ $-\lambda_{ez}\frac{\partial T}{\partial z}=u\rho C_p(T_0-T)$ at $z=0,\ r>0$ $\frac{\partial T}{\partial z}=0$ at $z=L, r>0$ $\lambda_{er}\frac{\partial T}{\partial r}=h_w(T-T_w)$ at $z=0,\ r=R$

Source: Doraiswamy, L.K. and Sharma, M.M., *Heterogeneous Reactions: Analysis, Examples, and Reactor Design*, Wiley, New York, 1984.

[a] The equations are given in dimensional form and can be easily recast in nondimensional form.

model equations can still be used with an acceptable degree of accuracy. Thus, the models used in the developments presented above are not based on thermodynamic properties, but on *effective properties* that can be used to analyze transport phenomena within the continuum, as in any homogeneous system.

Effective thermal conductivity

Effective diffusivity

The most important of these models is the *effective thermal conductivity* and *effective diffusivity*. It is important to note that each of these has a radial and an axial component. A number of correlations have been developed to estimate these (see Kulkarni and Doraiswamy, 1980, for extensive tables).

Another equally important property is the *heat transfer coefficient* for the bed to control the fluid. If there is no radial temperature gradient within the bed, it is assumed that the entire resistance inside the tube is localized within the film adjacent to the wall. This is the value to use in 1-D beds. On the other hand, if there is a radial gradient, a 2-D model must be used and heat transfer coefficient adjacent to the wall, and effective radial thermal conductivity within the bed are involved in the overall transport process. Several correlations for these have been listed by Kulkarni and Doraiswamy (1980).

1D pseudo-homogeneous NINA

1D pseudo-homogeneous nonisothermal, nonadiabatic flow (model A1-a in Table 8.1) Consider a simple reaction

$$A \rightarrow R \tag{R1}$$

occurring in a single tube of the multitubular packed-bed reactor. Taking a differential volume dV of the reactor, the material and energy balances can be readily written as

$$\frac{d[A]}{dt} + (-r_A) = 0 \tag{8.1}$$

$$\rho C_p \frac{dT}{dt} + U A_h (T - T_w) - (-\Delta H_r)(-r_A) = 0 \tag{8.2}$$

This is an initial value problem with the boundary conditions

$$\begin{aligned} [A] &= [A]_0 \quad \text{at } z = 0 \\ T &= T_0 \quad \text{at } z = 0 \end{aligned} \tag{8.3}$$

The second term in the heat balance accounts for heat removal by a circulating liquid at temperature T_W through a heat transfer area A_h, on the assumption of an overall heat transfer coefficient U across the tube wall. Note that r_A represents the rate per unit volume of the reactor. Hence,

$$-r_A\left(\text{or} - r_{VA}\right) = \left(1 - f_B\right)\left(-r_{vA}\right)$$
$$k\left(\text{or}\, k_V\right) = \left(1 - f_B\right)k_v \tag{8.4}$$

where r_{vA} and k_v represent, respectively, the rate and rate constant based on the catalyst volume, and f_B is the bed voidage. We now recast these equations in the dimensionless form as

$$\frac{d[\hat{A}]}{d\bar{t}} + \left(-r_A\right) = 0 \tag{8.5}$$

$$\frac{d\hat{T}}{d\bar{t}} + \frac{UA_h}{\rho C_p}\left(\hat{T} - \hat{T}_w\right) - \gamma_H\left(-r_A\right) = 0 \tag{8.6}$$

where

$$\left[\hat{A}\right] = \frac{[A]}{[A]_0},\quad \hat{T} = \frac{T}{T_0},\quad \hat{T}_w = \frac{T_w}{T_0},\ \gamma_H = \frac{-\Delta H_r}{\rho C_p T_0} \tag{8.7}$$

The parameters γ_M and γ_H are, respectively, the mass and heat transfer groups (with units of inverse rate) defined in terms of the common physical properties. The initial conditions are

$$\left[\hat{A}\right] = 1 \quad \text{at}\ \bar{t} = 0$$
$$\hat{T} = 1 \quad \text{at}\ \bar{t} = 0 \tag{8.8}$$

The solution to these equations is straightforward.

Reduction to isothermal operation If the reactor is assumed to be isothermal, only Equation 8.1 needs to be solved. An analytical solution can be found, which for a first-order reaction ($-r_{vA} = k_v[A]$) is given by

$$\ln\frac{[A]_0}{[A]} = \left(1 - f_B\right)\frac{k_v L}{u} = \frac{kV_T}{Q_T} \tag{8.9}$$

The reactor length L can be readily found for any given feed velocity u and for conversion.

Momentum balance Recall that we have only used material and energy balance equations in the design described above. The momentum balance should also be logically considered. Let us examine this situation a little further. Any significant change in momentum occurs only when the fluid velocity in the reactor increases (or decreases) axially. For this to happen, either there should be a steady increase in temperature or a volume increase with the reaction. The extent to which these changes normally occur are not significant enough to include a momentum balance. We shall have an occasion later when we consider radial reactors to

Momentum balance

Brinkmann–Forchheimer–Darcy equation

necessarily include a momentum balance in the set of design equations. Brinkmann–Forchheimer–Darcy equation is used to obtain $v_z(r)$ as a function of $\in(r)$

$$0 = -\frac{dP}{dz} + \mu_{\text{eff}}\left(\frac{\partial^2 v_z}{\partial r^2} + \frac{1}{r}\frac{\partial v_z}{\partial r}\right) - \frac{\mu_f}{K} v_Z - \frac{F\rho}{\sqrt{K}} v_Z^{\;2} \tag{8.10}$$

where

$$K = \frac{\epsilon^3(r) d_p^2}{150\left(1-\epsilon(r)\right)^2}, \quad F = \frac{1.75}{\sqrt{150\epsilon^3(r)}} \tag{8.11}$$

The solution of this equation gives $v_z(r)$, which is then incorporated into the energy balance and mass balance equations for the reactor, along with reaction terms (Dixon and Nejemeisland, 2001).

In designing the packed-bed reactors, the pressure-drop issues are efficiently handled through the Ergun (1952) equation:

Ergun equation

$$\frac{\Delta P}{L} = \frac{150\mu}{\varphi_p d_p^2}\left[\frac{(1-\epsilon)^2}{\epsilon^3} v + \frac{1.75\rho_f}{\varphi_p d_p}\frac{(1-\epsilon)}{\epsilon^3} v^2\right] \tag{8.12}$$

The basic model (model A2-a): 2D pseudo-homogeneous nonisothermal, nonadiabatic with no axial diffusion This is perhaps the most useful model and hence we refer to it as the *basic model* (see Table 8.1 for governing equations). Example 8.1 illustrates the application of the basic model to an industrially important organic reaction.

Example 8.1: Catalytic hydrogenation of nitrobenzene to aniline: Application of the basic model

Aniline is produced by the catalytic hydrogenation of nitrobenzene in a fixed-bed reactor according to the reaction

$$C_6H_5NO_2 + 3\text{–}H_2 \longrightarrow C_6H_5NH_2 + 2\text{–}H_2O$$

$$A + 3B \rightarrow R + 2S \tag{R2}$$

The rate equation is given by

$$-r_{wA} = \frac{k_w[A]}{\left(1 + K_B[B]\right)^2}, \text{ mol/gcat s} \tag{E8.1.1}$$

with

$$k_w = k_w^0 \exp\left(-\frac{E}{RT}\right), \text{cm}^3/\text{gcat s} \qquad \text{(E8.1.2)}$$

$$K_B = K_B^0 \exp\left(\frac{\Delta H_{ad}}{RT}\right), \text{cm}^3/\text{mol} \qquad \text{(E8.1.3)}$$

Assuming a pseudo-homogeneous 2-D reactor with plug flow of fluid and constant properties, calculate the axial concentration and temperature profiles at several radial positions along the axis of a single tube. Use the following property/parameter values:

Material Balance-Related Data	Energy Balance-Related Data	Physical Properties	Process Data
$[A]_0 = 7.1 \times 10^{-4}$ mol/L	$-\Delta H_r = 180$ kcal/mol	$\rho_s = 2.18$ g/cm^3	$f_B = 0.312$
$K_B^0 = 10.7$ cm^3/mol, $\Delta H_{ad} = 8039$ cal/mol	$C_p = 0.49$ cal/g (reactant gases)°C at 200°C	$\rho = 0.0944$ g/L	$u = 40$ cm/s
$k_w^0 = 9.46 \times 10^{-3}$ L/gcatalyst s, $E = 2631$ cal/mol	$\lambda = 1.4$ cal/m°C s	$D_{er} = 4.74 \times 10^{-4}$ cm^2/s	$T_0 = 160$°C
	$h_w = 9.5 \times 10^{-4}$ cal/cm^2°C s		$T_w = 100$°C

The calculations were done using the commercial CFD tool, COMSOL (formerly FEMLAB), and the following equations of model A2-a of Table 8.1 are solved

$$D_{er}\left(\frac{\partial^2 [A]}{\partial r^2} + \frac{1}{r}\frac{\partial [A]}{\partial r}\right) - u\frac{\partial [A]}{\partial z} - (-r_A) = 0$$

$$\lambda_{er}\left(\frac{\partial^2 T}{\partial r^2} + \frac{1}{r}\frac{\partial T}{\partial r}\right) - u\rho C_p\frac{\partial T}{\partial z} + (-\Delta H_r)(-r_A) = 0$$

with the boundary conditions

$$\frac{\partial [A]}{\partial r} = 0 \quad \text{at} \quad z > 0, \quad r = 0 \quad \text{and at} \quad z > 0, \quad r = R$$

$$[A] = [A]_0 \quad \text{at} \quad z = 0, \quad r > 0$$

$$\frac{\partial [A]}{\partial z} = 0 \quad \text{at} \quad z = L, \quad r > 0$$

$$\frac{\partial T}{\partial r} = 0 \quad \text{at} \quad z > 0, \quad r = 0$$

$$-\lambda_{er}\frac{\partial T}{\partial r} = h_w(T - T_w) \quad \text{at} \quad z > 0, \quad r = R$$

$$T = T_0 \quad \text{at} \quad z = 0, \quad r > 0$$

$$\frac{\partial T}{\partial z} = 0 \quad \text{at} \quad z = L, \quad r > 0$$

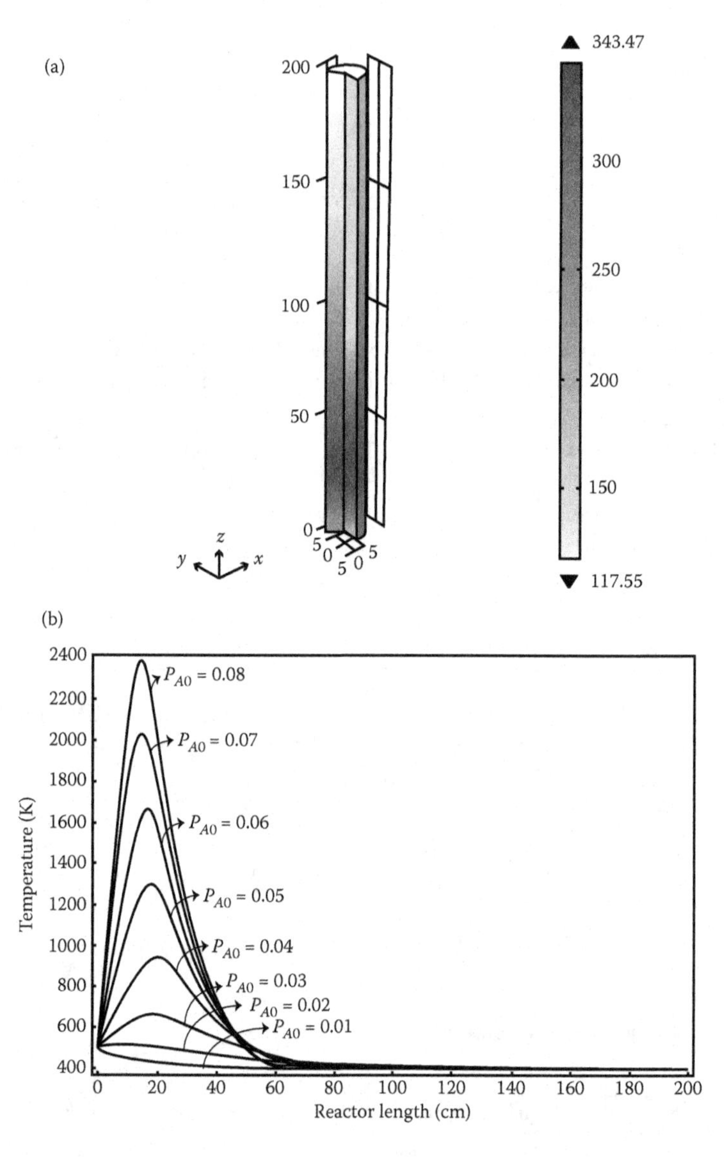

Figure 8.5 2D nonisothermal solution without axial dispersion. The axisymmetric solution plane is rotated 210° along the *z*-axis. (a) The variation of temperature along with the reactor length at a low inlet partial pressure; (b) the effect of the feed amount on the level of hot spot at the inlet of the reactor.

Note that in describing the procedure for a simpler model above, we used the dimensionless forms of the equations. One can also use the dimensional forms given in the table, as we do in this example. We used both the methods to illustrate that the choice is largely personal. A three-dimensional (3 D) display of the result is given, showing the hot spot at the entrance (Figure 8.5).

The conclusion from this contour plot is self-evident. It is interesting to note that a temperature maximum occurs very close to the reactor entrance and then the heat is lost rapidly to the cooling jacket. For higher partial pressures, the peak can reach 1000°C. In the case shown, the temperature falls from a high of 420°C and approaches to the wall temperature gradually that is held constant at 100°C.

Extension to nonideal models with and without heterogeneity The continuity and energy equations for a number of models are included in Table 8.1 along with the methods of solution to be used. All the models listed in the table can also be expressed in dimensionless form. For this purpose, the effective diffusivities and thermal conductivities in the axial and radial directions are expressed as corresponding Peclet numbers. The definitions of these Peclet numbers along with their values for the extremes of plug flow and full mixing are given in Table 8.2.

Table 8.2 Peclet Numbers for Mass and Heat Transfer

	Peclet Number for Mass Transfer	Peclet Number for Heat Transfer
Axial		
Based on reactor length, L	$Pe_{mz} = \dfrac{uL}{D_{ez}}$	$Pe_{hz} = \dfrac{u\rho C_p L}{\lambda_{ez}}$
Based on pellet size, d_p	$Pe'_{mz} = \dfrac{ud_p}{D_{ez}}$	$Pe'_{hz} = \dfrac{u\rho C_p d_p}{\lambda_{ez}}$
Value for PFR	∞	∞
Value for MFR	0	0
Radial		
Based on reactor radius, r_T	$Pe_{mr} = \dfrac{ur_T}{D_{er}}$	$Pe_{hr} = \dfrac{u\rho C_p r_T}{\lambda_{er}}$
Based on pellet size, d_p	$Pe'_{mr} = \dfrac{ud_p}{D_{er}}$	$Pe'_{hr} = \dfrac{u\rho C_p d_p}{\lambda_{er}}$
Value for PFR	0	0
Value for LFR[a]	∞	∞

[a] This represents the laminar flow reactor that is characterized by a parabolic radial velocity profile. There is no radial diffusion as in PFR; hence, the radial diffusivity is zero and the corresponding Peclet number is infinity.

Adiabatic reactor

The approach

As there is no heat removal or addition in an adiabatic reactor, radial transport of heat is absent and that of mass can usually be neglected. Hence, in modeling an adiabatic reactor, a simple 1 D model would be adequate. What is important, however, is the height of the bed for a given diameter (as determined by the production rate), that is, the allowable conversion within the bed before the product mixture is taken out as the final product. Alternatively, remembering that one has no control over the temperature in an adiabatic situation, the product may have to be cooled (for an exothermic reaction) before it becomes too hot or heated (for an endothermic reaction) before it becomes too cold, and is introduced into a second bed. It may happen that the height of each bed would be so small that a number of beds (stages) with interstage cooling or heating would be necessary to achieve a desired final conversion. This means that two decisions must be taken for each bed: the inlet temperature and the outlet conversion to be achieved in it. In other words, if there are N beds, $2N$ decision variables must be simultaneously varied to arrive at an optimum policy for the reactor that would maximize profit. This leads to an enormous amount of computation. A particularly attractive technique for reducing the amount of computation is *dynamic programming* introduced by Bellman (1957).

Dynamic programming

The principle of dynamic programming can be exploited in the adiabatic reactor design in two ways:

1. By using a rigorous mathematical procedure or one that is partly graphical (Lee and Aris, 1963).
2. By using a simple trial-and-error graphical method suggested by Levenspiel (1993).

We shall first consider a straightforward single bed for the entire reaction with no heating or cooling and then explain both the methods mentioned above for multiple-bed reactors. Since the basis for all these methods is the unique conversion–temperature relationship that exists for the adiabatic reactor, we begin by a consideration of this plot.

A unique conversion–temperature relationship Consider the reaction

$$A \underset{k_+/k_-}{\leftrightarrow} R \qquad \text{(R3)}$$

for which the rate equation can be written as

$$-r_A = [A]_0 \left(\frac{T_0 P}{T P_0}\right)\left[k_+\left(\frac{1-X_A}{1+\varepsilon_A X_A}\right) - k_-\left(\frac{X_A}{1+\varepsilon_A X_A}\right)\right] \qquad (8.13)$$

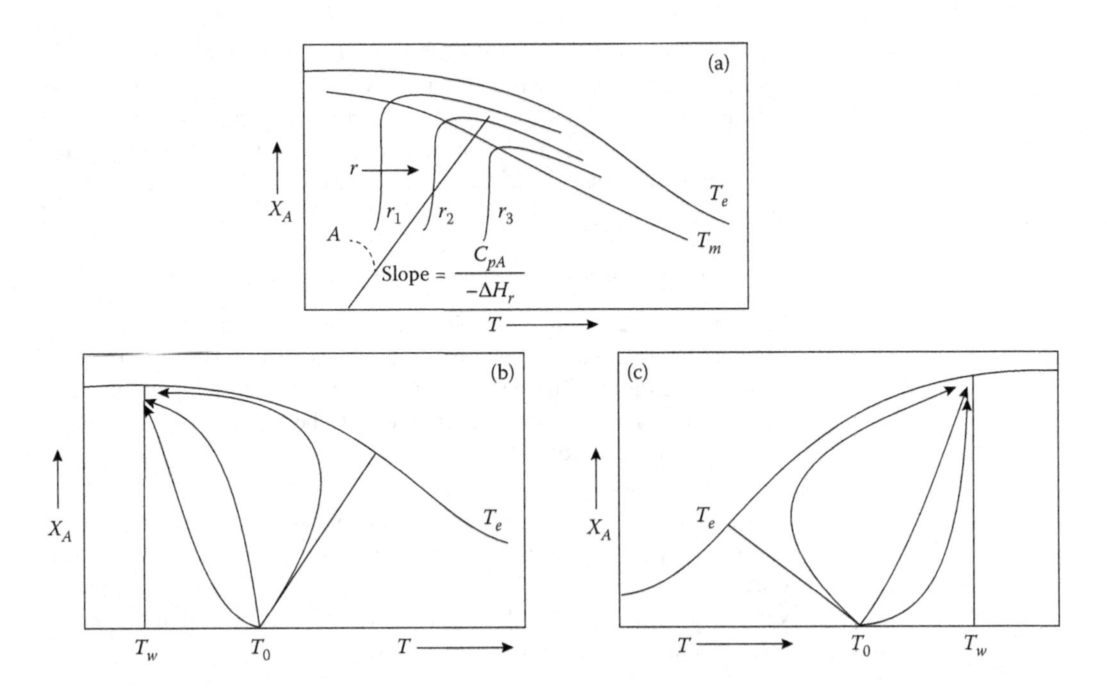

Figure 8.6 Trajectories of (a) adiabatic, (b) wall-cooled (for exothermic), and (c) wall-heated (for endothermic) reactions.

If the reaction is exothermic, the equilibrium curve relating the conversion to temperature will have the shape shown in Figure 8.6a and is designated as T_e. The figure also shows a number of constant rate curves, that is, rate contours. The locus of the maxima observed in these contours is designated as the T_m curve. Now, reverting to reaction R3, the following energy balance can be readily written for an adiabatic reactor:

$$(X_A - X_{A0})(-\Delta H_r) = C_{pm}(T - T_0) \tag{8.14}$$

giving (since $X_{A0} = 0$)

$$X_A = \left(\frac{F_{T0}C_{pm}}{F_{A0}(-\Delta H_r)}\right)(T - T_0) \tag{8.15}$$

or (since $\Delta[A] = X_A[A]_0$)

$$\Delta[A] = J\Delta T \tag{8.16}$$

where for a pure feed of A (when $F_{T0} = F_{A0}$)

$$J = \frac{C_{pm}[A]_0}{-\Delta H_r} \tag{8.17}$$

It will be recalled that we already defined J earlier in Chapter 1. We note from Equation 8.17 that a trajectory of slope $C_{pm}/-\Delta H_r$ on an X_A–T plot, as shown by line A in Figure 8.6a, represents the path of an adiabatic reaction. The T_e and T_m curves along with the adiabatic line A and the rate contours shown in the X_A–T plane are central to the design of an adiabatic reactor. Note that this relationship between conversion and temperature is unique to adiabatic reactors and cannot be used for any nonadiabatic situation (i.e., where heat is supplied or removed).

If a constant wall temperature is maintained by cooling, then the reactor is no longer adiabatic, and the trajectories will lie between the limits shown in Figure 8.6b where T_w represents the wall temperature. The behavior of an endothermic reaction (where heat has to be supplied) is sketched in Figure 8.6c.

Single-bed reactor As mentioned earlier, we shall first illustrate a method for designing a single-bed reactor using the unique X_A–T relationship of Equation 8.17. To provide a common basis for explaining various designs, we use the running reaction scheme of the chapter: hydrogenation of nitrobenzene to aniline.

Example 8.2: Design of a single-bed adiabatic reactor for the hydrogenation of nitrobenzene to aniline

A fixed-bed adiabatic reactor is to be designed to produce 6000 tons per annum (300 working days) of aniline (B). Vertical tubes, 0.8, 1.0, or 1.5 m diameter, packed with catalyst pellets, are proposed to be used. The desired conversion at the reactor outlet is 99.7% and there is no heat exchange between the reactor and the surroundings. The rate equation for the disappearance of nitrobenzene (A) is the same as in Example 8.1 and is reproduced below in different units:

$$-r_{wA} = \frac{k_{wP} P_A}{\left(1 + K_B p_B\right)^2}, \text{ mol/gcat h}$$

$$k_{wP} = 8.77 \exp\left[-2631/R_g T\right], \text{ mol/h g atm}$$

$$K_B = 2.77 \times 10^{-3} \exp\left[8040/R_g T\right], \text{ atm}^{-1}$$

Data

Material Balance-Related Data	Energy Balance-Related Data	Process Parameters
H_2:nitrobenzene = 60:1	ΔH_r = 180 kcal/mol	T = 200°C
H_2:nitrobenzene = 80:1	$\bar{C}_p$ = 7.5 cal/mol°C	$T_{ambient}$ = 25°C
		P = 1 atm
		ρ_{bulk} = 2.18 g/cm³
		ϕ_p = 0.312

Calculate the reactor length, with conversion and temperature profiles, for each of the tube sizes for both the ratios of H_2 to nitrobenzene.

SOLUTION

The production rate of aniline is

$$\frac{6000 \times 10^3}{300 \times 24 \times 93} = 8961 \text{ mol/h}$$

From the stoichiometry of the reaction, the feed rate of nitrobenzene (A) is

$$F_{A0} = \frac{8961}{0.997} = 8987 \text{ mol/h}$$

Also, for the ratio of 60:1, for instance, the rate equation can be recast as

$$-r_{wA} = \frac{8.77 \exp(-2631/2T)(1 - X_A/61 - X_A)}{\left(1 + 2.77 \times 10^{-3} \exp(8040/2T)(X_A/61 - X_A)\right)^2} \tag{8.18}$$

The material balance on A gives

$$F_{A0}\, dX_A = (-r_{wA})\, dW \tag{8.19}$$

or

$$F_{A0}\, dX_A = (-r_{wA})\, \rho_B (1 - f_B) A_c\, dz \tag{8.20}$$

and energy balance gives

$$F_{A0}(-\Delta H_r)\, dX_A = F_{t0} C_{pm}\, dT \tag{8.21}$$

where F_{t0} is the total flow rate of the inlet gases. In view of the large excess of hydrogen, F_{t0} can be assumed to be constant.

From Equations 8.20 and 8.21, we can write

$$dX_A = \left[\frac{\rho_B (1 - f_B) A_c (-r_{wA})}{F_{A0}}\right] dz \tag{8.22}$$

$$dT = \left[\frac{(-\Delta H_r)\, \rho_B (1 - f_B) A_c (-r_{wA})}{F_{t0} C_{pm}}\right] dz \tag{8.23}$$

or

$$dX_A = \bar{\alpha}\, dz \tag{8.24}$$

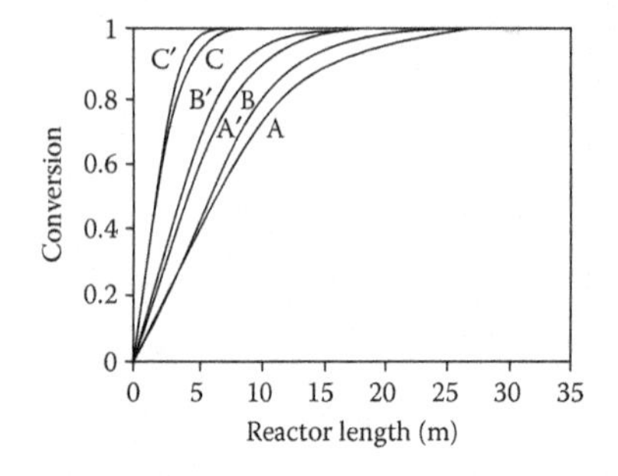

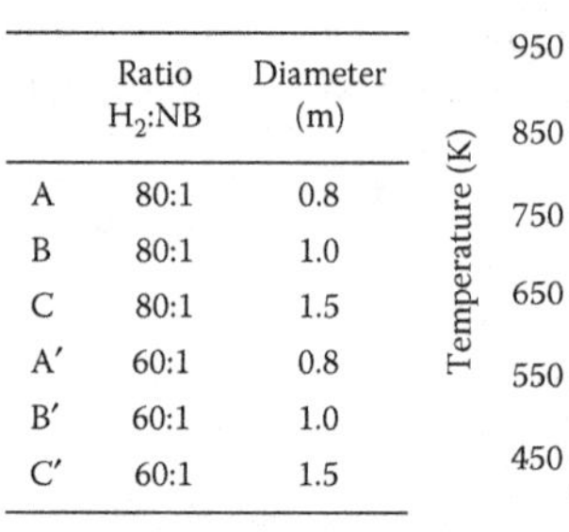

	Ratio H_2:NB	Diameter (m)
A	80:1	0.8
B	80:1	1.0
C	80:1	1.5
A′	60:1	0.8
B′	60:1	1.0
C′	60:1	1.5

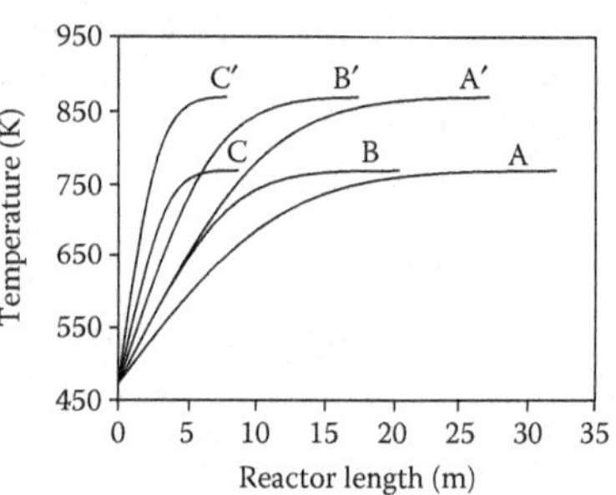

Figure 8.7 Conversion and temperature profiles for a single adiabatic fixed-bed reactor for aniline production.

$$dT = \bar{\beta} dz \tag{8.25}$$

Equations 8.24 and 8.25 are coupled as ordinary differential equations in conversion and temperature. They can be solved for the initial conditions

$$z = 0,\ X_A = 0,\ T_0 = 473\,\text{K}$$

using Runge–Kutta fourth-order method for the different tube diameters and ratios given (note that Equation 8.40 will be slightly different for the ratio 80:1). The results are presented in Figure 8.7.

Multiple-bed reactor We first describe a rigorous method based on the principle of dynamic programming, and follow it up with a simpler graphical method.

Dynamic programming

Design by dynamic programming Consider again the reaction of Example 8.1: hydrogenation of nitrobenzene. Noting the exothermicity of the reaction, it would be desirable to design a multiple-bed reactor with interstage cooling, as sketched in Figure 8.4. We start with the last bed, designated as stage 1, for which the inlet temperature and conversion are not known yet. The computation begins by optimizing this stage for a whole set of possible inlet conditions, the reasoning being that irrespective of the operating conditions of the preceding stage, the sequence as a whole cannot be optimal unless the last stage is optimal. We then add one more stage, designated as stage 2, and optimize the two-stage system as a whole. In this new optimal policy, stage 2 will not necessarily operate at the conditions optimal to its performance alone. However, for the combined policy to be optimal, stage 1 has to be optimal for the feed coming from stage 2. It is only necessary to find the inlet conditions for stage 2, which is really the optimal policy for stages 1 and 2 together, since stage 1 has already been optimized.

The same principle can be extended to three stages from the end. Here, we consider stages 1 and 2 as a single pseudostage for which the inlet conditions have already been optimized, and determine the inlet conditions for stage 3 added to this pseudostage. Thus, if there are N stages, the last optimization step would involve optimization of stage N added to an already optimized pseudo-single stage consisting of $N-1$ stages. This is Bellman's optimization policy and may be formally expressed as

$$\begin{bmatrix}\text{Maximum}\\ \text{profit}\\ \text{from } N\\ \text{stages}\end{bmatrix} = \text{Maximum of}\left[\begin{pmatrix}\text{Profit}\\ \text{from}\\ \text{stage } N\end{pmatrix} + \begin{pmatrix}\text{Maximum}\\ \text{profit from}\\ N-1 \text{ stages}\\ \text{with feed}\\ \text{from } N\text{th stage}\end{pmatrix}\right] \quad (8.26)$$

On the basis of this policy, a computer-aided graphical procedure has been developed by Lee and Aris (1963). This is an elegant procedure, but the geometric construction of the various adiabatic lines and other curves on the X–T plane can clutter, as can be seen from Figure 8.8 even for a far simpler graphical procedure.

Example 8.3: The dynamic programming method for multiple-bed adiabatic reactor design (Doraiswamy and Sharma, 1984)

Chartrand and Crowe (1969) used the following experimental data on the catalytic oxidation of sulfur dioxide in an existing four-bed converter of the Canadian Industries Ltd. to optimize the operation.

Reactor diameter, 5.18 m, total volume of catalyst available 35.94 m^3, total air available to add to the reactor, 159.8 kg/h, reactor feed rate 1328 kmol/h, reactor feed composition 9.5% SO_2, 11.5% O_2, 79% N_2, and reactor feed pressure 1.2 kg/cm^2 absolute. The decision variables were the temperature into each bed. The following assumptions were allowed to be made: plug flow, adiabatic walls, constant effectiveness factor, and ideal gas behavior. It was desired to optimize the variable for the four beds to obtain maximum conversion.

SOLUTION

$$\frac{dF_{SO_3}}{dl} = A_c r_{VSO_3}$$

$$\frac{dT}{dl} = \frac{A_c r_{VSO_3}(-\Delta H)}{\sum_i F_i C_{pm,i}}$$

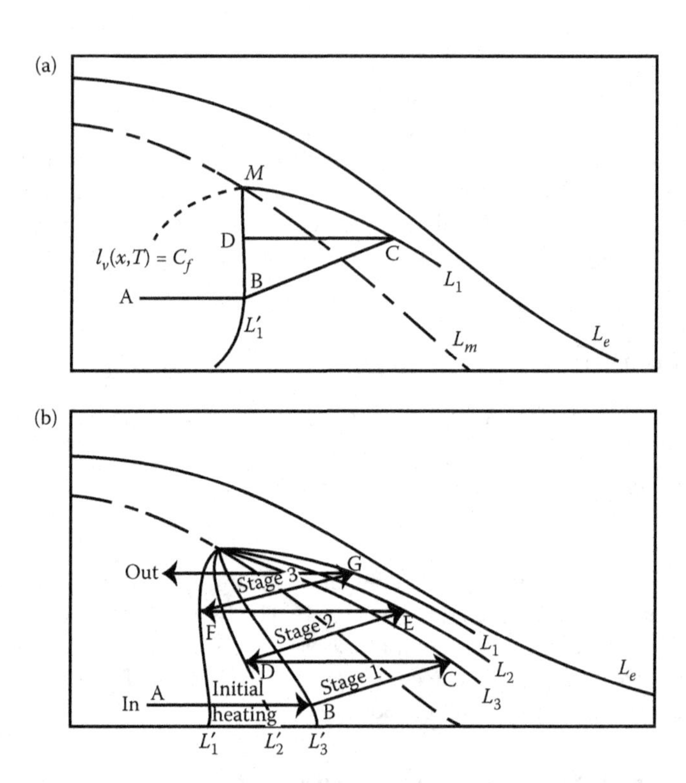

Figure 8.8 The Graphical method of Lee and Aris. Conversion versus temperature plots of equilibrium conversion and the constant rate curves for the design of (a) a single stage adiabatic reactor: after initial heating (A–B) reaction proceeds until the same value of initial rate (B–C) followed by cooling to the same rate (C–D); (b) three stage adiabatic reactor with interstage cooling. (Adapted from Doraiswamy, L.K. and Sharma, M.M., *Heterogeneous Reactions: Analysis, Examples, and Reactor Design,* Wiley, New York, NY, 1984.)

The kinetic equation used was that of Kubota et al. (1959), namely

$$r_{VSO_3} = \frac{\varepsilon k_p\left[p_{O_2} - \left(p_{SO_3}/K_{pSO_2}\right)\right]}{D^2}$$

with

$$D = \begin{cases} \dfrac{1 + K_a p_{SO_3}}{p_{SO_2}} & \text{for} \quad T < 450°\text{C} \\ \dfrac{1 + K_b p_{O_2}^{1/2}}{p_{SO_2}} & \text{for} \quad T < 500°\text{C} \end{cases}$$

A detailed analysis of the alternative optimization methods was given in Doraiswamy and Sharma (1984). Dynamic programming was the only method capable of finding a global maximum rather than local maxima, as was the case with the other methods. Generally, if more than one maximum exists, the global maximum can be found only by comparing the local ones.

The optimization parameters for the problem considered here are

Number of beds:	Four
Bed inlet temperature:	Varied
Catalyst bed depths:	Fixed
An addition to beds:	Fixed
Constraint on temperature:	a. None
	b. $T \leq 600°C$

A simple graphical procedure

The principle of dynamic programming as outlined above can be used, but without much of its rigor, to establish a simple trial-and-error procedure. The various steps in this procedure are shown in Figure 8.9 and adequately explained therein, for the case of cold-shot injection of feed (Levenspiel, 1993). Other reactors and situations are also possible, such as recycle flow reactors, mixed flow reactors, cooling by inert injection, combination cooling, and so on.

Strategies for heat exchange Heat exchange in an adiabatic reactor is one of the most important aspects of its design. Several strategies of heat

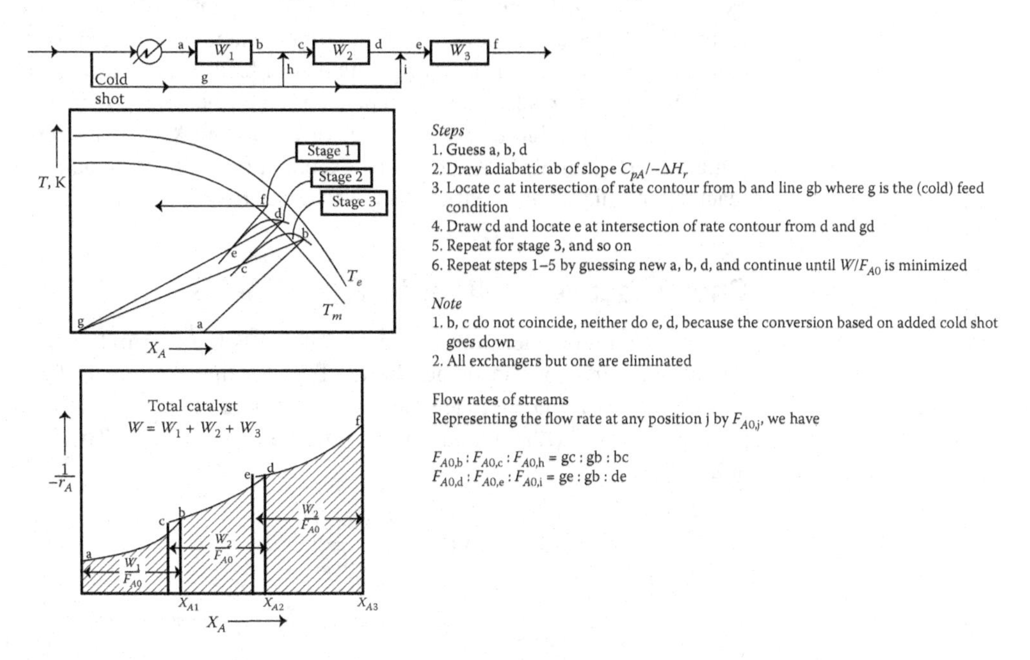

Figure 8.9 Graphical design of a PFR with cold-shot cooling.

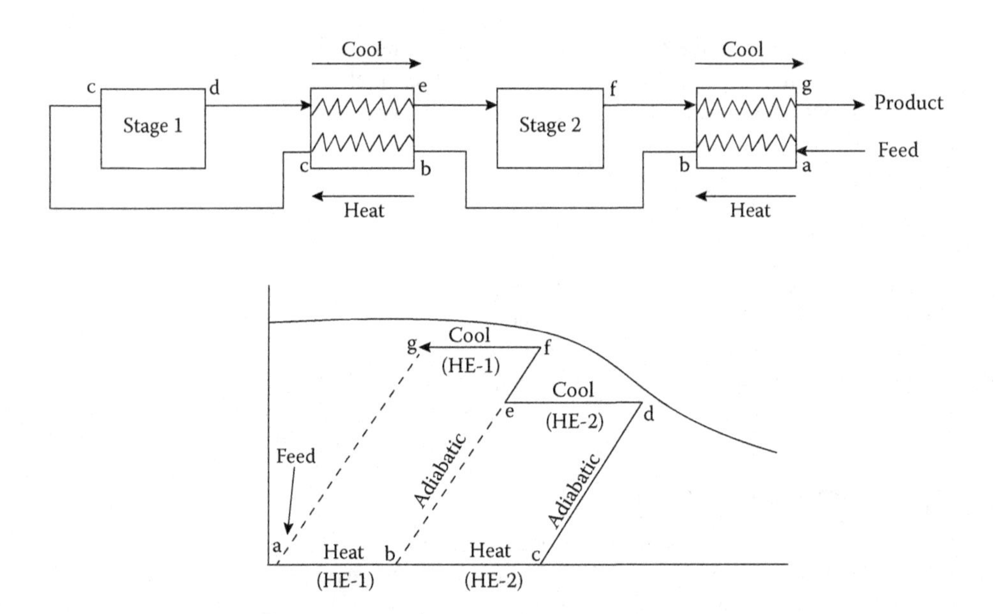

Figure 8.10 Heat exchange between the hot product and cold feed by countercurrent flow of the fluids in suitably located heat exchangers.

exchange have been used in which heat is transferred from a hot to a cold stream within the insulated system such that maximum use of the exothermic heat is accomplished (without removing it out of the system by external heat transfer media). Note that no heat is supplied or abstracted externally in these strategies. As an important example, consider a multistage plug-flow reactor operating with a cold feed. We can arrange the flows in several ways for maximum heat conservation. One possible scenario is shown in Figure 8.10 with the corresponding X_A–T plane representation. The ammonia reactor is an excellent example of an adiabatic reactor for which many heat-exchange schemes have been used.

Choice between NINA-PBR and A-PBR

An important consideration in reactor choice is the thermal mode of operation: nonadiabatic or adiabatic. Two parameters are useful in making a preliminary choice: adiabatic temperature change at complete conversion (ATC), and temperature sensitivity (TS). These are defined as

$$\text{ATC} = \frac{(-\Delta H_r)\, y_{A0}}{C_p} \tag{8.27}$$

$$\text{TS} = \left[\frac{d(-r_A)}{dT}\right]_{[A]} \frac{1}{(-r_A)} = \frac{E}{RT^2} \tag{8.28}$$

ATC is positive for exothermic reactions and negative for endothermic reactions. The most important is the value of TS. Very high values of the two reactions would require cooling during the reaction, and hence, adiabatic operation would be precluded. The values for a few representative reactions are:

	ATC	TS
Toluene dealkylation	162	5.2 (exo)
Phthalic anhydride from *o*-xylene	278	12.8 (exo)
Ethane cracking	−979	−29.4 (endo)

Some practical considerations

Backmixing or axial dispersion Let us now consider some important practical aspects of NINA-PBR design and operation. Backmixing or axial dispersion depends, to a large extent, on the value of L/d_p. The effect of backmixing can be assumed to be absent if this ratio is greater than 50 (variously reported as 30–70) for gaseous reactions and 200 for liquid reactions (see Carberry and Wendel, 1963; Carberry, 1964, 1976). Since in commercial reactors, this ratio is almost always much higher than the indicated values, backmixing is usually not a problem in their design. However, laboratory reactors are shorter so that the data obtained in these reactors cannot be directly applied to large-scale reactors. A more rigorous analysis shows that the following criterion must be satisfied for the data to be free of backmixing effects (Mears, 1971)

For the absence of backmixing

$L/d_p > 50$ in gas phase reactions

$L/d_p > 200$ in liquid phase reactions

$$\frac{L}{d_p} > \frac{20}{Pe_{ma}} \ln \frac{[A]_0}{[A]_e} \tag{8.29}$$

where Pe_{ma} is the axial Peclet number, $d_p u/D_{ea}$.

Nonuniform catalyst distributions between tubes Another problem can arise if the catalyst is not uniformly distributed (leading to nonuniform pressure drops) among the thousands of tubes that are normally present in NINA-PBR. If a particular tube has a lower pressure drop, more gas will flow through it, thus blowing over a part of the catalyst to the neighboring tubes. This process can continue till the tube becomes almost empty. Such a situation can be avoided by adopting the following catalyst-filling policy. Make incremental additions (say in 10% lots) of the catalyst and measure the pressure drop in each tube after each addition. The tubes must be tapped or vibrated between additions to ensure uniform pressure drop at each height. Another strategy is to provide for a much larger pressure drop at the tube entrance than that due to the catalyst bed itself, so that any fluctuations in the latter will not affect the flow distribution. This can be accomplished by using a tube sheet with a single nozzle at the bottom of each tube, with a high-pressure drop across it. However, this practice is not generally followed since tackling the

Nonuniform packing

problem at the source, as described earlier, is a much better alternative. The problem of fluid distribution is not critical except in the headers whose diameter corresponds to the shell diameter, which is very much larger than that of the tubes. Care should be taken to ensure uniform entry of the gas across the entire cross-section of the tube sheet.

Scale-up considerations

The rigorous procedures described above for fixed-bed reactor design notwithstanding, it is often necessary to carry out studies on a pilot plant scale to account for hidden scale-up factors that do not show up in the design. One way of ensuring good scale-up for multitubular reactors (the most common NINA-PBR) is to carry out the reaction in a single tube and commonly used feed velocities. Then calculate the number of tubes required for the desired production rate and design a multi-tubular reactor in which each tube would function equivalently to the single tube. This strategy was found to be eminently workable at the National Chemical Laboratory for the design of an aniline reactor. The predictions from this strategy were surprisingly close to those from the rigorous design.

On the basis of the above observations and other considerations, some working rules can be proposed for the design of conventional fixed-bed catalytic reactors in general. These rules are based on considerations of dynamic, thermal, and chemical similarities. The parameters used in determining them are the flow rate of feed (F), pellet diameter (d_p), tube diameter (d_t), reactor height (H), total desired production rate (P), and production rate in single-tube experiments (p). For the NINA-PBR: Same F, d_t, and d_p (dynamic similarity); same H, same H/F (chemical similarity); same heat transfer medium properties and temperature with high enough rates of circulation, or use of multiple heat exchangers, which would keep the temperature of the medium nearly constant (thermal similarity); and the number of tubes would be P/p.

Table 8.3 from Rase (1990) summarizes the scale-up rules for fixed-bed reactors.

Alternative fixed-bed designs

In addition to the two most common reactors described above, other designs have also been used. The most important of these reactors are the radial-flow and catalytic wire-gauze reactors, followed by the rather infrequently used spherical reactor.

Radial-flow reactor

Radial-flow reactors

New designs are often triggered by the needs of a particularly important industrial reaction. One outstanding example of this is ammonia synthesis, which has consistently occupied a critically important position in the chemical industry—chiefly because it is a basic chemical for the

TABLE 8.3 Scale-Up Rules for Gas Phase Fixed-Bed Reactors

Reactor Type	Relation between Small and Large Sizes	Constraints
Adiabatic	Dynamic similarity, same G and d_p Chemical similarity, same L/G and temperature, same L and the inlet temperature, and same catalyst size and type Reactor diameter $D = \left(\frac{4W_F}{G\pi}\right)$ where D is the diameter of the reactor and W_F is the feed flow rate in mass/time	The proven methods for assuring good distribution on the bed at the inlet and the outlet must be used $D/d_p < 30$
Nonadiabatic	*Dynamic similarity:* Same G, D, and d_p *Chemical similarity:* Same L/G and temperature, same L and inlet temperature *Thermal similarity:* Same heat transfer media, temperature and properties *Number of tubes:* $N_{\text{tube}} = \frac{\text{Total desired production (mass/time)}}{\text{Production/tube (mass/time)}}$	Each tube must be packed in an identical manner so that ΔP will be the same across each tube, thereby assuring equal distribution among the tubes The temperature of the heat transfer fluid must ideally be essentially constant, a goal reachable with boiling water reactors or high rates of circulation of the heat transfer fluid Many high-temperature endothermic reactors require heating by hot gases flowing transversely to the tubes and low heat capacity means large changes in gas temperature. Design multipass exchange for a gas to assure close constant T at any particular axial position. This axial gas temperature profile must be mimicked in the single-tube pilot reactor

Source: Rase, H.F., *Fixed Bed Reactor Design and Diagnostics,* Butterworths, Boston, MA, 1990.

production of fertilizers. The scale of manufacture is so immense that ordinary fixed-bed reactors tend to be uneconomical. One way of overcoming this problem is to move away from the general practice of letting the reactants flow axially down a reactor tube packed with the catalyst. Instead, they are made to flow radially over the entire axial length of the reactor through holes in the reactor wall. A typical segment of such a reactor for ammonia synthesis is shown in Figure 8.11. The chief features of this flow configuration are that the travel distance for the fluid is reduced and the average cross sectional area offered is higher, both of which help reduce the pressure drop. Furthermore, radial flow reactors compensate the pressure drop due to the volume change during reactions. With this background to radial-flow packed reactors (RFPRs), let us now set up the external field equations that would mathematically define the reactor. We shall assume that intraparticle diffusion effects are absent and the pseudo-homogeneous assumption fully holds.

Radial-flow packed reactors

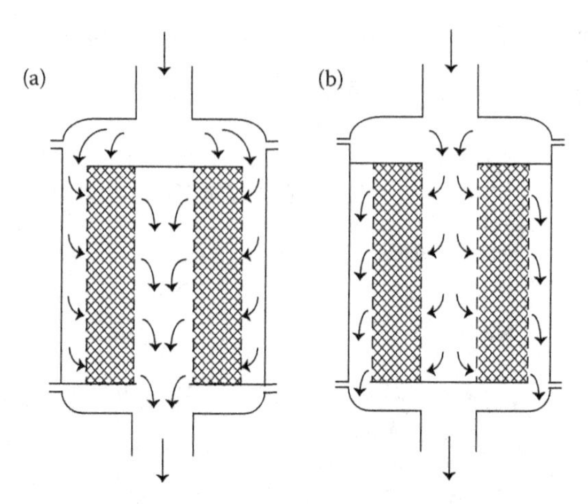

Figure 8.11 Segment of a typical (a) inward flow and (b) outward flow radial reactor.

Before we do that, it may be noted that a simple procedure effectively used for RFPR design is to assume that the equations for a regular axial flow packed-bed reactor described earlier can be directly used for RFPR as well, with the simple change of replacing the axial coordinate z in the equations by the radial coordinate r. Among many others, a program called PFRSIM is available for performing the calculations (Rase, 1990). The inlet flow velocity will change with the changing cross-sectional area, which is typical of an RFPR, but the variation is small enough to be neglected. The interested reader will find an illustration of this procedure in Rase's (1990) solution of the styrene dehydrogenation.

As pointed out under regular fixed-bed reactors, there are situations where the fluid velocity inside a packed reactor can increase with axial distance. A typical example is the RFPR, a configuration that was developed specifically to meet the high tonnage demands of basic chemicals such as ammonia (50–100 tons/h). In this reactor (first proposed by Haldor Topsøe), the catalyst is placed between coaxial cylinders and the gas is allowed to flow either from or to the center, as shown in Figure O.2 of the Overview chapter. Several basic studies can be found in the literature. We will guide the curious reader to *Ullman's Encyclopedia* as a starting point (Li, 2007).

Clearly, as more and more of the feed is introduced down the axial direction, the fluid velocity increases correspondingly, leading to a change in momentum. Thus, it is necessary in the design of an RFPR to include a momentum balance equation. With this in mind, we shall now formulate the design equations for an RFPR.

Material, momentum, and energy balances

Material balance The complete material balance must include

- Forced convection axial flow
- Forced convection radial flow (an additional feature)
- Radial dispersion
- Axial dispersion
- Reaction
- Accumulation

A model that includes all the above terms is normally called the radial axial flow packed reactor (RAFPR) and represents the most sophisticated fixed-bed reactor model. The coupled differential equations for momentum balance, component balance, and for a nonisothermal reactor, the energy balance must be simultaneously solved (Li, 2007; Singh, 2005).

Mass balance

$$\varepsilon\left[\frac{\partial[A_i]}{\partial t}\right]_{z,t}+\left[\frac{\partial([A_i]u_z)}{\partial z}\right]_{t,r}+\left[\frac{\partial([A_i]v_r)}{\partial r}\right]_{t,z}-\varepsilon D_a\left[\frac{\partial^2[A_i]}{\partial z^2}\right]_{t,r}$$
$$-\varepsilon D_r\left[\frac{\partial^2[A_i]}{\partial r^2}+\frac{1}{r}\frac{\partial[A_i]}{\partial r}\right]_{t,z}=\rho_b\sum_1^n v_{in}R_n$$

Momentum balance

r-component

$$\rho\left(\frac{\partial v_r}{\partial t}+v_r\frac{\partial v_r}{\partial r}+u_z\frac{\partial v_r}{\partial z}\right)=-\frac{\partial p}{\partial r}+\mu\left(\frac{\partial}{\partial r}\left(\frac{1}{r}\frac{\partial r v_r}{\partial r}\right)+\frac{1}{r^2}\frac{\partial^2 v_r}{\partial\theta^2}+\frac{\partial^2 v_r}{\partial z^2}\right)$$

z-component

$$\rho\left(\frac{\partial u_z}{\partial t}+v_r\frac{\partial u_z}{\partial r}\right)=-\frac{\partial p}{\partial z}+\mu\left(\frac{1}{r}\frac{\partial}{\partial r}\left(r\frac{\partial u_z}{\partial r}\right)+\frac{1}{r^2}\frac{\partial^2 u_z}{\partial\theta^2}+\frac{\partial^2 u_z}{\partial z^2}\right)$$

These expressions must be combined with appropriate source terms for flow through porous media, such as Darcy's law. The solution of these highly coupled equations is nontrivial and will not be dealt with here.

Some important observations

1. The plug-flow assumption can be misleading; a simple and approximate criterion for plug flow is

$$Pe > 50 \tag{8.30}$$

that is not easily satisfied.

2. For reactions with no volume change, the outward flow of gas yields a higher conversion than inward flow for positive-order kinetics, whereas inward flow is superior for negative-order kinetics.
3. For a first-order reaction with increase in volume, outward flow gives higher conversions, whereas for a reaction with decrease in volume, the inward flow is superior.
4. For ideal plug flow, the direction of flow is inconsequential. The difference in behavior between the two directions of flow is therefore entirely due to the dispersion effect. For example, for the ammonia reactor, which satisfies the plug-flow criterion given by Equation 8.30, the change of direction makes no difference.

Catalytic wire-gauze reactors

Catalytic wire-gauze reactors

In certain reactions involving precious metals such as platinum, rhodium, or silver as catalysts, the catalyst is used in the form of wire gauze or filament. Examples of reactions that use wire-gauze catalysts are: oxidation of ammonia to nitric oxide on Pt wire, oxidation of methanol to formaldehyde on Ag–Cu screen, oxidation of ethylene to ethylene oxide on Ag gauze, oxidation of some hydrocarbons on Pt screen, and the Andrussow process for manufacturing hydrogen cyanide (HCN) on Pt gauze with NH_3, air, and CH_4. The most important consideration in the design is the loss prevention of the precious metal by vaporization. The *L/D* ratio for these reactors is usually very low, of the order of 1/200.

Catalyst flicker

A condition known as *catalyst flicker* is mainly responsible for catalyst loss. Several studies on catalyst flicker have been reported, for example, Schmidt and Luss, 1971; Luss and Erwin, 1972; Edwards et al., 1973, 1974. Flicker is the result of nonuniform surface temperatures on the gauze caused by fluctuating transport coefficients. The ratio of the characteristic time for changes in wire temperature to that for changes in surface concentration is an important parameter in the analysis of flicker, but much more work needs to be done before reliable predictions can be made. Studies on mass and heat transfer from/to wire gauzes have also been reported, for example, Satterfield and Cortez, 1970; Shah and Roberts, 1974; Rader and Weller, 1974. The most important application of wire-gauze reactors is in the oxidation of ammonia. The catalyst employed consists of 20–60 layers of shining Pt wire screen along with other catalytic and support screens (Powell, 1969; Gillespie and Kenson, 1971). A typical reactor is sketched in Figure 8.12.

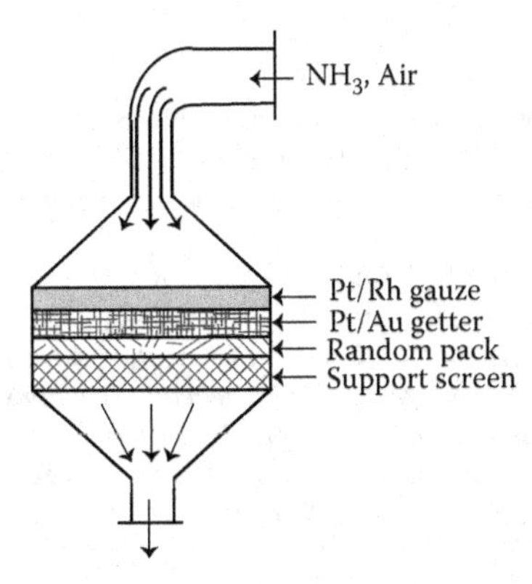

Figure 8.12 A typical catalytic wire-gauze reactor. (Adapted from Joshi, J.B. and Doraiswamy, L.K. *Chemical Reaction Engineering, in Albright's Chemical Engineering Handbook,* CRC Press, Albright, Boca Raton, FL, 2009.)

Explore yourself

1. List as many models used to design/analyze packed-bed reactors as possible. Can you differentiate the main features of these models? Can you choose the most suitable model for your design situation?
2. Outline a methodology for the selection of the most plausible model for the solid catalyzed reactor design.
3. Why is it important to have criteria for temperature runaways in packed-bed reactors?
4. What is the difference between the quasi-continuum model and the cell model in designing packed-bed reactors both in terms of the physical picture and the mathematics involved?
5. What is the main advantage of radial flow reactors? List some industrial examples.
6. How much did computational fluid dynamics (CFD) enter into the actual design of the packed-bed reactors? Perform a thorough literature survey and find a solid evidence of the use of CFD tools in packed-bed reactor design.
   a. Following the previous question, make a list of the unknowns as well as the factors to be known before *a priori* design of reactors from first principles becomes possible. The list you generate should have the potential to provide research topics for future students.
   b. Make a list of recent and developing reactor technologies. In what ways can these technologies replace fixed-bed reactors? Which deficiency in the present-day technology should be eliminated for the new technology to replace the fixed-bed reactors?

References

Bellman, R., *Dynamic Programming*, Princeton University Press, Princeton, NJ, 1957.
Carberry, J.J., *Ind. Eng. Chem.*, **56**, 39, 1964.
Carberry, J.J., *Chemical and Catalytic Reactor Engineering*, McGraw-Hill, New York, NY, 1976.
Carberry, J.J. and Wendel, M., *AIChE J.*, **9**, 129, 1963.
Chatrand, C. and Crowe, C.M., *Can. J. Chem. Eng.*, **47**, 296, 1969.
Dixon, A.J. and Nejemeisland, M., *Ind. Eng. Chem. Res.* **40**, 5246, 2001.
Doraiswamy, L.K. and Sharma, M.M., *Heterogeneous Reactions: Analysis, Examples, and Reactor Design,* Wiley, New York, NY, 1984.
Edwards, W.M., Worley, F.L. Jr., and Luss, D., *Chem. Eng. Sci.*, **28**, 1479, 1973.
Edwards, W.M., Zuniga-Chaves, J.E., Worley, F.L. Jr., and Luss, D., *AIChE J.*, **20**, 571, 1974.
Ergun, S., *Chem. Eng. Progr.*, **48**, 89, 1952.
Gillespie, G.R. and Kenson, R.E., *Chem. Tech.*, **1**, 627, 1971.
Hlavacek, V. and Votruba, J., in *Chemical Reactor Theory—A Review* (eds., Joshi, J.B. and Doraiswamy, L.K.), *Chemical Reaction Engineering,* in *Albright's Chemical Engineering Handbook,* edited by Lyle F., CRC Press, Albright, Boca Raton, FL, 2009.
Kubota, H., Ishizawa, M., and Shindo, M., *Sulphuric Acid, Japan*, **12**, 243, 1959.
Kulkarni, B.D. and Doraiswamy, L.K., *Catal. Rev. Sci. Eng.*, **22**, 431, 1980.
Lee, K.U. and Aris, R., *Ind. Eng. Chem. Proc. Des. Dev.*, **2**, 300, 1963.
Levenspiel, O., *The Chemical Reactor Omnibook*, Oregon St University Bookstores, Corvallis, OR, 1993.
Li, L.F.C., *Radial-Flow Packed-Bed Reactors*, Ullmann's *Encyclopedia of Industrial Chemistry*, DOI: 10.1002/14356007.122_101, 2007.
Luss, D. and Erwin, M. A., *Chem. Eng. Sci.*, **27**, 315, 1972.
Mears, D.E., *Chem. Eng. Sci.*, **26**, 1361, 1971.
Powell, R., *Nitric Acid Technology: Recent Developments*, Vol. 30, Chemical Process Review Series, Elsevier Science and Technology Books, Amsterdam, 1969.
Rader, C.G. and Weller, S.W., *AIChEJ*, **20**, 515,1974.
Rase, H.F., *Fixed Bed Reactor Design and Diagnostics,* Butterworths, Boston, 1990.
Satterfield, C.N. and Cortez, D.H., *Ind. Eng. Chem. Fundam.*, **9**, 613, 1970.
Schmidt, L.D. and Luss, D., *J. Catal.*, **22**, 269, 1971.
Shah, M.A. and Roberts, D., *Adv. Chem. Ser.*, **133**, 259, 1974.
Singh, A.K. *Modeling of Radial Flow Dynamics for Fixed Bed Reactor*, MS thesis, Oklahoma State University, 2005.

Bibliography

Aris, R., *The Optimal Design of Chemical Reactors—A Study in Dynamic Programming*, Academic Press, New York, NY, 1961.
Aris, R., *Discrete Dynamic Programming*, Blaisdell, MA, 1964.
Aris, R., *Introduction to the Analysis of Chemical Reactors*, Prentice-Hall, Englewood Cliffs, NJ, 1965.
Aris, R., *Elementary Chemical Reactor Analysis*, McGraw-Hill, New York, NY, 1969.
Aris, R., *Chemical Reaction Engineering* (eds. Doraiswamy, L.K. and Mashelkar, R.A.), Wiley Eastern, New Delhi, India, 1984.
Butt, J.B., *Reaction Kinetics and Reactor Design*, Prentice-Hall, Englewood Cliffs, NJ, 1980.
Carberry, J.J., *Chemical and Catalytic Reaction Engineering*, McGraw-Hill, New York, NY, 1976.

Carberry, J.J. and Varma, A., *Chemical Reaction and Reactor Engineering*, Marcell Dekker, New York, NY, 1987.
Doraiswamy, L.K., *Recent Advances in the Engineering Analysis of Chemical Reacting Systems*, Wiley Eastern, New Delhi, 1984.
Doraiswamy, L.K. and Mashelkar, R.A. (eds.), In *Frontiers in Chemical Reaction Engineering*, Vols. 1 and 2, Wiley Eastern, New Delhi, 1984.
Doraiswamy, L.K. and Mujumdar, A.S., *Transport in Fluidized Particle Systems*, Elsevier, New York, NY, 1989.
Gavalas, G.R., *Non-Linear Differential Equations of Chemically Reacting Systems*, Springer-Verlag, New York, NY, 1968.
Joshi, J.B. and Doraiswamy, L.K., *Chemical Reaction Engineering,* in *Albright's Chemical Engineering Handbook,* edited by Lyle F., CRC Press, Albright, Boca Raton, FL, 2009.
Lee, H.H., *Heterogeneous Reactor Design*, Butterworths, Boston, MA, 1985.
Nauman, E.B., *Chemical Reactor Design*, Wiley, New York, NY, 1987.
Nauman, E.B. and Buffham, B.A., *Mixing in Continuous Flow Systems*, Wiley, New York, NY, 1983.
Rase, H.F., *Chemical Reactor Design for Process Plants*, Wiley, New York, NY, 1977.
Rase, H.F., *Fixed Bed Reactor Design and Diagnostics*, Butterworths, Boston, MA, 1990.
Roberts, S.M., *Dynamic Programming in Chemical Engineering and Process Control*, Academic Press, New York, NY, 1964.
Rose, L.M., *Chemical Reactor Design in Practice*, Elsevier, Amsterdam, 1981.
Schmidt, L.D., *Engineering of Chemical Reactions*, Oxford University Press, New York, NY, 1998.
Smith, J.M., *Chemical Engineering Kinetics*, McGraw-Hill, New York, NY, 1956.
Tarhan, M.O., *Catalytic Reactor Design*, McGraw-Hill, New York, NY, 1983.

Chapter 9 Fluidized-bed reactor design for solid catalyzed fluid-phase reactions

Chapter objectives

After successful conclusion of the chapter, students must be able to

- Articulate the advantages of the fluidized-bed reactors over other types of reactors.
- Select the most suitable model for a specific reaction system among various models of fluidized-bed reactors.
- Choose the design parameters for fluidized-bed reactors.
- Size a fluidized-bed reactor for a specific reaction.
- Determine the conversion and selectivity for a given fluidized-bed reactor.
- Identify the factors associated with the operation of a fluidized-bed reactor in the case of a deactivating catalyst.

General comments

While fluidization by liquids results in an expansion of the bed with smooth internal movement of the individual particles of the suspension in a flowing stream of the liquid, fluidization by gas breaks up the flowing gas into bubbles (i.e., individual voids) at or soon after the onset of fluidization. These two patterns of behavior are referred to as *particulate* or homogeneous and *aggregative* or heterogeneous fluidization, respectively. As the gas density increases, and/or the solid density decreases, the behavior approaches that of particulate fluidization (even if the fluid is not a liquid), and vice versa. The following criteria can be used to roughly distinguish between the two modes of fluidization (i.e., for defining the so-called quality of fluidization):

$$(Fr_{mf})(Re'_{mf})\left(\frac{\rho_s - \rho_g}{g}\right)\left(\frac{L_{mf}}{d_T}\right) < 100,\ \text{particulate;} \quad > 100,\ \text{aggregative} \tag{9.1}$$

where $Fr_{mf} = u^2_{mf}/gd_p$ (Froude group at minimum fluidization) and $Re'_{mf} = d_p u_{mf} \rho_G/\mu$ (Reynolds number at minimum fluidization).

Although a number of studies, of varying degrees of rigor, have added to our understanding of fluidization over the last 60 years, six developments of a very general nature stand out as significant—providing the basic structure of fluid-bed reactor analysis and design.

1. Davidson's fluid dynamic approach to fluid-bed reactor design (see Davidson and Harrison, 1963).
2. Geldart's classification (1973) of solids in terms of their fluidization behavior.
3. Grace's explicit recognition (1986) of different regimes of fluidization (drawing from other similar previous studies, e.g., Yerushelmi and Cankurt, 1978; Li and Kwauk, 1980; Werther, 1980; Squires et al. 1985; Horio et al., 1986), through a comprehensive map that demarcates the different regimes.
4. Kunii and Levenspiel's modification (see their book, 1991, for original references) of the Davidson model and formulation of reactor design procedures for the different categories of Geldart's particles.
5. The finding by Lewis and Gilliland (see Kunii and Levenspiel, 1991) that solids circulation between two fluidized beds (usually a reactor and a regenerator) and in the transport lines connecting them can occur stably.
6. The finding that a fluidized-bed reactor can operate at more than one steady state (Elnashaie and Cresswell, 1973; Bukur and Amundson, 1975a,b; Furusaki et al., 1978; de Lasa et al., 1981), in particular, the Kulkarni–Ramachandran–Doraiswamy criterion in 1980 for multiple solutions for a first-order reaction.

Fluidization: Some basics

Minimum fluidization velocity

The velocity at which the constituent particles of a bed begin to behave as independent entities (and not as a single bed) is designated as the *minimum fluidization velocity* u_{mf}. The pressure drop in the bed remains practically constant thereafter. This velocity can be easily determined in a laboratory experiment. Many correlations are also available for estimating it (see Couderc, 1985, for a review; also Yang et al., 1985), and the following are recommended (Kunii and Levenspiel, 1991):

Coarse particles (Chitester et al., 1984):

$$Re'_{mf} = [(28.7)^2 + 0.0494Ar]^{1/2} - 28.7 \tag{9.2}$$

Fine particles (Wen and Yu, 1966):

$$Re'_{mf} = [(33.7)^2 + 0.0408Ar]^{1/2} - 33.7 \tag{9.3}$$

where

$$Re'_{mf} = \frac{d_p u_{mf} \rho_G}{\mu}, \quad Ar(\text{Archimedes number}) = \frac{d_p^3 \rho_G (\rho_S - \rho_G) g}{\mu^2} \tag{9.4}$$

Two-phase theory of fluidization

The entire edifice of fluidization theory rests on the concept, amply supported by experiment, that the gas–solid fluid bed can broadly be divided into two phases:

1. The bubble phase formed by gas in excess of that required for the onset of fluidization. The bubble is usually surrounded by a *cloud* of gas–solid mixture and is characterized by an indentation caused by suction due to the upward movement of the bubble. The solids that fill up this region are called the *wake*. The bubbles are usually large and move faster than the surrounding emulsion gas flowing at u_{mf}, thus giving rise to the cloud. This behavior is usually characteristic of Geldart A and B particles (defined in the next section). Thus, the original two-phase theory has undergone some modification, and our present understanding of it, as outlined above, is depicted in Figure 9.1.
2. The gas at incipient fluidization percolates through the solid particles, creating a liquid-like phase referred to as the emulsion phase. Although this so-called two-phase theory (Toomey and Johnstons, 1952) is not entirely accurate, it is generally valid (within acceptable error limits) and has served as the basis for a variety of fluidized-bed reactor models. The main feature of these models is that conversions are increased due to mass transfer between the emulsion and bubble phases. Without such mass transfer, in view of the large quantities of gas rising through the bed, there would be short-circuiting of the bubbles resulting in lowered conversions.

Geldart's classification

Based on extensive studies involving a variety of solid particles and fluids with a wide range of properties, Geldart (1973) classified the fluidization behavior of systems under four categories of particles: A, B, C, and D (Figure 9.2). Class B particles conform strictly to the two-phase theory of fluidization described earlier: the gas in excess of that needed

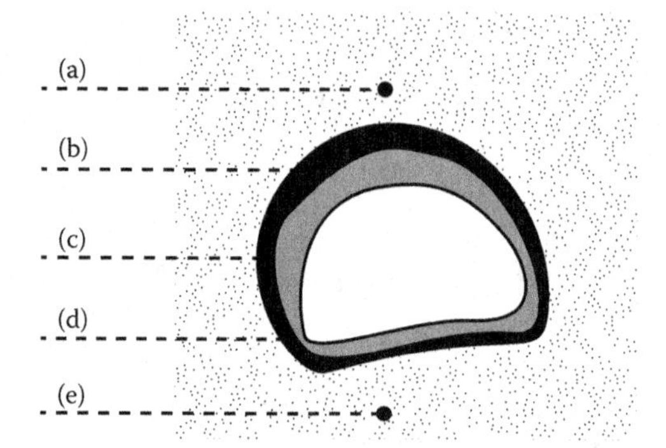

Figure 9.1 Basic features of a bubble in a fluidized bed: (a) emulsion, (b) cloud, (c) void, (d) wake, and (e) emulsion.

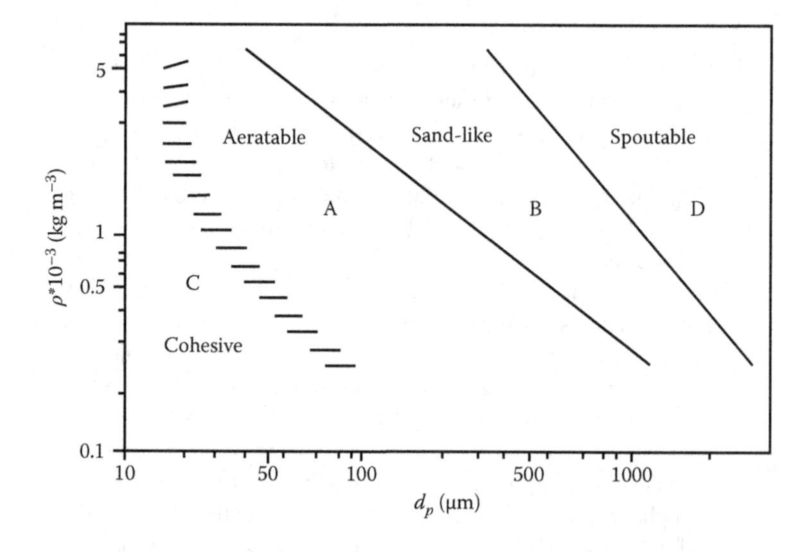

Geldart's classification

Figure 9.2 Geldart's classification of fluidization behavior of the particles.

for the onset of fluidization immediately breaks into bubbles. In the case of class A particles, bubbling commences only after a certain amount of gas has passed through the bed beyond incipient fluidization. Our concern in this section is essentially restricted to class A and B particles. The smaller C particles are too cohesive and fluidize poorly, whereas the larger class D particles are more relevant to reactions in which the solid also reacts such as coal combustion.

Classification of fluidized-bed reactors

Several categories of fluidized-bed reactors are possible, depending on the mode of operation. The chief features of these reactors are summarized in Table 9.1 and sketched in Figure 9.3. As can be seen from this table, bed behavior (and category) is essentially determined by the

Table 9.1 Principal Features of Different Types of Fluidized-Bed Reactors

Fluidization Regime	Main Features	Examples
1. Incipiently fluidized bed (stationary)	Upward flow of gas at about u_{mf} for A and B particles; no solids mixing; gas mostly in plug flow; solids mixing by a stirrer is sometimes useful; $u_o \leq 1.2u_{mf}$ with no bubbles; $\varepsilon_s \approx$ 0.5–0.6 [b] throughout bed	Methylchlorosilanes
2. Bubbling fluidized bed (stationary)	Upward flow of gas through a wide range of A and B particles; onset of bubbling depends on particle size, ranging roughly from $u_b = 40u_{mf}$ to $70u_t$ for small particles to a very narrow range ($u_{mf} < u_b \leq 2\ u_{mf}$) for large particles; $\varepsilon_s \approx$ 0.6 (bottom)–0.4 (top)	Polymerization of ethylene to LD polyethylene, ethylene dichloride, vinyl acetate
3. Turbulent bed[a] (stationary)	Starts gradually at $u_o >> u_t$ for small particles and $u_o \approx 0.5u_t$ for large particles and merges smoothly into fast fluidized-bed region at higher velocities in each case; as u_o is not very high, internal cyclone is usually adequate; solids entrainment is usually high and, instead of bubbles, clusters of solids and voids of gas move through the bed; $\varepsilon_s \approx$ 0.3–0.4 (bottom) to 0.2–0.3 (top); the void lifetime is short so that, overall, the bed looks more uniform than in regime 2	Phthalic anhydride, o-cresol and 2–6 xylenol, acrylonitrile, chloromethanes
4. Fast fluidized bed[a] (circulating)	Continuous feed of both gas and solids; sufficiently high solids velocity—in excess of the upper limit for regime 5; the transition point (from the reverse direction) causes *choking* at the entry and collapse of the lean dispersion of that regime into the fluidized mass of regime 4; suitable gas distributor is used to ensure high density at bottom that merges smoothly with the low-density region at the top (corresponding roughly to the freeboard region of the bubbling bed); $\varepsilon_s \approx$ 0.5–0.2 (bottom) to 0.05–0.01 (top); essentially a circulating bed with plug flow of gas accompanied by slugs of emulsion; bed even more uniform than in regime 3	Fischer–Tropsch synthesis of hydrocarbons

[a] The distinction between bubbling-bed and turbulent-bed reactors is not always clear. Hence the classification of reactions under these categories is uncertain.
[b] ε_s: Solid porosity.

fluidizing gas (reactant) velocity and particle size. This transition from fixed to pneumatic bed is usually depicted in terms of a *fluidization map*, different renditions of which have been published from time to time (e.g., Čatipović et al., 1979; van Deemter, 1980; Werther, 1980; Squires et al., 1985; Horio et al., 1986; Grace, 1986). The version consolidated by Kunii and Levenspiel (1991) is reproduced in Figure 9.4. Our main concern is with the bubbling bed. The turbulent bed is only qualitatively different from the bubbling bed.

Fluidization map

Velocity limits of a bubbling bed

As already noted, the velocity for the onset of bubbling u_{mb} does not always coincide with that for the onset of fluidization u_{mf}. Depending on the nature of the solid, bubbling can occur at or beyond u_{mf}. The upper limit of velocity for bubbling is the velocity for the onset of slugging.

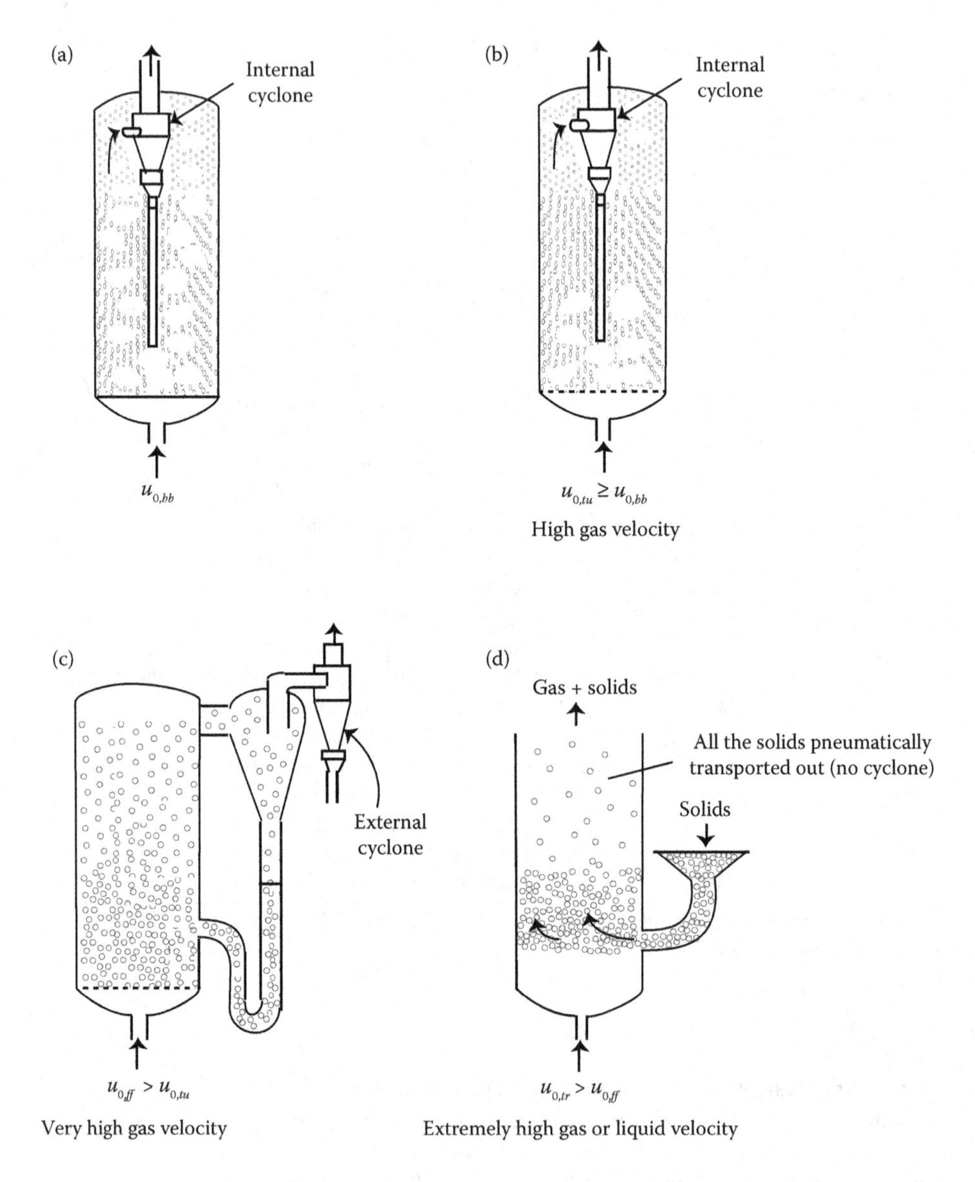

Figure 9.3 Classification of fluidized-bed reactors. (a) Bubbling-bed reactor; (b) turbulent- (or fluid-) bed reactor; (c) Fast fluidized-bed reactor; (d) transport reactor. (From Joshi, J.B. and Doraiswamy, L.K. In *Albright's Chemical Engineering Handbook* (Ed. Lyle, F.), CRC Press, Boca Raton, FL, 2009.)

Slugging defines a condition where the bubble size becomes nearly equal to the tube diameter, or (as in the case of group C particles) portions of the bed are bodily lifted, resulting in alternate zones of packed and void regions. We are usually concerned with the former, and the velocity for the onset of this condition is given by

$$u_{ms} = u_{mf} + 0.07(gd_T)^{1/2}, \text{cm/s} \tag{9.5}$$

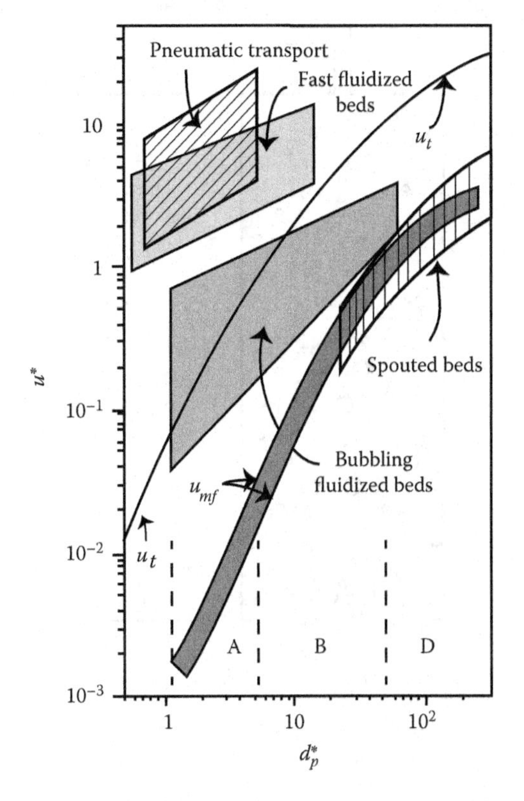

Fluidization map

Figure 9.4 Fluidization map. The parameters are $u^* = u[(\rho_s - \rho_G)_g/\mu]^{1/3}$, $d_p^* = [\rho_G^2]/\mu(\rho_s - \rho_G)^{1/3}$. (From Grace, J.R., In *Gas Fluidization Technology* (Ed. Geldart, D.), Wiley, New York, 1986; as modified by Kunii, D., and Levenspiel, O., *Fluidization Engineering,* Butterworth-Heinemann, Boston, 1991.)

Fluid mechanical models of the bubbling bed The analysis of the bubbling bed reactor is based largely on the fluid mechanical model first described fully by Davidson and Harrison (1963) and modified later by a number of investigators. Our description of fluidized-bed reactor modeling will be based on the Kunii–Levenspiel (K-L) adaptation (see Levenspiel, 1993). A few other important bubbling bed models will also be considered.

Complete modeling of the fluidized-bed reactor The bubbling bed constitutes the largest segment of the fluid-bed reactor as a whole, but not the entire reactor. To account for all aspects of the fluidized bed, it is necessary to recognize three regions of the bed, as shown in Figure 9.5: the grid region, the main fluid-bed region, and the dilute phase region. Much of the conversion occurs in the main fluid-bed region (under normal conditions of operation), commonly referred to as the *bubbling bed*, but the reaction occurring in the other regions cannot be ignored. We derive models for all the three regions of the reactor, with much greater

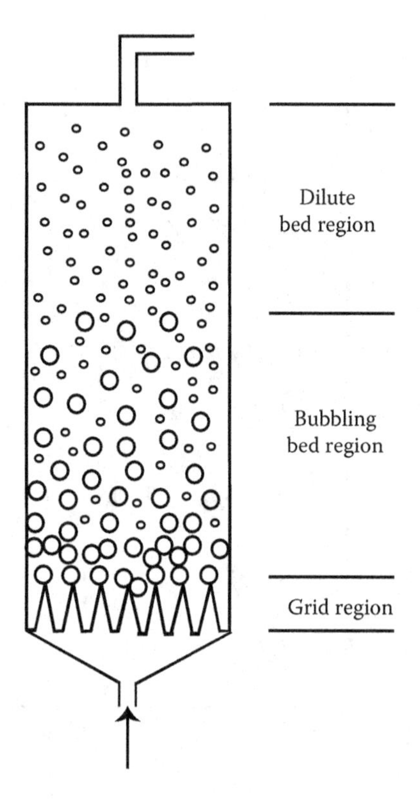

Figure 9.5 The three regions of a complete fluidized bed where reaction can occur.

emphasis on the bubbling segment and illustrate their application via worked examples.

Bubbling bed model of fluidized-bed reactors

Bubbling bed

Modeling of the bubbling bed region is based on several special characteristics/assumptions/definitions. These are outlined below along with the governing equations in each case. The equations will depend on the nature of particles used: fine, intermediate size, or large. We restrict our treatment to small particles.

It will be assumed that bubble growth occurs close to the distributor, so that a single effective bubble diameter for the entire bed can be assumed. This is a contentious assumption, and several studies accounting for bubble size distribution and other hydrodynamic features of the bed have been proposed. However, we persist with the constant bubble size assumption because it is sound enough for a preliminary design.

Based on this assumption, the distinguishing feature of the bubbling bed is the magnitude of the ratio α of the bubble rise velocity u_b (for a bubble diameter d_b) to the interstitial velocity u_i (equal to u_{mf}/f_{mf})

$$\alpha = \frac{u_b}{u_i} = \frac{u_b}{u_{mf}/f_{mf}} \tag{9.6}$$

Bubble rise velocity

We now derive an expression for the bubble rise velocity. This velocity consists of two contributions: (1) the free rise velocity u_{br} that is determined only by the properties of the bed and the gas and is independent of gas velocity and (2) the bulk flow of the bubble phase as a whole that is dependent on gas velocity and is given by (u_0-u_{mf}). Thus

$$u_b = u_{br} + (u_0 - u_{mf}) \tag{9.7}$$

The free rise velocity u_{br} may be assumed to be equal to the velocity of a bubble released from a single nozzle into an inviscid fluid, that is, the rise velocity when the bed is at u_{mf} and is given by

$$u_{br} = 0.711(gd_b)^{1/2}, \text{cm/s} \tag{9.8}$$

Equation 9.8 has been empirically modified to provide separate correlations for Geldart A and B particles (Kunii and Levenspiel, 1991), but this simple expression is considered quite adequate for fine particles.

Main features

The bubbling bed model holds when fast bubbles are rising through a bed of small particles, that is, $u_b >> u_e$. The situation is depicted in Figure 9.6a. The gas circulation is restricted to the bubble and a small region called the *cloud* surrounding it. In fact, the bubble gas is completely segregated from the rest of the gas passing through the bed. From simple fluid mechanical concepts, it can be shown that

$$\frac{r_c - r_b}{r_b} \cong \frac{u_e}{u_b} = \frac{1}{\alpha} \tag{9.9}$$

As an example, if the bubble rises 10 times as fast as the emulsion gas, the cloud thickness (r_c-r_b) will be just 1/10th of the bubble radius (i.e., 1/20th the bubble diameter).

A more rigorous equation for the ratio of cloud to bubble radii is

$$\frac{r_c}{r_b} = \left(\frac{\alpha + 2}{\alpha - 1}\right)^{1/3} \tag{9.10}$$

Also, the picture is more complicated than depicted in Figure 9.6a. As the bubble rises, it carries with it a small amount of the solids as *wake*.

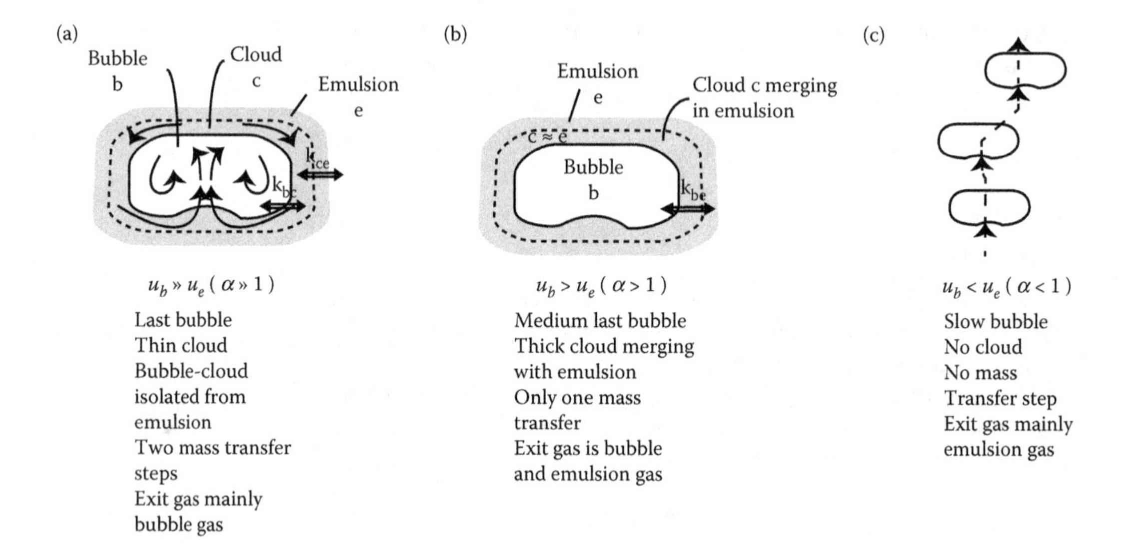

Figure 9.6 Bubbling bed models for (a) fine, (b) intermediate, and (c) large particle fluidized beds. (From Levenspiel, O., *Chemical Reactor Omnibook*, Oregon State University Bookstore, Corvallis, OR, 1993.)

Thus, a rigorous model should really recognize four regions: emulsion, bubble, cloud, and wake. In the K–L model, it is assumed that the wake solids are evenly distributed in the cloud phase. This simplifies the computations without seriously affecting the accuracy.

Figure 9.6b and c depict the situations for medium- and large-sized particles, which, as already noted, will not be considered here.

Mass transfer between bubble and emulsion An important feature of fluidized-bed reactors is mass transfer between bubble and emulsion. Several models have been proposed for this exchange. The Davidson model assumes no cloud, so that only one mass transfer coefficient k_{be} (for direct bubble–emulsion exchange) is involved. On the other hand, the K–L model is based on two successive mass transfer steps, leading to the coefficients k_{bc} (for bubble–cloud exchange) and k_{ce} (for cloud-emulsion exchange). The equations for the K–L model are given in Table 9.2.

Solids distribution

Since three phases are present, the extent of reaction in each phase must be computed. And since the reaction occurs only in the presence of solids, the distribution of solids in the three phases, emulsion, cloud, and bubble, must be known. These are expressed as fractions of the bubble phase, s_{bb}, s_{cb}, and s_{eb} for the bubble, cloud, and emulsion phases,

Table 9.2 Expressions for Estimating Important Fluid-Bed Properties

Bubble Rise Velocities	
Single bubble in a quiescent bed at u_{mf}	$u_{br} = 0.711(gd_b)^{1/2}$, cm/s
Bubbles in a bubbling bed	$u_b = u_{br} + (u_0 - u_{mf})$, cm/s
Bubble fraction	$\delta = \dfrac{u_0 - u_{mf}}{u_b}$
Solids distribution	$s_{bb} = 0.001 - 0.01$
	$s_{cb} = (1 - f_{mf})\left[\dfrac{3u_{mf}/f_{mf}}{u_{br} - u_{mf}/f_{mf}} + s_{wb}\right]$
	$s_{eb} = \left[\dfrac{(1 - f_{mf})(1 - \delta)}{\delta} - s_{cb} - s_{bb}\right]$
	$s_{wb} = 0.2 - 2.0$
Mass Transfer Coefficients	
Bubble cloud	$k_{bc} = 4.50\left(\dfrac{u_{mf}}{d_b}\right) + 5.85\left(\dfrac{D^{1/2}g^{1/4}}{d_b^{5/4}}\right)$, s^{-1}
Cloud–emulsion	$k_{ce} = 6.78\left(\dfrac{f_{mf}Du_{br}}{d_b^3}\right)^{1/2}$, s^{-1}
Fluid-bed height relationships	$L_{mf} = L_f(1 - \delta)$
	$L_m(1 - f_m) = L_{mf}(1 - f_{mf}) = L_f(1 - f_f)$

respectively. As already mentioned, every bubble carries with it a small amount of solids as wake. The volume fraction of the wake is given by s_{wb} = volume of wake/volume of bubble phase.

Estimation of bed properties

Knowledge of several properties and parameters of the fluidized bed (including those mentioned above) is necessary in fluidized-bed reactor design. A list of these properties along with the equations to estimate them is included in Table 9.2. Many of these equations will probably have to be revised in light of the recent observation (Gunn and Hilal, 1997) that u_{mf} is likely to be affected by the scale of equipment and by distributor design. Thus, correlations (such as for bed expansion) should more logically be based not on comparisons at the same gas velocity, as has been the practice so far, but at the same excess gas velocity (i.e., same u_0/u_{mf}).

Heat transfer

Good heat transfer is one of the most attractive features of the fluidized bed. From the standpoint of its use as a chemical reactor, the most important mode of heat transfer is that from a fluidizing bed to a bank of tubes (with a circulating fluid) immersed within it. The value of the heat transfer coefficient will depend on whether the tube bank is vertical or horizontal. A number of correlations are available for predicting these and other modes of heat transfer in a fluidized bed, and good reviews of these correlations can be found in Botterill (1966), Zabrodsky (1966), Muchi et al. (1984), and Kunii and Levenspiel (1991). Most of them are restricted to relatively narrow ranges of variables. Two useful correlations are listed in Table 9.3. It is important to note that there are reactions such as the chlorination of methane (Doraiswamy et al., 1972), which are entirely heat transfer controlled. The rate of heat removal and design of reactor internals become crucial considerations in such cases.

Recent studies have made it easier to design reactors with vertical tubular inserts. This arises from the observation (Gunn and Hilal, 1994, 1996, 1997) that the heat transfer coefficients for these systems are almost equal to those for the corresponding open fluidized beds of the same diameter operating with the same particles. Hence, correlations for the latter (which are readily available) can be used for vertical inserts without significant loss of accuracy. Vertical inserts have an additional advantage over horizontal inserts: in horizontal inserts, unlike in the vertical orientation, there is accumulation of particles on the top of the tubes and depletion of particles at the bottom, a situation that leads to a spatial variation in heat transfer coefficient.

Table 9.3 Recommended Correlations for Heat Transfer in Fluidized Beds

Bed-wall heat transfer	$h_w = 35.8\rho_s^{0.2}\lambda_G^{0.6}d_p^{-0.36}$ (mks units)
Bed-immersed tube heat transfer (vertical tubes)	$\dfrac{h_w d_p}{\lambda_G} = 0.0184(CF)(1-f_f)\left(\dfrac{C_{pG}\rho_G}{\lambda_G}\right)^{0.43}\left(\dfrac{d_p\rho_G u}{\mu}\right)^{0.23}\left(\dfrac{C_{ps}}{C_{pG}}\right)^{0.8}\left(\dfrac{\rho_s}{\rho_G}\right)^{0.66}$ for $\dfrac{d_p\rho_G u}{\mu} = 10^{-2}\text{–}10^{2}$
Bed-immersed tube heat transfer (horizontal tubes)	$\dfrac{h_w d_0}{\lambda_G} = 0.66\left(\dfrac{C_{pG}\mu_G}{\lambda_G}\right)^{0.3}\left[\left(\dfrac{d_0\rho_G u}{\mu}\right)\left(\dfrac{\rho_s}{\rho_G}\right)\left(\dfrac{1-f_f}{f}\right)\right]^{0.44}$ $\left(\text{for } \dfrac{d_0 u\rho_G}{\mu} < 2000\right)$ $\dfrac{h_w d_0}{\lambda_G} = 420\left[\left(\dfrac{C_{pG}\mu_G}{\lambda_G}\right)\left(\dfrac{d_0\rho_G u}{\mu}\right)\left(\dfrac{\rho_s}{\rho_G}\right)\left(\dfrac{\mu^2}{d_p^3\rho_s g}\right)\right]^{0.3}$ $\left(\text{for } \dfrac{d_0 u\rho_G}{\mu} > 2500\right)$

Calculation of conversion

We now come to the main element of the model: calculation of conversion. Here, considering the fact that the amount of emulsion gas is negligible compared to the bubble-phase gas, reaction in the emulsion phase can be neglected. Thus, conversion is based only on bubble-phase flow. A complete accounting of reactant A can be done as shown in Figure 9.7. Eliminating the intermediate concentration and by simple algebraic manipulations, the following expression can be derived:

$$-u_b \frac{d[A]_b}{dz} = K_F [A]_b \tag{9.11}$$

where

$$K_F = \left[s_{bb} k_v + \cfrac{1}{\cfrac{1}{k_{bc}} + \cfrac{1}{s_{cb} k_v + \cfrac{1}{\cfrac{1}{k_{ce}} + \cfrac{1}{s_{eb} k_v}}}} \right] \tag{9.12}$$

Integrating Equation 9.11 results in the following expression for reactant concentration as a function of height:

$$\frac{[A]_b}{[A]_0} = \exp\left(-K_F \frac{z}{u_b}\right) \tag{9.13}$$

Expressing this in terms of conversion at the reactor exit, we obtain

$$(1 - X_A) = \frac{[A]_{bf}}{[A]_0} = \exp(-K_F) = \exp\left(-K_F \frac{Z_F}{u_b}\right) \tag{9.14}$$

Notice that Equation 9.14 is identical to the plug-flow equation except that K_F is not a true rate constant but a composite constant consisting of the true constant k_v and the various physical parameters of the model.

End region models

Dilute bed region

Dilute bed region In all the models developed above, it was assumed that reaction is restricted to the bubbling bed but the data of Lewis et al. (1962) and Fan et al. (1962) show that an axial distribution of bed density exists. Further, it seems most likely that bubbles carry solid particles along with them through the central region of the bed and enter the dilute phase by a process of bursting on the emulsion surface (Miyauchi, 1974;

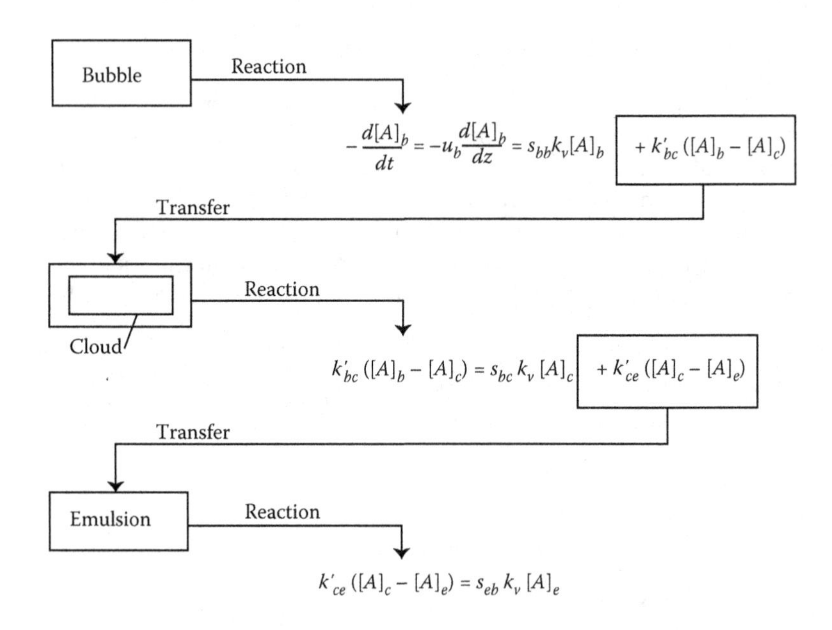

Figure 9.7 The Kunii–Levenspiel model.

Miyauchi and Furusaki, 1974). A bubble-free emulsion then flows down the bed peripherally. This situation clearly leads to some reaction in the dilute phase. An elegant model that accounts for reaction in both the bubbling and dilute regions of the bed has been proposed by Miyauchi (1974), and another by Kunii and Levenspiel (1991) (more in line with their fine particle model).

According to Miyauchi's model, the concentration at the exit of the dilute bed $[A]_{ft}$ is given by (see Figure 9.8)

$$\frac{[A]_{ft}}{[A]_0} = 1 - X_A = \exp[-(K' + K_b + K_d)] = \exp(-K'_R) \tag{9.15}$$

where

$$\frac{1}{K'} = \frac{1}{K_m} + \frac{1}{K_o(1-\delta)} \tag{9.16}$$

The various groups (K_m, K_o, K_b, K_d) are defined in the figure. Note that experimental determination of the bed density distribution $[(1 - \delta)dz_f]$ in the dilute region is necessary to estimate K_d.

Grid or jet region As the fluidizing gas enters the bed through the openings in the grid plate, it usually issues as fluid jets of velocities in the range 40–80 m/s (see, e.g., Behie et al., 1976; Mori and Wen, 1976). These jets penetrate a certain distance into the bed before they collapse

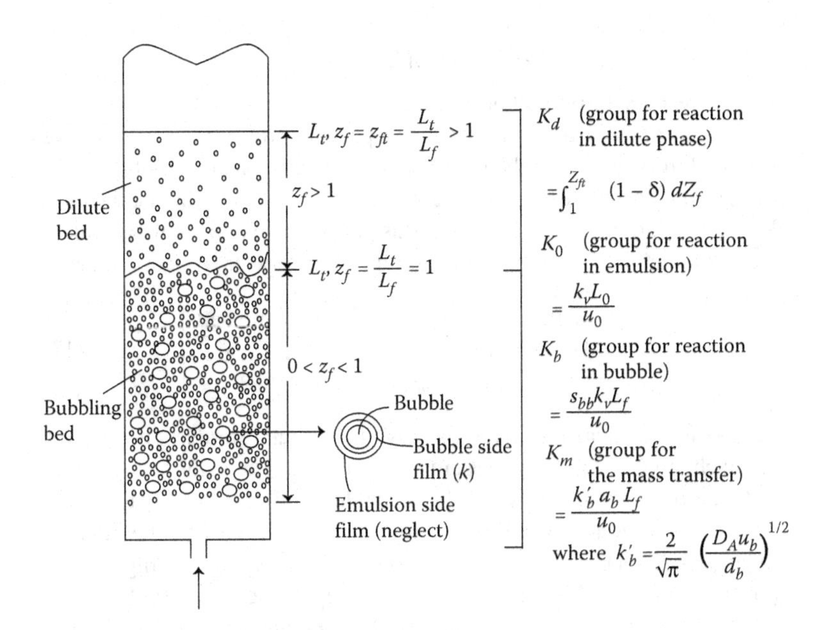

Figure 9.8 Model that accounts for reaction in both the bubbling and dilute regions of the bed.

into bubbles, and no reactor internals should be placed in the path of these jets to avoid damage. The jet region is significant in the case of fast reactions in very-large-scale productions, in which the jet velocities tend to be high, but is usually of no particular consequence in the production of organic intermediates.

Practical considerations

The modeling of fluidized-bed reactors provides a preliminary indication of the size and main characteristics of the reactor for a given reaction. However, there are so many uncertainties in the scale-up of fluidized beds that it is impossible at the present time to avoid pilot plant and semicommercial units before arriving at a suitable commercial design. Some of the problems encountered in scale-up are: violent circulating currents (of the order of 20–30 cm/s) known as "gulf streaming" in large beds, resulting in low bubble residence times in relation to those in smaller laboratory or pilot plant reactors (Davidson, 1973); "caking" of parts of the bed (sometimes the whole bed) due to stagnant pockets in exothermic reactions (Doraiswamy et al., 1968); defluidization of bed due to sudden changes in temperature or pressure, or coke deposition on particles leading to higher u_{mf}; and location of inlets of gases and choice of fluidizing gas in the case of reactions involving two reactants, which cannot be mixed outside the bed due to flammability considerations, as in the chlorination of methane (Doraiswamy et al., 1972). These are only indicative of the many problems involved.

Gulf streaming

Caking

Defluidization

Recommended scale-up procedure

Perhaps the most important problem is the lack of understanding of bubble behavior in beds of different sizes and particle properties. An effective way of addressing this problem is to operate a pilot plant reactor of a certain size and design the larger reactor in such a way as to simulate the bubble behavior of the pilot reactor. For this purpose, we define an equivalent reactor diameter d_e given by the hydraulic diameter (see Volk et al., 1962)

Equivalent diameter

$$d_e = 4\left[\frac{\text{Free cross-sectional area of the bed}}{\begin{array}{c}\text{Wetted perimeter of all vertical}\\ \text{surfaces exposed to the bed}\end{array}}\right] \quad (9.17)$$

and provide vertical internals (usually tubes that can also be used for heat exchange) in the larger reactor so that d_e has the same value as the pilot reactor diameter. This results in bubbles of approximately the same size in the two reactors (corresponding to 1–1.5 times the tube pitch in the larger reactor). This simple method will not apply to very high scale-up ratios, but should be useful in the scale-up of medium-sized organic chemicals production, as demonstrated by Doraiswamy et al. for the chlorination of ethylene to hexachloroethane (1968) and ethylation of aniline to monoethylaniline (1973).

Example 9.1: Design of a fluidized-bed reactor for the hydrogenation of nitrobenzene to aniline

Design a fluid-bed catalytic reactor for the commercial production of 6000 tons per annum of aniline by the hydrogenation of nitrobenzene. A conversion of 97% is to be achieved in the reactor. The following data are given:

Material Balance-Related Data	Energy Balance-Related Data	Physical Properties	Process Parameters
H_2: nitrobenzene = 20:1	$-\Delta H_r$ = 180 kcal/mol	$\mu = 1.3 \times 10^{-2}$ cP (at 200°C)	$T = 270$°C
$k = 1.2$ s^{-1}		$D_{eA} = 0.25$ cm^2/s (at 270°C)	Catalyst particle diameter $d_{ave} = 0.031$ cm $\rho_s = 2.2$ g/cm^3 Bulk density of solids Void fraction at minimum fluidization $f_{mf} = 0.50$ Wake fraction $s_{wb} = 0.43$ Bubble diameter $d_b = 12$ cm

SOLUTION

A broad methodology

1. Calculate the minimum fluidization velocity (u_{mf}), minimum bubbling velocity (u_{mb}), and minimum slugging velocity (u_{ms}). As the calculation of u_{ms} requires a value of the reactor diameter, which is not known, assume a value and go through the calculations. This velocity represents the maximum permissible velocity and therefore the minimum diameter. From the given volumetric flow rate, calculate the maximum diameter by assuming a velocity of, say, $3u_{mf}$. Choose a suitable diameter between these limits. Iteration may be needed.
2. Laboratory-scale data are best simulated by using a number of vertical internal tubes (which can also be used for heat transfer) in a reactor shell containing the fluidizing solids and ensuring that the equivalent diameter d_e calculated from Equation 9.17 is within normal limits, that is, 15–20 cm. The effective bubble size should be approximately 1–1.5 times the tube pitch in the larger reactor (which will depend on the size and number of the internal tubes used). Thus, design the grid plate such that the maximum and minimum sizes of the bubbles generated by the orifices will straddle the effective bubble size in a narrow range.
3. With the information now available, use Equation 9.14 to calculate the bed height z_f needed to obtain the desired conversion. The present example is primarily concerned with this crucial aspect of design.
4. Since cooling would be needed, see whether the number of internal tubes calculated in step 2 would be adequate for the purpose. If not, use more tubes, and repeat pertinent calculations using this new value. A good design would be one where the number of tubes calculated from Equation 9.17 would be approximately equal to the number based on heat transfer requirements. In any case, adequate number of tubes must be provided to ensure maintenance of the required temperature.

Bed height for 97% conversion

The rise velocity of bubble is calculated as

$$u_{br} = 0.711\left[gd_b\right]^{1/2} = 0.711(981 \times 12)^{1/2} = 77.14 \text{ cm/s}$$

The absolute rise velocity of the bubble is then

$$u_b = u_o - u_{mf} + u_{br} = 36 - 12 + 77.14 = 101.14 \text{ cm/s}$$

The fraction of the fluidized bed occupied by bubbles is given by

$$\delta = \frac{u_0 - u_{mf}}{u_b} = \frac{36 - 12}{101.1} = 0.237$$

The interchange coefficient between bubble and cloud is calculated from the appropriate equation given in Table 9.2.

$$k_{bc} = 4.5\left(\frac{12}{12}\right) + 5.85\left(\frac{0.25^{1/2}981^{1/4}}{12^{5/4}}\right) = 5.23\ \text{s}^{-1}$$

Similarly, the interchange coefficient between cloud/wake and emulsion phases is calculated as

$$k_{ce} = 6.78\left(\frac{0.5 \times 0.25 \times 77.14}{12^3}\right) = 0.506\ \text{s}^{-1}$$

Let us now calculate the fraction of solids dispersed in the bubble, cloud/wake, and emulsion phases per unit volume of bubble in the bed:

$$\left.\begin{aligned} s_{bb} &\approx 0.003 \\ s_{wb} &\approx 0.43 \end{aligned}\right\} \text{Assume}$$

$$s_{cb} = 0.798\ [\text{from Table 9.2}]$$

$$s_{eb} = 0.800\ [\text{from Table 9.2}]$$

From Equation 9.14

$$1 - X_A = \exp(\hat{K}_F)$$

$$1 - 0.97 = \exp(\hat{K}_F)$$

$$\hat{K}_F = 3.51$$

and from Equation 9.12

$$K_F = 1.09$$

giving finally, height of reactor

$$L_f = 3.24\ m$$

Strategies to improve fluid-bed reactor performance*

One of the chief drawbacks of the fluid-bed reactor is deviation from plug flow. This is also a feature of other reactor types such as for gas–liquid

* For a full coverage of the subject, see Doraiswamy and Sharma (1984).

and gas–liquid–solid reactors. A remedial measure introduced in those reactors is to provide for improved contact through development of more efficient packings. Another is staging of the reactor.

Packed fluidized-bed reactors

This strategy of using packings of different designs can, in theory, be used for fluidized beds also. This would involve packing the bed with stationary packing as in a regular fixed bed and introducing fluidizable particles into this bed. This is a highly favorable situation where design is concerned, as we shall see below, but suffers from a severe limitation. The particles can lodge in unaerated regions within a packing's interstices, leading to dead zones and gas maldistribution. However, such flow maldistribution can be minimized by using structured-grid packings with well-defined uniformly distributed flow areas. Several such packings have been listed by Papa and Zenz (1995) along with a corresponding equation for reactor design.

We shall now take a closer look at the packed fluidized bed.

Reactor model for packed fluidized beds As in the case of the conventional bubbling bed, the packed fluidized bed also consists of the bubble and emulsion phases, and the bubbles carry along with them solid particles in the wake. The size of the bubbles in the packed fluidized bed is, however, restricted by the packing, and any increase in the gas flow rate helps in increasing the number of gas bubbles per unit volume of the bed and the bubble rise velocity. A fluid flow model based on the size of the bubble in the bed thus seems unsuitable for the packed fluidized bed. Kato et al. (1974) have employed a fluid flow model similar to that of Mathis and Watson (1956) and Lewis et al. (1959) in calculating the conversion in packed fluidized-bed reactors.

In view of the large effective thermal conductivity realized in packed fluidized beds, these reactors are more nearly isothermal than the conventional fluidized beds. On the assumption of plug flow on the gas, in the bubble and emulsion phases, the following conservation equations can be written.

Bubble phase:

$$\frac{d[A]_b}{dz} + \frac{(R_B - 1)k_{gv}}{R_B(u - u_{mf})}([A]_b - [A]_e) + \frac{s_{bb}r_b}{R_B(u - u_{mf})} = 0 \tag{9.18}$$

Emulsion phase:

$$\frac{d[A]_e}{dz} + \frac{(R_B - 1)k_{gv}}{R_B u_{mf}}([A]_b - [A]_e) + \frac{(1 - s_{bb})r_e}{R_B u_{mf}} = 0 \tag{9.19}$$

Boundary condition:

$$z = 0, [A]_B = [A]_E = [A]_0 \tag{9.20}$$

where r_b and r_e denote the reaction rates in the bubble and emulsion phases, respectively. The concentration at the exit of the packed fluidized bed can be calculated as

$$[A] = \left(\frac{u_{mf}}{u}[A]_e + \frac{(u - u_{mf})}{u}[A]_b \right)_{z=L} \tag{9.21}$$

The mass balance Equations 9.18 and 9.19 require knowledge of two parameters: the fraction of solids contained in the bubble s_{bb} and the gas interchange coefficient k_{gv} between the two phases. The gas interchange coefficient can be obtained from the relationship (Kato et al. 1967)

$$k_{gv} = 4.2 \times 10^{-3} \left(\frac{(u - u_{mf})}{u_{mf}} \right)^{-0.6} \text{cm/s} \tag{9.22}$$

Equations 9.18 and 9.19 can be integrated analytically for a first-order reaction. Numerical solutions have been presented by Kato et al. (1974) for a general-order reaction with the reaction of solids in the bubble s_{bb} as the parameter. Their results indicate that the reactant concentrations in the bubble and emulsion phases are quite sensitive to s_{bb}, when the reaction rate constant is high. Increasing the fraction of solids in the bubble phase increases the conversion in the bubble phase, and the larger the rate constant, the larger the concentration difference between the bubble and the emulsion phases. Reactant conversion calculated from this model for a first-order reaction with reaction rate constant as a parameter is reproduced in Figure 9.9. The figure shows the conversion in a packed fluidized bed to be almost the same as that in the plug-flow reactor when the rate constant is small (~0.1 s^{-1}). For intermediate values of the rate constant (~1.0 s^{-1}), conversion lies between those in the plug-flow and perfectly mixed reactors. When the reaction rate constant is large (~10.0 s^{-1}), the conversion calculated from this model is smaller than that in the perfectly mixed flow reactor.

The fluid flow model is perhaps the most important single consideration in the design of a packed fluidized-bed reactor. The model as discussed above requires knowledge of two parameters, s_{bb} and k_{gv}. For a reaction with known kinetics (determined separately in a fixed bed), the model parameter s_{bb} can be calculated from the equations presented above. Since the bed diameter and the bed height have no influence on the fluid flow pattern, this parameter, calculated from a simple laboratory experiment, can be directly used in the design of a commercial packed fluidized-bed reactor. Thus, scaling-up problems are considerably simplified in those reactors.

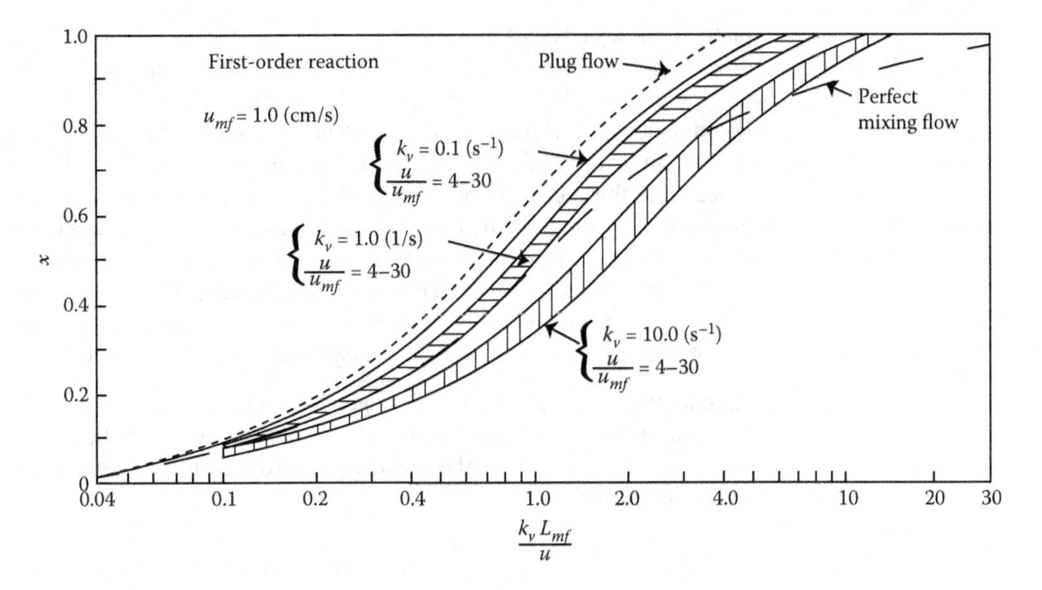

Figure 9.9 Performance of a packed fluidized-bed reactor in relation to that of plug-flow and fully mixed reactors. (From Kato, K., Arai, H., and Ito, U., *Adv. Chem. Ser.*, **133**, 270, 1974; adapted from Doraiswamy, L.K. and Sharma, M.M., *Heterogeneous Reactions: Analysis, Examples, and Reactor Design*, Vol. 1, Wiley, New York, 1984.)

Examples

This model has been successfully used by Kato et al. (1979) to interpret results for the reactant conversion in the case of catalytic cracking of cumene in a packed fluidized bed under conditions of no deactivation. When catalyst deactivation exists, the authors have shown that for reactions with high initial rate constants (>1.0 L/s), the reactant conversion can be calculated by assuming perfect mixing of particles in the bed. For reaction with lower initial rate constants (<0.3 L/s), the assumption of plug flow of particles within the bed seems adequate.

Another potential application of the packed fluidized bed is in the regeneration of spent activated carbon after its use in municipal waste water treatments and in the removal of odors from waste gases (Kato et al. 1980). The adsorption capacity of the regenerated catalyst as measured by the iodine number was found to agree well with the values predicted by the model described above.

Staging of catalyst

Vertical staging of catalyst can sometimes be advantageous because the gas flow would tend to approximate plug flow. The operation can be either countercurrent or cross-current. The region between two consecutive beds is obviously the freeboard region of the lower bed. The holes in the grid plates of these beds must be carefully designed to balance the upward and downward flows of solids from each bed. Thus, the holes in the plate of a given stage should be large enough to allow particles from the lower freeboard region to flow into this stage (thus preventing their accumulation at the bottom of the plate), but small enough to prevent particles from the bed to leak into this freeboard region and then into the lower stage. There is, however, always a through flow of solids, downward or upward. For countercurrent contacting of gas and solid, downflow of solids is necessary, as in fluidized-bed reduction of metal ores. For details of particle interchange at perforated plates and factors influencing particle leakage in staged reactors, reference Briens et al. (1978) (among others) may be consulted.

An energy balance for the single stage and multistage modes of operation can be readily written (Kunii and Levenspiel, 1991) and the following simple expressions for the efficiency improvement as a result of the added stages can be obtained:

N: number of stages

$$\eta_s = \eta_g = \frac{N}{N+1} \text{ for counter current flow} \tag{9.23}$$

and

$$\eta_s = N\phi_f \eta_g \text{ for cross-current flow} \tag{9.24}$$

where

$$\phi_f = \frac{A_c u \rho_g C_p}{F C_{ps}} \tag{9.25}$$

where η_s and η_g are the efficiencies for the solid and fluid phases, respectively.

An important practical shortcoming of the bubbling bed model considered in the previous sections is that fluid-bed reactors normally operate at velocities usually in excess of the limits of bubbling beds, that is, $u_0/u_{mf} > 15$ (Avidan and Edwards, 1986; Bolthrunis, 1988). Reactors of this type give rise to turbulent behavior as opposed to bubbling bed behavior (Squires et al., 1985). Increasing attention is being paid to these reactors in the recent literature. This takes us to a consideration of other regimes of fluid-bed operation, besides the bubbling bed. A practical overall strategy for fluidized-bed reactor scale-up has also been suggested (Jazayeri, 1995).

Extension to other regimes of fluidization types of reactors

From Figure 9.3, four different (but somewhat overlapping) categories of fluidized-bed reactors can be recognized: bubbling bed, turbulent (or fluid) bed, fast bed, and pneumatic (or transport) bed. Occasionally, reactor operation at velocities close to u_{mf} has been attempted.

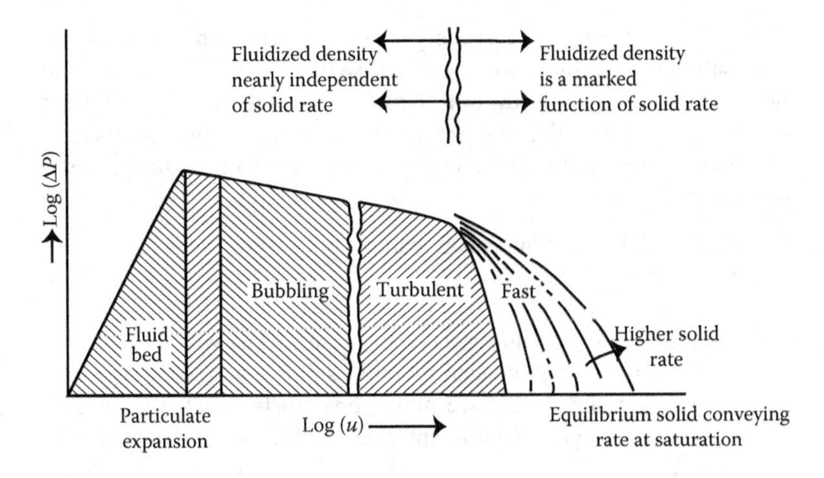

Figure 9.10 Different regimes of fluidization. (From Doraiswamy, L.K. and Sharma, M.M., *Heterogeneous Reactions: Analysis, Examples, and Reactor Design*, Vol. 1, Wiley, New York, 1984.)

The operation of a fluid bed in terms of demarcation into these regimes can be best understood by mapping the effect of velocity on the pressure gradient across the bed. Such a map is shown in Figure 9.10 as a logarithmic plot of ΔP versus velocity.

Turbulent bed reactor

This differs from the bubbling bed reactor mainly in respect of the happenings at the bed surface. Owing to the higher gas velocities (which make the reactor more suitable for high-throughput reactions), there is greater turbulence at the interface leading to more violent bursting of bubbles and splashing of emulsion clusters. Otherwise, it is not much different from the bubbling bed, despite the fact that the bubble and emulsion phases are not now as clearly demarcated. Thus, the general procedure described for the bubbling bed, including the dilute bed (Kunii and Levenspiel, 1991) or the Miyauchi–Marooka model described above can be used. Note that the porosity distribution data will now be different. In particular, the freeboard porosity will be higher. Strictly, new correlations for k_{bc} and k_{ce} will also be required, but available correlations for the bubbling bed can be used as a first approximation.

Fast fluidized-bed reactor

Here, the gas velocity is even higher than in the turbulent bed. What we have is a bottom-dense zone consisting of a mixture of bubbles with more solids in them than in the bubbling bed and clumps or packets of particles rising through most of the bed's cross-section, while some emulsion flows down at/near the wall. These packets naturally have a terminal velocity much higher than that of the particles and hence cannot be sustained by the rising gas. They would therefore fall back, with subsequent disintegration, giving rise to a higher slip velocity and hence backmixing. The overall coefficient K_f will be different from that defined by Equation 9.12 for the K–L model. At present, there is no equivalent equation. These observations are also true for the upper dilute region.

In conclusion, the most important features of the three beds can be succinctly stated as follows:

Bubbling regime

- In the bubbling regime, gas is brought in contact with the solids at a relatively high concentration of solids, and the carryover is not significant. A discernible upper surface is maintained where the bubbles burst to enter the bubble-free lean or dilute phase.

Turbulent regime

- In the turbulent regime, the solid's concentration in the bed is lower and the solid's carryover significantly more, but the upper surface continues to be relatively intact, in spite of being more turbulent. If the carryover is equivalently replenished by continuous addition, the fluid-bed density does not differ from that of the bubbling bed.

Fast fluidization regime

- As the velocity is further increased, a point is reached when the bed density becomes a strong function of the rate of solids feed. This corresponds to the fast fluidization regime and is characterized by the significant fact that fluid-bed densities as high as those in a bubbling bed can be maintained by adjusting the solids flow rate.

Transport (or pneumatic) reactor

FCC reactors

Fluid catalytic cracker (FCC) units are characterized by reaction in the transport lines between the reactor and the regenerator, in addition to that in the main reactor. Thus, the design of these reactors is very complicated, involving more than one regime of fluidization. Innumerable studies have been reported on the modeling of FCC reactors and they should be consulted for detail (e.g., de Croocq, 1984; Chuang et al., 1992). Some salient features are described here.

As shown in Figure 9.10, the fluidization regime changes with increase in gas velocity (sometimes also expressed as the ratio of the superficial velocity to the minimum fluidization velocity). FCC units use very fine particles. In the more recent designs based on highly active zeolite catalysts, the main cracking occurs almost entirely in the riser, that is, the section that transports the catalyst between the reactor and the regenerator or back into the same reactor. Hence, although the bubbling or turbulent regime may be involved within the reactor or regenerator, the transport line (or the riser) operates in the pneumatic regime. In the main reactor, bubbling does not commence immediately after u_{mf} is reached, and the turbulent regime sets in far beyond u_t. In this regime, bubble short-circuiting is much less prevalent and hence the conversions are higher. On the other hand, the regenerator operates in the bubbling regime even though the velocities involved (0.6 m/s) correspond to the turbulent regime. This is because of the absence of fines. The presence of fines plays an important role in the operation of a fluidized-bed reactor (see Yadav et al., 1994) as will be discussed in the next section.

Circulation systems

We already saw that certain types of fluidization (fast and pneumatic) involve solids recirculation. As shown in Figure 9.11, this type of operation essentially consists of solids circulating between two fluidized beds A and B. They are connected through two curves of a U-tube in such a way that the difference in static pressures drives the solids from one bed to the other. The use of a second U-tube completes the circulation between the beds. As there is a frictional resistance associated with solids flow (increasing with increasing flow rate), the rate of circulation is controlled by a balance between the frictional resistance and the static pressure difference mentioned earlier. The frictional resistance can be controlled by varying the average densities of the flowing gas–solid mixtures in the various sections of the circulation loop.

Different circulation systems have been devised over the years, two of which are shown in Figure 9.12.

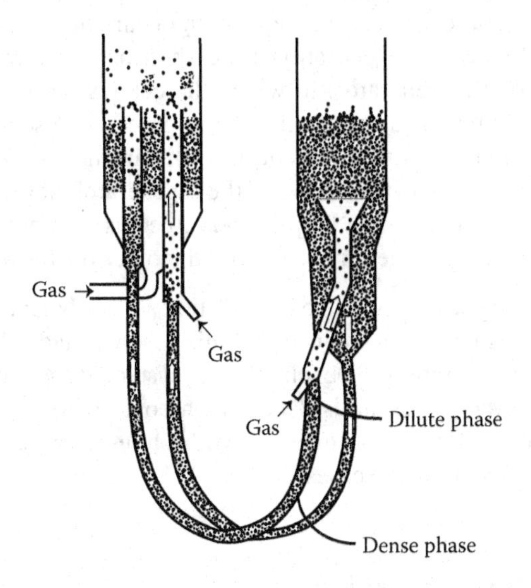

Figure 9.11 Main features of solids circulation in fluidized-bed reactors. (From Joshi, J.B. and Doraiswamy, L.K. In *Albright's Chemical Engineering Handbook* (Ed. Lyle, F.), CRC Press, Boca Raton, FL, 2009.)

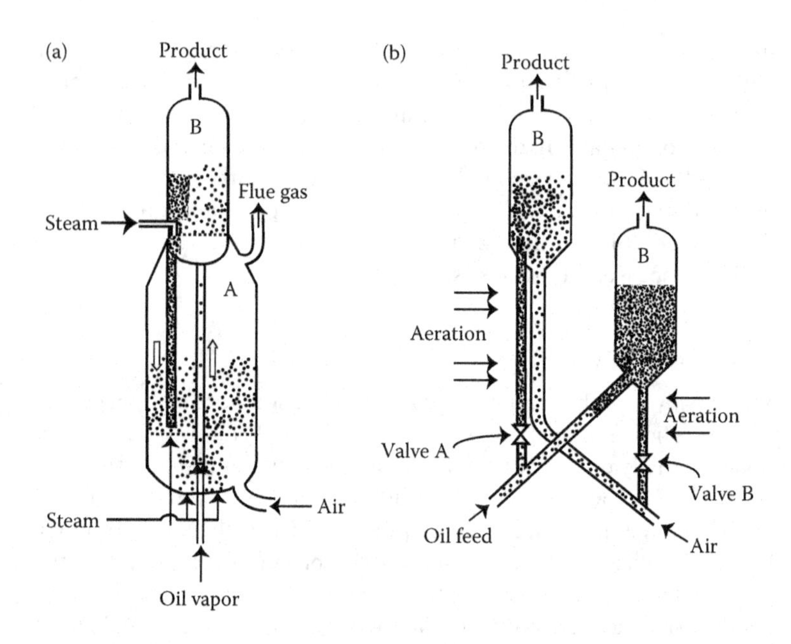

Figure 9.12 Two examples of circulation systems in fluidized-bed operation. (From Joshi, J.B. and Doraiswamy, L.K. In *Albright's Chemical Engineering Handbook* (Ed. Lyle, F.), CRC Press, Boca Raton, FL, 2009.)

Deactivation control

From the point of view of reactor design, solid circulation is most useful for a deactivating catalyst. Catalytic reaction including simultaneous deactivation by carbon deposition occurs in one location (the reactor) and catalyst regeneration by carbon burn-off with air in a second location (regenerator) to which the catalyst is transported by circulation. Although different fluidizing gases are used in the two reactors, the reactant (usually vaporized oil) in the case of the catalytic converter and air in the case of the regenerator, solid circulation can be restricted to a single loop for both the gases or can be accomplished in two loops, one for the oil and the other for the air.

Coupled endothermic and exothermic reactions

The second important aspect of circulation is the balancing of heat between its absorption in the endothermic reaction and release in the exothermic regeneration. The circulation system design depends on whether the overall scheme is deactivation controlled or heat transfer controlled. We give the final equations, with a brief reference to the principles, for the two extreme cases.

Deactivation control

To arrive at a balanced circulation rate of solids, let us first define the activity Ω of the catalyst as

$$\Omega = \frac{\text{rate of reaction on catalyst at a given condition}}{\text{rate of reaction on a fresh catalyst}} \tag{9.26}$$

Then, we assume that the catalyst undergoes deactivation in the reactor but that the activity is not fully recovered in the regenerator. This general situation is depicted in Figure 9.13, which shows an average activity $\overline{\Omega}_1$ in the reactor and $\overline{\Omega}_2$ in the regenerator. Assuming first-order deactivation, we have

$$\text{Reaction:} \quad -\frac{d[A]}{dt} = k_v[A]\Omega \tag{9.27}$$

$$\text{Deactivation:} \quad -\frac{d\Omega}{dt} = k_d\Omega \tag{9.28}$$

$$\text{Regeneration:} \quad \frac{d\Omega}{dt} = k_a(1-\Omega) \tag{9.29}$$

It is reasonable to assume that the solid flow is fully backmixed with an exponential distribution of residence times. Based on this assumption and by writing an expression for the average activity of the leaving catalyst stream (which contains particles of all ages with their corresponding activities), we can derive the following equations:

$$\text{Reactor:} \quad k_a\overline{t}_1 = \frac{\overline{\Omega}_2 - \overline{\Omega}_1}{\overline{\Omega}_1} \tag{9.30}$$

$$\text{Regenerator:} \quad k_a\overline{t}_2 = \frac{\overline{\Omega}_2 - \overline{\Omega}_1}{1 - \overline{\Omega}_1} \tag{9.31}$$

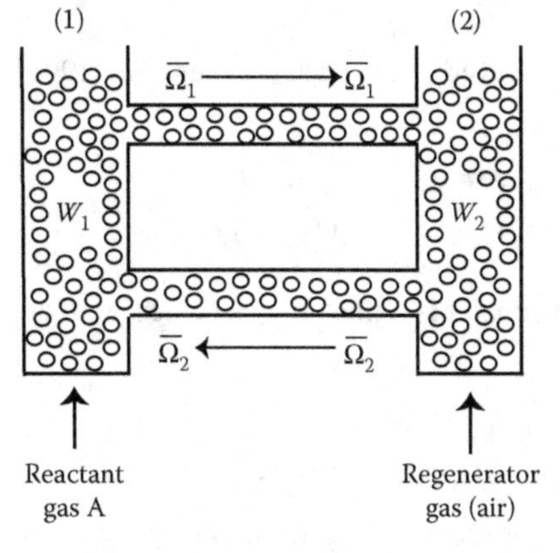

Note: When regeneration is complete (i.e. 100%) $\overline{\Omega}_2 = 1$, the equations get simplified

Figure 9.13 Fluidized-bed circulation system for catalyst deactivation control. (From Joshi, J.B. and Doraiswamy, L.K. In *Albright's Chemical Engineering Handbook* (Ed. Lyle, F.), CRC Press, Boca Raton, FL, 2009.)

where $\bar{t}_1 = W_1/F_S$, $\bar{t}_2 = W_2/F_S$. Eliminating $\bar{\Omega}_2$ from Equations 9.30 and 9.31, we obtain

$$\bar{\Omega}_1 = \frac{k_a\bar{t}_2}{k_d\bar{t}_1 + k_dk_a\bar{t}_2 + k_a\bar{t}_2} \tag{9.32}$$

This equation for average catalyst activity is then combined with the rate constant k_v to give $\bar{\Omega}_1 k_v$ in the usual reactor equations, which would normally use simply k_v. Thus, we obtain for the three common kinds of gas flow:

$$\text{Fully mixed:} \quad \frac{[A]}{[A]_0} = \frac{1}{1 + k_v\bar{\Omega}_1\bar{t}_1} \tag{9.33}$$

$$\text{Plug:} \quad \frac{[A]}{[A]_0} = \exp(-k_v\bar{\Omega}_1\bar{t}_1) \tag{9.34}$$

$$\text{Fluidized:} \quad \frac{[A]}{[A]_0} = \exp(-K_f\bar{\Omega}_1\bar{t}_1) \tag{9.35}$$

where K_f is the overall rate constant of the Kunii–Levenspiel model of the fluidized bed in the absence of deactivation. Solution of Equations 9.33 and 9.34 or 9.35 gives the circulation rate F_s and the bed weights for reactor (W_1) and regenerator (W_2).

Heat transfer controlled

An important aspect of solid circulation is that they carry heat from its source (usually exothermic regeneration) to its destination (usually endothermic oil cracking). This can be the controlling feature of a reaction–regeneration system, and the necessary circulation rate for balancing the two depends on the enthalpies at various points in the scheme as shown in Figure 9.14. The final equation derived by Kunii and Levenspiel (1991) assuming no heat loss is

$$F_s = \frac{F_1(-\Delta H_r + H_1 - H_3)}{C_{ps}(T_3 - T_4)} = \frac{F_2(H_4 - H_2)}{C_{ps}(T_3 - T_4)} \tag{9.36}$$

If the reaction is endothermic, the cooler is replaced by a heater, with no other change in the analysis.

Reactor choice for a deactivating catalyst

Catalyst deactivation is one of the most vexing problems in catalyst and catalytic reactor design. We shall not be concerned with this in the present chapter beyond using empirical equations to represent deactivation rates for use in design calculations (see, e.g., Weekman, 1968; Sadana and Doraiswamy, 1971; Doraiswamy and Sharma,

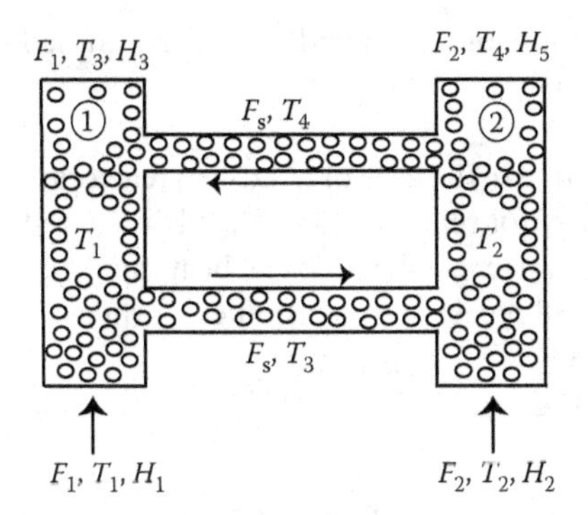

Figure 9.14 Fluidized-bed circulation system for heat transfer control. (From Joshi, J.B. and Doraiswamy, L.K. In *Albright's Chemical Engineering Handbook* (Ed. Lyle, F.), CRC Press, Boca Raton, FL, 2009.)

1984, for details). We have already seen an example of this while dealing with circulating solid beds. Here, we briefly touch upon the problem of reactor design and choice for a deactivating catalyst in general. An important parameter in considering the role of catalyst deactivation is the production or onstream time (which is the same as the catalyst decay time). This is the time t_p for which the reactor is run before subjecting the catalyst to regeneration. The total decay time for a given level of decay is t_{p1}. Assuming that reactant A is passing through a decaying bed of catalyst under diffusion-free conditions and is undergoing a simple reaction, the first step in the analysis is to write the continuity equation for the reactor. We therefore begin with this.

Basic equation Restricting the treatment to isothermal plug flow, the continuity equation for a reactor containing a time-decaying catalyst through which reactant A is passing and reacting under diffusion-free conditions may be written as

$$\left(\frac{\partial[A]}{\partial t_p}\right) + u\left(\frac{\partial[A]}{\partial z}\right) = -[-r_A([A], t_p)] \tag{9.37}$$

or

$$\frac{f_B \rho_G}{M}\left(\frac{\partial y_A}{\partial t_p}\right) + \frac{G_M}{M}\left(\frac{\partial y_A}{\partial z}\right) = -[-r_A(y_A, t_p)] \tag{9.38}$$

where t_p is the production or onstream time. It will be noticed that the rate has been written as a function of both concentration and onstream time. In other words, at a given concentration, the rate is also dependent on time (which it would not be for a nondecaying catalyst).

We now develop the governing equations for fixed and fluidized-bed reactors and compare their performances.

Fixed-bed reactor

Fixed-bed reactor The reaction time in the case of fixed-bed reactors is evidently the same as the total decay (or reaction) time t_{pl}, when viewed from the standpoint of catalyst decay. Hence, the time can be normalized with respect to t_{pl}, and Equation 9.38 recast in dimensionless form as

$$\alpha\left(\frac{\partial y_A}{\partial \hat{t}}\right)+\frac{\partial y_A}{\partial \hat{z}} = -\beta[-r_A(y_A,\hat{t})] \tag{9.39}$$

where $G_M = \rho_F S_{V,F} L$ (any feed) and $G_M = \rho_L S_{V,L} L$ (liquid feed), and

$$\hat{t} = \frac{t_P}{t_{pl}},\quad \hat{z} = \frac{z}{L}$$

$$\alpha = \frac{f_B \rho_G}{\rho_L S_{V,L} t_{p1}},\quad \beta = \frac{M}{\rho_F s_{V,F}} = \frac{M}{\rho_L s_{V,L}},\quad \frac{\text{cm}^3\text{s}}{\text{mol}} \tag{9.40}$$

Equation 9.39 can be simplified if the first term can be neglected. It is entirely reasonable to do so since the constant α in that term, which represents the ratio of feed transit time to decay time, is usually negligibly small. Thus, Equation 9.39 reduces to

$$\frac{dy_A}{dz} = -\beta[-r_A(y_A,\hat{t})] \tag{9.41}$$

We now express the rate as

$$-r_A(y_A, t_p) = k_v(t_p)(1 - f_B)y_A^m \tag{9.42}$$

If we assume that the catalyst decays exponentially with time, that is

$$k_v(t_p) = k_{v0}\exp(-d_c t_p) \tag{9.43}$$

then

$$\frac{dy_A}{dz} = -\beta'\exp(-\lambda\hat{t})y_A^m \tag{9.44}$$

where d_c is the decay constant (1/s)

$$\lambda = d_c t_{p1}$$

$$\beta' = \beta(1 - f_B)k_{vo} = \frac{M}{\rho_F s_{v,F}}(1 - f_B)k_{vo} \tag{9.45}$$

Equation 9.45 is the basic nondimensional equation describing the mole fraction of A in a fixed-bed reactor containing an exponentially decaying catalyst as a function of position and time in terms of two dimensionless parameters β' and λ. The performance of this reactor can be best judged by solving the equation for the reactor exit, that is, for $z = 1$. The solution for a first-order reaction ($m = 1$) is given by Sadana and Doraiswamy (1971). It is also possible to assume various other forms of catalyst decay. Solutions are included in Table 9.4 for two other forms, one of them linear.

When a decaying catalyst is used, it is important to estimate the average conversion over a given period of time. From an economic point of view, this quantity is much more important than the conversion at the exit, and is given by

$$\bar{X}_A = 1 - \bar{y}_A = 1 - \int_0^1 y_A d\hat{t} \tag{9.46}$$

Solutions to this equation (for $z = 1$) are included in the table for all the three forms of decay considered.

Fluidized-bed reactor

Fluidized-bed reactor A reasonable assumption of the fluidized-bed reactor is that the fluid is partially mixed, whereas the solid is fully mixed. If we assume that the fluid is in plug flow, the residence time distribution of the solids will be given by $\exp(-\hat{t})$. As in the case of the fixed-bed reactor, we shall first consider exponential decay. The average rate constant to be used in solving Equation 9.44 is therefore

$$[k_v(\hat{t})]_{av} = k_{vo} \int_0^\infty \exp(-\hat{t}) \exp(-\lambda \hat{t}) d\hat{t} \tag{9.47}$$

The solution to this equation is (Weekman, 1968)

$$[k_v(\hat{t})]_{av} = \frac{k_{vo}}{1 + \lambda} \tag{9.48}$$

Substituting this equation for the rate constant in Equation 9.42 gives

$$-r_A(y_A, \hat{t}) = \left[\frac{k_{vo}(1 - f_B)}{1 + \lambda}\right] y_A^m \tag{9.49}$$

for a reaction of order m taking place in a steady-state fluid-bed reactor. Then, upon incorporation of this equation in Equation 9.44, we obtain

$$\frac{dy_A}{dz} = -\left(\frac{B''}{\lambda + 1}\right) y_A^m \tag{9.50}$$

Solutions to this equation for $m = 1$ and $m \neq 1$ as well for the other decay forms considered for the fixed-bed reactor given by Sadana and Doraiswamy (1971) are presented in Table 9.4.

Moving bed reactor

Moving bed reactor In this reaction, catalyst decay is a function of position and the decay time at any position z is given by zt_p. Hence, the

Table 9.4 Expressions for Mole Fraction and Conversion of Reactant A for Various Decay Forms

	Expression for y_A		Expression for Conversion	
	mth Order ($m \neq 1$)	First Order	mth Order ($m \neq 1$)	First Order
$k_v = k_{v0}e^{-\lambda\hat{t}}$				
Fixed	$\left[\dfrac{1}{(m-1)B''e^{-\lambda\hat{t}}+1}\right]^c$ (1a)	$\exp\{-[B''\exp(-\lambda\hat{t})]\}$ (1b)	$1-\int_0^1\left[\dfrac{1}{(m-1)B''e^{-\lambda\hat{t}}+1}\right]^c d\hat{t}$ (1c)	$1+\dfrac{1}{\lambda}Ei^*(B'')-Ei^*(B''-\lambda)$ (1d)
Fluid	$\left[\dfrac{\lambda+1}{(m-1)B''+(\lambda+1)}\right]^c$ (2a)	$\exp\left(-\dfrac{B''}{\lambda+1}\right)$ (2b)	$1-\left[\dfrac{\lambda+1}{(m-1)B''+(\lambda+1)}\right]^c$ (2c)	$1-\exp\left(-\dfrac{B''}{\lambda+1}\right)$ (2d)
$k_v = k_{v0}-\lambda\hat{t}$				
Fixed	$\left[\dfrac{k_{v0}}{B''(m-1)(k_{v0}-\lambda\hat{t})+k_{v0}}\right]^c$ (3a)	$\exp\left[-\left(B''-\dfrac{B''\lambda\hat{t}}{k_{v0}}\right)\right]$ (3b)	$1-\int_0^1\left[\dfrac{k_{v0}}{B''(m-1)(k_{v0}-\lambda\hat{t})+k_{v0}}\right]^c d\hat{t}$ (3c)	$1-\int_0^1\exp\left[-\left(B''-\dfrac{B''\lambda\hat{t}}{k_{v0}}\right)\right]d\hat{t}$ (3d)
Fluid	$\left[\dfrac{k_{v0}}{B''(m-1)(k_{v0}-\lambda\hat{t})+k_{v0}}\right]^c$ (4a)	$\exp\left[\left(\dfrac{B''}{k_{v0}}\right)(k_{v0}-\lambda)\right]$ (4b)	$1-\left[\dfrac{k_{v0}}{B''(m-1)(k_{v0}-\lambda\hat{t})+k_{v0}}\right]^c$ (4c)	$1-\exp\left[\left(-\dfrac{B''}{k_{v0}}\right)(k_{v0}-\lambda)\right]$ (4d)
$k_v = k_{v0}-\lambda\hat{t}^d$				
Fixed	$\left[\dfrac{k_{v0}}{(B''k_{v0}-B''\lambda\hat{t}^d)(m-1)+k_{v0}}\right]^c$ (5a)	$\exp\left[\left(\dfrac{B''}{k_{v0}}\right)\lambda\hat{t}^d-B''\right]$ (5b)	$1-\int_0^1\left[\dfrac{k_{v0}}{(B''k_{v0}-B''\lambda\hat{t}^d)(m-1)+k_{v0}}\right]^c d\hat{t}$ (5c)	$1-\int_0^1\exp\left(\dfrac{B''}{k_{v0}}\lambda\hat{t}^d-B''\right)d\hat{t}$ (5d)
Fluid	$\left[\dfrac{k_{v0}}{B''(m-1)(k_{v0}-d!\lambda)+k_{v0}}\right]^c$ (6a)	$\exp\left[B''\left(\dfrac{d!\lambda}{k_{v0}}-1\right)\right]$ (6b)	$1-\left[\dfrac{k_{v0}}{B''(m-1)(k_{v0}-d!\lambda)+k_{v0}}\right]^c$ (6c)	$1-\exp\left[B''\left(\dfrac{d!\lambda}{k_{v0}}-1\right)\right]$ (6d)

Source: Adapted from Sadana, A. and Doraiswamy, L.K., *J. Catal.*, **23**, 147, 1971.

[a] $c = 1/(m-1)$. The last two decay forms are useful but dimensionally inconsistent.

reduced decay time is $zt_p/t_p = z$. The following equation can then be written for the mole fraction of A:

$$\frac{dy_A}{dZ} = B' \exp(-\lambda Z) y_A^m \tag{9.51}$$

and X_A is related to y_A by Equation 9.46.

A comparison of the performances of the three types of reactors for given values of the governing parameters is given in Figure 9.15 for a first-order reaction. The important role of the decay parameter λ is clearly evident from this figure and provides a strong point (among many other issues to be considered) in reactor choice. Sadana and Doraiswamy (1971) and Prasad and Doraiswamy (1974) have extended the treatment to non-first-order and complex reactions.

The procedures described above are relatively simple. More rigorous methods of design for both types of reactors have been documented in several books (e.g., Doraiswamy and Sharma, 1984; Kunii and Levenspiel, 1991).

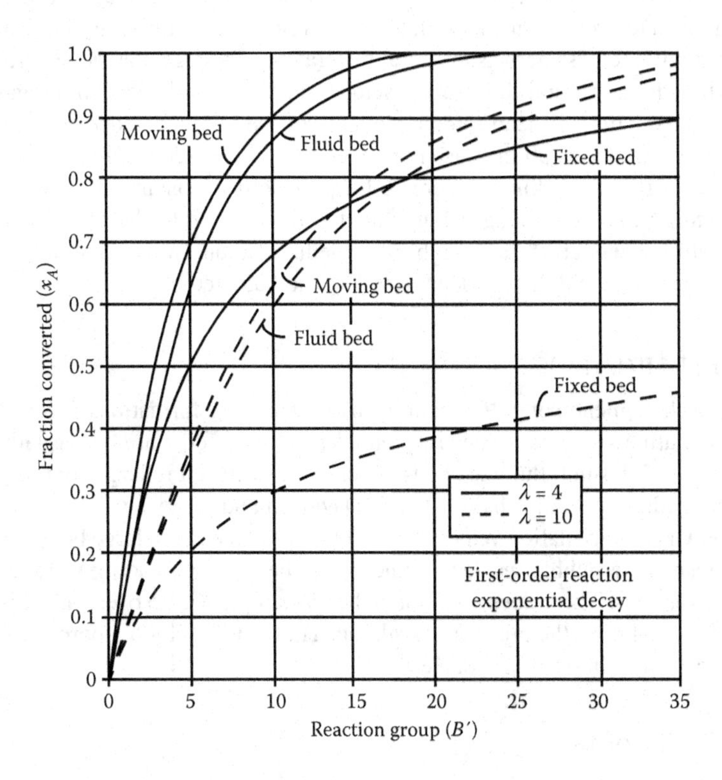

Figure 9.15 Comparison of performances of different reactor types for a reaction on a deactivating catalyst.

Some practical considerations

The fluidized-bed reactor is far more difficult to scale-up than the fixed-bed reactor, and there are also many practical considerations involved in its operations. Some of these are briefly outlined here.

Slugging

The bed is assumed to slug when the ratio of the bubble to tube diameter is greater than 0.8–0.9 (Stewart and Davidson, 1967). This criterion is however not valid beyond a certain height, and Baeyens and Geldart (1974) give a more elaborate criterion for such a situation. Generally, the conversion obtained in a narrow slugging bed is higher than in the scaled-up version. Hence, slugging is basically not an "unacceptable" mode of operation. Multitubular reactors operating under fluidizing conditions might provide a useful alternative.

Defluidization of bed: Sudden death

During the operation of a fluid-bed reactor, care should be taken to ensure that there is no sudden increase in pressure due to malfunctioning of certain valves or choking in downcomers by solid particles. If this is allowed to proceed unchecked, the mass velocity for minimum fluidization will increase at the same total flow rate, and a stage may be reached when the minimum fluidization velocity might actually exceed the gas velocity. This will lead to defluidization of the bed. Defluidization can also occur due to an increase in temperature or coke deposition of catalyst particles (leading to increased u_{mf}). Another reason is the switch from an inert fluidizing gas to start the fluidization to the actual gas. Disturbances at this stage might cause defluidization. Every effort should be made to avoid this "sudden death" of the fluidized bed.

Gulf streaming

The phenomenon of gulf streaming arises due to the formation of violent circulating currents induced by bubbles. The currents are not usually observed in small fluidized beds but can be significant in large commercial beds (especially shallow beds). Davidson and Harrison (1971) have shown that circulation velocities of 200 mm/s can exist in large beds. As a result, the bubble residence time would be smaller, leading to lower conversions. There does not seem to be any simple way to overcome this problem during the process of scale-up. Thus, a pilot plot of appropriate size seems almost unavoidable.

Effects of fines

It is always desirable to have a size distribution of particles rather than a single size in a fluidized bed. The two-phase theory of fluidized-bed operation is suspect when a bed contains a reasonable fraction of fines, and hence models based on this theory should be used with caution in the presence of fines. The dense phase in such cases should really be regarded as consisting of two phases: emulsion and clusters of fines (d_p < 40 μm). Indeed, the results of Yadav et al. (1994) on commercial propylene ammoxidation catalyst clearly show that the fines do agglomerate. An interesting observation is that there exists a critical level of fines (30%) at which the fluid-bed behavior in terms of bed expansion, aeratability, and cluster size is optimum. Yadav et al. (1994) have proposed a model that takes the two dense phase components (emulsion and cluster) into account. A practical consequence of adding fines is that it widens the limits of operable gas velocities. Also, carryover does not segregate the particles according to size, as in the absence of fines. Clearly, the utility of adding fines to improve performance should be an important practical consideration in fluid-bed reactor design.

Start-up

The start-up of a fluidized bed requires an initial burst of pressure to lift the solids and bring them to a state of fluidization. This can severely damage the bed internals, indeed the entire structural framework. It is essential therefore to begin the operation with an empty reactor (by removing the solids after each shutdown) and progressively increase the gas velocity and introduce the solids incrementally at the same time. As has been noted earlier, sudden death of the fluid bed is also dangerous and should be avoided.

Other practical considerations are attrition of particles; start-up of a fluidized bed; caking of catalyst from malfunctioning of the reactor due to the formation of tarry products (resulting sometimes in "cakes" as large as the reactor diameter); and the need to avoid premixing of reactants, particularly when they can form explosive mixtures and fix their relative locations within the bed (e.g., in the chlorination of methane and ammoxidation of propylene). A more detailed coverage of these topics can be found in Doraiswamy and Sharma (1984).

Fluidized-bed versus fixed-bed reactors

Following their first major success in 1942 in the refining/petrochemical industries, fluidized-bed reactors were hailed as the panacea for most reactor evils. This optimism was clearly too hasty in view of many spectacular failures. But more recently, there has been resurgence in their use with the development of fluidized-bed coal combustion processes and a better understanding of fluid-bed behavior in general. It is no longer restricted to deactivating catalysts requiring a second reactor for regenerating it or to really large-scale productions. Its main advantages are prevalence of near-isothermal conditions in the entire bed due to solids movement (and therefore absence of hot spots, a key drawback of fixed bed reactors), less danger of explosions and temperature excursions, better operational flexibility, and higher heat and mass transfer rates compared to other modes of fluid–solids contacting (often leading to smaller-sized heat exchangers). Because of these advantages, the traditional preference for fixed-bed reactors is no longer a common feature of reactor choice, in spite of their many advantages, namely relatively easy scale-up, minimum catalyst loss due to attrition, and the theoretical possibility of imposing an optimum temperature profile (taking advantage of the inherent nonisothermicity of the bed).

In fact, the bias today is toward fluidized-bed reactors, particularly toward pressurized fluidized-bed reactors in gas–solid noncatalytic systems such as coal gasification and coal combustion power plants (Dutta and Gualy, 1999). Higher pressures enhance productivity per unit reactor volume, and in many cases can significantly increase equilibrium conversion and selectivity. An important example of reactor choice for gas–solid noncatalytic reactions based on extensive studies on different alternatives is the hot gas desulfurization process using zinc titanate as the regenerative adsorbent. ZnO, the active component of this adsorbent, reacts with H_2S to form ZnS, which is regenerated by air oxidation and recycled back to the adsorber. This process is part of the integrated gasification combined cycle (IGCC) power plants. Experimental results on moving bed, bubbling bed, and circulating fluidized-bed reactors have clearly shown the circulating bed to be the preferred candidate. Details of these studies are discussed by Dutta (1994).

A classical example of the use of fluidized beds in catalytic reactions is the FCC. In view of its tremendous importance, improvements in modeling and operation of this unit are being continually reported. This is perhaps one of the few instances of a clear advantage of the fluidized bed over the fixed bed for catalytic processes.

Explore yourself

1. List as many models used to design/analyze fluidized-bed reactors as possible. Can you differentiate the main features of these models? Can you choose the most suitable model for your design situation?
2. Outline a methodology for the selection of the most plausible model for the solid catalyzed fluidized-bed reactor design.
3. What is the difference between the packed-bed and fluidized-bed reactors both in terms of the physical picture and the mathematics involved?
4. What are the main advantages of fluidized-bed reactors and the chief uncertainties in their design? Why is such a reactor with all its uncertainties so widely used in industry?

5. How much did computational fluid dynamics (CFD) enter into the actual design of the fluidized-bed reactors? Perform a thorough literature survey and find a solid evidence of the use of CFD tools in fluidized-bed reactor design.
   a. Following the previous question, make a list of the unknowns as well as the factors to be known before a *priori* design of reactors from first principles becomes possible. The list you generate should have the potential to provide research topics for future students.
   b. Make a list of recent and developing reactor technologies. In what ways can these technologies replace fluidized-bed reactors? Which deficiency in the present-day technology should be eliminated for the new technology to replace the fluidized-bed reactors?

References

Avidan, A. and Edwards, M., *Modeling and Scale-Up of Mobil's Fluid Bed MTG Process, 5th Int. Conf. Fluidization*, Elsinore, Denmark, May 18–23, 1986.

Baeyens, J. and Geldart, D., *Chem. Eng. Sci.*, **29**, 255, 1974.

Behie, L.A., Bergougnou, M.A., and Baker, C.G.J., *Fluidization Technology, Vol. 1* (Ed., Keairus, D.L.), Hemisphere, Washington, DC, 1976.

Bolthrunis, C.O., *Chem. Eng. Prog.*, **5**, 51, 1989.

Botterill, J.S.M., *BY Chem. Eng.*, **11**, 122, 1966.

Briens, C.L., Bergougnou, M.A., and Baker, C.G.J., *Fluid Proc. Eng. Found. Conf. 2nd*, 38, 1978.

Bukur, B.D. and Amundson, N.R., *Chem. Eng. Sci.*, **30**, 847, 1975a.

Bukur, B.D. and Amundson, N.R., *Chem. Eng. Sci.*, **30**, 1159, 1975b.

Ĉatipović, N. and Levenspiel, O., *Ind. Eng. Chem. Process Des. Dev.*, **18**, 558, 1979.

Chitester, D.C., et al., *Chem. Eng. Sci.*, **39**, 253, 1984.

Chuang, K.C. Young G.W., and Benslay R.M., *Advanced Fluid Catalytic Cracking Technology*, American Institute of Chemical Engineers, New York, NY, 1992.

Couderc, J.P., In *Fluidization*, 2nd ed. (Eds., Davidson, J.F., Clift, R., and Harrison, D.), Academic Press, New York, 1985.

de Croocq, D., *Catalytic Cracking of Heavy Petroleum Fraction*, Inst. Francais du Petrole, Technip., Paris, 1984.

Davidson, J.F., *AIChE Symp. Ser. (No. 128)*, **69**, 16, 1973.

Davidson, J.F., and Harrison, D., *Fluidized Particles*, Cambridge University Press, New York, 1963.

Davidson, J.F. and Harrison, D., *Fluidization*, Academic Press, London, 1971.

DeLasa, H.I., Errazu, A., Barreiro, E., and Solioz, S., *Can. J. Chem. Eng.*, 59, 549, 1981.

Doraiswamy, L.K. and Sharma, M.M., *Heterogeneous Reactions: Analysis, Examples, and Reactor Design*, Vol. 1, Wiley, New York, 1984.

Doraiswamy, L.K., Sadasivan, N., and Venkitakrishnan, G.R., *NCL Report*, Pune, India, 1972, 1973.

Doraiswamy, L.K., Sadasivan, N., Mukherjee, S.P., and Venkitakrishnan, G.R., *NCL Report*, Pune, India, 1968.

Dutta, S. *AIChE Symp. Ser.*, **301**, 157, 1994.

Dutta, S. and Gualy, R., *Hydrocarbon Process.*, **78**, 45, 1999.

Elnashaie, S.S.E.H. and Cresswell, D., *Proc. Int. Symp. Fluidization and Its Applications*, Tolouse, France, 1973.

Fan, L.T., Lee, C.J., and Bailie, R.C., *AIChE J.*, **8**, 239, 1962.
Furusaki, S., Takahasi, M., and Miyauchi, T., *J. Chem. Eng. Jpn.*, **11**, 309, 1978.
Geldart, D., *Powder Technol.*, **1**, 285, 1973.
Grace, J.R., In *Gas Fluidization Technology* (Ed. Geldart, D.), Wiley, New York, 1986.
Gunn, D.J. and Hilal, N., *Int. J. Heat Mass Transfer*, **37**, 2465, 1994.
Gunn, D.J. and Hilal, N., *Int. J. Heat Mass Transfer*, **39**, 3357, 1996.
Gunn, D.J. and Hilal, N., *Chem. Eng. Sci.*, **52**, 2811, 1997.
Horio, M., Nonaka, A., Hoshiba, M., Morisita, K., Kobukai, Y., Naito, J., Tashibana, O., Watanabe, K., and Yoshida, N., In *Circulating Fluidized Bed Technology* (Ed. Basu, P.), Pergamon Press, New York, 1986.
Jazayeri, B., *Chem. Eng. Prog.*, **91**, 26, 1995.
Joshi, J.B. and Doraiswamy, L.K. Chemical Reaction Engineering, In *Albright's Chemical Engineering Handbook* (Ed. Lyle, F.), CRC Press, Boca Raton, FL, 2009.
Kunii, D. and Levenspiel, O., *Fluidization Engineering*, Butterworth-Heinemann, Boston, 1991.
Kulkarni, B.D., Ramachandran, P.A., and Doraiswamy, L.K., In *Fluidization* (Eds. Grace, J.R., and Matsen, J.M.), Plenum Press, New York, 1980.
Kato, K., Imafuku, K., and Kubota, H., *Chem. Eng. Jpn.*, **31**, 967, 1967.
Kato, K., Arai, H., and Ito, U., *Adv. Chem. Ser.*, **133**, 270, 1974.
Kato, K., Ito, H., and Omura, S., *J. Chem. Eng. Jpn.*, **12**, 403, 1979.
Kato, K., Matsuura, K., and Hanzawa, T., In *Fluidization* (Eds. Grace, J.R., and Matsen, J.M.), Plenum Press, New York, p. 555, 1980.
Levenspiel, O., *Chemical Reactor Omnibook*, Oregon State University Bookstore, Corvallis, OR, 1993.
Lewis, W.K., Gilliland, E.R., and Glass, W., *AIChE J.*, **5**, 419, 1959.
Lewis, W.K., Gilliland, E.R., and Girouard, H., *Chem. Eng. Prog. Symp. Ser. (No. 38)*, **58**, 87, 1962.
Li, Y. and Kwauk, M., In *Fluidization* (Eds. Grace, J.R., and Matsen, J.M.), Plenum Press, New York, p. 537, 1980.
Mathis, J.F. and Watson, C.C., *AIChE J.*, **2**, 518, 1956.
Miyauchi, T., *Chem. Eng. Jpn.*, **7**, 201, 1974.
Miyauchi, T. and Furusaki, S., *AIChE J.*, **20**, 1087, 1974.
Mori, S. and Wen, C.Y., *Fluidization Technology* (Ed., Keairus, D.L.), Hemisphere, Washington, DC, 1976.
Muchi, I., Mori, S., and Horio, M., *Chemical Reaction Engineering with Fluidization* (in Japanese), Baifukan, Tokyo, 1984.
Papa, G. and Zenz, F.A., *Chem. Eng. Prog.*, **91**, 32, 1995.
Prasad, K. and Doraiswamy, L.K., *J. Catal.*, **32**, 384, 1974.
Sadana, A. and Doraiswamy, L.K., *J. Catal.*, **23**, 147, 1971.
Squires, A.M., Kwauk, M., and Avidan, A., *Science*, **230**, 1329, 1985.
Stewart, P.S.B. and Davidson, J.M., *Powder Technol.*, **1**, 61, 1967.
Toomey, R.D. and Johnston, H.F., *Chem. Eng. Prog.*, **48**, 220, 1952.
van Deemter, J.J., In *Fluidization III* (Eds. Grace, J.R. and Matsen, J.M.), Plenum Press, New York, 1980.
Van Swaaij, In *Fluidization*, 2nd ed. (Ed., Davidson, J.F.), Academic Press, New York, 1985.
Volk, W., Johnson, C.A., and Stotler, H.H., *Chem. Eng. Prog. Symp. Series*, **58**(38), 38, 1962.
Weekman, V.W., Jr. *Ind. Eng. Chem. Proc. Des. Dev.*, **7**, 90, 1968.
Wen, C.Y. and Yu, Y.H., *AIChE J.*, **12**, 610, 1966.
Werther, J., *Int. Chem. Eng.*, **20**, 529 1980.
Werther, J. and Schoessler, M., In *Proc. 16th Int. Symp. Heat Mass Transfer*, Dubrovnik, 1984.
Yadav, N.K., Kulkarni, B.D., and Doraiswamy, L.K., *Ind. Eng. Chem. Res.*, **33**, 2412, 1994.

Yang, W.C., Chitester, D.C., Kornosky, R.M., and Keairns, D.L., *AIChE J.*, **31**, 1085, 1985.
Yerushelmi, J. and Cankurt, N.T., *Chemtech.*, **8**, 564, 1978.
Zabrodsky, S.S., *Hydrodynamics and Heat Transfer in Fluidized Beds* (in Russian), translated by Zenz, F.A., MIT Press, Cambridge, MA, 1966.

Chapter 10 Gas–solid noncatalytic reactions and reactors

Chapter objectives

Upon successful completion of this chapter, the students should be able to

- Classify the gas–solid reactions according to the interaction domains.
- Select the most appropriate model for a gas–solid reaction at hand.
- Apply mass and energy conservation laws to derive the relevant differential equations and select the most appropriate boundary conditions.
- Recite the difference between a homogeneous model and a heterogeneous model, in mathematical domain and in physical domain.
- Understand the limitations of a particular model.
- List a number of industrial gas–solid reactions.

Introduction

Reactions between gases and solids are widespread and include operations such as combustion of solid fuels, environmental control (pollution abatement), energy generation, mineral processing, chemical vapor deposition, and catalyst manufacture and regeneration. Several classes of gas–solid reactions can be identified. Representing the solid by (s) and gas by (g), a number of categories are listed in Table 10.1 along with selected examples of each category. The analysis and modeling of these reactions obviously depend on the specific category at hand, but many common principles can be brought out by considering the most general case

$$\nu_A A\,(\text{g}) + \nu_B B\,(\text{s}) \rightarrow \nu_R R\,(\text{g}) + \nu_S S\,(\text{s}) \tag{R1}$$

Thus, our presentation will largely be confined to this class of reactions, although brief references will be made to a few other classes, notably: $A\,(\text{g}) + B\,(\text{s}) \rightarrow R\,(\text{g})$, represented by the gasification of coal.

Table 10.1 Industrially Important Examples of Different Types of Noncatalytic Gas–Solid Reactions

Type	General Reaction Scheme	Reaction	
A	Solid + fluid → solid + fluid	Roasting of zinc ore	$2ZnS\ (s) + 3O_2\ (g) \rightarrow 2ZnO\ (s) + 2SO_2\ (g)$
		Production of uranium tetrachloride by chlorination	$UO_2\ (s) + CCl_4\ (g) \rightarrow UCl_4\ (s) + CO_2\ (g)$
		Selective chlorination of iron in ilmenite	$FeTiO_3\ (s) + CO\ (g) + Cl_2\ (g) \rightarrow FeCl_2\ (g) + CO_2\ (g) + TiO_2\ (s)$
B	Solid + fluid → solid	Nitrogenation of calcium carbide to produce cyanamide	$CaC_2\ (s) + N_2\ (g) \rightarrow CaCN_2\ (s) + C\ (s)$
		Rusting reaction in iron	$2Fe\ (s) + O_2\ (g) \rightarrow 2FeO\ (s)$
		Absorption of SO_2 by dry limestone injection	$CaO\ (s) + SO_2\ (g) \rightarrow CaSO_3\ (s)$
C	Solid → fluid + solid	Calcination of limestone	$CaCO_3\ (s) \rightarrow CaO\ (s) + CO_2\ (g)$
		Decomposition of magnesium hydroxide	$Mg(OH)_2\ (s) \rightarrow MgO\ (s) + H_2O\ (g)$
D	Solid + fluid → fluid	Production of carbon disulfide	$C\ (s) + S_2\ (g) \rightarrow CS_2\ (g)$
		Chlorination of rutile to titanium tetrachloride	$TiO_2\ (s) + 2C\ (s) + 2Cl_2\ (g) \rightarrow TiCl_4\ (g) + 2CO\ (g)$
		Gasification of carbon	$C\ (s) + H_2O\ (g) \rightarrow CO\ (g) + H_2\ (g)$
E	Solid → fluid	Decomposition of ammonium chloride	$NH_4Cl\ (s) \rightarrow NH_3\ (g) + HCl\ (g)$
		Decompositions of ammonium sulfate	$(NH_4)SO_4\ (s) \rightarrow 2NH_3\ (g) + SO_3\ (g) + H_2O\ (g)$
F	Fluid → solid + fluid	Mond process for nickel production	$Ni(CO)_4\ (g) \rightarrow Ni\ (s) + 4CO\ (g)$
		Oxidation of silicon tetrachloride to silicon dioxide	$SiCl_4\ (g) + O_2\ (g) \rightarrow SiO_2\ (s) + 2Cl_2\ (g)$
		Burning of titanium tetrachloride to rutile	$TiCl_4\ (g) + O_2\ (g) \rightarrow TiO_2\ (s) + 2Cl_2\ (g)$

General reaction: $v_A A + v_B B \rightarrow v_R R + v_S S$

Source: Adapted from Joshi, J.B. and Doraiswamy, L.K., *Chemical Reaction Engineering in Chemical Engineers' Handbook*, CRC Press, Boca Raton, FL, 2009.

Modeling of gas–solid reactions

Shrinking core (or sharp interface) model

The first model, the shrinking core model (SCM) or the sharp interface model (SIM), was first proposed about half a century ago. Many other models have since been proposed to describe the behavior of the solid as it undergoes reaction with a gas (or is reacting by itself to produce a gas). These categories (many not mutually exclusive), include: shrinking (or expanding) core, volume reaction, reaction zone, particle–pellet (or grain), grain–micrograin, discrete, computational, and percolation models. The percolation models are based on the statistical physics of disordered media and include such phenomena as aggregation processes, scaling, network modeling of the pore space, discretization, and random walk representation of diffusion processes. An increasing number of papers are being published on percolation models (see Sahimi et al., 1990, for a review) but research on the other models continues because of their practical usefulness (see, e.g., Szekely et al., 1976; Ramachandran and Doraiswamy, 1982; Kulkarni and Doraiswamy, 1986; Doraiswamy and Kulkarni, 1987; Bhatia and Gupta, 1992).

The simplest of these models are those in which the internal structure of a pellet is not considered and its behavior as a whole is modeled. These are normally called the macroscopic models, but are also referred to as basic models. In all other models, the behavior of the distinctive elements of a pellet such as the grain, micrograin, or the pore constitutes the central feature. In essence, these models account for structural changes accompanying reaction. The so-called random pore models are the most commonly used structural models of this kind. Our discussion will be confined to macroscopic models followed by a brief consideration of structural models.

Shrinking core model

Shrinking core model

Figure 10.1 illustrates the basic features of SCM. The gas first diffuses through the film surrounding the pellet and reacts at the interface. As the reaction progresses, the interface moves inward leaving behind a shell of the exhausted solid (called ash). Thus, in effect, the unreacted core keeps shrinking till the entire solid has reacted. This behavior is possible only if the solid is nonporous. Otherwise, the gas will diffuse beyond the interface and the reaction will no longer be confined to the interface.

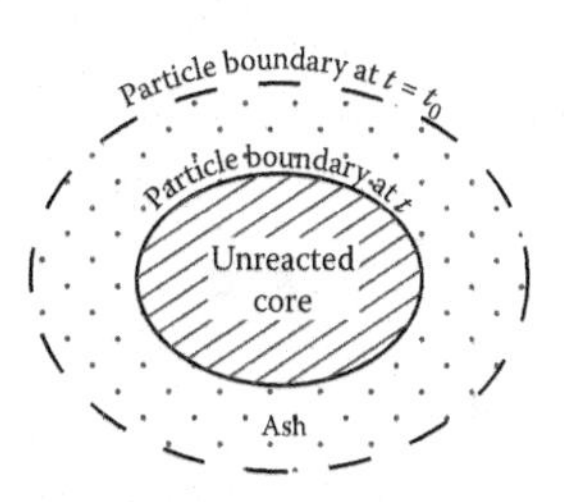

Figure 10.1 The SCM for gas–solid noncatalytic reactions.

The mathematical analysis of this model is greatly facilitated by the pseudo-steady-state (PSS) assumption, that is, the interface remains stationary while the mass flux equations are written. This is generally satisfied for gas–solid reactions. Thus, we can write the following equations for the rate of transport of A diffusing per unit time for a single pellet, r_A:

Diffusion through gas film:

$$r_A = 4\pi R^2 k_G \left([A] - [A]_s\right) \tag{10.1}$$

Diffusion through ash layer:

$$r_A = \frac{4\pi R r_i D_{e,As}}{R - r_i}\left([A]_s - [A]_i\right) \tag{10.2}$$

The chemical reaction at the interface

What is the volume ratio of a sphere with radius r_i and R? How is it related to conversion?

$$r_A = 4\pi r_i^2 k_S [A]_i \tag{10.3}$$

Noting that conversion of the solid B in a spherical pellet is related to the ratio of the initial and interface radii by the equation

$$X_B = 1 - \left(\frac{r_i}{R}\right)^3 \tag{10.4}$$

Equations 10.1 through 10.3 can be combined to give

$$r_A = \left(\frac{1}{4\pi R^2 k_G} + \frac{R - r_i}{4\pi R r_i D_{e,As}} + \frac{1}{4\pi r_i^2 k_s} \right)^{-1} [A] \tag{10.5}$$

Note that each of the terms in parentheses represents a resistance due to the corresponding kinetic phenomenon. To express the rate in terms of the solid reactant B, we write the following equation for the rate of movement of the sharp interface:

$$-\frac{d}{dt}\left(\frac{4}{3}\pi r_i^3 \frac{\rho_B}{M_B} \right) = \frac{v_B}{v_A} r_A = v r_A \tag{10.6}$$

Substituting Equation 10.5 for r_A into Equation 10.6 and integrating, we obtain an equation of the general form

$$\tau = f_1(X_A) + f_2(X_A) + f_3(X_A) \tag{10.7}$$

or

$$\tau = \tau_f + \tau_a + \tau_c \tag{10.8}$$

where τ_f, τ_a, and τ_c represent, respectively, the times required for complete conversion if film transfer, ash diffusion, or chemical conversion alone were to be the controlling step. The functions f_1, f_2, and f_3 and the various τ's assume different forms for different geometries of the pellet and are defined in Table 10.2. An interesting observation is that the dependence of τ on pellet size is different for different controlling regimes: first order in R for reaction control, second order for ash diffusion control, and 1.5–2.0 order for film diffusion control.

Table 10.2 Time Conversion Relationships for SIM for Different Particle Geometries

	$\nu_A A\ (g) + \nu_B B\ (s) \rightarrow \nu_R R\ (g) + \nu_S S\ (s)$			
	Functional Forms for			
Controlling Regime	**Flat Plate**	**Cylinder**	**Sphere**	τ
Film diffusion, $f_1(X_B)$	X_B[a]	X_B	X_B	$\frac{\rho_B R}{v k_g [A]_b}$
Ash diffusion, $f_2(X_B)$	X_B^2	$X_B + (1 - X_B)\ln(1 - X_B)$	$1 - (1 - X_B)^{2/3} + 2(1 - X_B)$	$\frac{\rho_B R^2}{2 v D_{e,As} [A]_b}$
Reaction, $f_3(X_B)$	X_B	$1(1 - X_B)^{1/2}$	$1(1 - X_B)^{1/3}$	$\frac{\rho_B R}{v k_s [A]_b^n}$

[a] Conversion $X_B = 1 - (r/R)^s$, where $s = 1, 2, 3$ for flat plate, cylinder, and sphere, respectively.

Improvements/modifications of the above procedure have been suggested. There can be situations where a porous film of a solid product is deposited on the ash layer, adding one more resistance to the overall process. For example, during the reduction of ilmenite with hydrogen, a porous film of iron is formed on the ash layer (Briggs and Sacco, 1991). Thus, one has to be cautious in routinely applying the additivity of resistances principle for treating combined control as was done in developing Equation 10.7 or Equation 10.8.

An alternative approach is to express the results in analogy with those for catalytic systems, in terms of an effectiveness factor (Ishida and Wen, 1968). Unlike in a catalytic pellet, here the rate changes with time. Hence, the effectiveness factor also changes with time, that is, with r_i, and the following equation can be derived for a first-order reaction in a sphere:

$$\varepsilon = \left[1 + \hat{R}_i Da\left(\frac{1}{Sh} + \frac{1-\hat{R}_i}{\hat{R}_i}\right)\right]^{-1} \tag{10.9}$$

where $Sh = k_G R/D_{e,As}$ and $Da = k_s R/D_{e,As}$. The equation takes an implicit form for reaction orders different from 1.

SCM is a phenomenological model that predicts the total conversion of a solid in a finite time and is well suited for many practical systems. However, it cannot account for such features as the leveling off of conversion at a value lower than the total conversion. Most importantly, it is not suitable for porous solids.

Volume reaction model

Volume reaction model

When the solid is porous, the reaction occurs throughout the pellet with no sharp interface. If diffusion is assumed to be fast, the gas concentration will be uniform throughout the pellet, leading to the so-called homogeneous model. The rate of reaction can then be simply written as

$$r_A = k_V [A]^m [B]^n \tag{10.10}$$

The general conservation equations for the solid and reactant species for the volume reaction model in dimensionless form are as follows:

$$\nabla^2[\hat{A}] = \phi^2[\hat{A}]^m[\hat{B}]^n \tag{10.11}$$

$$-\frac{d[\hat{B}]}{d\hat{t}} = [\hat{A}]^m[\hat{B}]^n \tag{10.12}$$

where

$$\phi = \left[\frac{k_V R^2 [A]_b^{m-1} [B]_0^n}{D_{e,As}}\right]^{1/2} \tag{10.13}$$

and

$$\hat{t} = \upsilon k_s [A]_b^m [B]_0^{n-1} t \tag{10.14}$$

The boundary conditions are

$$\hat{t} = 0,\ \hat{R} = 1:\ [\hat{B}] = 1$$

$$\hat{t} > 0,\ \hat{R} = 1: \frac{d[\hat{A}]}{d\hat{R}} = Sh(1 - [\hat{A}]) \text{ or } [\hat{A}] = 1 \tag{10.15}$$

$$\hat{t} > 0,\ \hat{R} = 0: \frac{d[\hat{A}]}{d\hat{R}} = 0$$

No analytical solution for this set of equations is possible for arbitrary values of m and n. However, analytical solutions can be found for some simplified situations. For example, for the homogeneous model corresponding to low values of ϕ (and hence uniform concentration of A throughout the pellet), the solution is

$$X_B = \begin{cases} 1 - \exp(-\hat{t}) & \text{for } n = 1 \\ \hat{t} & \text{for } n = 0 \end{cases} \tag{10.16}$$

A practically important case is when m =1. In this situation, Equations 10.11 and 10.12 can be combined into a single equation by defining a cumulative gas concentration as

$$\psi = \int_0^{\hat{t}} [A] d\hat{t} \tag{10.17}$$

The transformed equation for the case of $m = n = 1$ is (del Borghi et al., 1976; Dudukovic and Lamba, 1978; see also Ramachandran and Kulkarni, 1980)

$$\nabla^2 \psi = \phi^2 [1 - \exp(-\psi)] \tag{10.18}$$

with boundary conditions

$$\text{at } \hat{R} = 1,\ \ \psi(0,\hat{t}) = \hat{t}, \text{ at } \hat{R} = 0, \left(\frac{d\psi}{d\hat{R}}\right) = 0 \tag{10.19}$$

This transformation is an extremely useful tool for systems with no structural changes and for reactions with power-law kinetics.

The importance of reaction orders m and n has been examined at length (see, e.g., Doraiswamy and Sharma, 1984). The case of $m = 1$, $n = 0$ (i.e., zero order with respect to the solid reactant) is particularly important, since the gas concentration can drop to zero within the pellet, depending on the value of ϕ. In fact, a critical value given by

$$\phi_{cr} = \frac{6}{2/Sh + 1} \tag{10.20}$$

exists beyond which the concentration of A can fall to zero at some point within the pellet. For $\phi < \phi_{cr}$, the concentration would be finite at all points in the pellet and Equation 10.17 describes the conversion-time behavior.

Zone models

Zone models

The homogeneous model behaves in part as an SCM when the reaction–diffusion interaction is such that the outer layers become exhausted leading to the formation of an ash layer as in SIM. The difference, however, is that the reaction is not topochemical, that is, it is not confined to the interface, but occurs throughout the reactant matrix (core) as in the homogeneous model. Ishida and Wen (1968) have derived equations for this so-called two-zone model. A more general model is, however, one in which a reaction zone is sandwiched between the ash layer and the unreacted core (Bowen and Cheng, 1969). The model, sketched in Figure 10.2, is characterized by three stages (Mantri et al., 1976): (1) zone formation starting from the pellet surface till it has reached a thickness determined by the reaction–diffusion interaction for the system, (2) zone travel to the interior leaving a layer of ash as the shell, and (3) zone collapse as it merges with the core (thus becoming a two-zone model), the reaction continuing in the core till the entire solid is exhausted. The main features of this model have been studied by Mantri et al. (1976). The experimental results of Prasannan and Doraiswamy (1982) on the oxidation of zinc sulfide reveal all the three stages of the reaction. The zone width is clearly a function of the Thiele modulus. The chief feature of this model is that when the zone thickness is zero, it reduces to SIM, and when it is of the pellet dimension it reduces to the homogeneous model (Figure 10.3).

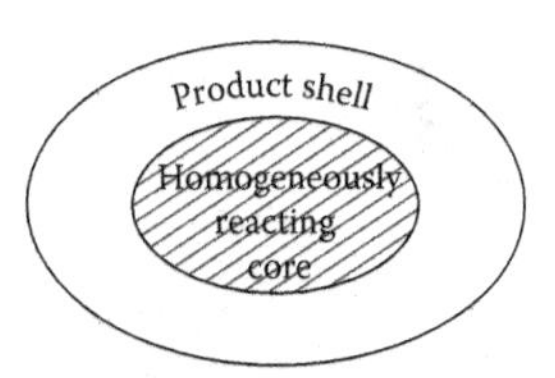

Figure 10.2 The two-zone model for gas–solid noncatalytic reactions.

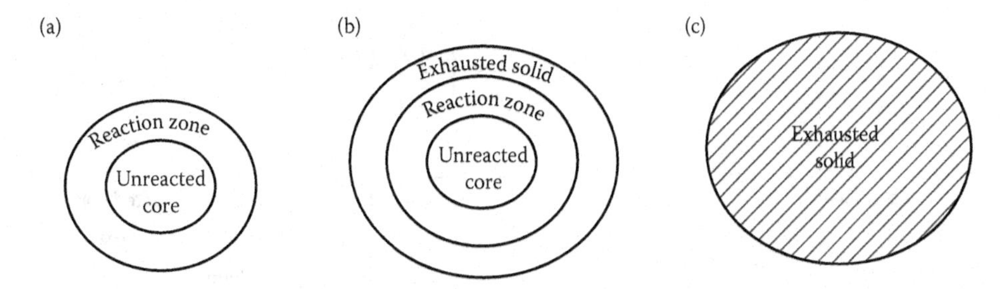

Figure 10.3 The three-zone model for gas–solid noncatalytic reactions. (a) Stage 1: Zone formation, (b) Stage 2: Shell of ash layer forms, (c) Stage 3: Reaction zone merges with core.

Grain model

The particle–pellet or grain models

Although this class of models is based on an explicit recognition of the granular structure of the pellet, the first version did not account for changes in the grain size. Hence they can broadly be considered as spanning the macroscopic models in which pore evolution is ignored and the structural models in which the progress of reaction is explicitly related to pore evolution with time (i.e., to structural changes). The basic feature of these models, sketched in Figure 10.4, is that the grains constituting a pellet are spherical and of the same size, that each grain reacts according to SIM and that the size of the grain does not change with reaction (thereby implying no voidage change with reaction and hence no pore evolution).

The mathematical formulation of the model requires consideration of the rate processes within an individual grain, and the overall mass balance for the gaseous reactant in the pellet and its stoichiometric relationship with the extent of solid consumed. As far as the individual grain is concerned, the rates of diffusion through the reacted portion of the grain and of reaction at the interface can be obtained in analogy with Equations 10.2 and 10.3 as

$$r_{GA} = \frac{4\pi D_{eG} r_{Gi} r_{G0}}{r_{Gi} - r_{G0}}([A]_s - [A]_i) \tag{10.21}$$

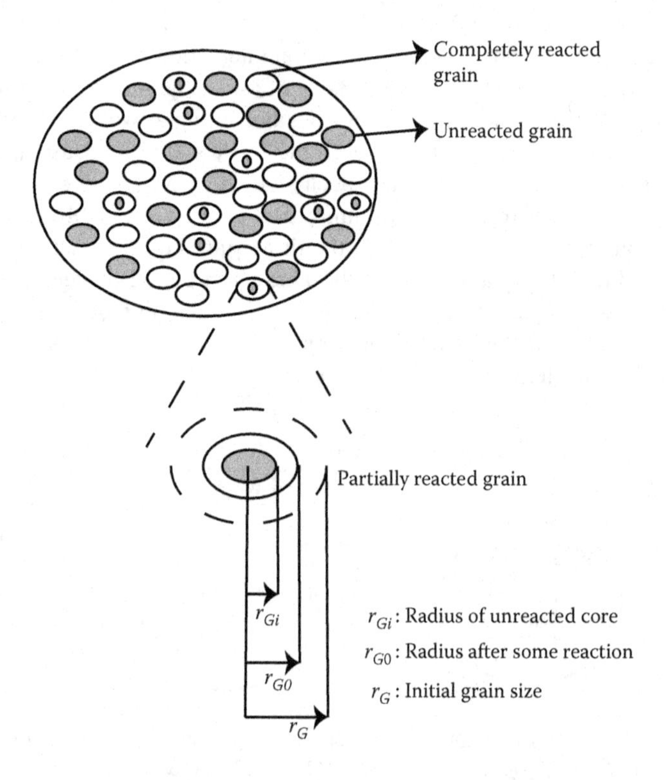

Figure 10.4 The particle pellet model for gas–solid noncatalytic reactions.

$$r_{GA} = 4\pi r_{Gi}^2 k_s [A]_i \tag{10.22}$$

Eliminating the unknown concentration $[A]_i$ at the interface, we obtain the overall rate per unit grain as

$$r_{GA} = \frac{4\pi r_{Gi}^2 k_s [A]}{\left(1 + \dfrac{r_{Gi} k_s}{D_{eG}}\right)\left(1 - \dfrac{r_{Gi}}{r_{G0}}\right)} \tag{10.23}$$

Once the rate of reaction for the individual grain is known, we can proceed to write the overall pellet equation as

$$\nabla^2 [A] = r_{GA} \frac{3(1-\varepsilon_p)}{4\pi r_{G0}^3} \tag{10.24}$$

where the term in parentheses refers to the grains in the pellet volume. The term r_{GA} involves a knowledge of the interfacial position r_{Gi} within each grain, which is a function of both the time and position within the pellet. To evaluate this, a stoichiometric balance on the solid reactant B can be written for an individual grain in analogy with Equation 10.7 as

$$-\frac{d}{dt}\left[\frac{4}{3}\pi r_{Gi}^3 \frac{\rho_B}{M_B}\right] = r_{GA} \tag{10.25}$$

The term $[A]$ appearing in r_{GA} (Equation 10.24) in this equation fixes the position of the individual grain in the pellet.

The appropriate boundary conditions to the problem are

$$\begin{aligned} r = R &: D_{e,As} \frac{d[A]}{dr} = k_G([A]_b - [A]_S) \\ r = 0 &: \frac{d[A]}{dr} = 0 \\ t = 0 &: r_{Gi} = r_{G0} \end{aligned} \tag{10.26}$$

In general, this set of equations requires numerical solution.

Considering the physical features of the model, two parameters are involved: τ_G, the time required for complete conversion of the grain in the $[A]$ environment; and τ_p, the time for complete conversion of the particle by diffusion if the grain conversion process is extremely fast. In the limiting case of grain diffusion controlling, the simple homogeneous model is recovered. The individual grains could follow the SCM with ash diffusion or reaction controlling. Because the processes within the grain determine the system behavior, the conversion–time relationship will be independent of the pellet dimensions. On the other extreme, when diffusion within the pellet controls, one would observe shrinking core behavior with ash diffusion control, and typically the system behavior will be dependent on the pellet dimensions ($t \propto R^2$). In the intermediate region where τ_G and τ_p are of the same order of magnitude, one could expect the pellet behavior to lie within the limiting cases of shrinking core with reaction and ash diffusion control.

The grain models are useful in cases where pellets are formed by compaction of particles in very fine sizes. This is not so in some naturally occurring minerals in which case fictitious grains will have to be invoked to apply the model. Also the model, in its simple form, does not explain S-shaped behavior and leveling off of conversion.

Other models

What does topochemical mean?

Several other models have also been proposed, such as the nucleation model. Nucleation effects are often significant in systems such as reduction of metallic oxides. In these systems, the process proceeds with the generation of nuclei, which subsequently grow and finally overlap. When the nuclei generation rate is faster, the whole surface gets covered with the metallic phase and the reaction proceeds topochemically. On the other hand, for a slow generation rate, the metal–oxide interface is irregular, and different considerations prevail in estimating the conversion–time relationships. The following empirical model, proposed by Avrami as far back as 1940, is still surprisingly valid:

$$X_B = 1 - \exp(-at^b) \tag{10.27}$$

where a and b are constants. Modified forms have been suggested by Erofeev (1961), Ruckenstein and Vavanellos (1975), Rao (1979), and Bhatia and Perlmutter (1980), and experimental data provided by Neuberg (1970) and El-Rahaiby and Rao (1979) to validate the model.

Extensions to the basic models

Some basic macroscopic models were described in the previous sections, including the slightly more rigorous particle–pellet model. These models ignored several complexities, mainly the effects of bulk flow, nonisothermicity, and variations in structure due to reaction. The first two can be included in the basic models and are hence regarded as extensions of these models. On the other hand, the effects of structural changes can be better brought out in newer models which incorporate them at a more basic level.

Bulk-flow or volume-change effects

In addition to diffusion, bulk flow can occur within a reacting pellet (Beveridge and Goldie, 1968; Gower, 1971; Sohn and Sohn, 1980). This effect is considerably magnified for reactions with volume change such as

$$C\,(s) + CO_2\,(g) \rightarrow 2CO\,(g) \tag{R2}$$

$$FeCl_2\,(s) + H_2\,(g) \rightarrow Fe\,(s) + 2HCl\,(g) \tag{R3}$$

Note that these reactions are different from those in which the molar volume of the solid itself changes leading to a structural change as the reaction progresses. Both the effects can occur simultaneously as in the second reaction shown above.

For reactions with a change in the gas volume, a continuity equation can be written with appropriate boundary conditions (in analogy with that for catalytic reactions by Weekman and Goring, 1965) and nondimensionalized to incorporate the effect through a dimensionless quantity for volume change. The final asymptotic solution (Sohn and Sohn, 1980) obtained is

$$\begin{aligned}\frac{\ln(1+\theta)}{\theta}\frac{\hat{t}}{\phi^2} &= 1 - \frac{(\beta+1)(1-X_B)^{2/(\beta+1)} - 2(1-X_B)}{(\beta-1)} \\ &= f(X_B)\end{aligned} \tag{10.28}$$

where β is the shape factor and ϕ and θ are, respectively, the Thiele modulus and a volume change modulus defined as

Note that the length parameter of the Thiele modulus is now related to the volume and the surface area of the particle!

$$\phi = \frac{sv_p}{A_p}\sqrt{\frac{k_v}{2D_{e,As}}} \tag{10.29}$$

$$\theta = \left(\frac{v_R}{v_A} - 1\right)X_A \tag{10.30}$$

Clearly, $\theta = 0$ for a reaction with no volume change.

A word of caution is necessary in extending the above analysis to nonisothermal situations. It will be recalled that the SIM equations are based on the applicability of the PSS assumption. This assumption is not valid for heat flow. Hence, these equations are not applicable to nonisothermal reactions.

Effect of temperature change

Although, in general, SIM cannot be applied to analyze a nonisothermal reaction, it has been found to be well suited for a certain class of reactions, that is, decomposition reactions (Narasimhan, 1961; Hills, 1968; Campbell et al., 1970) such as

$$A\,(s) \rightarrow R\,(s) + S\,(g) \tag{R4}$$

The process typically yields SIM behavior and is controlled either by heat or gas diffusion through the product layer. For heat transfer through the product layer controlling, the interface stays isothermal and the equation for SIM with $D_{e,As}$ replaced by the corresponding heat transfer parameters in the definition of τ represents the conversion–time behavior. Where gas diffusion is controlling, the variation of $D_{e,As}$ with temperature should be accounted for. This variation usually takes the form $D_{e,As} = T^{1.5-2.0}$ in the bulk diffusion regime with $D_{e,As} = T^{0.5}$ in the Knudsen regime. Luss and Amundson (1969) have provided a more rigorous analysis that incorporates the transient heat accumulation term and gives the interface temperature as a function of the interfacial position r_i.

Structural variations

Models that account for structural variations

The main structural changes that occur in a solid are those due to reaction and sintering.

Effect of reaction

The reaction effect is mainly due to the difference in molar volumes of the product and reactant solids, leading to voidage and therefore diffusivity changes as the reaction progresses. To incorporate these effects in any model, it is necessary to relate the overall solids conversion to voidage and diffusivity. An important feature of the structural effect is that when the porosity at the surface of the solid becomes zero (pore closure), the governing equations predict incomplete conversion, so often observed in gas–solid reactions (and not predicted by the basic models).

One way to account for structural changes is to allow for changes in the grain size in the particle–pellet model (Garza-Garza and Dudukovic, 1982a,b). A more useful way is to incorporate the effect through changes in pore size distribution. The simplest of such models is the single pore model of Ramachandran and Smith (1977a) and Chrostowski and Georgakis (1978). In the Ramachandran–Smith approach, changes in a single pore are assumed to reflect changes in the pellet as a whole. The pore contracts, expands, or remains unchanged depending on whether there is an increase, decrease, or no change in the solid volume due to reaction. The model yields a simple conversion–time relationship based on the knowledge of the pore radius and length and the radius of the associated solid. Ulrichson and Mahoney (1980) have extended this model to incorporate the effects of bulk flow and reversibility of the reaction.

Perhaps the most realistic model is the random pore model of Bhatia and Perlmutter (1980, 1981a,b, 1983). This model assumes that the actual reaction surface of the reacting solid B is the result of the random overlapping of a set of cylindrical pores. Surface development as envisaged in this model is illustrated in Figure 10.5. The first step in model development is therefore the calculation of the actual reaction surface, based on which the conversion–time relationship is established in terms of the intrinsic structural properties of the solid. *In the absence of*

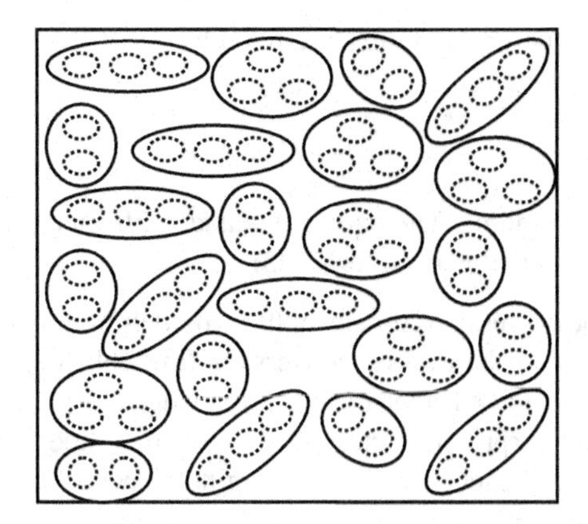

Figure 10.5 The random pore model of gas–solid noncatalytic reactions: stages in surface development.

intraparticle and boundary layer resistances, the following final relationship is obtained:

$$\hat{t} = \frac{k_s [A]_b^m S_0 t}{1 - \varepsilon_0} = \int \frac{dX_{B1}}{S^*(X_{B1})}$$

$$= \frac{\beta'}{2} \int_0^{X_{B1}} \int_0^{X'_B} \left[\frac{1}{S^{*2}(X'_B)} + \frac{Z_v - 1}{S_P^{*2}(X'_B)} \right] dX'_B \, dX_B \qquad (10.31)$$

where X_{B1} and X'_B are dummy variables, S^* and S_p^* refer, respectively, to dimensionless reaction surface area and pore surface area, and

$$\beta' = \frac{2 k_s v_A \rho_B (1 - \varepsilon_0)}{v_B M_B D_{e,As} S_0} \qquad (10.32)$$

This characterizes the diffusional resistance to the flow of gas (zero for kinetic control and infinity for product layer diffusion control). Expressions for S^* and S_p^* depend on the reaction model used. Thus, for the grain model,

$$S^*(X_B) = (1 - X_B)^g \qquad (10.33)$$

$$S_p^*(X_B) = [1 + (Z_v - 1) X_B]^g \qquad (10.34)$$

where g is the grain shape factor (2/3 for the sphere, 1/2 for the cylinder, and 0 for the flat plate). For the random pore model,

$$S^*(X_B) = (1 - X_B)\sqrt{1 - \psi' \ln(1 - X_B)} \qquad (10.35)$$

$$S_p^*(X_B) = [1 + (Z_v - 1) X_B]\sqrt{1 - \psi' \ln[1 + (Z_v - 1) X_B]} \qquad (10.36)$$

where ψ' is a structural parameter defined by

$$\psi' = \frac{1}{\ln(1/(1-\varepsilon_0))} \tag{10.37}$$

for uniform pore radius. The equation gets complicated for nonuniform radius (see Bhatia and Perlmutter, 1983).

Substituting the expressions for any of these models in Equation 10.30 and integrating leads to the desired conversion–time relationship. Although the random pore model appears more realistic, the predictions of the grain model are surprisingly close to those of this model. A number of improvements and extensions, many marginal, have been suggested (see Bhatia and Gupta, 1992).

Effect of sintering

The use of high temperatures in certain reactions such as those in gas cleaning using lime-based adsorbents, or generation of large amounts of heat in exothermic reactions, leads to sintering of the solid. The effect becomes more severe at higher temperatures (usually over 800 K). This can happen because of a decrease in the effective diffusivity of the solid or an increase in grain size leading to a lowering of its specific area. Also, there could be a decrease in porosity and an increase in the tortuosity factor, both leading to a lowering of the effective diffusivity. Simple empirical laws have been used to account for sintering, such as exponential decay for diffusivity and first-order decay for surface area (Ranade and Harrison, 1979, 1981). The combined effects of the two have been considered by Kim and Smith (1974), Chan and Smith (1976), and Ramachandran and Smith (1977b). The following equation is recommended:

$$D_e = \frac{1}{F(f_p)}\left[1-(1-\varepsilon_0)\left(\frac{r_{Gi}}{r_{G0}}\right)^3\right]^2 (1-f_p) \tag{10.38}$$

where f_p is the fraction of pores removed and is given by

$$\frac{df_p}{dt} = k_p(1-f_p) \tag{10.39}$$

where k_p is the rate constant for pore removal.

A general model that can be reduced to specific ones

We now develop a general mathematical formulation that should be applicable to most of the models described in the earlier sections. For this, we begin with the volume reaction model described by Equations

Table 10.3 Functional Forms of $f(X_B)$ for Different Gas–Solid Noncatalytic Reaction Models

Reaction Model	**Functional Form $f(X_B)$**
Volume reaction model	$(1 - X_B)^{-n}$
Grain model	$-1 - \frac{(1-X_B)^{1/3} + (1-X_B)^{2/3}}{Sh}$
Grain model with structural variations	$-1 - \frac{(1-X_B)^{1/3}}{Sh} + \frac{(1-X_B)^{2/3}}{Sh\left[Z_v + (1-Z_v)(1-X_B)^{1/3}\right]}$
Nucleation model	$\frac{1}{n}\left(\frac{1}{1-X_B}\right)\ln\left(\frac{1}{1-X_B}\right)^{(1-n)/n}$

10.11 through 10.15. Then, for a reaction first-order in the gaseous component, we recast these equations as

$$\frac{1}{\xi^\beta}\frac{\partial}{\partial\xi}\left(\alpha\xi^\beta\frac{\partial[\hat{A}]}{\partial\xi}\right) = \phi^2\frac{\partial X_B}{\partial\hat{t}} \quad (10.40)$$

$$\frac{dX_B}{d\hat{t}} = \frac{[\hat{A}]}{f(X_B)} \quad (10.41)$$

where α refers to the diffusivity ratio $D_{e,As}/D_{e,As0}$, X_B is the solid conversion, and ϕ is the Thiele modulus (Equation 10.29). Using the cumulative gas concentration defined by Equation 10.17, the following final equation can be developed:

$$f(X_B)\frac{d^2X_B}{d\xi^2} + \beta\frac{f(X_B)}{\xi}\frac{dX_B}{d\xi} + f'(X_B)\left(\frac{dX_B}{d\xi}\right)^2 - k(X_B) = 0 \quad (10.42)$$

The solution of this equation directly gives the conversion profiles. The equation is sufficiently general, since several functional forms representing different models can be chosen. Typical forms of $f(X_B)$ for some of the more frequently used models are presented in Table 10.3.

Prasannan et al. (1986) have employed a collocation procedure to solve this equation for the models considered in the table.

Gas–solid noncatalytic reactors

As in the case of gas–solid catalytic reactors, here also it is common practice to use fixed-bed, fluidized-bed, or moving-bed reactors. However, since all gas–solid noncatalytic reactions are inherently time-dependent, time becomes an unavoidable parameter in the analysis. We briefly outline the procedures for the three reactor types mentioned, and also touch upon a few other types. For a comparative evaluation of fixed- and fluidized-bed reactors, the reader is referred to Dutta and Gualy (1999).

Fixed-bed reactors

Examples of the use of fixed-bed reactors are roasting and sintering of ores, incineration of solid wastes, reduction of metal oxides, and production of light-weight aggregates. They can be of the conventional types as shown in Figure 10.6a in which a tube is packed with the reactant pellets and the gas passed through the bed. An alternative design, not used in catalytic reactors, is shown in Figure 10.6b. Here, the solids are continuously fed on a moving grate and the gas is blown through the bed. The reaction front (which moves with the reaction progress) is marked in both the cases. The conventional design, by its very nature, is restricted to batch reactions and hence is not applicable to large-scale production.

In an ideal fixed-bed reactor, plug flow of gas is assumed. This is not a good assumption for reactive solids because the bed properties would vary with position, mainly due to changing pellet properties (and dimensions in most cases), and hence the use of nonideal models is often necessary. The dispersion model, with all its limitations, is still the most practical model to use. The equations involved are quite cumbersome, but the asymptotic solutions to these equations are simple, particularly for systems conforming to pellet reaction control. Many reactions can, under the practical conditions of operation, be assumed to correspond to this limiting case. Further, their use is especially appropriate for the following reasons. As brought out earlier, the controlling regime in a reactive pellet changes as the reaction progresses. For example, in the case of SIM, the

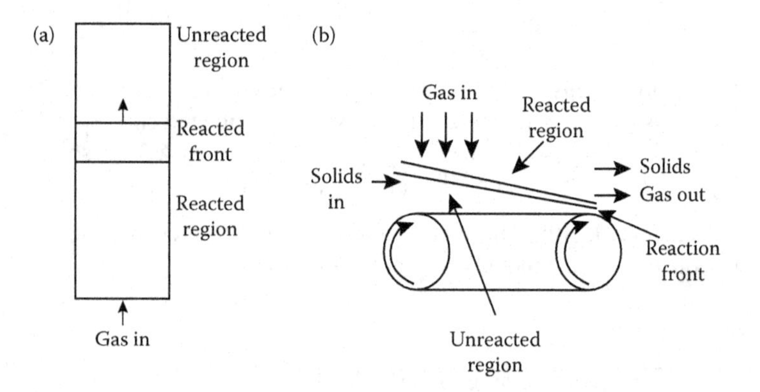

Figure 10.6 Reactors for gas–solid noncatalytic reactions. (a) Packed bed and (b) moving bed.

reaction usually starts with reaction control but eventually becomes diffusion-controlled. This makes it necessary to consider individual pellets in the reactor explicitly. Thus, pseudo-homogeneous models cannot be used and recourse to one- or two-dimensional heterogeneous models becomes necessary. In such cases, the choice of the gas–solid reaction model has an important bearing on reactor simulation (Sotirchos and Zarkanitis, 1989). Mutasher et al. (1989) have simulated a fixed-bed reactor using the zone model. We use the particle–pellet model because of its greater generality. Consider a typical reaction

$$A\,(\mathrm{g}) + B\,(\mathrm{s}) \rightarrow R\,(\mathrm{g}) + S\,(\mathrm{s}) \qquad \text{(R5)}$$

It can be assumed that the pellet itself is isothermal even for reactions involving large heat effects. One can also reasonably assume that axial diffusion effects are absent (particularly for long beds), and hence the one-dimensional model can be used. We shall also assume that the heat generated by reaction is lost through the sensible heat carried by the gas leaving the system and by convection at the outside surface of the reactor wall. The nonisothermicity of the reaction greatly affects the linear velocity and density of gas along the reactor, and these affects must be taken into account. However, in our relatively simple approach, we ignore these effects.

For a complete derivation of the model equation, reference may be made to the original article (Evans and Song, 1974). The following equations are written: material and energy balances for the gas phase, and the rate equations for the pellet assuming any one of the models described in this chapter earlier. As mentioned above, we use the particle–pellet model. Appropriate initial and boundary conditions are also written. Assuming isothermal behavior, the material balance equation is

$$-G_A \frac{\partial c}{\partial z} = \frac{(1-\varepsilon_P)(1-\varepsilon_b)\rho_{Bm}}{\upsilon}\frac{dX^x}{dt} + (\varepsilon_P + \varepsilon_b)[\bar{A}]\frac{\partial c}{\partial t} \qquad (10.43)$$

where $[\bar{A}]$ is the external field concentration at some point in the bed and c is the dimensionless external field concentration. The changing value of the concentration $[\bar{A}]$ with position allows for a more realistic simulation of the reactor.

Note that the reactor equation is general and independent of the rate equation, which depends on the gas–solid reaction model used. It can also be adapted to other reactors with appropriate changes/simplifications. However, this involves unacceptable assumptions for the fluid-bed model, like ignoring the bubble phase. We therefore do not apply this method to the fluid-bed reactor.

We begin by writing the particle–pellet equation in terms of a generalized effectiveness factor defined as

$$\varepsilon = \frac{\int_0^1 \xi^{s-1} g \eta^{g-1} [\hat{A}]_p d\xi}{\int_0^1 \xi^{s-1} d\xi} \tag{10.44}$$

where s and g are, respectively, the shape factors for the pellet and grain (3 for the sphere, 2 for the cylinder, and 1 for the flat plate), defined by

$$s = \frac{\text{Volume of the shape}}{\text{External area of the shape}}$$

and ξ and η are the dimensionless positions in the pellet and grain, respectively. We use this in the material balance Equation 10.43 recast in dimensionless form to give

$$-\frac{\partial c}{\partial Z^*} = \frac{dX^*}{dt^*} = c\varepsilon \tag{10.45}$$

and then manipulate it into the form

$$\frac{dX^*}{dt^*} = \varepsilon \exp\left[-\int_0^{Z^*} \varepsilon \, d\lambda\right] \tag{10.46}$$

where λ is a dummy variable, and t^* and Z^* are dimensionless time and distance given by

$$t^* = \left(\frac{\upsilon k_s [\bar{A}]_0 S_g}{\rho_{Bm} g V_g}\right) t \tag{10.47}$$

$$Z^* = \frac{k_s [\bar{A}]_0 S_g}{G_A g V_g} (1 - \varepsilon_p)(1 - \varepsilon_b) L \tag{10.48}$$

and X^* is the extent of reaction of the pellet defined by

$$X^* = \frac{\int_0^1 \xi^{s-1} (1 - \eta^g) d\xi}{\int_0^1 \xi^{s-1} d\xi} \tag{10.49}$$

The calculation procedure is straightforward. We compute ε for different values of X_B and a reaction modulus σ defined as

$$\sigma = \frac{sV_p}{A_p} \left[\frac{(1 - \varepsilon_p) k_s}{D_{e,As}} \frac{S_g}{V_g}\right]^{1/2} \frac{1}{(2sg)^{1/2}} \tag{10.50}$$

and may be considered to be a dimensionless pellet size. The precise table given by Evans and Song can be used to estimate σ, but the following approximate equation for a sphere ($s = 3$) is usually adequate (Sohn and Szekely, 1972):

$$\varepsilon \cong \frac{1}{s}\left(1 - X^*\right)^{1/g-1} + \sigma\left[2(1 - X)^{-1/3} - 2\right]^{-1} \tag{10.51}$$

Knowing ε from Equations 10.44 and 10.45 can be integrated and X^* plotted as a function of t^* for a given value of σ.

Several complicating features can be added to these equations such as a different bed porosity at the wall by using the correlation of Chandrasekhar and Vortmeyer (1979), allowing for nonisothermicity of the bed (Sampath et al., 1975), accounting for propagation of the reaction front (Bagajewicz, 1992), and the variation of gas properties with temperature in the bed. These are not necessarily additions to the model used by us but in some cases to other available models, and will not be considered here.

Moving-bed reactors

We use the same approach here as for the fixed-bed reactor but by making allowance for the special features of the moving-bed reactor (Figure 10.7). The main difference is that the solid is also moving and a mass balance

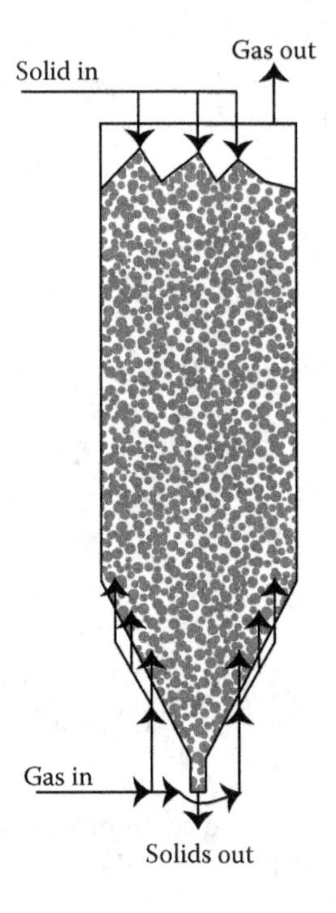

Figure 10.7 A typical moving-bed reactor. (Adapted from Joshi, J.B. and Doraiswamy, L.K., *Chemical Reaction Engineering* in *Chemical Engineers' Handbook*, CRC Press, Boca Raton, FL, 2009.)

equation for the solid phase is therefore needed—both for plug flow and complete mixing of the solids. The following development is based on the more likely plug-flow behavior.

The equations for the gas and solid phases are

$$\text{Gas:} \quad \frac{dc}{dZ^*} = -c\varepsilon \tag{10.52}$$

$$\text{Solid:} \quad \frac{dX^*}{dZ^*} = \upsilon\left(\frac{G_A}{G_B}\right)c\varepsilon = R'c\varepsilon \tag{10.53}$$

The boundary conditions are

$$c = 1, \; Z^* = 0 \tag{10.54}$$

and

$$\begin{aligned} &\text{for co-current flow: } X^* = 0, Z^* = 0 \\ &\text{for counter-current flow: } X^* = 0, Z^* = L^* \end{aligned} \tag{10.55}$$

where L^* is a dimensionless reactor length defined for co-current flow defined as

$$L^* = \left[\frac{k_s[\bar{A}]_0 S_g(1-\varepsilon_p)(1-\varepsilon_b)}{G_A g V_g}\right]L \tag{10.56}$$

From stoichiometry

$$X^* - X_0^* = R'(c-1) \tag{10.57}$$

Substituting Equation 10.57 into 10.53 and integrating, we obtain

$$\text{for co-current flow: } Z^* = \int_0^{X^*} \frac{d\chi}{(R'-1)\varepsilon(x)} \tag{10.58}$$

where χ is a dummy variable, and

$$\text{for counter-current flow: } L^* = \int_0^{X_0^*} \frac{dX^*}{(R' - X^* + X_0^*)\varepsilon(X^*)} \tag{10.59}$$

Using Equation 10.44 for ε, these equations can be solved to give c and X^* as functions of Z^* or L^* for given values of σ.

A comparison of the plots (not included) shows that counter-current operation gives significantly higher conversions. This is in consonance with the superior performance of counter-current flow in general.

Fluidized-bed reactors

As in the case of the other two reactors considered so far, the prerequisite for the design of a fluidized-bed reactor is a reaction model for the single pellet. Single-pellet models were already discussed in this chapter. The main features of the fluidized bed was outlined in detail in Chapter 9 and hence is excluded here. What is peculiar to the fluidized-bed reactor for gas–solid noncatalytic reactions is the particle size distribution, which changes with reaction. This happens mainly because the particle density changes with reaction. Thus, an important design feature is the prediction of particle size distribution in the product solids from a knowledge of the distribution in the reactant solids. It is also necessary to make allowance for elutriation of fines and their partial return to the reactor. Particle growth or shrinkage during reaction will also affect the particle size distribution and fluidization characteristics of the bed and add further complexity to the design. This complexity can also arise in fixed- and moving-bed reactors but are not as crucial. In the foregoing treatment of fixed- and moving-bed reactors, it was assumed that the particle size remains constant. Nondecaying catalytic systems can be operated in the batch mode with respect to the solids, but the consequences of reaction and solids consumption often require that noncatalytic gas–solid reactions be operated in the continuous mode.

References

Avrami, M., *J. Chem. Phys.*, **8**, 212, 1940.
Bagajewicz, M., *Chem. Eng. Commun.*, **112**, 145, 1992.
Beveridge, G.S.G. and Goldie, P.J., *Chem. Eng. Sci.*, **23**, 912, 1968.
Bhatia, S. and Gupta, J.S., *Rev. Chem. Eng.*, **8**, 177, 1992
Bhatia, S.K. and Perlmutter, D.D., *AIChE J.*, **26**, 379, 1980.
Bhatia, S.K. and Perlmutter, D.D., *AIChE J.*, **27**, 226, 1981a.
Bhatia, S.K. and Perlmutter, D.D., *AIChE J.*, **27**, 247, 1981b.
Bhatia, S.K. and Perlmutter, D.D., *AIChE J.*, **29**, 287, 1983.
Bowen, J.H. and Cheng, C.K., *Chem. Eng. Sci.*, **24**, 1829, 1969.
Briggs, R.A. and Sacco, A., Jr., *J. Mater. Res.*, **6**, 574, 1991.
Campbell, R.R., Hills, A.W.D., and Paulin, A., *Chem. Eng. Sci.*, **25**, 929, 1970.
Chan, S.F. and Smith, J.M., *Indian Chem. Eng.*, **18**, 42, 1976.
Chandrashekar, B.C. and Vortmeyer, D., *Warme-Stoff.*, **12**, 105, 1979.
Chrostowski, J.W. and Georgakis, C., *ACS Symp. Ser.*, 65 (*Chem. React. Eng.*), Houston, TX, 225, 1978.
del Borghi, Dunn, J.C., and Bischoff, K.B., *Chem. Eng. Sci.*, **31**, 1065, 1976.
Doraiswamy, L.K. and Kulkarni, B.D., in *Chemical Reaction and Reactor Analysis* (Eds. Carberry, J.J., and Varma, A.), Marcel Dekker, New York, NY, 1987.
Doraiswamy, L.K. and Sharma, M.M., In *Heterogeneous Reactions: Analysis, Examples, and Reactor Design*, Vol. 1: *Gas–Solid and Solid–Solid Reactions*, Wiley, New York, NY, 1984.
Dudukovic, M.P. and Lamba, H.S., *Chem. Eng. Sci.*, **33**, 303, 1978.
Dutta, S. and Gualy, R., *Hydrocarbon Processing*, **78**, 45, 1999.
El-Rahaiby, S.K. and Rao, Y.K., *Metall. Trans. B*, **10**, 257, 1979.
Erofeev, B.V., in J.H. deBoer (Ed.) *Reactivity of Solids*, Proc. 4th International Symposium on Reactivity of Solids, pp. 273–282, Elsevier, Amsterdam, 1960.

Evans, J.W. and Song, S., *Ind. Eng. Chem. Process. Des. Dev.*, **13**, 146, 1974.
Garza-Garza, O. and Dudukovic, M.P., *Comp. Chem. Eng.*, **6**, 131, 1982a.
Garza-Garza, O. and Dudukovic, M.P., *Chem. Eng. J.*, **24**, 35, 1982b.
Gower, R.C., *A Determination of the Transport-limited Reaction Rate for a Gas–Solid Reaction Forming a Porous Reaction Product*, Lehigh University, 1971.
Hills, A.W.D., *Chem. Eng. Sci.*, **23**, 297, 1968.
Ishida, M. and Wen, C.Y., *AIChE J.*, **14**, 311, 1968.
Joshi, J.B. and Doraiswamy, L.K., *Chemical Reaction Engineering* in *Chemical Engineers' Handbook* (Ed. L.F. Albright), CRC Press, Boca Raton, FL, 2009.
Kim, K.K. and Smith, J.M., *AIChE J.*, **20**, 670, 1974.
Kulkarni, B.D. and Doraiswamy, L.K., *Mass Transfer and Reactor Design* (Ed. Cheremisinoff, N.P.), Gulf Publishing House, Houston, TX, 1986.
Luss, D. and Amundson, N.R., *AIChE J.*, **13**, 759, 1967; *AIChE J.*, **15**, 194, 1969.
Mantri, V.B., Gokarn, A.N., and Doraiswamy, L.K., *Chem. Eng. Sci.*, **31**, 779, 1976.
Mutasher, E.I., Khan, A.I., and Bowen, J.H., *Ind. Eng. Chem. Res.*, **28**, 1150, 1989.
Narasimhan, G., *Chem. Eng. Sci.*, **16**, 7, 1961.
Neuberg, H.J., *Ind. Eng. Chem. Proc. Des. Dev.*, **9**, 285, 1970.
Prasannan, P.C. and Doraiswamy, L.K., *Chem. Eng. Sci.*, **37**, 925, 1982.
Prasannan, P.C., Ramachandran, P.A., and Doraiswamy, L.K., *Chem. Eng. J.*, **33**, 19, 1986.
Ramachandran, P.A. and Doraiswamy, L.K., *AIChE J.*, **28**, 881, 1982.
Ramachandran, P.A. and Kulkarni, B.D., *Ind. Eng. Chem. Proc. Des. Dev.*, **19**, 717, 1980.
Ramachandran, P.A. and Smith, J.M., *Chem. Eng. J.*, **4**, 137, 1977a.
Ramachandran, P.A. and Smith, J.M., *AIChE J.*, **23**, 353, 1977b.
Ranade, P.V. and Harrison, D.P., *Chem. Eng. Sci.*, **34**, 427, 1979.
Ranade, P.V. and Harrison, D.P., *Chem. Eng. Sci.*, **36**, 1079, 1981.
Rao, Y.K., *Met. Trans. B.*, **10**, 243, 1979.
Ruckenstein, E. and Vavanellos, T., *AIChE J.*, **21**, 756, 1975.
Sahimi, M., Gavalas, G.R., and Tsotsis, T.T., *Chem. Eng. Sci.*, **45**, 1443, 1990.
Sampath, B.S., Ramachandran, P.A., and Hughes, R., *Chem. Eng. Sci.*, **30**, 125, 1975.
Sohn, H.Y. and Sohn, H.J., *Ind. Eng. Chem. Process Des. Dev.*, **19**, 237, 1980.
Sohn, H.Y. and Szekely, J., *Chem. Eng. Sci.*, **27**, 763, 1972.
Sotirchos, S.V. and Zarkanitis, S., *AIChE J.*, **35**, 1137, 1989.
Szekely, J., Evans, J.W., and Sohn, H.Y., *Gas–Solid Reactions*, Academic Press, New York, NY, 1976.
Ulrichson, D.L. and Mahoney, D.J., *Chem. Eng. Sci.*, **35**, 567, 1980.
Weekman, W.W., Jr. and Gorring, R.L., *J. Catal.*, **4**, 260, 1965.

Chapter 11 Gas–liquid and liquid–liquid reactions and reactors

Chapter objectives

After successful conclusion of this chapter, the students must be able to

- Classify the gas–liquid reactions according to the relative rates of the reaction and mass transfer.
- Use Hatta number to determine the controlling regime in a fluid–fluid reaction system.
- Identify the controlling regimes in the gas–liquid reaction systems in terms of film mass transfer, reaction, and diffusion rates.
- Select the most suitable reaction equipment to minimize the dominant resistance or controlling regime.

Introduction

Fluid–fluid (gas–liquid and liquid–liquid) reactions are of great industrial importance and contribute to more than 70% of industrial applications. For gas–liquid, liquid–liquid, gas–liquid–solid (noncatalytic and catalytic), and solid–liquid systems, examples of industrial importance have been given by many authors.

Now, let us identify the problems:

1. Two distinct phases create two distinct reaction environments.
2. These distinct regimes may act like, and should be treated as independent reactors.
3. The rates depend on mass transfer through the films and the kinetics: the unknown regimes make the design process complicated.

These problems also prevail in the liquid–liquid reactions and to some extent in gas–liquid–solid reactors.

The general physical picture characterizing fluid–fluid reactions is sketched in Figure 11.1. A is the solute in phase 1 (gas or liquid) and

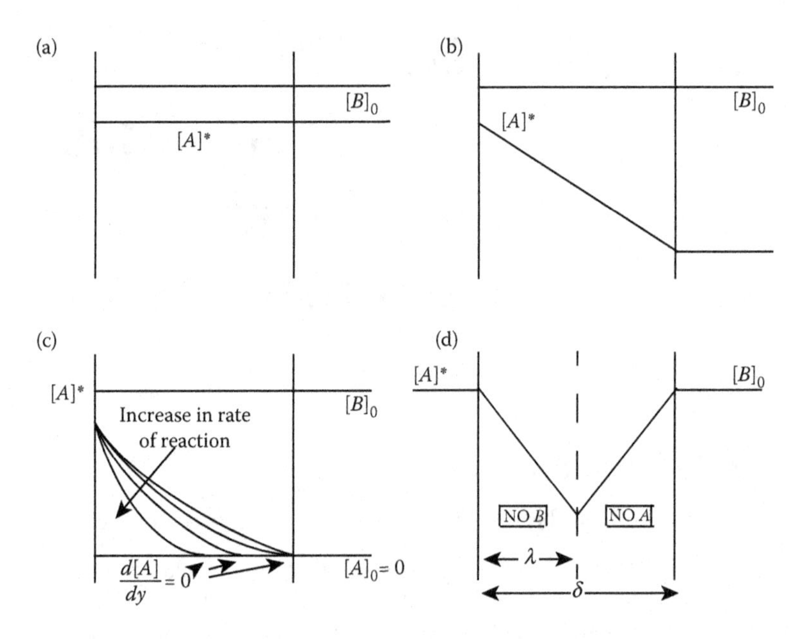

Figure 11.1 The interface behavior of gas–liquid reactions: (a) regime 1, (b) regime 2, (c) regime 3, and (d) regime 4. (From Joshi, J.B. and Doraiswamy, L.K., *Chemical Reaction Engineering in Chemical Engineer's Handbook* (Ed. L.F. Albright), CRC Press, Boca Raton, FL, 2009.)

is slightly soluble in phase 2, which is always a liquid. Upon entering phase 2, A reacts with B present in that phase. Where phase 1 is a gas, the reaction is almost always restricted to phase 2 (except in the rare case of a desorbing product reacting in the gas phase); but where phase 1 is a liquid or a solid, the reaction can occur in both the phases. We consider in this chapter the case where the reaction is confined to phase 2, that is, gas–liquid reactions. Since two phases are present, mass transfer across the interface is clearly an important consideration. Therefore, the basis of the analysis is the interaction between mass transfer and reaction, leading to the formulation of conditions and rate expressions for reactions with varying roles of the two processes (i.e., with different controlling regimes).

Consider a reaction of the general form

$$\upsilon_A A\ (g) + \upsilon_B B\ (l) \rightarrow R \qquad \text{(R1)}$$

with $\upsilon_A = 1$. Our objective is to examine the effect of the chemical reaction on mass transfer. Depending on the relative rates of mass transfer and chemical reaction, the following regimes of control can be postulated:

Very slow reactions:	regime 1
Slow reactions:	regime 2
Fast reactions:	regime 3
Instantaneous reactions:	regime 4

The distinction between these regimes reflects itself in the mathematical description of the problem. Of the very slow and slow reactions, the reaction rate is comparable to the rate of mass transfer by diffusion; therefore, it is accounted for in the differential equation describing the physical situation. As the reactions get faster, the kinetics prevails at the interface between the phases. Under these conditions, the reaction rate must be defined in the boundary conditions. We dealt with the mathematics of these in Chapter 6 in detail. We will focus on the engineering aspects in this chapter.

Obviously, the distinction between regimes cannot be as sharp as the above classification would indicate; but this deficiency can be overcome by accounting for overlaps of regimes and formulating appropriate conditions for different overlaps, such as between regimes 1 and 2, 2 and 3, and 1, 2, and 3.

Where the phase 1 reactant is a pure gas, no gas film resistance is involved. On the other hand, where it is a mixture, gas film resistance must also be accounted for. The film theory of mass transfer is the simplest and most extensively used, in spite of its many limitations. Our treatment too will be based on this theory (see Chapter 6), although a reference will be made to the penetration theory in some cases.

In the absence of the reaction, we have pure mass transfer of A across the film. The specific rate of this reaction is given by

$$r_A' = k_L'([A]^* - [A]_b) \tag{11.1}$$

where $[A]^*$ is the concentration of A at the interface of phases 1 and 2, and is equal to the solubility of A in phase 2; and $[A]_b$ is the concentration in the bulk of phase 2. The rate r_A' has the units of moles per centimeter square second (or other consistent units), and k_L' is the conventional phenomenological mass transfer coefficient with units of centimeter per second. If the rate r_A' is multiplied by the interfacial area per unit volume a_L, then we have

$$r_A' a_L = r_A = k_L' a_L([A]^* - [A]_b) \tag{11.2}$$

where $r_A = r_A' a_L$ is the volumetric rate of mass transfer with units of moles per centimeter cube second, and $k_L' a_L$ $(= k_L)$ has the unit liter per second.

The chemical reaction rate for a reaction of arbitrary order both with respect to the gas and liquid phase components is given by

$$-r_A = bk_{mn}[A]_b^m[A]_b^n \tag{11.3}$$

where b is the liquid phase holdup, that is, the volume fraction of phase 2 in the gas–liquid system, and k_{mn} is the volumetric rate constant for a reaction of orders m and n with respect to A and B, respectively. The entire treatment of gas–liquid reactions, including the determination of the regime of a particular reaction, revolves around the relative values of the rates expressed by Equations 11.2 and 11.3 in one form or the other. The analysis can be practically applied to any multicomponent gas–liquid reaction.

The simple case of an irreversible reaction in a gas–liquid system provides the best basis for a clear exposition of the principles underlying the analysis. The basic treatment of a gas–liquid reaction was given in Chapter 6. There, we derived the balance equations and solved these differential equations for the fast reactions and for the slow reactions, the two extreme cases. For the slow reactions, the rate expression was a part of the differential equation, whereas for the fast reaction, the rate was a part of the boundary condition as the reaction took place at the interface. In the following sections, we present the intermediate cases.

Diffusion accompanied by an irreversible reaction of general order

Diffusion and reaction in series with no reaction in film: Regimes 1 and 2 (very slow and slow reactions), and regimes between 1 and 2

Here, we are concerned with regimes in which the reaction occurs exclusively in the bulk and is controlled either by chemical reaction (regime 1) or diffusion, that is, mass transfer across the liquid film (regime 2), as well as with the intermediate regime in which there is a mass transfer resistance in the film but the reaction still occurs exclusively in the bulk.

Regimes 1 and 2: Very slow and slow reactions Consider the case where the reaction is so slow that the liquid bulk is saturated with diffusing A before any measurable reaction occurs between A and B. For such a situation, the condition to be satisfied is given by

$$k_L' a_L [A]^* \gg b k_{mn} [A]^{*m} [B]_b^n \tag{11.4}$$

and the rate is given by Equation 11.3. This is designated as regime 1 or *very slow-reaction regime* (also known as *kinetic regime*) and is shown in Figure 11.1.

It can also happen that first A diffuses in the film without the reaction and is then fully consumed immediately on reaching the bulk ($[A]_b = 0$). This situation is represented by regime 2 of Figure 11.1. Thus, Equation 11.1 for mass transfer becomes

$$-r_A = k_L' a_L [A]^* \tag{11.5}$$

The reaction is controlled by film diffusion although it is still quite slow and occurs only in the bulk. This is referred to as regime 2 or the *slow-reaction regime*, the condition for which is

$$k_L' a_L [A]^* \ll b k_{mn} [A]^{*m} [B]_b^n \tag{11.6}$$

Another condition for regime 2 is that the amount of A that reacts in the film before reaching the bulk be negligible. For the very slow and slow reactions, the kinetic term resides along with the differential equation describing the transport of the species in the film, such that the definition of Hatta modulus is necessary as was derived in Chapter 6 for a pseudo-first-order reaction. In Chapter 6, we defined the Hatta modulus as

$$M_H = \left(\frac{\text{Characteristic diffusion time}}{\text{Characteristic reaction time}} \right)^{1/2}$$

For a reaction of general order, the Hatta number takes the form

$$M_H = \frac{\sqrt{2/(m+1) D_A k_{mn} [A]^{*(m-1)} [B]_b^n}}{k_L'} \tag{11.7}$$

Thus, the condition for very slow to slow reaction is

$$M_H \ll 1 \tag{11.8}$$

When reaction is very slow, $M_H \ll 1$

Recall that the *enhancement factor* η, as defined in Chapter 6 is a measure of enhancement in mass transfer rate due to reaction:

$$\eta = \frac{\text{Rate of mass transfer in the presence of reaction}}{\text{Rate of mass transfer in the absence of reaction}} = \frac{-r_A}{k_L' a_L [A]^*} \tag{11.9}$$

For a pseudo-first-order reaction, the enhancement factor (see Chapter 6)

$$\eta = \frac{M_{H,1}}{\tanh M_{H,1}} \tag{11.10}$$

For a pseudo-mth-order reaction, the enhancement factor takes the following form:

$$\eta = \frac{M_{H,m}}{\tanh M_{H,m}} \tag{11.11}$$

Note that unlike in the definition of effectiveness factor for catalytic reactions where the normalizing rate was the rate of reaction, here, the normalizing rate is the rate of mass transfer. Thus, the reaction is considered as the *intruder* (albeit a benevolent or *enhancing* one), while for catalytic reactions, diffusion was the intruder (often, but not always, a retarding one).

Regimes between 1 and 2 Let us now visualize a situation where the residual concentration of A in phase 2, $[A]_b$, does not fall to zero, that is, Equation 11.5 does not hold. In this situation, the residual concentration of A is now lying between $[A]^*$ and $[A]_b$. The rate equation for this regime may be obtained by a simultaneous solution of Equations 11.2 and 11.3. The nature of the solution obviously depends on the orders m and n. For a reaction that is of first order in both A and B, the solution is given by

$$-r_A = k'_{LR}a_L[A]^* = \frac{[A]^*}{1/k'_L a_L + 1/bk[B]_b} \tag{11.12}$$

where c is an overall constant expressed as the sum of resistances due to mass transfer and reaction:

$$\frac{1}{k'_{LR}a_L} = \frac{1}{k'_L a_L} + \frac{1}{bk[B]_b} \tag{11.13}$$

Diffusion and reaction in film, followed by negligible or finite reaction in the bulk: Regime 3 (fast reaction), and regime covering 1, 2, and 3

For a conceptual analogy of G–L and G–S reactions, see Kulkarni and Doraiswamy (1975)

We now move to a case where the reaction is so fast that it occurs even while A is diffusing through the film. If it is not completed within the film, the rest of the reaction is completed in the bulk. The situation is similar to that considered for gas–solid catalytic reactions in Chapter 9.

Reaction entirely in film The condition for the reaction to occur entirely in the film would be

When $M_H \gg 1$, the reaction occurs entirely in the film

$$M_H \gg 1 \tag{11.14}$$

Also, if the condition

$$M_H \ll \frac{[B]_b}{\upsilon_B[A]^*}\sqrt{\frac{D_B}{D_A}} \tag{11.15}$$

is satisfied, it can additionally be assumed that the concentration of B is uniform throughout phase 2, that is, there is no depletion of B in the film even though it reacts with A , that is, $[B]^n$ is invariant. This situation is sketched in Figure 11.1 as regime 3. When the reaction occurs in the entire film, it is called a *pseudo-mth-order reaction*, and when it is completed within the film at a distance $x < \delta$, it is usually designated as *fast pseudo-mth order.*

The governing equation for slab geometry is

$$D_A\frac{d^2[A]}{dx^2} = k_{mn}[B]_b^n[A]^m = k_m[A]^m \tag{11.16}$$

where

$$k_m = k_{mn}[B]_b^n \quad (11.17)$$

is the pseudo-mth-order rate constant. The boundary conditions can be readily seen to be

$$x = 0, \quad [A] = [A]^*, \quad d[B]/dx = 0, \quad (11.18a)$$

$$x = \delta, \quad [A] = 0 \quad (11.18b)$$

With some manipulation, the solution in its final form may be found as

$$[A] = \frac{[A]^* \sinh[M_H(1 - (x/\delta))] + [A]_L \sinh(M_H(x/\delta))}{\sinh(M_H)} \quad (11.18c)$$

Under the circumstances, $M_H > 3$ such that the enhancement factor is equal to M_H (see Equation 11.11) and thus, the following equations prevail:

$$-r_A' = k_L'[A]^* M_H = [A]^* \sqrt{2/(m+1) D_A k_{mn} [A]^{*(m-1)} [B]_b^n} \quad (11.19a)$$

For a pseudo-mth-order reaction, $[B]$ does not change during the reaction progress and thus is included in the rate constant such that

$$-r_A' = k_L'[A]^* M_{H,m} = [A]^* \sqrt{2/(m+1)\, D_A k_m [A]^{*(m-1)}} \quad (11.19b)$$

Finally for the pseudo-first-order reaction, the equations take the following form:

$$-r_A' = k_L'[A]^* M_{H,1} = [A]^* \sqrt{D_A k_1} \quad (11.19c)$$

where $M_{H,m}$ and $M_{H,1}$ represent, respectively, the film-to-bulk reaction ratios for pseudo-mth-order and pseudo-first-order reactions. Equations 11.19b and c may also be written as

$$M_{H,m} = \frac{\sqrt{2/(m+1) D_A k_m [A]^{*(m-1)}}}{k_L'} \quad (11.20a)$$

$$M_{H,1} = \frac{\sqrt{D_A k_1}}{k_L'} \quad (11.20b)$$

Reactions both in film and bulk (regimes 1–2–3) Mathematically, Equation 11.16 continues to be valid, but the boundary conditions will be different

$$\begin{aligned} x &= 0, \quad [A] = [A]^* \\ x &= \delta, \quad [A] = [A]_b \end{aligned} \quad (11.21)$$

Measurement of mass transfer coefficients

It must be clear from the various equations developed above that gas–liquid interfacial area is a very important parameter in determining the rate of mass transfer. Any precise measurement of mass transfer coefficient is only possible if the area is correctly known. This is best accomplished by using a stirred cell with a fixed gas–liquid interfacial area, although other experimental reactors such as the wetted wall column, laminar jet, and disk contactor have also been used (see Danckwerts, 1970; Doraiswamy and Sharma, 1984). The two commonly used cell designs are those of Danckwerts (1970) and Levenspiel and Godfrey (1974).

A sketch of the Levenspiel–Godfrey cell is shown in Figure 11.2. It consists of two flanged sections of tubing, one for the upper part (gas phase) and the other for the lower part (liquid phase), with an interface plate between the two sections. Ports are provided for introducing the gas and liquid phases separately. A distinctive feature of this cell is that the actual interfacial area can be independently varied with a high level of precision by providing holes in the plate and controlling their number. The two phases are independently stirred to produce gradientless conditions in each section. As a result, the measured rates directly give single-point values.

Microfluidic devices

The vast advantages offered by microfluidic devices in chemical reaction engineering can also be seen in the multiphase reactor operations. In microfluidic devices, it is possible to create monodisperse bubble sizes. As a result, uniform contact interfaces are possible. Furthermore,

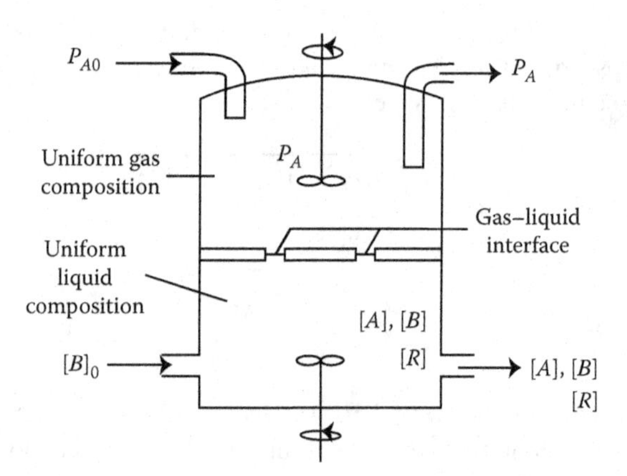

Figure 11.2 A gradientless cell for measuring the mass transfer coefficients with provision for independently varying the operating parameters including the interfacial area. (From Levenspiel, O. and Godfrey, J.H., *Chem. Eng. Sci.*, **29**, 1723, 1974.)

surface and interface forces are no longer negligible, and can assist mixing, such as in Taylor flow reactors. The ability to manipulate the bubble size and mixing within the bubbles can lead to selectivity tuning. This makes the microfluidic devices as very viable options to carry out the gas–liquid and liquid–liquid reactions. A recent review article summarizes the various aspects of using microfluidic devices especially hydrodynamics and the mass transfer (Sobieszuk et al., 2012).

Taylor flows generated by the differences in the flow rates of the different phases provide excellent mixing characteristics. Taylor flow domains shown schematically in Figure II.5 (of Interlude II) eliminates the interfacial gradients and enhances the rates of mass transfer that was inhibited by the interfacial gradients. Indeed, several orders of magnitude greater rates of mass transfer were measured in microfluidic devices (Sobieszuk et al., 2012).

Well-defined flow characteristics in a microfluidic device can also reveal new aspects of mass transfer. For example, it was possible to elucidate the relative rates of the mass transfer at the bubble cups and at the lubricating fluid interface between the gas bubble and the solid wall (see the sidebar figure). Van Baten and Krishna (2004) have demonstrated through computational fluid dynamics (CFD) simulations that the mass transfer rates at the hemispherical bubble cups and the lubricating liquid interfaces of the rising Taylor gas flows contribute differently to the rates of the mass transfer.

One final factor we will mention in this descriptive part of the chapter for improved mass transfer rates in microfluidic devices is the Marangoni effect. Widely known as the "tears of wine," Marangoni effect is observed if there is a significant difference between the surface tensions of the components making up of a mixture. The tendency for heat and mass to travel to areas of higher surface tension within a fluid can create spontaneous interfacial convection. These effects become dominant especially in microfluidic systems and under microgravity (Zhang and Zheng, 2012).

Reactor design

For a gas–liquid reactor, the major input variables for the design are

- Flow rates of reactants
- Regime of operation and corresponding rate equations
- Mixing characteristics of gas and liquid phases as determined by the type of contactor

So far, we have formulated a number of regimes with the corresponding conditions and governing rate equations. Now, we recast the rate equations in a general form that would be indicative of the relative roles of reaction, liquid film diffusion, and gas film diffusion. We then briefly discuss the design principles of the more common classes of fluid–fluid reactors.

A generalized form of equation for all regimes

The most convenient way to define a rate process is by the ratio

$$\text{Rate} = \frac{\text{Driving force}}{\text{Resistance}} = \left[\frac{\text{Driving force}}{1/\text{Rate coefficient}}\right] \quad (11.22)$$

The rate coefficient can be for reaction (k), for any of the mass transfer steps (k_G or k_L), or for combinations thereof. Equation 11.22 can, therefore, be written as

$$\text{Rate} = \frac{\text{Driving force}}{\text{Resistance}} = \frac{\text{Driving force}}{\sum 1/k_i} \quad (11.23)$$

where k_i represents any rate coefficient. The overall rate equations for each of the regimes considered so far can thus be recast in this form. It is often convenient to express gas phase concentrations in terms of partial pressures. Additionally, if the gas phase is not a pure gas but a mixture (i.e., A is mixed with an inert gas), then gas film resistance should also be included.

Regime 1: Very slow reaction

Since in this regime, the reaction is kinetically controlled throughout, Equation 11.23 becomes

$$-r_A = b\frac{p_A[B]_b}{H'_A/k_2} \quad (11.24)$$

where the concentrations have been expressed in terms of partial pressures, k_2 is the second-order reaction rate constant with units of meter cube per meter second, and H'_A is the reciprocal of the Henry's law constant H_A.

Regime 2 and regime between 1 and 2: Diffusion in film without and with reaction in the bulk

The controlling resistance in regime 2 is due to the liquid film. In addition, as already stated, there can also be a gas film resistance, resulting in the equation

$$-r_A = \frac{p_A}{1/k_{GP} + H'_A/k_L} = \frac{p_A}{1/k'_{GP}a_L + H'_A/k'_L a_L} \quad (11.25)$$

If the reaction in the bulk is also a contributing resistance, we add this resistance from Equation 11.24 to the above expression, with the result

$$-r_A = \frac{p_A}{1/k'_{GP}a_L + H'_A/k'_L a_L + H'_A/bk_2[B]_b} \quad (11.26)$$

Regime 3: Fast reaction

Referring to Equation 11.9, the reaction in the film (that occurs simultaneously with diffusion in this regime) can be written in terms of pure mass transfer in the liquid film multiplied by the enhancement factor:

$$-r_A = (k'_L a_L [A]^*)\eta = \left[k'_L a_L \left(\frac{p_A}{H'_A}\right)\right]\eta \tag{11.27}$$

When this rate is added to the rate of gas film diffusion, we obtain

$$-r_A = \frac{p_A}{1/k'_{GP}a_L + H'_A/k'_L a_L \eta} \tag{11.28}$$

Regime between 2 and 3

In this regime, the reaction occurs both in the film and in the bulk. A simple way of accounting for the reaction in bulk would be to add the resistance corresponding to this additional reaction to Equation 11.26, giving

$$-r_A = \frac{p_A}{1/k'_{GP}a_L + H'_A/k'_L a_L \eta + H'_A/bk_2[B]_b} \tag{11.29}$$

Regime 4: Instantaneous reaction

We observed in Chapter 6 that the enhancement factor for this regime is an asymptotic value since it corresponds to the extreme case of enhancement due to an instantaneous reaction. The enhancement in the case of no reaction is obviously one and that corresponding to any other regime will lie between these two asymptotes. The reaction rate within the liquid film may be written as

$$-r_A = \frac{[A]^*}{1/k'_L a_L \eta} \tag{11.30}$$

where

$$\eta_a = 1 + \frac{D_B[B]_b}{D_A v_B [A]^*} = 1 + \frac{D_B[B]_b H'_A}{D_A v_B p_A} \tag{11.31}$$

Adding the gas film resistance to the rate equation gives

$$-r_A = \frac{p_A}{1/k'_{GP}a_L + H'_A/k'_L a_L \eta} \tag{11.32}$$

Note that since the reaction occurs instantaneously and completely in the film, there is none in the bulk. Hence, Equation 11.32 does not contain a term for the reaction in the bulk.

A special case of instantaneous reaction is when it occurs entirely at the gas–liquid interface. Clearly, when the reaction occurs so fast as to not permit any penetration of A into the film, the controlling step would be the rate at which gas A is supplied to the interface. In other words, the reaction would be gas film-controlled, with the rate given by

$$-r_A = k'_{GP}a_L\left(\frac{1}{H'_A}\right)p_A = \frac{p_A}{H'_A/k'_{GP}a_L} \tag{11.33}$$

Classification of gas–liquid contactors

There are two broad methods of classifying gas–liquid contactors:

1. On the basis of the manner in which contact between gas and liquid is achieved
2. On the basis of the manner in which power for dispersion of one phase into the other is delivered

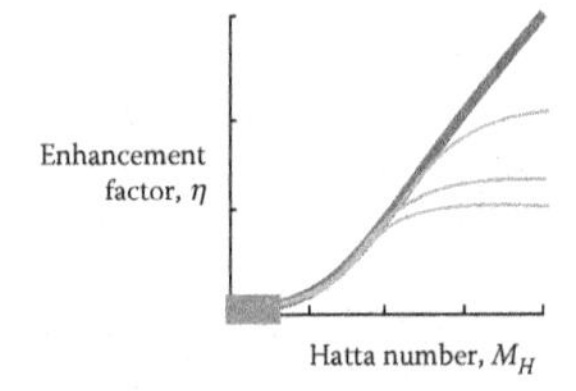

A classical description of the selection of the equipment is given by Krishna and Sie (1994). The qualitative enhancement factor versus Hatta number diagram shown in the sidebar figure is classified for the reaction rate (slow, fast, or instantaneous) and a dispersion ratio β, the ratio of the liquid phase volume to the volume of the diffusion layer.

Classification-1 (based on manner of phase contact)

Using classification-1, most industrial contactors may be classified under three broad categories, with a number of variations in each category (Van Krevelen, 1950).

1. Contactors in which the liquid flows as a thin film
   a. Packed columns
   b. Thin-film reactors
   c. Disk reactors
2. Contactors with dispersion of gas into liquids
   a. Plate columns (including controlled cycle contactors)
   b. Mechanically agitated contactors
   c. Bubble columns
   d. Packed bubble columns
   e. Sectionalized bubble columns
   f. Two-phase horizontal cocurrent contactor
   g. Coiled reactors
   h. Vortex reactors
3. Contactors where liquid is dispersed in the gas phase
   a. Spray columns
   b. Venturi scrubbers

The title of each category clearly describes the nature of the contact, whereas the subtitles indicate the manner in which such contact is achieved.

Classification-2 (based on the manner of energy delivery)

The rate and the manner in which energy is delivered influence such important fluid–fluid parameters as dispersed-phase holdup, interfacial area, and mass transfer coefficient (Calderbank, 1958; Nagel et al., 1972, 1973; Kastaněk, 1976; Zahradnik et al., 1982; Oldshue, 1983). Thus, an energy-based classification would be both useful and appropriate. In this classification, the contactors are essentially divided into three broad groups:

1. Contactors in which energy is supplied through the gas phase
2. Contactors in which energy is supplied through the liquid phase
3. Contactors in which energy is delivered through mechanically agitated parts

This classification is not always unambiguous. For example, in reactors belonging to groups 2 and 3, some energy is also delivered by the gas phase.

The principal features of the major contactors irrespective of their classification have been described by many authors at considerable length, for example, Laddha and Degaleesan (1976), Schügerl (1977, 1980, 1983), Shah et al. (1982), Doraiswamy and Sharma (1984), Shah and Deckwer (1985), and Kastaněk et al. (1993). A few common designs are sketched in Figure 11.3.

Mass transfer coefficients and interfacial areas of some common contactors

The values of the mass transfer coefficients and interfacial areas for the more common contactors (packed columns, plate columns, bubble columns, mechanically agitated contactors, and static mixers) are usually known or can be estimated from correlations published in the literature or supplied by the manufacturers. The typical values are given in Table 11.1.

Role of backmixing in different contactors

A particularly important consideration in designing contactors for gas–liquid reactions is the validity of the ideal flow assumption [plug flow (PF) or mixed flow (MF)] normally used in the design. Numerous studies have been reported on the role of nonideality (i.e., backmixing) in gas–liquid contactors, and based on these, some qualitative guidelines

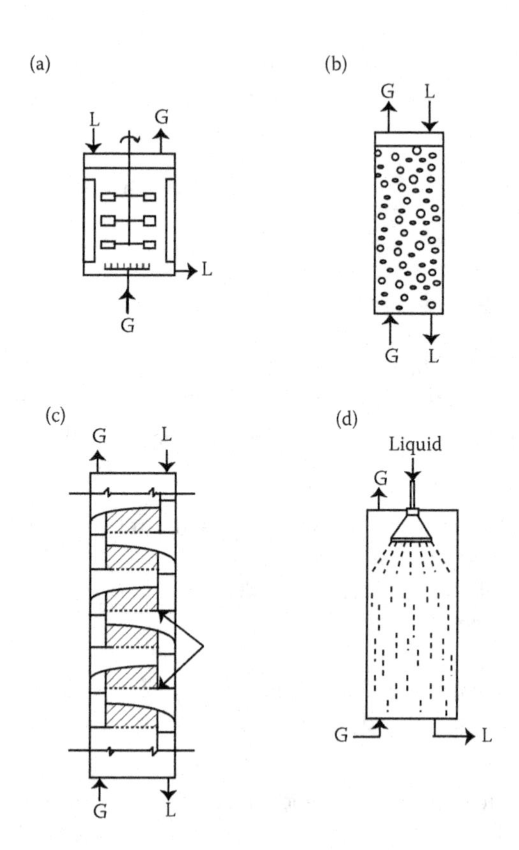

Figure 11.3 Sketches of some common gas–liquid contactor designs. (a) Mechanically agitated contactor, (b) bubble column, (c) plate column with downcomer, and (d) spray column.

Table 11.1 Flow and Mass Transfer Features of the More Common Gas–Liquid Contactors

Type of Contactor	Typical Range of Superficial Gas Velocities (cm/s)	Mixing Characteristics		k_L' (cm/s)	a_L (cm²/cm³)	Liquid Holdup
		Gas Phase	Liquid Phase			
Packed column	10–100	PF	PF	0.3–2.0	0.2–3.5	0.05–0.1
Bubble column	1–30	PF	PMF	1.0–4.0	0.25–10	0.6–0.8
Packed bubble column	1–20	PF	PMF	1–4	1–3	0.5–0.7
Plate column (without downcomers)	50–300	PF	MF[a]	1–4	1–2	0.5–0.7
Mechanically agitated contactors	0.1–2.0	MF (or PMF)	MF	1–5	2–10	0.8–0.9
Spray column	5–300	PMF	PF	0.5–1.5	0.2–1.15	0.05

Source: Doraiswamy, L.K. and Sharma, M.M., *Heterogeneous Reactions—Analysis Examples and Reactor Design*, Vol. 2, Wiley, New York, NY, 1984.

Note: PMF, partial mixed flow.

[a] Good mixing on plates, no mixing between plates.

can be formulated. Such an attempt has been made in several reviews and books, for example, Doraiswamy and Sharma (1984), Shah and Deckwer (1985), and more exhaustively by Kastaněk et al. (1993).

Reactor design for gas–liquid reactions

The overall strategy

A knowledge of the regime of operation, that is, location of the reaction region within the gas–liquid system is important in selecting the type of reactor to be used. Depending on whether the reaction occurs in the bulk or in the film, the "effective amount of liquid" or liquid holdup will be different and forms an important parameter in reactor choice.

For purposes of design, the reaction regime can be determined from laboratory experiments carried out within the framework of the analytical methods described so far. The basic objectives of the design are then to minimize the reactor volume and the energy required for a given production rate. The overall strategy can be stated as follows:

1. Make a preliminary list of reactors and flow arrangements (plug or mixed for the two phases in countercurrent or cocurrent flow) that deserve to be considered.
2. Calculate the reactor volume for each of the preselected reactors.
3. Using the parameter values available from calculations in (2), and make a more rational choice of the reactor based on volume minimization.
4. Assess each reactor for its energy requirement and select the one with minimum requirement. It is important to note that the minimum energy criterion serves more as a basis for reactor selection for achieving a given reaction rate than as a design procedure.

Calculation of reactor volume

For making the calculations in step 2, it is convenient to assume that each phase is either in PF or in MF. One can readily visualize several combinations of these flows (Table 11.2 and Figure 11.4). The reactor types that closely approximate these combinations are also included in the table.

The first step in the design (for volume calculation) is a mass balance (differential for PF and overall for MF) written in terms of transfer of a selected component (say, A) from one phase to the other followed by the reaction. Thus, for the reaction

$$A\text{ (g)} + \upsilon_B B\text{ (l)} \rightarrow \text{Products} \qquad \text{(R2)}$$

Table 11.2 Reactors That Can Be Represented as Combinations of Ideal Plug- and Mixed-Flow Configurations

Type of Flow			
Liquid	**Gas**	**Type of Operation**	**Type(s) of Reactor**
PF	PF	Steady state	Bubble column ($H/d_r > 10$), packed column
MF	MF	Steady state	Mechanically agitated contactor, packed column ($H_r/d_r < 3$), spray column
MF	PF	Steady state	Plate column
PF	MF	Steady state	Spray column
Uniform conc.	MF	Unsteady state	Batch reactor

Note: H_r, reactor height; d_r, reactor diameter.

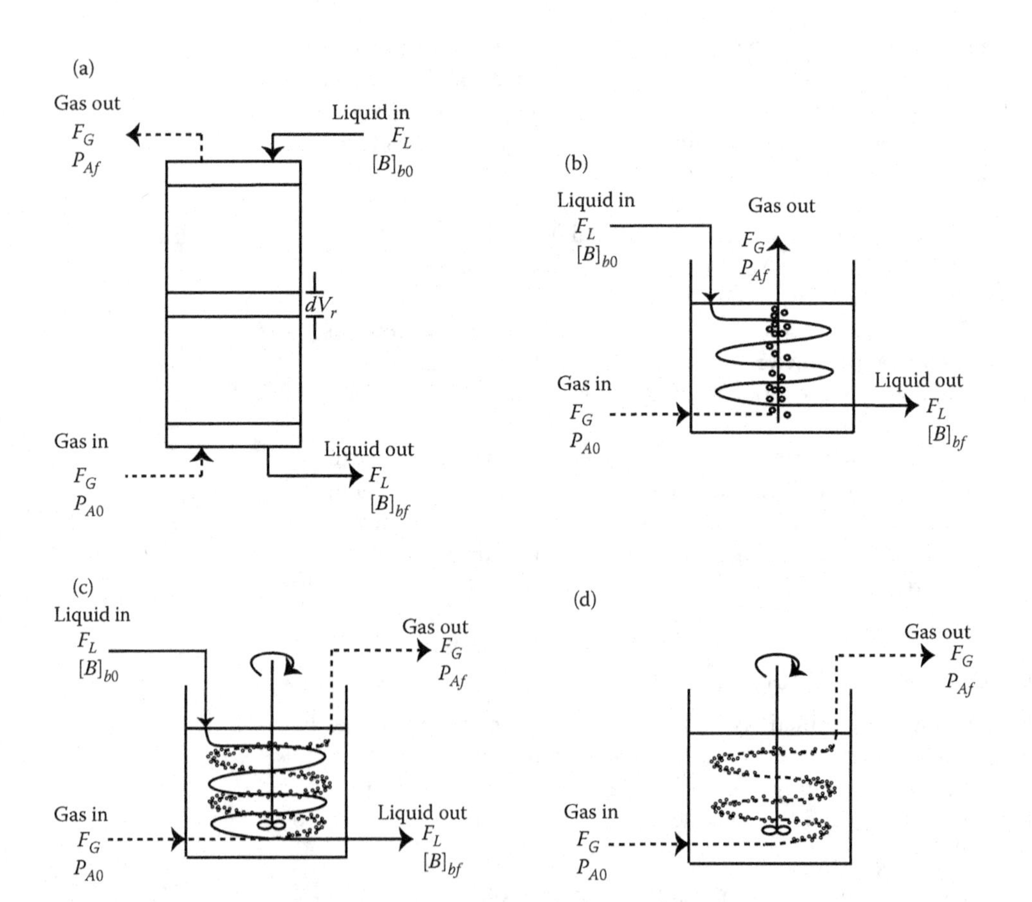

Figure 11.4 Four ideal flow configurations. (a) Plug-flow *G*–plug-flow *L*, (b) plug-flow *G*–mixed-flow *L*, (c) mixed-flow *G*–mixed-flow *L*, and (d) mixed-flow *G*–batch *L*.

we have

$$\begin{bmatrix}\text{Rate of loss of}\\ A \text{ by the gas}\end{bmatrix} = \begin{bmatrix}\text{Rate of reaction}\\ \text{of } A \text{ in the liquid}\end{bmatrix} \quad (11.34)$$

The equation can be recast to produce the now-familiar $1/-r_A$ versus X_A plots. The volume can then be determined graphically or numerically. However, the limitation of Equation 11.34 should be noted. It assumes that all A from phase 1 that enters phase 2 reacts in the latter phase without any accumulation. Thus, it cannot account for the regime between 1 and 2 or the regime overlapping 1, 2, and 3 where A exists in a finite concentration in phase 2 or is carried away by the liquid.

We now consider reaction R2 again and formulate the design equations for all the five cases presented in Table 11.2 (see Levenspiel, 1993, for a more detailed discussion).

Case 1: Plug gas, plug liquid, and countercurrent steady state Since both gas and liquid are in PF, we write the following differential balance (Figure 11.4a):

$$F_{GU}dy'_A = -\frac{F_{LU}dz'_B}{\upsilon_B} = (-r_A)dV_r = (-r'_A a_L)dV_r \quad (11.35)$$

Applicable to packed columns

where y'_A and z'_A represent the mole ratios of A to the inert in the gas and liquid phases, respectively, that is,

$$y'_A = \frac{p_A}{p_U},\quad z'_A = \frac{[A]}{[U]},\quad z'_B = \frac{[B]}{[U]} \quad (11.36)$$

and F_{GU} and F_{LU} are the flow rates (mol/s) of inerts in the gas and liquid phases, respectively. The integration of the equation leads to

$$V_r = F_{GU}\int_{y'_{A1}}^{y'_{A2}} \frac{dy'_A}{(-r'_A)a_L} = \frac{F_{LU}}{\upsilon_B}\int_{z'_{A2}}^{z'_{A1}} \frac{dz'_B}{(-r'_B)a_L} \quad (11.37)$$

with

$$F_{GU}(y'_{A2} - y'_{A1}) = \frac{F_{LU}}{\upsilon_B}(z'_{B1} - z'_{B2}) \quad (11.38)$$

Now, it only remains to introduce the appropriate equation for the rate in Equation 11.37 and solve it for the volume.

Case 2: Same as case 1 but with cocurrent flow The mass balance equations are the same as for case 1, but the signs of the flow terms F_{GU} and F_{LU} will change. If both streams flow down the column, F_{GU} becomes $-F_{GU}$, and when both streams flow up the column, F_{LU} becomes $-F_{LU}$.

Case 3: Plug gas, mixed liquid, and steady state In this case, sketched in Figure 11.4b, a differential balance must be written for the gas phase since it is in PF, but an overall balance for the liquid phase must be written since it is in MF. Thus, for the gas phase, by equating A lost by the gas to the amount disappearing by the reaction, we have

Applicable to bubble columns

$$F_{GU} dy'_A = (-r'_A)_{\text{liq exit}} a_L dV_r \tag{11.39}$$

Then for the liquid phase, by equating A lost by the gas to B lost by the liquid, we obtain

$$F_{GU}(y'_{A0} - y'_{Af}) = \frac{F_{LU}}{\upsilon_B}(z'_{B0} - z'_{Bf}) \tag{11.40}$$

The integration of Equation 11.39 for the gas phase and combining with Equation 11.40 leads to

$$V_r = F_{GU} \int_{y'_{Af}}^{y'_{A0}} \frac{dy'_A}{(-r'_A)_{\text{liq exit}} a_L} \tag{11.41}$$

Since the liquid phase is mixed, Equation 11.41 can be directly used to calculate V_r provided the exit conditions are known. Alternatively, if V_r is known, y'_{Af} and z'_{Bf} (and therefore $[B]_{bf}$) can be found by repeating the calculation for several assumed values of y'_{Af} or z'_{Bf} till the given value of V_r is matched.

Case 4: Mixed gas, mixed liquid, and steady state This case, sketched in Figure 11.4c, corresponds to mixed conditions in the entire reactor. Thus, the following mass balance can be written for the vessel as a whole:

Applicable to mechanically agitated contactors

$$F_{GU}(y'_{A0} - y'_{Af}) = \frac{F_{LU}}{\upsilon_B}(z'_{B0} - z'_{Bf}) = (-r'_A)_{\text{at liq gas exit}} a_L V_r \tag{11.42}$$

The solution of this equation for V_r is straightforward.

Case 5: Mixed gas, batch liquid, and unsteady state In this semibatch operation, a batch of liquid is taken in a reactor and a stream of gas is passed through it (Figure 11.4d). The problem here is usually one of calculating the time needed for a given conversion in a reactor of known volume.

The following material balance can be written as

$$F_{GU}(y'_{A0} - y'_{Af}) = -\frac{V_L}{\upsilon_B}\frac{d[B]_b}{dt} = (-r'_A) a_L V_r \tag{11.43}$$

An iterative procedure must be used for calculating the time needed for a given conversion. The steps involved are

1. Assume a value of $[B]_b$.
2. Guess a value of y'_{Af} (i.e., p_{Af}) for this $[B]_b$.
3. Calculate the rate from the appropriate rate equation for the reaction.
4. Compare the first and third terms of Equation 11.43 and iterate till they match.
5. Repeat the above steps for another value of $[B]_b$ and prepare a plot of reciprocal rate versus $[B]_b$.
6. Integrate to obtain the time

$$t = \frac{V_L}{\upsilon_B} \int_{[B]_{bf}}^{[B]_{b0}} \frac{d[B]_b}{(-r'_A)a_L} \tag{11.44}$$

Comments The following observations should serve as useful guidelines in assessing the applicability of the design equations:

1. MF of any phase is, in general, not desirable. However, this should be of no consequence in cases where the reaction is controlled by diffusion or diffusion is accompanied by the reaction in the film and is completed within it, that is, in regimes 2 and 3.
2. In writing Equation 11.35, PF was assumed both for the gas and liquid phases. Thus, this equation should be valid for packed columns.
3. Mechanically agitated contactors are characterized by complete backmixing of both liquid and gas phases; thus, Equation 11.42 should be applicable to these contactors. A deviation from MF is possible at low stirrer speeds. If the reaction is in regime 2, accurate values of k_L are needed. Since predictive correlations for this are unreliable, the values should be obtained by conducting experiments in a geometrically similar laboratory reactor (20–30-cm diameter). It is recommended that, to obtain such similarity, the ratio of tip speed to tank diameter should be kept constant (Juvekar, 1976).
4. In the widely used bubble column, PF for the gas phase and MF for the liquid phase are passable assumptions. Thus, Equations 11.39 through 11.41 should normally be valid for these reactors. However, this should be reviewed in light of more detailed information on backmixing.
5. Design equations can be developed for cases of partial mixing in either or both phases. The first step in such cases is to experimentally determine the residence time distribution for each phase. But the procedures tend to be quite involved and are outside the scope of this treatment (see Kastaněk et al., 1993, for a comprehensive exposition).

Example 11.1: Design of a semibatch reactor for the oxidation of *N*-butyraldehyde

It is proposed to oxidize a batch of 1000 L of *n*-butyraldehyde dissolved in *n*-butyric acid in the presence of 0.1% manganese acetate as the catalyst at essentially atmospheric pressure in a mechanically agitated contactor by passing air continuously. Given the following data, calculate the time required for obtaining a conversion corresponding to a drop in the aldehyde concentration $[B]_0$ from 6.0 to 0.85×10^{-3} mol/cm^3.

Reaction

$$O_2 + 2CH_3CH_2CH_2CHO \rightarrow 2CH_3CH_2CH_2COOH \quad (R3)$$

$$A \qquad B \qquad \rightarrow \qquad R$$

$$m \text{ (order in } O_2) = 1,\ n \text{ (order in aldehyde)} = 2,\ \upsilon_B = 2$$

Temperature = 20°C
Initial concentration of *n*-butyraldehyde, $[B]_{bi} = 6 \times 10^{-3}$ mol/cm^3
Final concentration of *n*-butyraldehyde, $[B]_{bf} = 0.85 \times 10^{-3}$ mol/cm^3
Diameter of contactor = 100 cm
Diameter of impeller = 35 cm
Rate of inert (nitrogen) feed F_N [=F_{GU}], 0.13 mol/s
Gas–liquid interfacial area $a_L = 12.0$ cm^2/cm^3 of liquid
True liquid side mass transfer coefficient $k_L' = 2 \times 10^{-2}$ cm/s

The following average properties may be used at 20°C:

Henry's law constant of oxygen in the liquid, $H_A = 1.043 \times 10^{-5}$ mol/cm^3 atm
Diffusivity of oxygen in the liquid, $D_A = 1.8 \times 10^{-5}$ cm^2/s
Third-order constant for the reaction
$k_3 = 4 \times 10^6$ (cm^3/mol)^2/s (Ladhabhoy and Sharma, 1969).

SOLUTION

It is assumed that the gas phase is completely backmixed and losses of *n*-butyraldehyde due to vaporization are negligible. (Although there will be a significant change in the viscosity of the liquid when it is oxidized, for the sake of illustration, average property values are assumed).

The regime of the reaction is not known. Hence, we will assume a relatively common regime, 2–3, and verify its applicability as part of the calculations. The rate equation for this regime (Table 11.1) is given by

$$-r_A' a_L = a_L [A]^* \sqrt{D_A k_3 [B]_b^2 + k_L'^2} \text{ (with } k_1 = k_3[B]_b^2) \quad \text{(E11.1.1)}$$

The conditions to be satisfied are

$$\frac{\sqrt{D_A k_3 [B]_b^2}}{k_L'} \cong 1 \quad \text{(E11.1.2)}$$

and

$$\frac{\sqrt{D_A k_3 [B]_b^2}}{k_L'} \ll \frac{[B]_b}{\upsilon_B [A]^*} \sqrt{\frac{D_B}{D_A}} \quad \text{(E11.1.3)}$$

Since the gas is completely mixed, the solubility of oxygen in the liquid should be based on the outlet partial pressure of oxygen in air, that is,

$$[A]^* = H_A P \left(\frac{y_{Af}'}{1 + y_{Af}'} \right) \quad \text{(E11.1.4)}$$

where y_{Af}' is the molar ratio of oxygen to nitrogen in air. If $y_{Af}' \ll 1$, then Equation E11.1.4 can be approximated as

$$[A]^* = H_A P y_{Af}' \quad \text{(E11.1.5)}$$

The material balance on oxygen for this semibatch reactor gives

$$F_N (y_{A0}' - y_{Af}') = \frac{V_r}{\upsilon_B} \frac{d[B]_b}{dt} = (-r_A') a_L V_r \quad \text{(E11.1.6)}$$

From Equations E11.1.1, E11.1.5, and E11.1.6, we obtain

$$F_N (y_{A0}' - y_{Af}') = H_A P y_{Af}' V_r a_L \sqrt{D_A k_3 [B]_b^2 + k_L'^2} \quad \text{(E11.1.7)}$$

The outlet mole ratio y_{Af}' can be obtained from the inlet ratio y_{A0}' by recasting Equation E11.1.7 as

$$y_{Af}' = \frac{y_{A0}'}{1 + \alpha' \sqrt{[B]_b^2 + \beta'^2}} \quad \text{(E11.1.8)}$$

where

$$\alpha' = \frac{a_L H_A P V_r \sqrt{D_A k_3}}{F_N}, \quad \beta' = \frac{k_L'}{\sqrt{D_A k_3}} \quad \text{(E11.1.9)}$$

Then by substituting Equation E11.1.8 for y'_{Af} into Equation E11.1.6, we obtain

$$-\frac{\gamma'}{y'_{A0}}\frac{d[B]_b}{dt} = \frac{\alpha'\sqrt{[B]_b^2 + \beta'^2}}{1 + \alpha'\sqrt{[B]_b^2 + \beta'^2}} \tag{E11.1.10}$$

where

$$\gamma' = \frac{V_r}{\upsilon_B F_N} \tag{E11.1.11}$$

The integration of Equation E11.1.10 gives the semibatch time as

$$t_{SB} = \frac{\gamma'}{y'_{A0}a'}\left[\ln\left(\frac{[B]_{bi} + \sqrt{[B]_{bi}^2 + \beta'^2}}{[B]_{be} + \sqrt{[B]_{be}^2 + \beta'^2}}\right) + \alpha'([B]_{bi} - [B]_{be})\right] \tag{E11.1.12}$$

We now check to see whether conditions (E11.1.2) and (E11.1.3) for the validity of the assumed rate equation are satisfied both at the beginning and at the end of the operation.

At the beginning:

$$\frac{\sqrt{D_A k_3 [B]_{bi}}}{k'_L} = 2.55 \cong 1$$

To apply the second condition (E11.1.3), we combine it with Equation E11.1.5 to give

$$\frac{\sqrt{D_A k_3 [B]_{bi}}}{k'_L} \ll \frac{[B]_{bi}}{\upsilon_B H_A P (y'_{Af})_i} \tag{E11.1.13}$$

Since $(y'_{Af})_i$, the outlet mole ratio at initial conditions is unknown, we calculate it from Equation E11.1.8. The parameter values to be used are $\alpha' = 8.25 \times 10^3$, $\beta' = 2.35 \times 10^{-3}$, $y'_{A0} = 0.266$, and $[B]_{bi} = 6 \times 10^{-3}$ mol/cm³. Thus, we obtain $(y'_{Af})_i = 0.0142$, giving

$$\frac{[B]_{bi}}{\upsilon_B H_A P (y'_{Af})_i} = 4 \times 10^3$$

Hence, both the conditions are satisfied.

At the end:

By repeating the above calculations for conditions at the end of the operation [when $y'_{Af} = (y'_{Af})_e$], it can be shown that the two conditions continue to be valid.

It may, therefore, be concluded that the rate equation, with the corresponding regime 2–3, is valid for the entire duration of the run. Equation E11.1.12 derived from this rate equation can thus be used for calculating t_{SB}.

Substituting the values of the relevant parameters in Equation E11.1.12, $F_N = 0.13$ mol N_2/s, $V_r = 10^3$ cm^3, $\alpha' = 8.25 \times 10^3$, $\beta' = 2.35 \times 10^{-3}$, $\gamma' = 3.88 \times 10^{-6}$, $y'_{A,\text{in}} = 0.266$, we obtain the semibatch time:

$$t_{SB} = 21.6 \text{ h}$$

Reactor choice

The criteria

As must be evident now, gas–liquid reactions are characterized by a large number of reactor types. This is also true of other multiphase reactions in which a liquid phase is involved. For other reactions such as gas–solid, catalytic, or noncatalytic, the choice of the reactor is confined to a lesser number of variations. Therefore, while the reactor choice is an important consideration for all reactions, particularly heterogeneous reactions, it is more so for gas–liquid, liquid–liquid, and slurry systems, all of which are widely used in industrial chemical synthesis. We discuss below the cost minimization criteria for a rational choice of the reactor for gas–liquid reactions.

Minimization of reactor volume is clearly one such criterion. In addition to directly reducing the reactor cost, this also lowers the energy cost. The latter derives from the reduction in energy for agitating or pumping the fluids. A second criterion (actually a set of two criteria) can also be formulated based exclusively on minimization of energy supply. Here, two situations can be visualized: minimization of the overall energy input to the reactor to initiate and sustain interfacial contact; and minimization of energy delivered to the system for maximizing mass transfer rates. These criteria are summarized in Figure 11.5.

Volume minimization criterion

General discussion Any criterion for volume minimization must essentially be based on the extent to which reaction occurs in the film. The criterion for reaction to occur entirely in the film is given by the inequality

$$M_H \gg 1 \tag{11.45}$$

$M_H \gg 1$, reaction entirely in the film

and for reaction to occur entirely in the bulk by the reverse of this criterion.

A more elaborate criterion (Kastaněk et al., 1993) can also be used based on a parameter N defined as

$$N = \left[\frac{M_H^2}{q^*} + M_H^2 - \frac{M_H^2}{q^*}(a + \eta)\right]^{1/2} \tag{11.46}$$

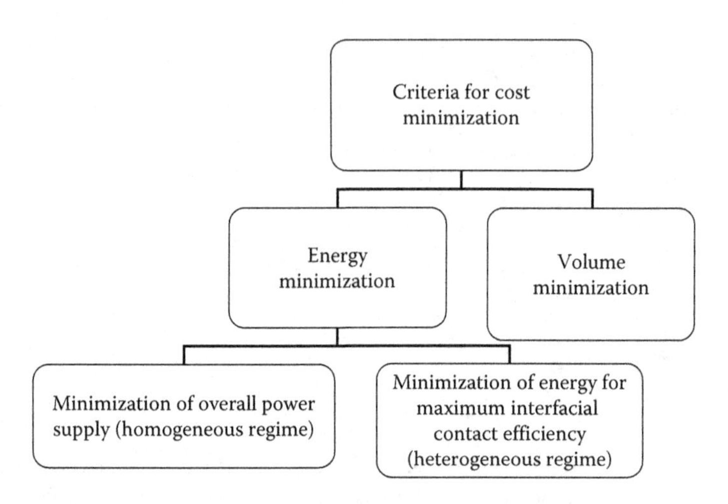

Figure 11.5 Classification of criteria for cost minimization. (Adapted from Kastaněk, F. et al., *Reactors for Gas–Liquid Systems*, Ellis Horwood, New York, NY, 1993.)

where

$$a = \frac{[A]_0}{[A]^*}, \quad q^* = \frac{[B]_b D_B}{\upsilon_B [A]^* D_A} \quad \text{(for the film model)}$$
$$\left[\text{or } q^* = \frac{[B]_b}{\upsilon_B [A]^*}\sqrt{\frac{D_B}{D_A}} \quad \text{(for the penetration model)}\right] \tag{11.47}$$

The magnitude of N provides a valuable guideline for reactor selection. Table 11.3 presents a broad indication of the types of contactors recommended for values of N ranging from $N \ll 0.03$ for reaction in the bulk to $N \gg 1.3$ for reaction in the film. These correspond to $M_H \ll 0.03$ and $M_H > 3$.

Table 11.3 Guidelines for Reactor Selection

Criterion Based on Value of N	Regime	Type of Reactor
$N \ll 0.03$	1, 2, 1–2	Bubble column
$0.03 \ll N \ll 1.3$	3, 2–3, 1–2–3	Packed column
		Plate column
		Mechanically agitated contactor
$N \gg 1.3$	3, 4	Plate column with shallow liquid layers on plates
		Spray column
		Wetted wall column

Source: Adapted from Kastaněk, F. et al., *Reactors for Gas–Liquid Systems*, Ellis Horwood, New York, NY, 1993.

Limitations of volume minimization It is useful to note that certain types of reactors get automatically eliminated for certain situations, such as the packed-bed reactor for dust-laden gas. Further, reactor selection is usually straightforward for extreme situations such as reaction occurring entirely in the bulk or in the film. In both cases, the quantity of bulk liquid is determined by the rate of production, but the extent to which the bulk phase is utilized for the reaction is minimal when reaction is confined to the film and maximal when it occurs fully in the bulk.

On the other hand, when reaction occurs both in bulk and the film, the reactor should combine good interfacial area generation with efficient bulk liquid utilization. A measure of the *efficiency of bulk liquid utilization* is the parameter γ_{bu} given by (Kramers and Westerterp, 1963):

$$\gamma_{bu} = \frac{1}{M_H \psi}\left[\frac{M_H[\psi - 1] + \tanh M_H}{M_H[\psi - 1]\tanh M_H + 1}\right] \tag{11.48}$$

where

$$\psi = \frac{V_c}{a_L \delta}$$

The parameter ψ is a direct measure of the ratio of the continuous phase fraction to film thickness. In other words, it represents the relative roles of reactor volume and interfacial area in the reaction. Equation 11.48 is graphically displayed in Figure 11.6 as plots of γ_{bu} versus ψ for different values of M_H.

The unshaded regions in the figure represent clear cases of reaction in bulk (regime 1) and film (regimes 3 and 4). For large values of M_H corresponding to regime 3 or 4, reactors that can generate large interfacial areas are desirable. Further, liquid phase holdup is negligibly small in such cases. On the other hand, for regime 1, that is, for reaction in the bulk, a large holdup is a primary requirement with no emphasis on area generation.

Reactor selection tends to be complicated in the range $0.03 < M_H < 0.1$ where large values of both interfacial area and liquid holdup are required. This is also true for regime 2 in spite of its very low value of M_H, since pure mass transfer through the film is the controlling resistance, so that interfacial area generation becomes an important consideration. In all such cases, the minimum volume criterion for reactor selection becomes inadequate, and the choice has to be based only on the minimum energy criterion to be discussed later.

Steps in volume minimization Procedures for calculating the reactor volume for various combinations of PF and MF of the liquid and gas

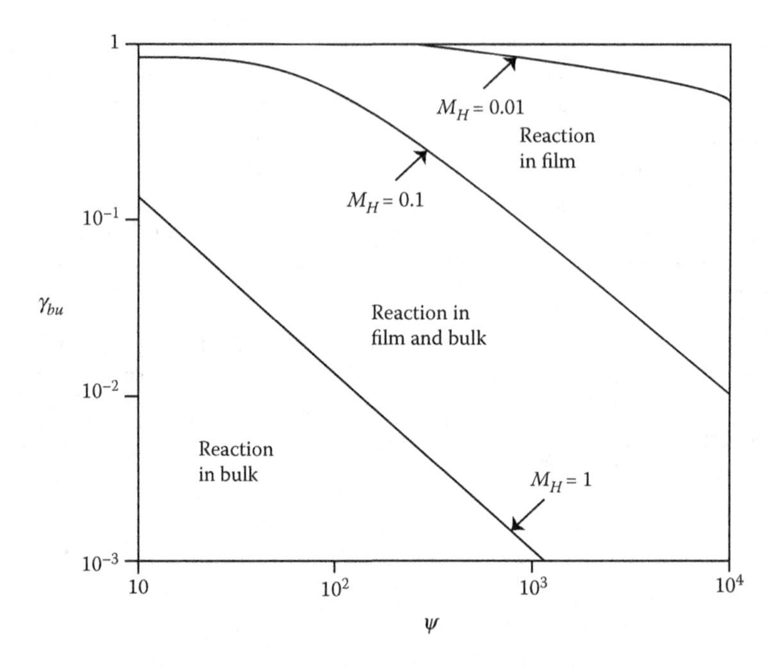

Figure 11.6 Efficiency of bulk phase utilization γ_{bu} as a function of composite parameter ψ for different values of M_H. (Adapted from Kramers, H. and Westerterp, K.R., *Elements of Chemical Reactor Design and Operation*, Chapman & Hall, London, 1963.)

phases were already described. These, however, constitute only one step in the overall strategy for the minimization of reactor volume for a given rate of production. The full strategy involving a number of steps is outlined in Figure 11.7. It will be noticed from the table that volume minimization does not always lead to a unique optimal design, and a second set of criteria based on energy minimization may frequently be necessary.

Energy minimization criteria

The energy minimization criteria depend on the regime of operation. We broadly classify these as *homogeneous* and *heterogeneous* regimes, 2(a) and 2(b).

2(a). Minimization of overall power supply to the contactor.
2(b). Maximization of interfacial contact efficiency.

Criterion 2(a): Homogeneous regime (regime 1) Regime 1 pertains to reactions which are so slow that they can be regarded as homogeneous reactions. For such reactions, the only energy consideration is the minimum energy required to establish contact between phases in order for reactions to occur, irrespective of the efficiency of this contact. In other words, criterion 2(a) will hold. Information on the minimum specific

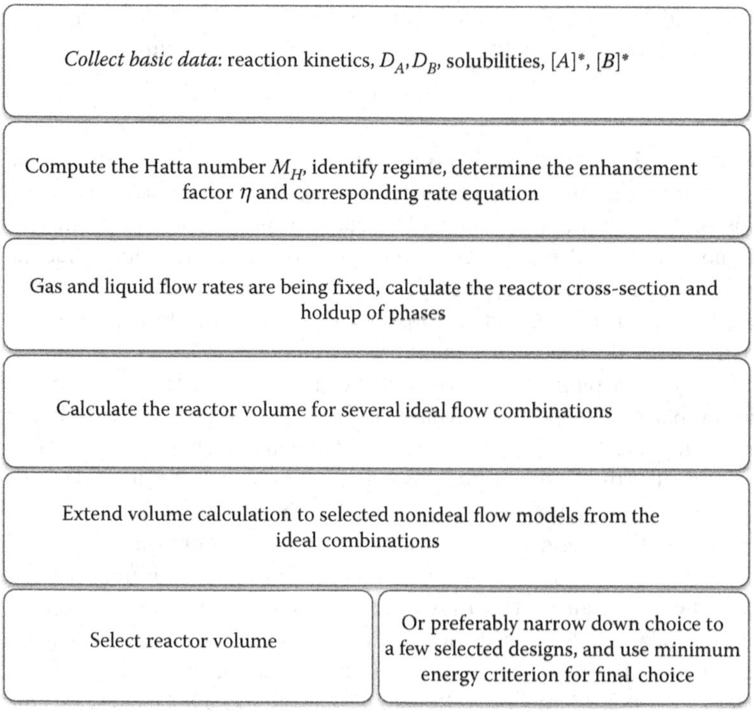

Collect basic data: reaction kinetics, D_A, D_B, solubilities, $[A]^*$, $[B]^*$

Compute the Hatta number M_H, identify regime, determine the enhancement factor η and corresponding rate equation

Gas and liquid flow rates are being fixed, calculate the reactor cross-section and holdup of phases

Calculate the reactor volume for several ideal flow combinations

Extend volume calculation to selected nonideal flow models from the ideal combinations

Select reactor volume

Or preferably narrow down choice to a few selected designs, and use minimum energy criterion for final choice

Figure 11.7 Steps in the minimization of the gas–liquid reactor volume.

energy required (i.e., power input per unit volume of reactor) for each type of contactor has been reported by Viesturs et al. (1986), and Sittig and Heine (1977). Table 11.4 lists the values based on the data reported in these studies. Since in the case of bubble columns some of the energy is supplied by the natural movement of the bubbles themselves, the specific energy requirement is minimum. Agitated contactors clearly require much more energy—which further increases as more agitators are added.

Table 11.4 Comparison of Energy Minimization Criteria 2(a) and 2(b)

Reactor Type	**Criterion 2(a): Minimum Specific Energy Required to Achieve Contact, P_a (kW/m³)**	**Criterion 2(b): Reciprocal of Minimum Energy for a Required Rate, Ω (kWh/m³)**
1. Agitated contactor with multiple stirrers	10	0.5
2. Agitated contactor with multiple stirrers and a central draught tube	11	0.5
3. Bubble column	2.5	≈0.014
4. Bubble column with forced liquid circulation	4	≈0.02
5. Multistage bubble column	3.5	0.01–0.003
6. Tower reactor with forced gas supply	5	0.007

Source: Adapted from Kastaněk, F. et al., *Reactors for Gas–Liquid Systems*, Ellis Horwood, New York, NY, 1993.

It should be noted that for deciding on the most favorable reactor, the specific energy P_a should be multiplied by the reactor volume.

$$P = P_a V_r \tag{11.49}$$

Criterion 2(b): Heterogeneous regime (regimes 2–4) In all regimes other than regime 1, the efficiency of interfacial contact becomes an important consideration. Thus, the quantity needed to be minimized is not the overall energy as used for homogeneous reactions, but the energy required to be supplied per unit area of interfacial contact generated or material transported. This includes: (1) energy losses in motors and moving devices such as pumps and agitators, (2) losses due to pressure drops in pipings and distributor elements, and (3) loss of energy at the place of dispersion formation. The last of these is obviously the most important and depends on the mechanism of dispersion. A detailed treatment of these energy losses is outside the scope of the present effort. Items (1) and (2) are usually "external" to the actual energy evaluation of a specific system. However, they can often be quite important, for example, proper plate design in a plate column can minimize energy losses due to transport and pressure drop in the column, or a properly designed gas distributor device can equally enhance energy utilization efficiency (Bohmer and Blenke, 1972; Zahradnik et al., 1982, 1985; Henzler, 1982; Mann, 1983; Rylek and Zahradnik, 1984). However, the "internal" source represented by item (3) constitutes the main element of criterion 2(b), the final effect of which is reflected in the value of the power input per unit reactor volume per unit interfacial area generated. Thus, a parameter

$$\Omega = \frac{P/V_r}{k_L' a_L} \tag{11.50}$$

can be used as a measure of the energy effectiveness of a given contactor.

Obviously, the mass transfer coefficient $k_L' a_L$ is the most important parameter in determining energy effectiveness, and several correlations have been reported for estimating it, for example, Meister et al. (1979), Botton (1980), and Heijnen and van't Riet (1984). Correlations have also been proposed for interfacial area a_L directly as a function of energy dissipation rate, for example, Nagel et al. (1972), Buchholz et al. (1983), Bisio and Kabel (1985), and Sisak and Hung (1986). Typical value ranges for k_L' and a_L for the more common reactors are listed in Table 11.5. An important conclusion from all these studies is that the energy effectiveness of a contacting device decreases with increasing demand on the intensity of interfacial contact (Kastaněk et al., 1993). In other words, each additional increase of mass flow rate requires a disproportionately larger increase of the specific energy dissipation rate.

Comparison of criteria Values of Ω for different contactors, again calculated from the data of Sittig and Heine (1977), are included in

Table 11.5 Flow and Mass Transfer Features of the More Common Liquid–Liquid Contactors

Type of Contactor	Mixing Characteristics: Continuous Phase	Mixing Characteristics: Dispersed Phase	Residence Time of Continuous Phase	Fractional Dispersed Phase Holdup	k_L' (cm/s)	a_L (cm²/cm³)	$k_L' a_L \times 10^2$ (s⁻¹)
Mechanically agitated contactors	MF	PMF	Variable over a wide range	0.05–0.40	0.3–1.0	1–800	0.3–800
Packed columns	PF	PF	Limited	0.05–0.10	0.3–1.0	1–10	0.3–10
Spray columns	MF	PF	Limited	0.05–0.10	0.3–1.0	1–10	0.1–10
Air (inert gas)-agitated contactors	MF	PMF	Variable over a wide range	0.05–0.30	0.1–0.3	10–100	1.0–30
Horizontal mixers	PF	PF	Limited	0.05–0.20	0.1–1.0	1–25	0.1–25

Table 11.4. It is clear from the table that the two criteria 2(a) and 2(b) lead to two entirely different choices of contactors. As already noted, criterion 2(b) is applicable to a much wider range of reactions than criterion 2(a) which is essentially restricted to reactions in the kinetic regime.

Liquid–liquid contactors

As already noted, the volume of a liquid–liquid reactor for a given output is calculated using the same principles as outlined for gas–liquid reactors. One major factor to be considered in liquid–liquid reactions, however, is the much smaller density difference between the two phases as compared to gas–liquid reactions. While this is, in general, conducive to creating greater interfacial area, too small a difference is likely to cause problems of phase separation. This often leads to emulsion formation and the consequent attention to designs that promote emulsion breaking.

Classification of liquid–liquid reactors

Liquid–liquid extractors can be classified as follows (adapted from Treybal, 1963; Warwick, 1973):

1. Gravity-operated extractors
   a. No mechanical moving parts
      Spray columns
      Packed columns
      Sieve plate columns

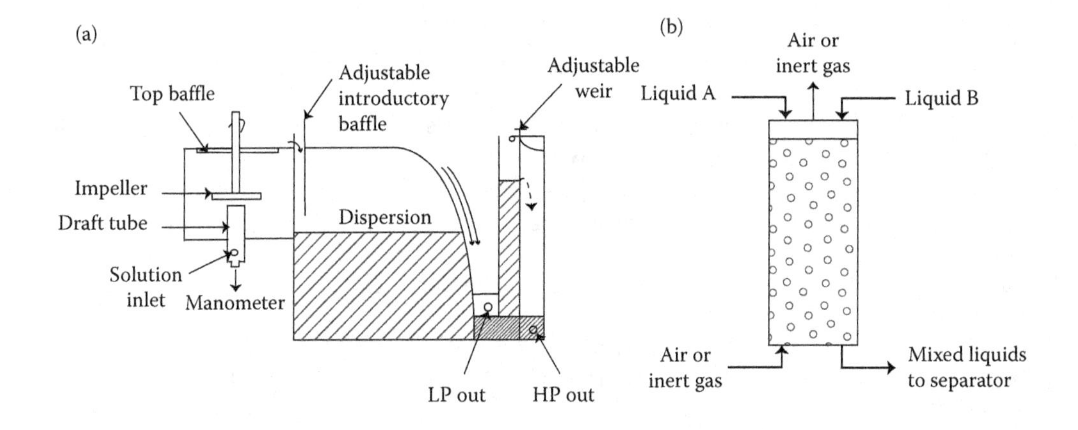

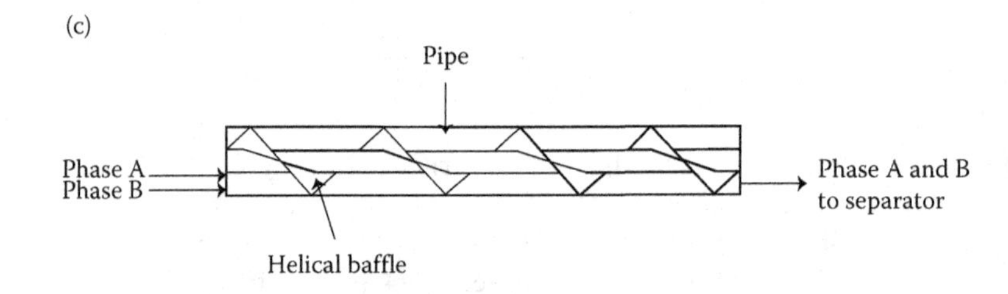

Figure 11.8 Sketches of some common liquid–liquid contactor reactor designs. HP, heavy phase; LP, light phase. (a) Mixer-settler, (b) horizontal pipeline mixer (static mixer), and (c) air agitated contactor.

b. Mechanically agitated contactors
   Mixer-settlers; inert gas agitated contactors
   Columns agitated with rotating stirrers
   Pulsed columns
2. Pipeline (horizontal) contactors; static mixers
3. Centrifugal extractors

It will be noticed that contactors under category A are common to gas–liquid and liquid–liquid reactions. The main difference stems from the need for phase separation in liquid–liquid reactions. A few common designs are shown in Figure 11.8. The design of liquid–liquid extractors in general has been well treated by Laddha and Degaleesan (1976) and the general subject of mass transfer by Taylor and Krishna (1993).

Values of mass transfer coefficients and interfacial areas for different contactors

Since k_L' and a_L are important parameters in contactor selection, approximate values of these for the more important contactors are listed in Table 11.5, along with some other features of these contactors.

Calculation of reactor volume/reaction time

The procedures are essentially similar to those for gas–liquid reactions, except that reaction in both the phases must be considered. As may often happen, reaction may well be confined to a simple phase, but such an assumption can only be made after experimental verification, or if the concerned parameter values distinctly rule out reaction in the other phase. We illustrate a typical procedure in Example 11.2.

Example 11.2: Design of a batch reactor for *m*-chloroaniline by liquid–liquid reduction of *m*-nitrochlorobenzene with disodium sulfide

A batch of *m*-chloroaniline is to be produced by reducing *m*-nitrochlorobenzene (MNCB) with an aqueous solution of sodium disulfide (Na_2S_2) in a fully baffled 2000 L mechanically agitated batch contactor at 105°C. The reaction is known to be restricted to the continuous aqueous phase (i.e., phase 2). The regime of this reaction (for 100% conversion with stoichiometric quantities of reactants) was shown to shift from 3 (fast) at the start to 1–2 at the end. Using this information and the following data, calculate the reaction time.

Data:

Amount of sodium disulfide, Na_2S_2 = stoichiometric requirement for 100% conversion of MNCB.
Final concentration of aqueous sodium disulfide, $[B]_{f,2} = 2.96 \times 10^{-5}$ mol/cm^3.
Volume of continuous phase, V_2 or $V_c = 8.805 \times 10^5$ cm^3.
Volume of dispersed phase, V_1 or $V_d = 3.125 \times 10^5$ cm^3.
Effective interfacial area, $a_L = 58.95$ cm^2/cm^3 of dispersion.

SOLUTION

Since the regime shifts from 3 to 1–2, and since this shift would probably be quite gradual, it is reasonable to consider the reaction as occurring in regimes 1–2–3. Also, the reaction is known to be restricted to phase 2.

Equation 11.29 gives the reaction rate for regimes 1–2–3. This equation can be recast as follows for a liquid–liquid reaction, with the dispersed phase (liquid phase 1) replacing the gas phase:

$$-r_A = \frac{k_{2,c}[B]_{b,c}a_L[A]_d^*\sqrt{D_{A,c}k_{2,c}[B]_{b,c} + k_L'^2}}{k_{2,c}[B]_{b,c} + (a_L/b)\sqrt{D_{A,c}k_{2,c}[B]_{b,c} + k_L'^2}} \quad \text{(E11.2.1)}$$

Note that there is no gas phase resistance involved here, hence the term $1/k'_{GP}a_L$ drops out. The various other terms are as defined under Notation, except that for phase 1 the symbol d is used instead of 1, and for phase 2 the symbol c is used instead of 2. For example, $k_{2,c}$ represents second-order rate constant in phase 2 (continuous phase).

The material balance for this case may be written as

$$-V_d \frac{d[A]_{b,d}}{dt} = -\frac{V_c}{\upsilon_B}\frac{d[B]_{b,d}}{dt} \tag{E11.2.2}$$

which on integration gives

$$[B]_{b,c} = ([B]_{b,c})_i - \frac{\upsilon_B V_d}{V_c}\{([A]_{b,d})_i - [A]_{b,d}\} \tag{E11.2.3}$$

where $([A]_{b,d})_i$ represents the initial concentration of A (MNCB) in the dispersed phase and $([B]_{b,c})_i$ that of B (Na_2S_2) in the continuous phase. We also have

$$-V_d \frac{d[A]_{b,d}}{dt} = (-r_A')a_L V_c \tag{E11.2.4}$$

Further, the interfacial concentration of A for this case can be written as

$$[A]_d^* = \frac{m_A}{[A]_{b,d}} \tag{E11.2.5}$$

Combining Equations E11.2.1, E11.2.3 through E11.2.5 and rearranging results in

$$-\frac{d[A]_{b,d}}{dt} = \frac{\alpha''\rho''\left(\gamma'' + [A]_{b,d}\right)\beta''\sqrt{\gamma'' + [A]_{b,d} + \delta''}[A]_{b,d}}{\alpha''\left(\gamma'' + [A]_{b,d}\right) + \eta''\sqrt{\gamma'' + [A]_{b,d} + \delta''}} \tag{E11.2.6}$$

where

$$\alpha'' = \frac{k_{2,c}\upsilon_B V_d}{V_c}, \quad \beta'' = a_L m_A \sqrt{\frac{D_A k_{2,c}\upsilon_B V_d}{V_c}}$$

$$\gamma'' = \frac{V_c([B]_{b,c})_i}{\upsilon_B V_d} - ([A]_{b,d})_i \tag{E11.2.7}$$

$$\delta'' = \frac{k_{2,c}^2 V_c}{D_A k_{2,c}\upsilon_B V_d}, \quad \eta'' = \frac{a_L}{b}\sqrt{\frac{D_A k_{2,c}\upsilon_B V_d}{V_c}}$$

$$\rho'' = \frac{V_d V_c}{V_d}$$

The initial and final conditions are at:

$$t = 0, [A]_{b,d} = ([A]_{b,d})_i$$
$$t = t_B, [A]_{b,d} = ([A]_{b,d})_f \tag{E11.2.8}$$

Also, since the reactants are present in stoichiometric amounts in this case, the term γ'' becomes zero, and Equation E11.2.6 reduces to

$$-\frac{d[A]_{b,d}}{dt} = \frac{\rho''\alpha''\beta''[A]_{b,d}^2\sqrt{[A]_{b,d} + \delta''}}{\alpha''[A]_{b,d} + \eta''\sqrt{[A]_{b,d} + \delta''}} \quad \text{(E11.2.9)}$$

Integration of this equation with boundary conditions (E11.2.8) gives

$$t_B = \frac{2}{\rho''\beta''\sqrt{\delta''}} \ln \frac{\operatorname{cosec}\left\{\tan^{-1}\sqrt{([A]_{b,d})_i/\delta''} - \cot\left[\tan^{-1}\sqrt{([A]_{b,d})_i/\delta''}\right]\right\}}{\operatorname{cosec}\left\{\tan^{-1}\sqrt{(([A]_{b,d})_f)/\delta''} - \cot\left[\tan^{-1}\sqrt{(([A]_{b,d})_f)/\delta''}\right]\right\}}$$

$$+ \frac{\eta''}{\rho''\alpha''\beta''}\left[\frac{1}{([A]_{b,d})_f} - \frac{1}{([A]_{b,d})_i}\right] \quad \text{(E11.2.10)}$$

The numerical values of the various parameters can be calculated using the given data. Thus,

$$\alpha'' = 1.228 \times 10^3,\ \beta'' = 9.88 \times 10^{-3}$$

$$\delta'' = 6.53 \times 10^{-4},\ \eta'' = 15.635$$

$$\rho'' = 1193 \times 10^3/312.5 \times 10^3 = 3.818$$

$$([A]_{b,d})_i = 8.454 \times 10^{-3},\ ([A]_{b,d})_f = 8.454 \times 10^{-5}$$

Substituting these into Equation E11.2.10 gives the batch time as

$$t_B = 7008 \text{ s}$$

Stirred tank reactor: Some practical considerations

The mechanically agitated contactor, also referred to as the stirred tank reactor (STR), is the most commonly used reactor for small- and medium-volume productions in organic synthesis. We shall, therefore, summarize the important practical features of this reactor, which should also be applicable to a CSTR.

1. Several types of impellers are used, the most common being the simple marine impeller and the highly efficient double-motion stirrer. The reactors are always suitably baffled.
2. An important consideration is mixing or dispersion (see point 5). As a rule of thumb, for homogeneous reactions or for reactions

Table 11.6 Scale-Up Criteria for a Stirred Reactor $N_2/N_1 = (V_2/V_1)^c$, where N Represents the Rotations of the Stirrer and V Represents the Volume of Reactors

Criterion	c	Remarks
Constant stirrer power/unit volume	−2/9	To be used when the main object is mixing; criterion 1
Constant heat transfer coefficient	−0.15	To be used when heat transfer is the main problem; additional heat exchange capacity must be added when needed
Constant tip speed	−1/3	Usually for gas liquid reaction; maintains the same gas distribution in the two reactors, satisfies criterion 2(b).
Constant pumping rate/unit volume	0 that is, $N_1 = N_2$	

in the kinetic regime of fluid–fluid reactions, a tip speed of 2.5–3.3 m/s is required for good mixing, whereas for fluid–fluid reactions in the fast regime a tip speed of 5–6 m/s is necessary. The power input for homogeneous reactions is of the order of 0.1 kW/m^3, while for fluid–fluid systems it is 2.0 kW/m^3.

3. For reactions with a sizable heat effect, heat is abstracted in several ways: reflux condenser, internal coil, external heat exchanger, cooling jacket, and half-round pipes wound on the reactor body. The overall heat transfer coefficient with water in the tank for a water-cooled or steam-heated-jacketed vessel is 0.15–1.7, for a tank with water-cooled half-round pipe is 0.3–0.9, and for a tank with water-cooled internal coil is 0.5–1.2 kW/m^2 K^{-1} (Rose, 1981).
4. Reactors are usually available only in standard sizes. It is much cheaper to buy an oversized tank conforming to the nearest standard than to have a reactor of actual calculated dimensions specially fabricated.
5. Scale-up criteria. In addition to design by modeling as discussed in previous sections, it is also possible to use simple scale-up criteria based on laboratory results. For this purpose, all the available criteria are consolidated in a common form in Table 11.6, with different values of the exponent c for the different criteria used. This kind of scale-up should be acceptable for the scales of production normally encountered in fine chemicals and intermediates synthesis, but could be misleading for bulk chemicals production.

References

Bisio, A. and Kabel, R.L., *Scale Up of Chemical Processes: Conversion from Laboratory Scale Test to Successful Commercial Scale Design*, Wiley, New York, NY, 1985.

Bohmer, K. and Blenke, H., *Verfahrenstechnik*, **6**, 50, 1972.

Botton, R., *Chem. Eng. J.*, **20**, 87, 1980.

Buchholz, R., Tsepetonides, J., Steinemannn, J., and Onken, U., *Ger. Chem. Eng.*, **6**, 105, 1983.

Calderbank, P.H., *Trans. Inst. Chem. Eng.*, **36**, 443, 1958.

Danckwerts, P.V., *Gas–Liquid Reactions*, McGraw-Hill, New York, NY, 1970.

Doraiswamy, L.K. and Sharma, M.M., *Heterogeneous Reactions—Analysis Examples and Reactor Design*, Vol. 2, Wiley, New York, NY, 1984.

Heijnen, J.J. and Van't Riet, K., *Chem. Eng. J.*, **B21**, 42, 1984.

Henzler, H.J., *Chem. Ing. Tech.*, **54**, 8, 1982.

Joshi, J.B. and Doraiswamy, L.K., *Chemical Reaction Engineering in Chemical Engineer's Handbook* (Ed. L.F. Albright), CRC Press, Boca Raton, FL, 2009.

Juvekar, V.A., *Studies in Mass Transfer in Gas–Liquid and Gas–Liquid–Solid Systems*, PhD (Tech.) thesis, University of Bombay, India, 1976.

Kastaněk, F., *Coll. Czech. Chem. Commun.*, **41**, 3709, 1976.

Kastaněk, F., Zahradnik, J., Kratochvil, J., and Cermak, J., *Reactors for Gas–Liquid Systems*, Ellis Horwood, New York, NY, 1993.

Kramers, H. and Westerterp, K.R., *Elements of Chemical Reactor Design and Operation*, Chapman & Hall, London, 1963.

Kulkarni, B.D. and Doraiswamy, L.K., *AIChE J.*, **21**, 501, 1975.

Laddha, G.S. and Degaleesan, T.E., *Transport Phenomena in Liquid Extraction*, McGraw-Hill, New Delhi, India, 1976.

Ladhabhoy, M.E. and Sharma, M.M., *J. Appl. Chem.*, **19**, 267, 1969.

Levenspiel, O., *Chemical Reactor Omnibook*, Oregon State University Bookstore, Corvallis, OR, 1993.

Levenspiel, O. and Godfrey, J.H., *Chem. Eng. Sci.*, **29**, 1723, 1974.

Mann, R., *Gas–Liquid Contacting in Mixing Vessels*, Institute of Chemical Engineers, Rugby Warks, England, 1983.

Meister, D., Post, T., Dunn, I.J., and Bourne, J.R., *Chem. Eng. Sci.*, **34**, 1367, 1979.

Nagel, O., Kurten, H., and Sinn, R. *Chem. Ing. Tech.*, 44, 899, 1972.

Nagel, O., Kurten, H., and Hegner, B., *Chem. Ing. Tech.*, **45**, 913, 1973.

Oldshue, J.Y., *Fluid Mixing Technology*, McGraw-Hill, New York, NY, 1983.

Krishna R. and Sie S.T., *Chem. Eng. Sci.*, 49, 4029, 1994.

Rose, L. M., *Chemical Reactor Design in Practice*, Elsevier, Amsterdam, 1981.

Rylek, M. and Zahradnik, J., *Coll. Czech. Chem. Commun.*, **49**, 1939, 1984.

Schügerl, K., *Chem. Ing. Tech.*, **49**, 605, 1977.

Schügerl, K., *Chem. Ing. Tech.*, **52**, 951, 1980.

Schügerl, K. In *Mass Transfer with Chemical Reaction in Multiphase Systems* (Ed., Alper, E.), Vol. 1, Martin Nijhoff, The Hague, 1983.

Shah, Y.T. and Deckwer, W.D. In *Scale-up of Chemical Processes Conversion from Laboratory Scale Tests to Successful Commercial Size Design* (Eds., Bisio, A., and Kabel, R.L.), Wiley, New York, NY, 1985.

Shah, Y.T., Kelkar, B.G., Godbole, S.P., and Deckwer, W.D., *AIChE J.*, **28**, 353, 1982.

Sisak, C. and Hung, J., *Ind. Chem.*, **14**, 39, 1986.

Sittig, W. and Heine, H., *Chem. Ing. Tech.*, **49**, 595, 1977.

Sobieszuk, P., Aubin, J., and Pohorecki, R., *Chem. Eng. Technol.* **35**(8), 1346, 2012.

Taylor, R. and Krishna, R., *Multicomponent Mass Transfer*, Wiley, New York, NY, 1993.

Treybal, R.E., *Liquid Extraction*, 2nd ed., McGraw-Hill, New York, NY, 1963; *Mass Transfer Operations*, McGraw-Hill, New York, NY, 1968.

van Baten J.M. and Krishna R., *Chem. Eng. Sci.*, **59**, 2535, 2004.

Van Krevelen, D.W., *Research*, **3**, 106, 1950.

Viesturs, U.E., Kuznecov, A.M., and Samenkov, V.V., *Fermentation Systems (in Russian)*, Zinatne, Riga, 1986.

Warwick, G.C.I., *Chem. Ind.*, **5**(May), 882, 1973.

Zahradnik, J., Kastaněk, F., and Kratochvil, *J. Coll. Czech. Chem. Commun.*, **15**, 27, 1982.

Zahradnik, J., Kratochvil, and Rylek, M. *Coll. Czech. Chem. Commun.*, **50**, 2535, 1985.

Zhang, Y. and Zheng, L., *Chem. Eng. Sci.*, **69**, 449, 2012.

Chapter 12 Multiphase reactions and reactors

Chapter objectives

After the completion of this chapter, the successful student must be able to

- Select the most suitable multiphase reactor for the reaction at hand.
- Determine mass transfer coefficients and gas (or liquid) bubble sizes from physical properties and flow parameters of the fluids.
- Differentiate various multiphase reactor types such as mechanically agitated slurry reactors (MASRs), bubble column slurry reactors (BCSRs), loop slurry reactors (LSRs), and trickle bed reactors (TBRs).
- Collect and interpret laboratory data for multiphase reactions.

Introduction

The thrust of this chapter is on reactions and reactors involving a gas phase, a liquid phase, and a solid phase, which can be either a catalyst (but not a phase-transfer catalyst) or a reactant, with greater emphasis on the former.

Design of three-phase catalytic reactors

The approach

Our emphasis in this chapter will be on slurry reactors (various designs of which are extensively used), but the other types of reactors will also be briefly considered.

Three-phase reactors are operated either in the semibatch or continuous mode, and batch operation is almost never used since the gas phase is invariably continuous. The general principles of design are the same for all types of reactors for a given mode of operation, that is, semibatch or continuous. They differ with respect to their hydrodynamic features, particularly mass and heat transfer. The rate constant k_{w1} would be the same for all the reactors, but what would be specific to each reactor type is the mass transfer term $\bar{k}_A$. Hence, we first consider the design of semibatch and continuous reactors, in general, and then briefly outline the mass transfer and other significant features of the more common types of reactors.

Semibatch reactors: Design equations for (1,0)- and (1,1)-order reactions

Consider reaction R1, for which the rate equation for both the cases can be written as

$$A\,(\mathrm{g}) + \nu_B B\,(\mathrm{l}) \xrightarrow{\text{solid catalyst}} R \tag{R1}$$

$$-r_A = k_1[A] \tag{12.1}$$

where k_1 is the true first-order rate constant for the (1,0)-order reaction, but for the (1,1) case, it is equal to $k_{11}[B]_b$ and is hence a function of $[B]_b$. The final equations for semibatch time for (1,0)- and (1,1)-order reactions are (Ramachandran and Chaudhari, 1980a) are as follows.

(1,0)-Order reaction:

$$t_{\mathrm{SB}} = \frac{[B]_{bi} X_B}{\nu_B [A]^*}\left(\frac{1}{\bar{k}_A} + \frac{1}{\varepsilon w k_{w1}}\right) \tag{12.2}$$

where

$$\varepsilon\ (\text{for a sphere}) = \frac{1}{\phi}\left[\coth(3\phi) - \frac{1}{3\phi}\right] \tag{12.3}$$

$$\phi = \frac{R}{3}\left(\frac{\rho_c k_{w1}}{D_{eA}}\right)^{1/2} \tag{12.4}$$

(1,1)-Order reaction:

$$t_{\mathrm{SB}} = \frac{[B]_{bi} X_B}{\nu_B [A]^*}\left(\frac{1}{\bar{k}_A}\right) + \frac{[B]_{bi} R^2 \rho_c I}{3 w D_{ea}} \tag{12.5}$$

$$I = \frac{2}{\phi_i}\ln\left[\frac{\phi_i \cosh\phi_i - \sinh\phi_i}{\phi_i\sqrt{1-X_B}\cosh(\phi_i\sqrt{1-X_B}) - \sinh(\phi_i\sqrt{1-X_B})}\right] \tag{12.6}$$

$$\phi_i = \frac{R}{3}\left[\frac{r_c k_{wi}[B]_{bi}}{D_{eA}}\right]^{1/2} \tag{12.7}$$

Equations 12.5 and 12.6 can be simultaneously solved to predict the semibatch time as a function of conversion and the Thiele modulus ϕ_i (for initial conditions) defined by Equation 12.7.

Continuous reactors

Three-phase reactions can also be run continuously in several types of reactors, for example, packed bed reactors, fluidized-bed reactors, BCSRs, and continuous stirred tank or sparged slurry reactors. These reactors differ mainly with respect to the flow patterns of the individual phases, and the performance equations will differ accordingly. However, a general equation applicable to all types of flow of the two phases can be developed if the gas phase concentration is assumed to be constant (thus eliminating gas phase hydrodynamics from consideration) and a suitable mixing model, such as the dispersion model, is used to characterize liquid phase flow. The equation can then be reduced to plug or mixed flow of liquid by appropriate simplification. The assumption of constant gas phase concentration is justified when pure gas is used, or when high rates of gas flow are used, thus minimizing concentration changes. The first assumption is usually valid for hydrogenation, carbonylation, or hydroformylation reactions, while for oxidation reactions (where air is normally used), conditions must be adjusted to validate the second assumption. Equations can also be developed for varying gas phase concentrations (Goto and Smith, 1978) but will not be attempted here.

The variables affecting continuous reactor operation and their dimensionless forms are explained in Figure 12.1. Considering an isothermal reaction with pseudo-first-order kinetics (that is, Equation 12.6 with $m = 1$), and accounting for backmixing by using the dispersion model characterized by the Peclet number

$$Pe = \frac{u_L L}{D_{el}} \tag{12.8}$$

the following material balance equation can be written:

$$\begin{pmatrix}\text{Moles of } A\\ \text{reacted}\end{pmatrix} = \begin{pmatrix}\text{Moles of } A\\ \text{in entering}\\ \text{liquid}\end{pmatrix} - \begin{pmatrix}\text{Moles of } A\\ \text{in leaving}\\ \text{liquid}\end{pmatrix} + \begin{pmatrix}\text{Moles of } A\\ \text{transferred from}\\ \text{the gas phase}\end{pmatrix} \tag{12.9}$$

$$V_r(-r_A) = Q_L[A]_{L0} - Q_L[A]_{Lf} + A_c\int_0^L k'_{GL,A}a_L\left([A]^* - [A]_L\right)dz \tag{12.10}$$

This equation can be recast in terms of a *reactor efficiency* E_R defined as

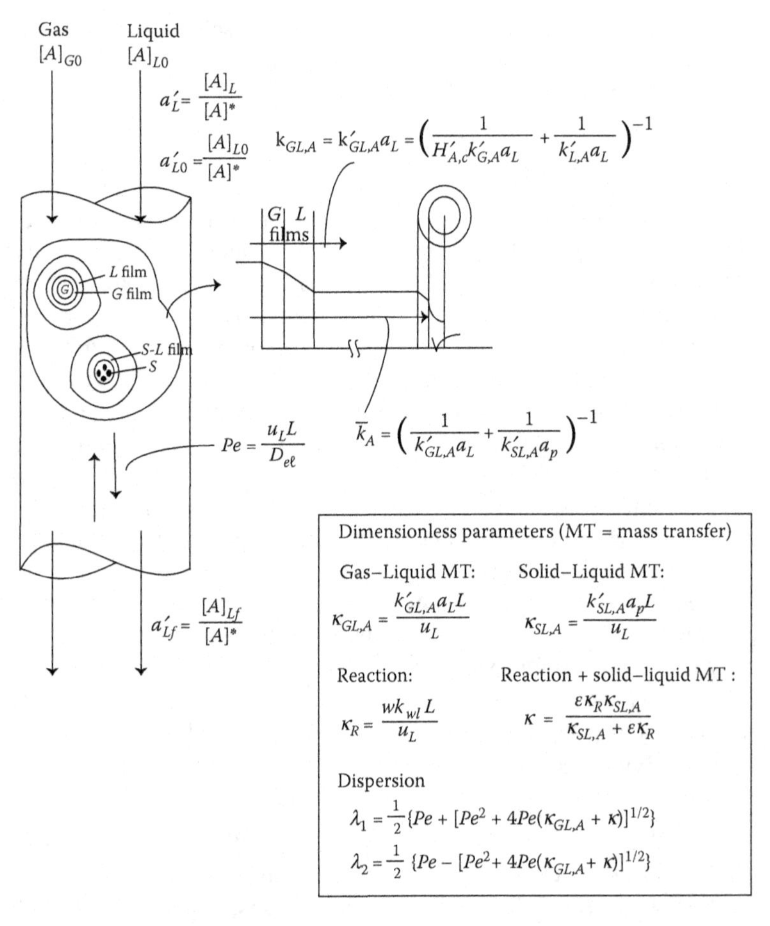

Figure 12.1 The variables and dimensionless groups used in the analysis of continuous three-phase slurry reactors.

$$E_R = \frac{\left(\begin{array}{c}\text{Actual rate of reaction of } A \text{ over} \\ \text{the entire catalyst in the reactor}\end{array}\right)}{\left(\begin{array}{c}\text{Rate based on the assumption that the} \\ \text{entire catalyst in the reactor is exposed} \\ \text{to a uniform concentration of } [A]^*\end{array}\right)} \tag{12.11}$$

to give

$$E_R = \frac{a'_{L0} - a'_{Lf}}{\kappa_R} + \frac{\kappa_{GL,A}}{\kappa_R}\int_0^1 (1 - a'_L)\,dz \tag{12.12}$$

where $a'_{L0} = [A]_{L0}/[A]^*$ and $a'_{Lf} = [A]_{Lf}/[A]^*$. The dimensionless quantities appearing in the above equation are defined in Figure 12.1.

The only single variable in Equation 12.12 is the reduced concentration a'_L, and equations for this can be developed for the PF, MF, and dispersion models. The following final expressions can then be derived for E_R for the three models (Ramachandran and Chaudhari, 1983):

Plug-flow model

$$E_R = \frac{\kappa}{\kappa_R(\kappa_{GL,A} + \kappa)}\left[\kappa_{GL,A} + \left(a'_{L0} - \frac{\kappa_{GL,A}}{\kappa_{GL,A} + \kappa}\right)\right]\{1 - \exp[-(\kappa_{GL,A} + \kappa)]\}$$

(12.13)

For the symbols in the equations, see Figure 12.1

Mixed flow model

$$E_R = \frac{\kappa(a'_{L0} + \kappa_{GL,A})}{\kappa_R(1 + \kappa_{GL,A} + \kappa)} \tag{12.14}$$

Dispersion model

$$E_R = \frac{\kappa}{\kappa_R(\kappa_{GL,A} + \kappa)}\left[\kappa_{GL,A} + \left(a'_{L0} - \frac{\kappa_{GL,A}}{\kappa_{GL,A} + \kappa}\right)\right]\left[1 - \frac{Pe(\lambda_1 - \lambda_2)e^{Pe}}{\lambda_1^2 e^{\lambda_1} - \lambda_2^2 e^{\lambda_2}}\right]$$

(12.15)

An indication of the effect of the dimensionless variables representing mass transfer at the gas–liquid and solid–liquid interfaces ($\kappa_{GL,A}$, $\kappa_{SL,A}$), chemical reaction (κ_R), and axial diffusion (Pe) can be had from the simulation results of Goto et al. (1976) (see also Shah and Paraskos, 1975; Shah et al., 1976; Goto and Smith, 1978). These may be summarized as follows:

1. Backmixing has very little effect, and at high values of the groups representing mass transfer at the two interfaces [($\kappa_{GL,A} + \kappa_{SL,A} > 3$)], the predictions of the two extreme models are almost identical.
2. Representing the group [$\kappa_{GL,A}/(\kappa_{GL,A} + \kappa)$] by ξ, the following conclusions may be drawn: for $a'_{L0} < \xi$, MFR performs better; for $a'_{L0} > \xi$, PFR performs better; and for $a'_{L0} = \xi$, the models predict identical results. This is a significant conclusion since in two-phase reactors, PFR is always superior to MFR.

   The fractional conversion of B can be predicted from

$$X_B = q\kappa_R E_R \tag{12.16}$$

   where q is defined as $v_B\,[A]^*/[B]_{bi}$.

Types of three-phase reactors

The four most important three-phase reactor types are the mechanically agitated slurry reactor (MASR), bubble column slurry reactor (BCSR), loop recycle slurry reactor (LRSR), and trickle bed reactor (TBR). The first three, sketched in Figure 12.2, are slurry reactors, whereas the fourth is a fixed-bed reactor. The features most relevant to a preliminary design of these reactors in organic synthesis/technology are briefly described below.

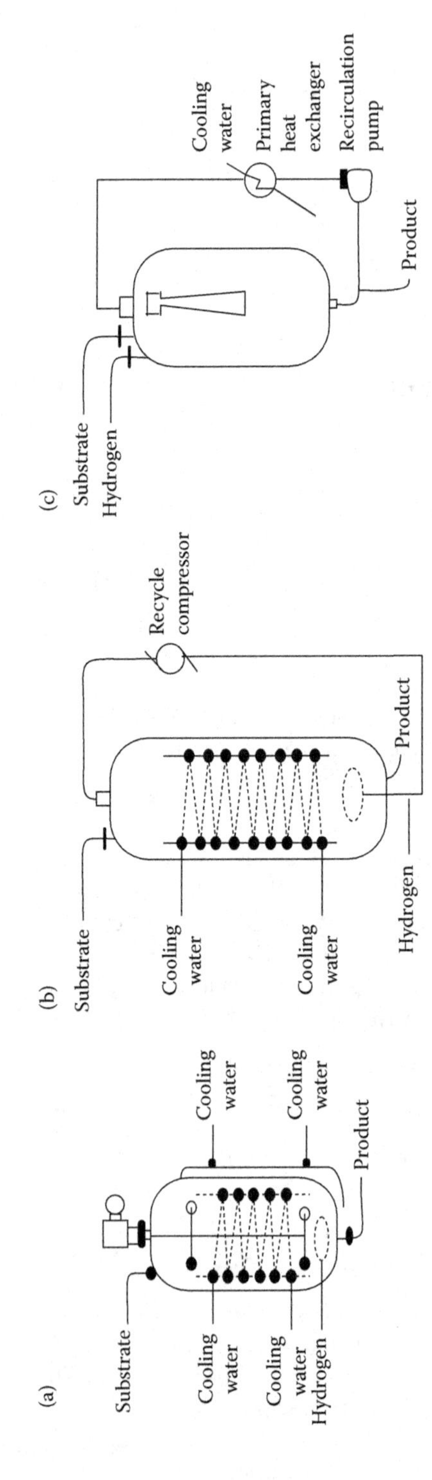

Figure 12.2 Schematic diagrams of different types of three-phase slurry reactors. (a) MASR; (b) BCSR; (c) jet LSR.

Mechanically agitated slurry reactors

This class of reactors is the most commonly used in organic synthesis. A cascade of stirred reactors can also be used. For example, a cascade with a column configuration as shown in Figure 12.3 is used in the production of the vitamin, carbol (Wiedeskehr, 1988).

Mass transfer Mills and Chaudhari (1997) recommend the following correlations of Yagi and Yoshida (1975) (Equation 12.17) and Bern et al. (1976) (Equation 12.18) for gas–liquid mass transfer:

$$\frac{k_L' a_L d_s^2}{D} = 0.06\left(\frac{d_s^2 N \rho_L}{\mu_L}\right)^{1.5}\left(\frac{d_s N^2}{g}\right)^{0.19}\left(\frac{\mu_L}{\rho_L D}\right)^{0.5}\left(\frac{\mu_L u_G}{\sigma_L}\right)^{0.6}\left(\frac{N d_s}{u_G}\right)^{0.32} \qquad (12.17)$$

$$k_L' a_L = 1.1 \times 10^{-2} N^{1.16} d_s^{1.98} u_G^{0.32} V_L^{-0.52} \qquad (12.18)$$

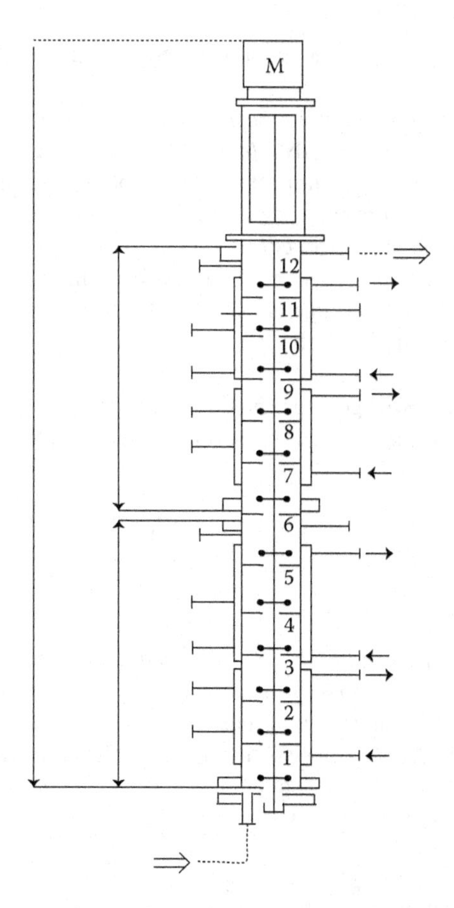

Figure 12.3 A cascade reaction column with multiple agitators for the synthesis of the vitamin "carbol." (After Wiedeskehr, H., *Chem. Eng. Sci.*, **43**, 1783, 1988.)

Liquid–liquid mass transfer depends on whether the transfer is from continuous to dispersed phase or vice versa. The liquid–liquid interfacial area a_{LL} can be estimated from

$$a_{LL} = \frac{6h_L}{d_0} \tag{12.19}$$

where h_L is the dispersed liquid phase holdup and d_0 is the average size of the dispersed droplets that can be determined from a correlation given by Okufi et al. (1990).

Solid–liquid mass transfer coefficients can be calculated from the following correlation proposed by Sano et al. (1974):

$$\frac{k'_{SL}d_p}{DS_F} = 2 + 0.4\left(\frac{ed_p^4\rho_L^3}{\mu_L^3}\right)^{1/4}\left(\frac{\mu_L}{\rho_L D}\right)^{1/3} \tag{12.20}$$

where $e = P/\rho_L V_L$, where P is the power.

The above correlation is also recommended for liquid–liquid systems.

Minimum speed for complete suspension It is necessary to ensure complete suspension of catalyst (see Joshi, 1982; Beenackers and Van Swaaij, 1993). The most important single parameter that influences complete suspension is the speed of agitation. This can be calculated from the early but still valid correlation of Zwietering (1958). In practice, the minimum agitation speed for uniform distribution of catalyst in the common loading range of $5 < w' < 30$ g/100 g liquid and particle size range of $10 < d_p < 150$ μm is 150–600 rpm.

Gas holdup Knowledge of gas holdup is important since it is an indication of the gas residence time. A useful correlation is that of Yung et al. (1979):

$$h_G = 6.8 + 10^{-3}\left(\frac{Q_G}{Nd_s^3}\right)^{0.5}\left(\frac{\rho_L N^2 d_s^3}{\sigma_L}\right)^{0.65}\left(\frac{d_s}{d_t}\right)^{1.4} \tag{12.21}$$

Controlling regimes in an MASR One can identify at least five major controlling regimes in an MASR: gas–liquid mass transfer, liquid–solid mass transfer of A, liquid–solid mass transfer of B, pore diffusion, and surface reaction. Table 12.1 lists the effects of different variables on the reaction for all these regimes of control. These effects have been classified as major and minor effects.

Bubble column slurry reactors

The BCSR is used mainly for large-volume productions, but in a few cases, it is also used for specialty chemicals, particularly where a

Table 12.1 Controlling Regimes in Slurry Reactors: Effect of Pertinent Variables on r_A

	Variables Whose Influence Is		
Controlling Resistance	**Major**	**Minor**	**Negligible**
Gas–liquid mass transfer, $k_L' a_L'$	$[A]^*$, type of impeller and stirring speed (mechanically agitated contactor), gas velocity (sparged contactor)	Temperature	$[B]_b$, d_p, w
Liquid–solid mass transfer of A, $k_{SL}' a_P$ (for A)	d_p, w, $[A]^*$, stirring speed (mechanically agitated contactor, turbulent regime)	Temperature, gas velocity (sparged reactor, fine particles)	$[B]_b$
Liquid–solid mass transfer of B, $k_{SL}' a_P$ (for B)	d_p, w, $[B]_b$, stirring speed (mechanically agitated contactor, turbulent regime)	Temperature, gas velocity (sparged reactor, fine particles)	$[A]^*$
Surface reaction (pore diffusion negligible)	d_p, w, temperature, $[A]^*$, $[B]_b$; independent of $[A]^*$ for $m=0$; independent of $[B]_b$ for $n = 0$		Type of impeller and stirring speed (mechanically agitated contactor), gas velocity (sparged contactor)
Surface reaction with pore diffusion	d_p, w, $[A]^*$, $[B]_b$, temperature	Pore structure	Type of impeller and stirring speed (mechanically agitated contactor), gas velocity (sparged contactor)

Source: Adapted from Doraiswamy, L.K. and Sharma, M.M., *Heterogeneous Reactions—Analysis Examples and Reactor Design*, Vol. 2, Wiley, New York, 1984.

gaseous product is produced and corrosion problems prohibit the use of agitated systems. The extent of backmixing can be controlled by using a draft tube. Such sparged reactors are particularly useful for chlorination, sulfonation, and phosgenation reactions.

Regimes of flow The principal operating feature of a BCSR is that the catalyst is kept in suspension by the turbulence induced by the gas flow. Depending on the gas and liquid velocity ranges, BCSR can operate in different regimes of flow, as shown in Figure 12.4. In the homogeneous (bubbly) flow regime observed at very low velocities, gas bubbles of uniform sizes are equally distributed in both the axial and radial directions. At higher velocities, bubbles tend to coalesce, leading to nonuniform sizes. This regime of flow is termed heterogeneous churn turbulent flow regime. Reactor scale-up is difficult in this regime and hence should be avoided. As the velocity is further increased, the so-called heterogeneous slug flow regime is observed—especially in small-diameter columns. This regime is also not of practical interest.

Mass transfer In the bibliography, the sources of many proposed correlations for both gas–liquid and solid–liquid mass transfer are listed. The recommended correlations are as follows.

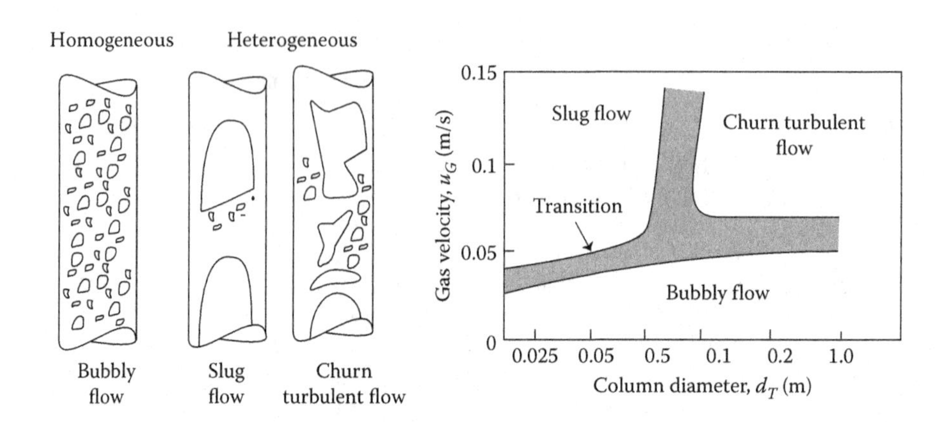

Figure 12.4 Flow regimes and transition gas velocities in BCSR. (Adapted from Shah, Y.T. et al., *AIChE J.*, **28**, 353, 1982.)

Gas–liquid mass transfer coefficient (Akita and Yoshida, 1973, 1974):

$$k_L' a_L = 0.6 D^{0.5} \left(\frac{\mu_L}{\rho_L} \right)^{-0.12} \left(\frac{\sigma_L}{\rho_L} \right)^{-0.62} d_T^{0.17} g^{0.93} h_G^{1.1} \tag{12.22}$$

Solid–liquid mass transfer coefficient:

Equation 12.20 for the MASR is especially valid for the BCSR, but with $e = u_G g$.

Unlike in MASRs, where liquid mixing is always considered to be complete, in this case, allowance must be made for partial mixing. Thus, it may often be necessary to use the dispersion model given by Equation 12.15. The liquid phase axial diffusion coefficient for estimating the Peclet number appearing in this equation may be calculated from the correlations of Hikita and Kikukawa (1975) or Mangartz and Pilhofer (1980).

Minimum velocity for complete solids suspension The following correlation proposed by Roy et al. (1964) for the maximum amount of solids that can be kept in complete suspension (w_{max}) may be used:

$$\frac{w_{max}}{\rho_L} = 6.8 \times 10^{-4} \frac{C_\mu d_T u_G \rho_G}{\mu_G} \left(\frac{\sigma_L h_g}{u_G \mu_L} \right)^{-0.23} \left(\frac{h_G u_{TS}}{u_G} \right)^{-0.18} \omega_F^{-3} \tag{12.23}$$

where C_μ is the viscosity correlation factor that may be estimated from $C_\mu = 0.23 - 0.20(\log \mu_L) + \log \mu_L^2$, u_{TS} the terminal velocity of the particles, and ω_F the wettability factor (which can usually be assumed to be 1).

Gas holdup Of the many correlations proposed (see Hikita et al., 1980a,b, for a review), those of Akita and Yoshida (1973) and Hikita et al. (1980b) are useful. One of the first correlations is that of Akita and Yoshida (1973), which continues to be the recommended one.

$$\frac{h_G}{(1-h_G)^4} = 0.2\left(\frac{gd_T'\rho_L}{\sigma_L}\right)^{1/8}\left(\frac{gd_T^2\rho_L^2}{\mu_L^2}\right)^{1/12}\left(\frac{u_G}{\sqrt{gd_T}}\right) \tag{12.24}$$

A more useful but slightly less accurate correlation is that of Hikita et al. (1980b).

$$h_g = 0.672 g^{-0.131} u_G^{0.578} \rho_G^{0.062} \mu_G^{0.107} u_L^{-0.053} \sigma_T^{-0.185} \tag{12.25}$$

Example 12.1: Reactor choice for hydrogenation of aniline

Aniline can be hydrogenated to cyclohexylamine in the presence of Raney nickel according to the reaction

$$\underset{A}{3H_2(g)} + \underset{B}{C_6H_5NH_2(l)} \rightarrow \underset{R}{C_6H_{11}NH_2(l)}$$

It is desired to choose between MASR and BCSR for carrying out the reaction on a large scale. To enable rational choice, calculate the conversion of aniline in a 50-cm bench-scale reactor with a liquid height of 50 cm. In addition, agitation speed is also a variable for MASR. The following data of Ramachandran and Chaudhari (1983) may be used:

$k_1 = 51.49$ cm^3/gs
$\rho_G = 6.05 \times 10^{-4}$ g/cm^3, $\rho_L = 1$ g/cm^3, $\rho_C = 1.75$ g/cm^3
$\mu_G = 1.1 \times 10^{-4}$ g/cm s, $\mu_L = 5 \times 10^{-3}$ g/cm s
$\sigma_T = 30$ dyne/cm^2, $[A]^* = 4.46 \times 10^{-6}$ mol/cm^3
$D = 1.16 \times 10^{-4}$ cm^2/s, $[B]_{bi} = 1 \times 10^{-3}$ mol/cm^3
$D_{eA} = 1.93 \times 10^{-5}$ cm^2/s, $d_p = 0.01$ cm, $d_s = 25$ cm
$u_G = 0.5$ cm/s, $w = 3.0 \times 10^{-2}$ g/cm^3, $a'_{L0} = 0$,
$u_L = 0.2$ cm/s

SOLUTION

The various dimensionless groups to be considered are (see Figure 12.1)

$$\kappa_{GL,A} = \frac{k'_L a_{GL} L}{u_L}, \quad \kappa_{SL,A} = \frac{k'_{SL} a_p L}{u_L}$$

$$\kappa_R = \frac{wk_1 L}{u_L}, \quad \kappa = \frac{\varepsilon\kappa_p\kappa_{SL,A}}{\kappa_{SL,A} + \varepsilon\kappa_R} \tag{E12.1.1}$$

The conversion can be calculated from the relation

$$X_B = \xi \kappa_R E_B \tag{E12.1.2}$$

where

$$\xi = \frac{\nu_B [A]^*}{[B]_{bi}} \tag{E12.1.3}$$

Calculations for an MASR

a. *Gas–liquid mass transfer coefficient*:
Liquid volume in the reactor, $V_L = 98{,}000\ \text{cm}^3$. From Equation 12.18

$$k_L' a_L = 1.1 \times 10^{-2} \times 10^{1.16} \times 25^{1.98} \times 0.50^{0.32} \times 98000^{-0.52}$$
$$= 0.187\,\text{s}^{-1}$$

b. *Liquid–solid mass transfer coefficient*:
Assume $e = 3.033 \times 10^5$ erg/s (calculated from the correlations given by Ramachandran and Chaudhari, 1983). Then, from Equation 12.20

$$k_{SL}' = 0.226\ \text{cm/s} \quad \text{and since}$$

$$a_p = \frac{6w}{\rho_c d_p} = \frac{6 \times 3 \times 10^{-2}}{1.75 \times 0.01} = 10.28\ \text{cm}$$

$$k_{SL}' a_P = 2.324\ \text{s}^{-1}$$

c. *Dimensionless mass transfer coefficient and reaction rate parameters*:

$$\kappa_{GL,A} = 467.5, \quad \kappa_{SL,A} = 5810, \quad \kappa_R = 3862$$

Catalytic effectiveness factor ε is calculated from

$$\varepsilon = \frac{1}{\phi}\left(\coth 3\phi - \frac{1}{3\phi}\right)$$

where the Thiele modulus ϕ for a first-order reaction is obtained from

$$\phi = \frac{d_p}{6}\left(\frac{k_1}{D_{eA}}\right)^{0.5} = \frac{0.01}{6}\left(\frac{51.49 \times 1.75}{1.93 \times 10^{-5}}\right)^{0.5} = 3.6$$

giving $\varepsilon = 0.252$ and $\kappa = 833.5$.

d. *Conversion*:
Overall reactor efficiency:
Assuming full mixing

$$E_R = \frac{833.5(0 + 467.5)}{3861.75(1 + 467.5 + 833.5)} = 0.0775$$

Assuming plug flow

$$E_R = \frac{833.5}{3861.75(467.5 + 833.5)} \times \left[467.5 + \left(0 - \frac{467.5}{467.5 + 833.5}\right)\{1 - \exp[-(467.5 + 833.5)]\}\right] = 0.0775$$

Both fully mixed and plug-flow calculations result in the same overall reactor efficiency.
Concentration ratio q:

$$q = \frac{1/3 \times 4.46 \times 10^{-6}}{1 \times 10^{-3}} = 1.487 \times 10^{-3}$$

Conversion of aniline (from Equation 12.16):

$$X_B = 1.487 \times 10^{-3} \times 3861.75 \times 0.0775 = 0.445$$

that is, 44.5%.

Calculations for a BCSR

Calculate the gas–liquid mass transfer coefficient from Equation 12.22 and the solid–liquid mass transfer coefficient from Equation 12.20 with $e = u_G g$.

Knowledge of the gas holdup is needed for calculating the gas–liquid mass transfer coefficient. This can be calculated from Equation 12.24 or 12.25.

The following sample calculations are given for the conditions: gas velocity $u_G = 0.5$ cm/s, liquid velocity $u_L = 0.2$ cm/s, and catalyst loading $w = 0.05$ g/cm^3.

a. *Gas holdup*:
From Equation 12.25,

$$h_G = 0.672 \times 981^{-0.131} \times 0.5^{0.578} \times (6.05 \times 10^{-4})^{0.062} \times 1^{0.069} \times (1.1 \times 10^{-4})^{0.107} \times (5 \times 10^{-3}) \times 30^{-0.185} = 0.0307$$

b. *Gas–liquid mass transfer coefficient*:
From Equation 12.21

$$k_L' a_L = 0.6 \times (1.16 \times 10^{-4})^{0.5} \times \left(\frac{5 \times 10^{-3}}{1}\right)^{-0.12} \times \left(\frac{30}{1}\right)^{-0.62}$$
$$\times\, 50^{0.17} \times 981^{0.93} \times 0.0307^{1.1} = 0.0387\ \mathrm{s}^{-1}$$

c. *Liquid–solid mass transfer coefficient*:
e = 0.5 × 981 = 490.5 cm²/s³, and a_P = 10.28 cm⁻¹. Then, from Equation 12.19, $k'_{AL} a_P = 0.657\,\mathrm{s}^{-1}$.

d. *Dimensionless mass transfer coefficient and reaction rate parameters*:
These are calculated in the same manner as for MASR. The results are as follows:

$$\kappa_{GL,A} = 94.5, \quad \kappa_{SL,A} = 1642.3$$

$$\kappa_R = 3861.75, \quad \kappa = 611$$

e. *Conversion*:
Calculations are identical to those for MASR.
Overall reactor efficiency:
Assuming fully mixed condition, $E_R = 0.0212$
Assuming plug flow condition, $E_R = 0.0212$
Concentration ratio: $q = 1.487 \times 10^{-3}$
Aniline conversion for both fully mixed and plug flow:
$X_B = 0.122$, that is, 12.2%.
Clearly, under the conditions given, MASR performs better than BCSR.

Loop slurry reactors

Types of loop reactors

There are essentially three types of LSR: jet LSR, gas lift LSR, and propeller LSR. Of these, the jet loop reactor is the most commonly used in chemical synthesis/technology (Figure 12.2c) due to the following reasons: higher heat and mass transfer rates, rapid dissipation of heat leading to precise temperature control, controlled residence time in the mixing nozzle where most of the reaction occurs, and easier scale-up to commercial size for mass transfer-controlled reactions (see Chaudhari and Shah, 1986). In actual operation, the liquid phase containing the catalyst particles is injected at a very high velocity (>20 m/s) through a nozzle from either the top or the bottom of the reactor, and hence an important factor in the scale-up of this reactor is the nozzle configuration. Many catalytic hydrogenations traditionally carried out in agitated slurry reactors have been switched to jet loop reactors (Leuteritz, 1973). Table 12.2 lists typical products made in LSR along with remarks on corresponding performances of mechanically agitated reactors.

Table 12.2 Comparison of LSR Performance with Mechanically Agitated Reactor (MASR) Performance under Corresponding Conditions

Hydrogenation of	Catalyst Concentration Relative to That in MASR	Hydrogenation Time and Operating Conditions Relative to Those in MASR
2,5-Dichloronitrobenzene	No basis for comparison because of different catalysts	Hydrogenation time about one-sixth as long; milder conditions
3,4-Dichloronitrobenzene	10% less	Hydrogenation time twice as long
p-Chloronitrobenzene	No basis for comparison as type of catalyst changed	Shorter time and milder conditions
o-Chloronitrobenzene	25% less	Shorter time
p-Nitroaniline	25% less	Shorter time
p-Nitroxylene	Same	Shorter time
o-Nitroethylbenzene	Same	Shorter time
o-Nitroaniline	25% less	Shorter time
Bisphenol A	25% more	Longer time
o-Nitroanisole	30% less	Shorter time

Source: Adapted from Leuteritz, G., *Process Eng.*, **54**, 62, 1973.

Mass transfer In a highly turbulent system like the jet loop reactor, the value of k_L' is not strongly dependent on hydrodynamics. Indeed, it has been found (Blenke and Hirner, 1974) that k_L' has an average value of 4.6×10^{-2} cm/s over a wide range of gas velocities (0.4–6.5 cm/s). The interfacial area can be calculated from the following correlation (Mills and Chaudhari, 1997):

$$a_L = 5.4 \times 10^3 u_G^{0.4} \left(\frac{P}{V_L} \right)^{0.66} \tag{12.26}$$

Trickle Bed Reactors (TBRs)

There are essentially two main classes of three-phase fixed bed catalytic reactors: trickle bed reactors and bubble bed reactors. The class of reactors characterized by cocurrent downflow of gas and liquid are called TBRs.

Regimes of flow

These reactors are in many ways similar to packed bed absorption columns but operate at much lower gas and liquid velocities, usually 0.1–2.0 and 10–300 cm/s, respectively. Depending on the flow rates of the individual phases, four regimes of operation can be identified (Charpentier and Favier, 1975): *trickle (film) flow*, where the liquid flows at a low rate as a laminar film over the packing in the continuous gas phase; *pulse flow*, corresponding to higher gas and liquid rates, with alternate gas-rich and liquid-rich elements passing through the column; *bubble flow*, corresponding to even higher rates, where the liquid phase becomes the continuous phase and the gas flows through it in the form of bubbles; and *spray flow*, where the liquid is dispersed in the form of fine droplets in the continuous gas phase, but a part of it continues to flow as a film over the packing.

Several flow maps have been proposed in which the various regimes are demarcated (e.g., Sato et al., 1973b; Charpentier and Favier, 1975; Chou et al., 1977; Fukushima and Kusaka, 1977; Specchia and Baldi, 1977; Talmor, 1977). These are useful in understanding TBR performance.

Mass transfer

The recommended correlations for gas–liquid and solid–liquid mass transfer for the different regimes are summarized in Table 12.3. The effect of liquid backmixing is usually unimportant.

Controlling regimes in TBRs

In addition to the flow regimes characteristic of TBR, there are also the usual controlling regimes as described earlier for MASRs. We summarize in Table 12.4 the effects of different variables on TBR performance in these regimes.

Collection and interpretation of laboratory data for three-phase catalytic reactions

Experimental methods

Of the different types of reactors that can be used, the stirred basket reactor is the most amenable to manipulation in terms of regimes (see Kenney and Sedriks, 1972). Such a reactor for gas–solid (catalytic) reactions was considered in Chapter 7. Typically, to operate under chemical control conditions, say in a fully baffled 15-cm diameter reactor provided with an 8-cm turbine agitator, the speed of agitation should be in the range 1000–5000 rev/min (corresponding to an impeller tip speed of 24,000–120,000 cm/min).

Effect of temperature

The overall effect of temperature would increase the mass transfer and the decrease the solubility of gases. However, this general conclusion should be viewed with caution since there are exceptions. An important one is hydrogen, whose solubility increases with temperature (e.g., the solubility in soybean increases by 60% as the temperature is raised from 20°C to 100°C). Thus, the overall effect in the case of hydrogenation, perhaps the most frequent "user" of the slurry reactor, will be that the rate will increase significantly with increase in temperature.

Table 12.3 Correlations for TBRs

Flow Regime	Correlation	Reference
Gas–Liquid Mass Transfer		
Trickle flow	$\frac{k'_L a_L d_p^2}{D(1-h_L/f_B)} = 2\left(\frac{S_e}{d_p^2}\right)^{0.2} Re_L^{0.73} Re_G^{0.2} \left(\frac{\mu_L}{\rho_L D}\right)\left(\frac{d_p}{d_T}\right)^{0.2}$	Fukushima and Kusaka (1977)
Pulse flow	$\frac{k'_L a_L d_p^2}{D(1-h_L/f_B)} = 0.11\, Re_L Re_G^{0.4} \left(\frac{d_p}{d_T}\right)^{-0.3}$	Fukushima and Kusaka (1977)
Gas–Liquid Interfacial Area		
Trickle flow	$\frac{k'_L a_L d_p}{1-h_L/f_B} = 3.4\times 10^{-1}\left(\frac{S_e}{d_p^2}\right)^{-1.0} Re_L^{0.4}$	Fukushima and Kusaka (1977)
Pulse flow	$\frac{a_L d_p}{1-h_L/f_B} = 5.9\times 10^{-2}\left(\frac{S_e}{d_p^2}\right)^{-0.3} Re_L^{0.67} Re_G^{0.2}$	Fukushima and Kusaka (1977)
Dispersed bubble flow	$\frac{a_L d_p}{1-h_L/f_B} = 4.5\times 10^{-4}\left(\frac{S_e}{d_p^2}\right)^{-0.9} Re_L^{1.8}$	Fukushima and Kusaka (1977)
Solid–Liquid Mass Transfer		
Trickle flow	$\frac{k'_{SL} d_p}{D}\frac{a_w}{a_p} = 0.815\, Re_L^{0.82} Sc_L^{0.33}$	Satterfield et al. (1978)
Pulse flow	$\frac{k'_{SL} d_p}{D}\frac{a_w}{a_p} = 0.334\left(\frac{e_L \rho_L^2 d_p^4}{h_L u_L^3}\right)^{0.2} Sc_L^{0.3}$	Satterfield et al. (1978)
Liquid Holdup		
Static holdup	$\frac{h_{Ls}}{f_B} = 3.7\times 10^{-2} Bd^n$ where $n = -0.07$ for $Bd < 1$ $= -0.65$ for $Bd > 1$	Mersmann (1972)
Dynamic holdup	$\frac{h_{Ld}}{f_B} = 3.86\, Re^{0.545} Ga^{-0.62}\left(\frac{a_p d_p}{f_B}\right)^{0.65}$	Specchia and Baldi (1977)

Note: $Re = d_p \rho_u \mu$, $Sc = \mu/\rho D_b$, Ga = Galileo number = $d_p^3 \rho_L (g\rho_L + \Delta P')$, $\Delta P'$ = pressure drop/unit bed height (dynes/cm^3), Bd = Bond number = $\rho_L g/\sigma_L a_p^2$.

Interpretation of data

Consider a simple first-order reaction. The total resistance to a first-order reaction can be expressed as the sum of resistances

$$-r_A = [A]^*\left(\frac{1}{k'_{GL}a_L} + \frac{1}{k'_{SL}a_p} + \frac{1}{\varepsilon w k_{w1}}\right)^{-1} \qquad (12.27)$$

on the assumption that $k'_{GL}a_L = k'_L a_L$. If it is assumed that the catalyst particles and the gas bubbles are spherical, the interfacial areas a_L and a_p can be expressed as

Table 12.4 Controlling Mechanisms in TBRs (Film Flow Regime): Effect of Pertinent Variables on r_A

	Variables Whose Influence Is		
Controlling Resistance	**Major**	**Minor**	**Negligible**
Gas–liquid mass transfer, $k_L'a_L$	$[A]^*$, d_p, superficial liquid velocity	Temperature	$[B]_0$, superficial gas velocity (at lower values), replacement of active by inactive catalyst particles
Liquid–solid mass transfer, $k_{SL}'a_p$ (for A)	d_p, $[A]^*$, superficial liquid velocity	Temperature	$[B]_b$, superficial gas velocity (at lower values), replacement of active by inactive catalyst particles
Liquid–solid mass transfer of $k_{SL}'a_p$ (for B)	d_p, $[B]_b$, superficial liquid velocity	Temperature	$[A]^*$, superficial gas velocity[a]
Surface reaction (pore diffusion negligible)	d_p, $[A]^*$, $[B]_b$ temperature, replacement of active by inactive catalyst particles	Superficial liquid velocity (above certain minimum)	Superficial gas velocity[a]
Surface reaction with pore diffusion	d_p, $[A]^*$, $[B]_b$, temperature,[b] replacement of active by inactive catalyst particles	Superficial liquid velocity (above certain minimum)	Superficial gas velocity[a]

Source: Adapted from Doraiswamy, L.K. and Sharma, M.M., *Heterogeneous Reactions—Analysis Examples and Reactor Design*, Vol. 2, Wiley, New York, 1984.

[a] It is assumed that this does not cause any change in the flow regime.

[b] The effect of temperature on r_A is less pronounced in this case than for surface reaction where pore diffusion is negligible.

$$a_L = \frac{6h_G}{d_b(1-h_G)}, \quad a_P = \frac{6w}{\rho_c d_p} \tag{12.28}$$

Substituting these in Equation 12.26 gives

$$\frac{[A]^*}{-r_A}\left[\frac{d_b(1-h_g)}{6h_G k_{GL}'}\right] + \frac{\rho_c d_p}{6w}\left(\frac{1}{k_{SL}'} + \frac{1}{\varepsilon k_{w1}}\right) \tag{12.29}$$

Thus, a plot of $[A]^*/(-r_A)$ versus $1/w$ gives straight lines with slope and intercept as shown in lines 1, 2, and 3 of Figure 12.5. An intercept of zero as in line 1 indicates negligible gas–liquid mass transfer resistance, a slope of zero (line 2) indicates gas–liquid mass transfer control, and finite values of slope and intercept (line 3) indicate combined control by all three resistances (gas–liquid, solid–liquid, and reaction).

By operating under conditions of high agitation where gas–liquid mass transfer effects are absent, line 1 is produced. From the slope and estimated value of k_{SL}', the rate constant can be obtained. Alternatively, if the rate constant is known, the correct value of k_{SL}' for the system at hand can be extracted from the slope. At very low agitation, line 2 is produced, and k_L' may be obtained from the intercept.

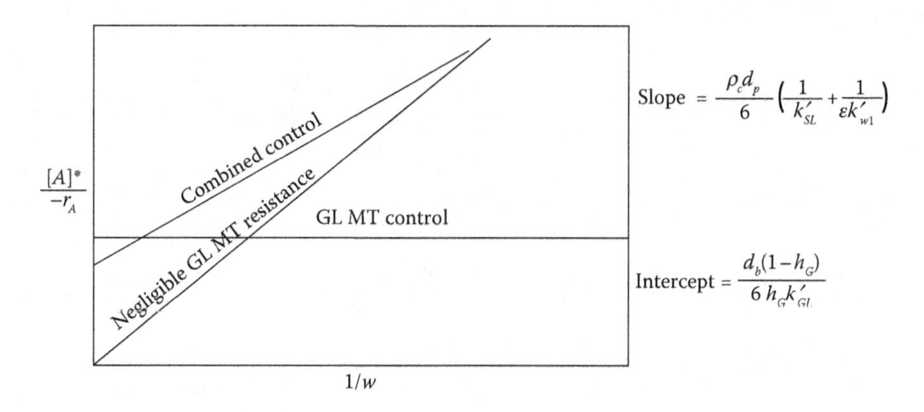

Figure 12.5 Interpretation of data for linear kinetics.

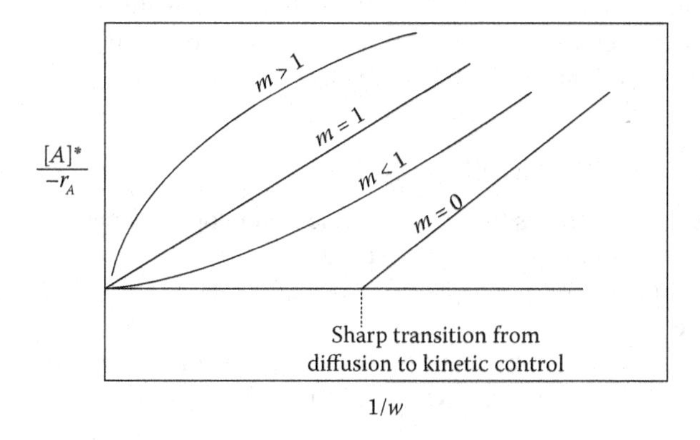

Figure 12.6 Plots of $[A]^*/-r_A$ for different orders (nonlinear kinetics).

If the kinetics is nonlinear, the simple additivity of resistances principle used in the above analysis will not apply. As a result, the $[A]^*/(-r_A)$ versus $1/w$ plots will no longer be straight as in Figure 12.5, but will be curves as sketched in Figure 12.6 for different orders. Even in these cases, it is possible to obtain $k'_L a_L$ by extrapolating the curves to $1/w = 0$, but a large number of experimental points very close to zero would be required. The value thus obtained, though not very accurate, should be adequate for most calculations.

In Figure 12.7, a broad classification of three-phase catalytic reactors is summarized.

Three-phase noncatalytic reactions

Here, we consider the reaction between solid B suspended in a liquid and gas A bubbled through it. If solid B is slightly soluble in the liquid, reaction occurs between dissolved A and B in the liquid phase. If B is

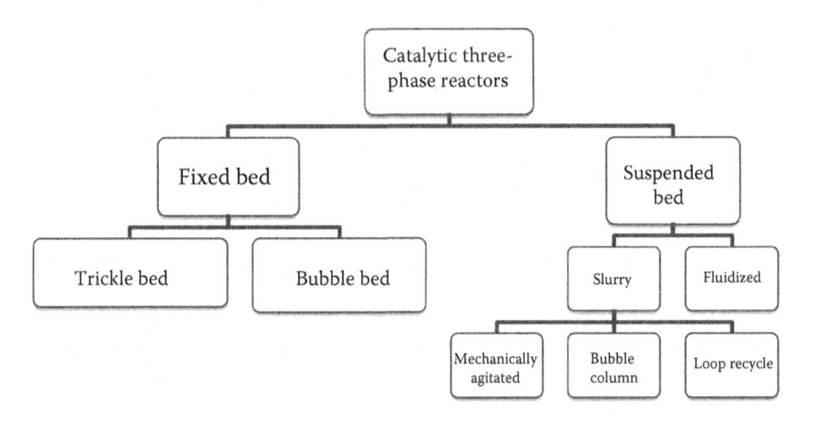

Figure 12.7 Classification of catalytic three-phase reactors.

insoluble, dissolved A diffuses and reacts with B within the solid. The situation represented by the first case is similar to gas–liquid reactions, but with provision for solid dissolution.

Solid slightly soluble

This is an interesting situation in organic technology/synthesis. Examples are the dissolution of acetylene in an aqueous slurry of CuCl as a step in the manufacture of propylene oxide, and alkylation of naphthalene with ethylene in a liquid medium in the presence of BF_3–phosphoric acid as dissolved catalyst. Note that the presence of the dissolved catalyst does not, in any way, alter the physical features of the system.

Such reactions can typically be represented as

$$\begin{aligned} &A\ (g) \rightarrow A\ (l) \\ &B\ (s) \rightarrow B\ (l) \\ &A\ (l) + \nu_B B\ (l) \rightarrow \text{products} \end{aligned} \qquad \text{(R2)}$$

Clearly, two liquid films are involved here: one surrounding the gas (which we designate F_1), and the other surrounding the solid (F_2). We consider two cases, one with negligible dissolution and the other with significant dissolution in F_2.

Negligible dissolution of solid in the gas–liquid film This represents a relatively simple situation where no solid dissolves in the gas–liquid film F_1. The following steps are involved: diffusion of gas A through the film; dissolution of solid B; and diffusion and simultaneous reaction of B with dissolved A in film F_1. The last step can occur in regime 3 (fast reaction, pseudo-order), regimes 3–4 (fast reaction, with depletion), or regime 4 (instantaneous reaction). The first two cases are sketched in Figure 12.8a and the third in 12.8b. The conditions and rate equations for the three cases are summarized in Table 12.5.

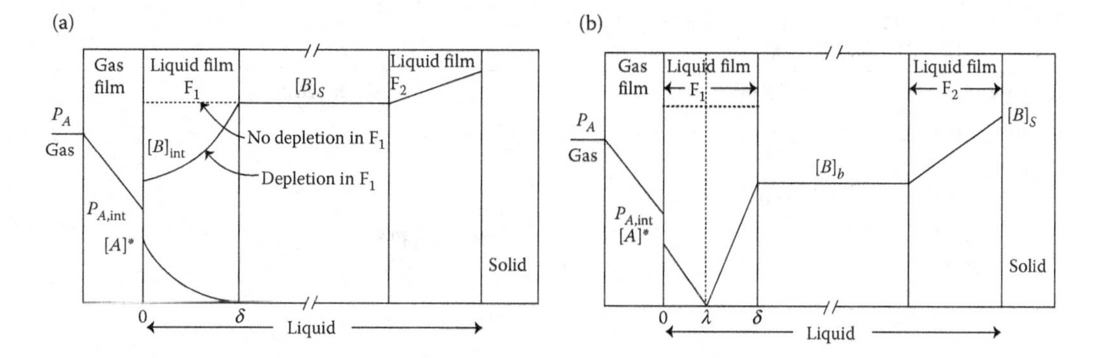

Figure 12.8 Gas–solid–liquid reaction: no solid dissolution in film (F_1) next to gas. (a) Fast reaction; (b) instantaneous reaction.

Significant dissolution of solid in the gas–liquid film If solid dissolution in the gas–liquid film is significant, the analysis becomes more involved. If the reaction occurs in the fast pseudo-first-order regime in film F_1, the bulk and film concentrations of B would be the same. Thus, dissolution of B in film F_1 will not make any difference to the reaction. On the other hand, an instantaneous reaction occurring in film F_1 will be further enhanced by this dissolution. The ultimate result would be that the reaction plane approaches the gas–liquid interface (i.e., $\lambda \rightarrow 0$). The condition and rate equation for this situation are included in Table 12.5.

Solid insoluble A typical example is the Kolbe–Schmitt carbonation of the sodium salt of β-naphthol in an inert liquid medium in the production of β-oxynaphthoic acid (commonly known as BON acid), a useful intermediate in the manufacture of dyes and other chemicals. Chlorination of wood pulp suspended in water is another example.

The only additional features in the analysis when compared to that for solid–liquid reactions are the mass transfer resistances (if any) associated with the gas–liquid films. We demonstrate the procedure by considering the Kolbe–Schmitt reaction modeled by Phadtare and Doraiswamy (1965, 1969).

Example 12.2: Kolbe–Schmitt carbonation of β-naphthol in an inert medium

β-Oxynaphthoic acid (BON acid), a useful organic intermediate used mainly in the manufacture of dyes, is produced by the Kolbe–Schmitt carbonation of sodium naphthonate prepared by reacting naphthalene with sodium hydroxide. The carbonation step is normally carried out in a dry atmosphere, but the possibility of conducting the reaction in a liquid medium such as kerosene has also been explored. The reaction scheme may be represented as

Table 12.5 Controlling Regimes in Gas–Liquid–Solid Noncatalytic Reactions: $A(g) \rightarrow A(l)$; $B(g) \rightarrow B(l)$; $A(l) + \nu_B B(l) \rightarrow R$

Regime	Condition(s)	Rate Equation
1. Negligible solid dissolution in liquid film F_1 next to gas	$\frac{k_{SL}'a_P}{4k_L'^2}\frac{D_A^2}{D_B} \ll 1$; $d_P \gg \delta$	
a. Fast reaction, no depletion		$r_A = \frac{a_L''H_Ap_A\sqrt{D_Ak_2'[B]_b}}{1+((H_A\sqrt{D_Ak_2'[B]_b})/k_G')}$ where $[B]_b = [B]_s - \nu_B(-r_A)/k_{SL}'a_P$
b. Fast reaction, depletion		$-r_A = \frac{a_L''H_Ap_A\sqrt{D_Ak_2'[B]_{int}}}{1+((H_A\sqrt{D_Ak_2'[B]_{int}})/k_G')}$ where $[B]_{int} = [B]_s - \frac{\nu_B D_A}{D_B}H_AP_A - \nu_B r_A \times \left[\left(\frac{H_A}{k_G'a_L''}+\frac{1}{k_L'a_L}\right)\frac{D_A}{D_B}+\frac{1}{k_{SL}'a_P}\right]$
c. Instantaneous reaction	$p_{A\,int} \ge 0$; $[B]_b \ge 0$	$-r_A = \frac{H_Ap_A + (D_B/D_A)([B]_s/\nu_B)}{(H_A/k_G'a_L'') + (1/k_L'a_L'') + (D_B/D_A)(1/k_{SL}'a_p)}$ where $p_{A\,int} = p_A - \frac{(-r_A)}{k_G'a_L''}$; $[B]_b = [B]_s - \frac{\nu_B(-r_A)}{k_{SL}'a_P}$
d. Gas film control	$p_{A\,int} \rightarrow 0,\ [B]_b > 0$	$-r_A = k_{GP}'a_L''p_A$
e. Solid dissolution control	$p_{A\,int} > 0,\ [B]_b \rightarrow 0$	$-r_A = \frac{k_{SL}'a_P[B]_s}{\nu_B}$
2. Significant solid dissolution in liquid film F_1 next to gas	$\frac{k_{SL}'a_P}{4k_L'^2}\frac{D_A^2}{D_B} \gg 1$; $d_P < \frac{\delta}{5}$	$-r_A' = \frac{[B]_s}{\nu_B}\sqrt{D_Bk_{SL}'a_P}\coth\left[(\delta-1)\sqrt{\frac{k_{SL}'a_p}{D_B}}\right] + k_{SL}'a_P[B]_s\frac{\lambda}{\nu_B}$
a. Instantaneous reaction at gas–liquid interface ($0 < \lambda < \delta$)	$\frac{\sqrt{D_Ak_2[B]_s}}{k_L} \gg$	Solve the following two equations for $-r_A'$ by assuming various values of λ and δ till they are satisfied:
	$\frac{[B]_s}{\nu_B[A]^*}\left(1+\frac{k_{SL}'a_p}{4k_L'^2}\frac{D_A^2}{D_B}\right)$	$-r_A' = \frac{D_A[A]^*}{\lambda} + k_{SL}'a_P[B]_s\frac{\lambda}{\nu_B}$ (1)
		$-r_A' = \frac{[B]_s}{\nu_B}\sqrt{D_Bk_{SL}'a_P}\coth\left[(\delta-1)\sqrt{\frac{k_{SL}'a_p}{D_B}}\right] + k_{SL}'a_P[B]_s\frac{\lambda}{\nu_B}$ (2)
b. Instantaneous reaction at gas–liquid interface ($\lambda \rightarrow 0$)	$\sqrt{\frac{k_{SL}'a_P}{D_B}}\delta > 5$	$-r_A' = \frac{[B]_s}{\nu_B}\sqrt{D_Bk_{SL}'a_P}$
	$\frac{[B]_s}{\nu_B} \gg [A]^*$	i.e., rate $\propto \sqrt{a_p} \propto \sqrt{(\text{particle loading}, w)}$

$$\text{2-naphthol (OH)} + NaOH \longrightarrow \text{sodium 2-naphthoxide (ONa)} + H_2O \quad \text{(E12.2.1)}$$

$$2\ \text{sodium 2-naphthoxide (ONa)} \xrightarrow[\text{(with } CO_2)]{\text{Carbonation}} \underset{\text{Na–salt of BON acid}}{\text{(ONa, COONa)}} + \text{naphthalene} \quad \text{(E12.2.2)}$$

When carried out in kerosene medium, the carbonation occurs in a three-phase system in which CO_2 first dissolves in the liquid and then diffuses to the solid naphthonate present as a particulate suspension in the liquid. Since the solid is insoluble in kerosene, reaction occurs only in the solid phase. Using the data of Phadtare and Doraiswamy (1965, 1969), reproduced in Table E12.2.1, formulate a model to predict the conversion of sodium naphthonate to BON acid as a function of time.

SOLUTION

As briefly outlined in Chapter 6, the shrinking core models are applicable to solids of very low porosity. The controlling resistance can be one of the following: (1) diffusion through the kerosene film surrounding the solid, (2) diffusion through an increasing layer of solid product (known as *ash* in gas–solid literature, but it is not an appropriate term for the solid product of a solid–liquid reaction), (3) chemical reaction on the surface of the receding reactant core, or (4) combinations of these.

In the case of solid–liquid reactions, the effective diffusivity of the fluid reactant through the solid product may be defined as

$$D_{eA} = \alpha D_{bA} \quad \text{(E12.2.3)}$$

where α is a factor that accounts for tortuosity and the density of particles in the product shell. Since D_{eA} cannot be estimated unlike in gas–solid catalytic reactions (Chapter 6) or even noncatalytic reactions, we express it entirely as a function of D_b and use Equation E12.2.3 in the analysis. The constant α is an unknown quantity and can only be determined by curve-fitting as part of a mass transfer group (as described below).

First, we examine the possibility of chemical control. This would require a linear dependence of the function $(1 - (1 - X_B)^{1/3})$ on time. The actual plot (not shown) is, however, distinctly nonlinear. Hence, it can be concluded that the reaction is not kinetically

Table E12.2.1 Kolbe–Schmitt Carbonation of β-Naphthol: Experimental Time-Conversion Data

Reaction Time	Total Conversion X_B		
t (min)	30 psig	70 psig	100 psig
Temperature, 230°C			
5	0.1798	0.1930	0.4733
15	0.4036	0.3610	0.3220
30	0.5544	0.5184	0.4500
40	0.6580	0.5920	0.5360
60	0.7180	0.6562	0.6220
120	0.6600	0.6890	0.6750
180	0.7360	0.7490	0.7390
Temperature, 250°C			
5	0.2308	0.1771	0.2500
15	0.3430	0.4200	0.4102
30	0.4998	0.5740	0.5616
40	0.5960	0.6800	0.6480
60	0.6750	0.702	0.7360
120	0.7260	0.7690	0.7530
180	0.8070	0.8044	0.8140
Temperature, 270°C			
5	0.3020	0.3720	0.3620
15	0.4580	0.4720	0.4580
30	0.6400	0.6572	0.6100
40	0.7332	0.7650	0.7140
60	0.7728	0.7990	0.7760
120	0.7200	0.7860	0.7840
180	0.7800	0.8260	0.8590

controlled. Thus, we now consider the case where the controlling resistances are diffusion through the kerosene film and the solid product. Since the particles are nearly spherical, and the porosity of the solid reactant is very low, we can relate the solid conversion to the shrinking core radius by the equation

$$(1 - X_B) = (\text{Fraction solid unreacted}) = \frac{4/3\,\pi r_c^3}{4/3\,\pi R^3}$$

giving

$$X_B = 1 - \left(\frac{r_c}{R}\right)^3 \tag{E12.2.4}$$

Then, by writing rate equations for diffusion of CO_2 through the kerosene film and the product shell, the following expression can be derived for the present reaction (Phadtare and Doraiswamy, 1965, 1969):

$$\frac{t}{X_B} = \frac{A_M}{3} + \frac{A_M k_M}{3}\left[\frac{F(X_B)}{X_B}\right] \tag{E12.2.5}$$

where

$$A_M = \frac{2\rho_s R}{k_L'[C]_0}, \quad k_M = \frac{R}{\delta\alpha}, \quad F(X_B) = [1.5 - X_B - 1.5(1 - X_B)] \tag{E12.2.6}$$

Thus, a plot of t/x_B versus $F(X_B)/X_B$ should give a straight line with

$$\text{Slope} = \frac{A_M k_M}{3}$$
$$\text{Intercept} = \frac{A_M}{3} \tag{E12.2.7}$$

Equation E12.2.5 can now be tested by plotting the data of Table E12.2.1 in the manner mentioned above. Plots at 230°C and different pressures are shown in Figure E12.2.1. The values of A_M and k_M obtained from the slopes and intercepts of such plots for all the temperatures are given in Table E12.2.2. The values of A_M were also determined by Phadtare and Doraiswamy (1969) from an independent series of mass transfer experiments and are included in the table. They are quite close to the kinetically determined constants at some temperatures and pressures, and of the same order of magnitude at others.

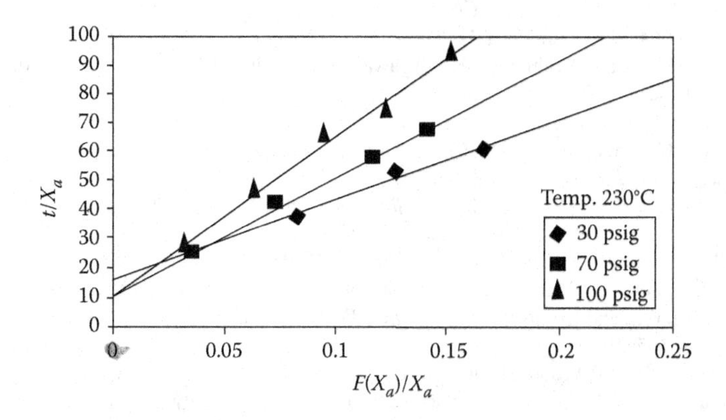

Figure E12.2.1 Validation of the mass transfer model for the Kolbe-Schmidt.

Table E12.2.2 Comparison of Experimental Values of Model Parameters with Those from Independent Mass Transfer Studies

		A_m		k_m
Temperature (°C)	**Pressure (psig)**	**From Kinetic Data**	**From Mass Transfer Studies**	**From Kinetic Data**
230	30	54.48	55	14.48
	70	36.86	35	32.19
	100	35.97	13	45.74
250	30	17.15	—	82.76
	70	58.02	39	11.82
	100	6.38	20	183.33
270	30	34.67	—	18.99
	70	41.08	74	13.47
	100	10.71	26	84.03

References

Akita, K. and Yoshida, F., *Ind. Eng. Chem. Proc. Des. Dev.,* **12**, 76, 1973.
Akita, K. and Yoshida, F., *Ind. Eng. Chem. Proc. Des. Dev.,* **13**, 84, 1974.
Beenackers, A.A.C.M. and Van Swaaij, W.P.M., *Chem. Eng. Sci.,* **48**, 3109, 1993.
Bern, L., Lidefelt, J.O. and Schoon, N.H., *J. Am. Oil Chem. Soc.,* **53**, 463, 1976.
Blenke, H. and Hirner, W., *VDI Ber.,* **218**, 549, 1974.
Charpentier, J.C. and Favier, M., *AIChE J.,* **21**, 1213, 1975.
Chaudhari, R.V. and Shah, Y.T., In *Concepts and Design of Chemical Reactors* (Eds. Whitaker, S. and Cassano, A.E.), Gordon and Breach, New York, 1986.
Chou, T.S., Worley, F.L., Jun., and Luss, D., *Ind. Eng. Chem. Proc. Des. Dev.,* **16**, 424, 1977.
Doraiswamy, L.K. and Sharma, M.M., *Heterogeneous Reactions—Analysis Examples and Reactor Design,* Vol. 2, Wiley, New York, 1984.
Fukushima, S. and Kusaka, K., *J. Chem. Eng. Jpn.,* **10**, 461, 1977.
Goto, S. and Smith, J.M., *AIChE J.,* **24**, 294, 1978.
Goto, S., Watabe, S., and Matsubara, M., *Can. J. Chem. Eng.,* **54**, 551, 1976.
Hikita, H. and Kikukawa, H., *Chem. Eng. J.,* **8**, 412, 1975.
Hikita, H., Asai, S., Ishikawa, H., and Uku, J., *Chem. Eng. Commun.,* **5**, 315, 1980a.
Hikita, H., Asai, H., Tanigawa, K., Segawa, K., and Kitao, M., *Chem. Eng. J.,* **20**, 59, 1980b.
Joshi, J.B., *Chem. Eng. Sci.,* **24**, 313, 1982.
Kenney, C.N. and Sedriks, W., *Chem. Eng. Sci.,* **27**, 2029, 1972.
Leuteritz, G., *Process Eng.,* **54**, 62, 1973.
Mangartz, K.H. and Pilhofer, TH., *Verfahrenstechnik,* **14**, 40, 1980.
Mersmann, A., *Verfahrenstechnik,* **6**, 203, 1972.
Mills, P.L. and Chaudhari, R.V., *Catal. Today,* **37**, 367, 1997.
Okufi, S., Perez de Ortiz, E.S., and Sawistowski, H., *Can. J. Chem. Eng.,* **68**, 400, 1990.
Phadtare, P.G. and Doraiswamy, L.K., *Ind. Eng. Chem. Proc. Des. Dev.,* **4**, 274, 1965; **8**, 165, 1969.
Ramachandran, P.A. and Chaudhari, R.V., *Chem. Eng. J.,* **20**, 75, 1980a.
Ramachandran, P.A. and Chaudhari, R.V., *Three Phase Catalytic Reactors,* Gordon and Breach, New York, 1983.
Roy, N.K., Guha, D.R., and Rao, M.N., *Chem. Eng. Sci.,* **19**, 215, 1964.
Sano, Y., Yamaguchi, N., and Adachi, T., *J. Chem. Eng. Jpn.,* **7**, 255, 1974.

Sato, Y., Hiroshe, T., and Ida, T., *Kagaku Kagaku*, **38**, 543, 1973.
Satterfield, C.N., Van Eck, M.W., and Bliss, G.S., *AIChE J.*, **24**, 709, 1978.
Shah, Y.T. and Paraskos, J.A., *Chem. Eng. Sci*, **30**, 465, 1975.
Shah, Y.T., Mhaskar, R.D., and Paraskos, J.A., *Ind. Eng. Chem. Proc. Des. Dev.*, **15**, 400, 1976.
Shah, Y.T., Kelkar, B.G., Godbole, S.P., and Deckwer, W.D., *AIChE J.*, **28**, 353, 1982.
Specchia, V. and Baldi, G., *Chem. Eng. Sci*, **32**, 515, 1977.
Talmor, E., *AIChE J.*, **23**, 868, 1977.
Wiedeskehr, H., *Chem. Eng. Sci.*, **43**, 1783, 1988.
Yagi, H. and Yoshida, F., *Ind. Eng. Chem. Proc. Des. Dev.*, **14**, 488, 1975.
Yung, C.N., Wong, C.W., and Chang, C.L., *Can. J. Chem. Eng.*, **57**, 672, 1979.

Bibliography

Below you will find a collection of relevant literature. The book by Ramachandran and Chaudhari (1983) is a particularly valuable one on three-phase catalytic reactions.

Beenackers, A.A.C.M. and Van Swaaij, W.P.M., *Chem. Eng. Sci.*, **48**, 3109, 1993.
Concordia, J.J., *Chem. Eng. Prog.*, **50**, 1990.
Crine, M. and L'Homme, G.A., In *Mass Transfer with Chemical Reaction in Multiphase Systems*, Vol. II (Ed. Alper, E.), Martinus Nijhoff, The Hague, 1983.
Hofmann, H., In *Mass Transfer with Chemical Reaction in Multiphase Systems*, Vol. II (Ed. Alper, E.), Three-Phase Systems, NATO-ASI Series, Series E, Applied Sciences No. 72-73, Nijhoff, Hingam, MA, 1983.
Joshi, J.B., Shertukde, P.V., and Godbole, S.P., *Rev. Chem. Eng.*, **5**, 71, 1988.
Kohler, M.A., *Appl. Catal.*, **22**, 21, 1986.
Mills, P.L., Ramachandran, P.A., and Chaudhari, R.V., *Rev. Chem. Eng.*, **8**, 1, 1992.
Ramachandran, P.A. and Chaudhari, R.V., *Three Phase Catalytic Reactors*, Gordon and Breech, New York, 1983.
Shah, Y.T. and Deckwer, W.D., In *Scale-Up of Chemical Processes Conversion from Laboratory Scale Tests to Successful Commercial Size Design* (Eds. Bisio, A., and Kabel, R.L.), Wiley, New York, 1985.
Tarmy, B., Chang, M., Coulaloglou, C., and Ponzi, P., *Chem. Eng.*, **407**, 18, 1984.
Villermaux, J. In *Multiphase Chemical Reactors VI. Fundamentals* (Eds. Rodriguez, A.E., Calo, J.M., and Sweed, N.H.), Sijthoffet Noordhoff, Alphen aan den Rijn, 1981.

Mass transfer in mechanically agitated slurry reactors:

Boon-Long, S., Laguerie, C., and Couderc, P.J., *Chem. Eng. Sci.*, **33**, 813, 1978.
Levins, D.M. and Glastonbury, J.R., *Chem. Eng. Sci.*, **27**, 537, 1972a.
Levins, D.M. and Glastonbury, J.R., *Trans. Inst. Chem. Eng.*, **50**, 132, 1972b.
Patil, V.K., Joshi, J.B., and Sharma, M.M., *Chem. Eng. Res.*, **62**, 247, 1984.
Teshima, H., and Ohashi, Y., *J. Chem. Eng. Jpn.*, **10**, 70, 1977.

Mass transfer in bubble column slurry reactors:

Arters, D.C. and Fan, L.S., *Chem. Eng. Sci.*, **41**, 107, 1986.
Arters, D.C. and Fan, L.S., *Chem. Eng. Sci.*, **45**, 965, 1990.
Hikita, H., Asai, S., Kikuwa, H., Zaike, T., and Masahiko, O., *Ind. Eng. Chem. Proc. Des. Dev.*, **20**, 540, 1981.
Kikuchi, K., Tadakuma, Y., Sugawara, T., and Ohashi, H.J., *Chem. Eng. Jpn.* **20**, 134, 1988.
Kikuchi, K., Takashi, H., and Sugarawa, T., *Can. J. Chem. Eng.*, **73**, 313, 1995.
Lemcoff, N.O. and Jameson, G.J., *Chem. Eng. Sci.*, **30**, 363, 1975a.
Lemcoff, N.O. and Jameson, G.J., *AIChE J.*, **21**, 730, 1975b.
Patil, V.K., and Sharma, M.M., *Chem. Eng. Res. Des.*, **61**, 21, 1983.

Chapter 13 Membrane-assisted reactor engineering

Introduction

We commenced our treatment with an introduction to the exploitable features of membrane reactors (without attempting to describe membrane synthesis) earlier in this book. We gave a brief introduction for the membrane reactors in *Interlude I* following Chapter 2. In this chapter, we describe the main variations in design and mode of operation of these reactors, develop performance equations for the more important designs, and compare the performances of some important designs with those of the traditional mixed and plug-flow reactors. Finally, we present a summary of the applications of membrane reactors in enhancing the rates of organic reactions.

General considerations

Major types of membrane reactors

The many attractive features of membrane reactors described in the interlude following Chapter 2 underscore the potential of these reactors in chemical technology/synthesis. A broad classification of these reactors is given in Figure 13.1 with sketches of a few specific ones in Figure 13.2.

Part (a) of the figure shows a design consisting of a permselective membrane tube placed coaxially inside an outer shell. Reaction occurs in the inner tube which is filled with a catalyst. One (or more) of the products from the inner tube permeates through the catalytically inert membrane wall into the outer shell from where it is swept away by an inert gas (usually argon). When the catalyst is a packed bed as shown in part (a), the reactor is designated as the *packed bed inert selective* membrane reactor* (IMR-P).

A useful variation of this design, shown in Figure 13.2b, consists of three concentric tubes (Oertel et al., 1987). The inner of the two annular spaces formed is filled with the catalyst, and selective permeation of products

* Use of the word "selective" denoting the permselective nature of the membrane is optional.

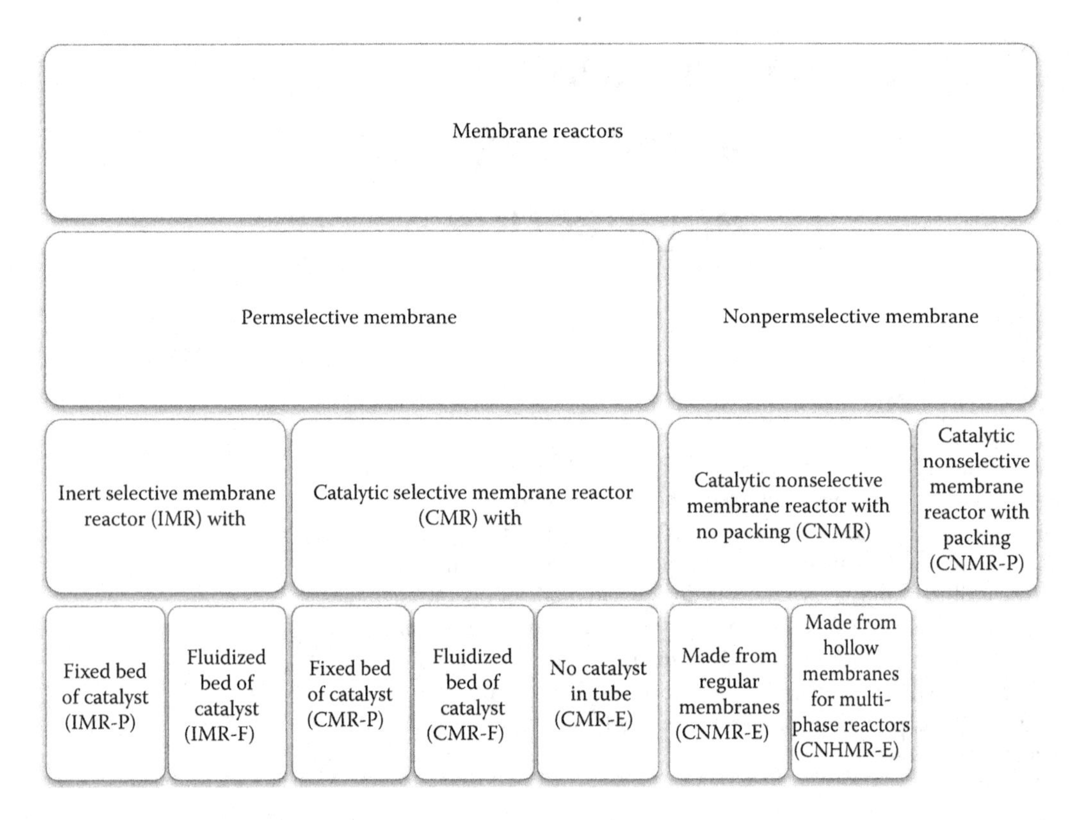

Figure 13.1 Broad classification of membrane reactors. E = empty, F = fluidized bed, P = packed bed.

to the central (product) tube is achieved by placing a number of tubular membranes inside this packed volume. It is, therefore, called the *packed bed inert selective multimembrane reactor* (IMMR-P). In yet another version of the IMR-P, the membrane is supported on the inner surface of a hollow fiber membrane tube and the catalyst is loaded around the hollow fiber [part (c)].

When the catalyst in design (a) is fluidized, it is designated as the *fluidized-bed inert selective* membrane reactor* (IMR-F). It is, however, more common to have the fluidized bed in the shell and sweep out the products through the membrane tubes, usually placed horizontally, as shown in part (d).

Part (e) shows a design that is similar to part (a), but with the catalyst supported on the membrane wall (and not placed inside the membrane tube). In some cases, the membrane itself acts as the catalyst. In this kind of reactor, as the reactant passes through the inner tube, it permeates across the catalytic membrane, with diffusion and reaction occurring simultaneously inside the membrane wall. We refer to this reactor as the *catalytic selective* membrane reactor* (CMR-E). When it is packed with the catalyst, the result is a *packed-bed catalytic selective membrane* reactor

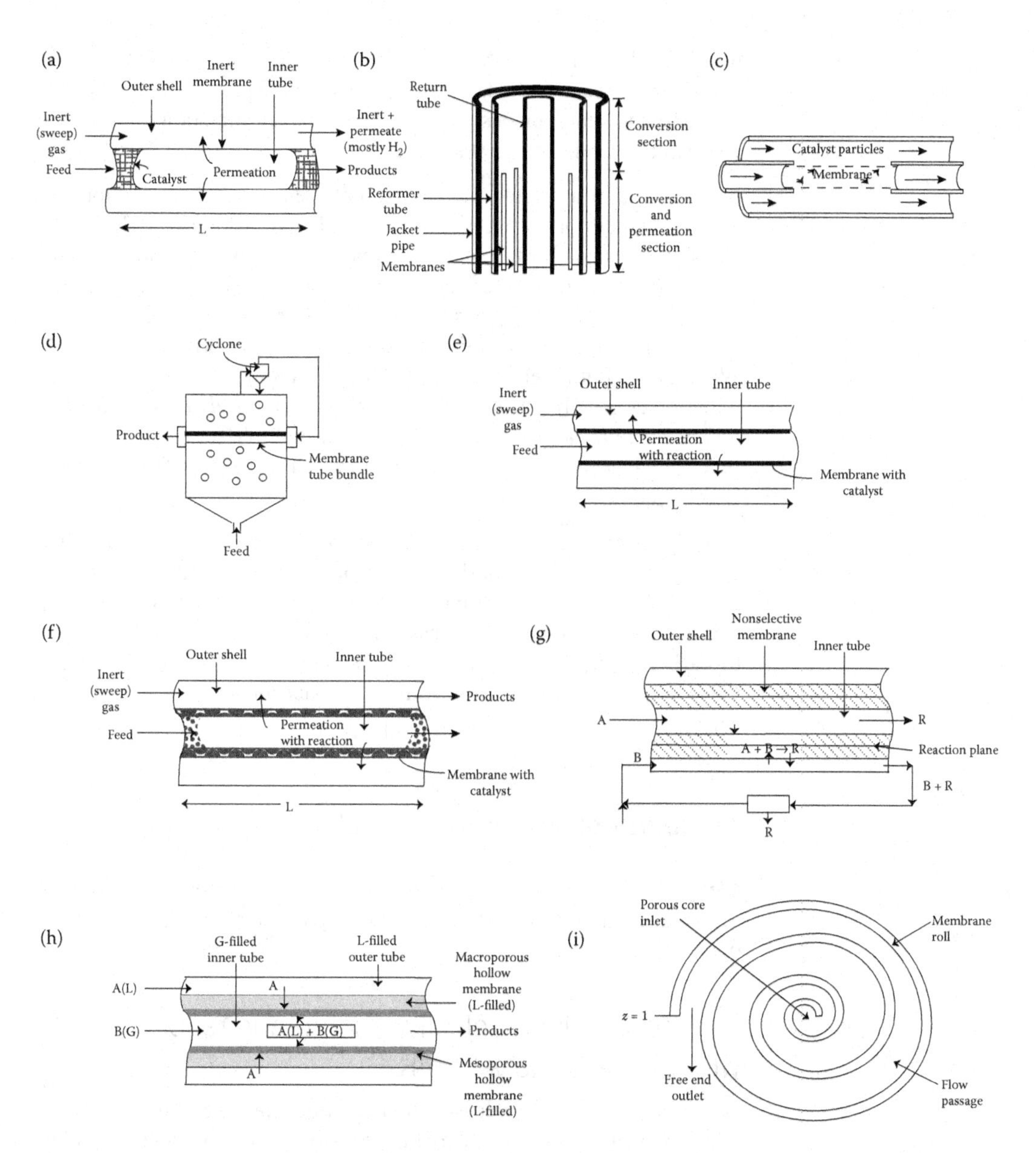

Figure 13.2 Types of membrane reactors. (a) IMR-P, (b) IMMR-P, (c) hollow membrane tube reactor with catalyst in shell (another version of IMR-P), (d) fluidized-bed inert selective membrane reactor (IMR-F), (e) CMR-E, (f) CMR-P, (g) catalytic nonreactive membrane reactor (CNMR-E), (h) catalytic nonselective hollow membrane reactor (CNHMR-E) for multiphase reactions: G = gas, L = liquid, and (i) immobilized-enzyme membrane reactor (IEMR). (Adapted from Shao, X., Xu, S., and Govind, R., *AIChE Symp. Ser.*, **268**, 1, 1989.)

(CMR-P) (Figure 13.2f). Modified designs have been used by Gryaznov (1986) and Gryaznov et al. (1974a,b) in which the reactor geometry is so as to provide simultaneously for a high degree of utilization of the vessel space and extended catalyst surface.

A schematic of a reactor made from a nonselective membrane for preventing the slip of an excess reactant is shown in Figure 13.2g. In the particular design shown, one of the reactants (A) is continuously recirculated on one side of the membrane so that complete conversion of B can be achieved on the opposite side without any slip. We refer to such a *catalytic nonselective** *membrane reactor* with no packing as CNMR-E. When packed, it is referred to as CNMR-P. Another nonselective membrane reactor with considerable potential is the *catalytic nonselective** *hollow membrane reactor* (CNHMR-E) for multiphase reactions. A useful design based on the principle explained in Figure I.2d *of the Interlude* is sketched in Figure 13.2h.

A design used in enzyme-catalyzed synthesis is shown in part (i) (Shao et al., 1989). This essentially consists of an enzyme immobilized in a sheet of microporous plastic (such as PVC) which is then rolled into a spiral and placed in a reactor vessel. Like the Gryaznov reactor referred to earlier, this *immobilized enzyme membrane reactor* (IEMR) provides for maximum utilization of reactor space and extension of catalyst surface.

Modeling of membrane reactors

We outline below the important features of the two most basic types of membrane reactors, the IMR-P and the CMR-E, supplemented by a brief discussion of some of their variations.

Packed-bed inert selective membrane reactor with packed catalyst (IMR-P)

Hydrogenation and dehydrogenation reactions have been found to benefit most from the use of membrane reactors, and a number of studies have been reported on the modeling of these systems. Thus, consider the following typical form of a dehydrogenation reaction:

$$A \leftrightarrow R + \nu_H \, H \qquad \text{(R1)}$$

where H is hydrogen. Figure 13.3 is a sketch of this reaction conducted in an isothermal IMR-P. Tubular reactor behavior will be assumed inside and outside the tubes.

* The word "nonselective" (abbreviated to N) should be retained to emphasize its difference from the permselective membrane.

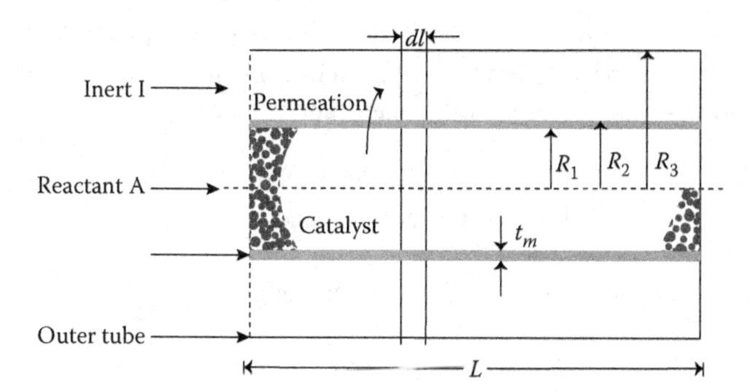

Figure 13.3 Packed-bed inert selective membrane reactor.

Model equations The general mass balance can be written as

$$\begin{pmatrix}\text{Rate of change of}\\ \text{component, mol/m}^3\text{s}\end{pmatrix} = \begin{pmatrix}\text{Rate of formation or}\\ \text{disappearance, mol/m}^3\text{s}\end{pmatrix} - \begin{pmatrix}\text{Rate of permeation}\\ \text{mol/m}^3\text{s}\end{pmatrix} \quad (13.1)$$

The last term is merely the specific rate of permeation r_i (mol/m^2 s) multiplied by the area per unit volume a, which for a tubular reactor is $2/R_1$ (1/m). The product $r_i a$ gives the rate of permeation in units of mol/m^3 s.

Inner (feed) tube:

The mass balance for this case becomes

$$\frac{dG_i^T}{dz} = r_i' - \frac{2\Pi_i}{R_1} \quad (13.2)$$

where G_i^T is the specific molar flow rate of species i on the tube (feed) side in mol/m^2 s; R_1 is the inner radius of the tube; and r_i' is the rate in mol/m^2 s, with $r_i' < 0$ for a reactant, > 0 for a product, and $= 0$ for an inert Π_i is the rate of permeation. The initial conditions are

$$z = 0, \quad G_{i=R,S}^T = 0, \quad G_A^T = G_{A0}^T, \quad G_I^T = \alpha\, G_{A0}^T \quad (13.3)$$

where α is the ratio of sweep gas-to-reactant gas rates:

$$\alpha\,(\text{sweep ratio}) = \frac{G_{I0}^S}{G_{A0}^T} = \frac{Q_{I0}^S}{Q_{A0}^T} \quad (13.4)$$

Outer (shell) side:

In writing the equations for the shell side, it must be noted that the reaction rate term in Equation 13.1 is zero. Thus, the only term we have on the right-hand side is the rate at which each component permeates into the annulus and adds on to the flow from the upstream side. Also, in formulating the permeation term, it must be remembered that the permeation area per unit volume of the outer tube is given by

$$a = \frac{\text{surface area of the inner tube}}{\text{volume of the annular space}} = \frac{2R_1}{R_3^2 - R_2^2} \tag{13.5}$$

Thus, the equations can be written in compact form as

$$\frac{dG_i^S}{dz} = \left(\frac{2R_1}{R_3^2 - R_2^2}\right)\Pi_i \tag{13.6}$$

where G_i^S is the specific molar flow rate in the outer shell (mol/m^2 s), and R_2 and R_3 are, respectively, the outer radius of the inner tube and the inner radius of the outer tube (Figure 13.3).

The initial conditions are

$$z = 0, \quad G_{i\neq 1}^S = 0, \quad G_1^S = G_{10}^S = \alpha\, G_{A0}^T \tag{13.7}$$

Permeation:

Since the membrane is assumed to be inert, the following simple mass balance can be written for the rate of permeation:

$$\Pi_i = \frac{D_i(p_i^T - p_i^S)}{t_m} \tag{13.8}$$

where D_i is the diffusivity, referred to as *permeability* in the membrane literature, of component i in units of mol/m atm s, p_i^T and p_i^S are the partial pressures of i in the inner tube and outer shell, respectively, and t_m is the membrane thickness.

The overall conversion equation:

The overall balance for reactant A in the inner and outer tubes leads to the following expression for its conversion:

$$X_A = \frac{\pi R_1^2(G_{A0}^T - G_A^T) - \pi(R_3^2 - R_2^2)G_A^S}{\pi R_1^2 G_{A0}^T}$$
$$= 1 - \frac{G_A^T}{G_{A0}^T} - \frac{G_A^S}{G_{A0}^T}\frac{(R_3^2 - R_2^2)}{R_1^2} \tag{13.9}$$

For the case of no permeation (i.e., $G_A^S = 0$), this reduces to

$$G_A^S = 0, \quad X_A = 1 - \frac{G_A^T}{G_{A0}^T} \tag{13.10}$$

We demonstrate below the use of the above equations in simulating an IMR-P for a specific reaction.

Example 13.1: Simulation of an IMR-P for the dehydrogenation of cyclohexane to benzene

The reaction

$$\text{(A)} \rightleftharpoons \text{(R)} + 3H_2 \qquad \text{(E13.1.1)}$$

is carried out in a porous Vycor membrane reactor with A = cyclohexane, R = benzene, and H = hydrogen (Figure 13.3). Cyclohexane is passed through the inner tube, and sweep gas argon (I) through the outer tube.

It is desired to determine whether conversions beyond the equilibrium limit corresponding to the temperature and pressure of the reaction can be achieved. Since membrane thickness controls the permeation rate, it is also desired to optimize the thickness for maximum performance. The input data for the simulation are given in Table E13.1.1.

Rigorous solution (isothermal operation)

Governing equations:

The material balance for the inner tube is given by Equation 13.2 which can be expanded into the following set:

$$\frac{dG_A^T}{dz} = -(-r_A) - \frac{2\Pi_A}{R_1} \qquad \text{(E13.1.2)}$$

$$\frac{dG_R^T}{dz} = (-r_A) - \frac{2\Pi_R}{R_1}, \quad [r_R = (-r_A)] \qquad \text{(E13.1.3)}$$

$$\frac{dG_H^T}{dz} = 3(-r_A) - \frac{2\Pi_H}{R_1}, \quad [r_H = 3(-r_A)] \qquad \text{(E13.1.4)}$$

Table E13.1.1 Input Parameters for Simulation

Reactor dimensions (cm)	$R_1 = 0.7$, $R_2 = 0.85$, $R_3 = 2.0$, $L = 20$
Gas flow rate at inlet (mol/m²/s)	$G_{I0}^T = 0.0797, G_{A0}^T = 0.0152, G_{I0}^S = 0.111$
Reaction temperature (K)	$T = 483$ K
Reaction pressure (Pa)	$P_t = 1.013 \times 10^5$
Kinetic parameters (from some previous studies)	$k^0 = 1.42 \times 10^{-5}$ mol m³ s/Pa, $E = 67.7$ kJ/mol

Source: Adapted from Itoh, N. et al., *Int. Chem. Eng.*, **25**(1), 139, 1985.

$$\frac{dG_I^T}{dz} = -\frac{2\Pi_I}{R_1} \qquad \text{(E13.1.5)}$$

Similarly, the material balance for the outer tube, Equation 13.6, can be expanded to give

$$\frac{dG_A^S}{dz} = \frac{2R_1\Pi_A}{R_3^2 - R_2^2} \qquad \text{(E13.1.6)}$$

$$\frac{dG_R^S}{dz} = \frac{2R_1\Pi_R}{R_3^2 - R_2^2} \qquad \text{(E13.1.7)}$$

$$\frac{dG_H^S}{dz} = \frac{2R_1\Pi_H}{R_3^2 - R_2^2} \qquad \text{(E13.1.8)}$$

$$\frac{dG_I^S}{dz} = \frac{2R_1\Pi_I}{R_3^2 - R_2^2} \qquad \text{(E13.1.9)}$$

The initial conditions are

$$\begin{aligned} G_A^T &= G_{A0}^T \\ G_R^T &= G_R^S = 0 \\ G_I^T &= \alpha G_{A0}^T \\ G_A^S &= G_R^S = G_H^S = 0 \\ G_I^S &= G_{I0}^S \end{aligned} \qquad \text{(E13.1.10)}$$

Parameter values:

The rate equation is given by

$$(-r_A) = k^0 \exp\left(\frac{E}{R_g T}\right)\left(p_A - \frac{p_R p_H^3}{K}\right) \qquad \text{(E13.1.11)}$$

The values of k^0, E, and K, and of the other parameters of the system, are listed in Table E13.1.1.

We now have all the data needed to solve Equations E13.1.2 through E13.1.10 along with E13.1.11. This can easily be done by any of the well-known numerical methods.

Discussion of results:

Figure 13.4a (recalculated results of Itoh et al., 1985) shows the mole fraction profiles of cyclohexane, benzene, and hydrogen in the inner (reaction) tube. The reactant (cyclohexane) profile shows a continuously falling trend, whereas the profiles of products benzene and hydrogen show maxima. The latter occur because of competition between the rate of reaction and that of permeation of hydrogen, one dominating the entrance region and the other the exit region.

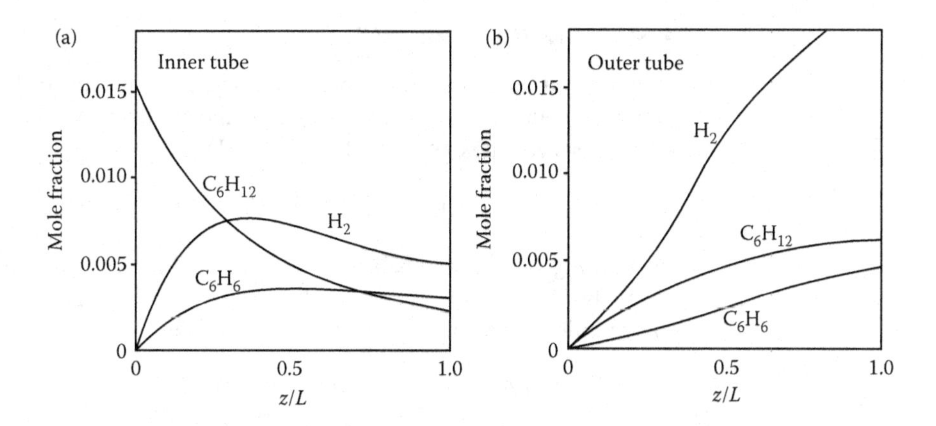

Figure 13.4 Dehydrogenation of cyclohexane: computed profiles in the (a) tube and (b) shell sides. (Adapted from the results of Itoh, N. et al., *Int. Chem. Eng.*, **25**(1), 139, 1985.)

As regards benzene, since the membrane is more selective to hydrogen, the permeation rate of benzene is slower than that of hydrogen, resulting in a slower rate of decrease of its concentration in the reaction tube. On the other hand, as expected, the mole fractions of all the three components rise steadily in the outer, separation tube [part (b)].

The effect of membrane thickness is shown in Figure 13.5. Opposing influences are exerted by reaction and permeation, resulting in an optimum thickness of 0.1 cm.

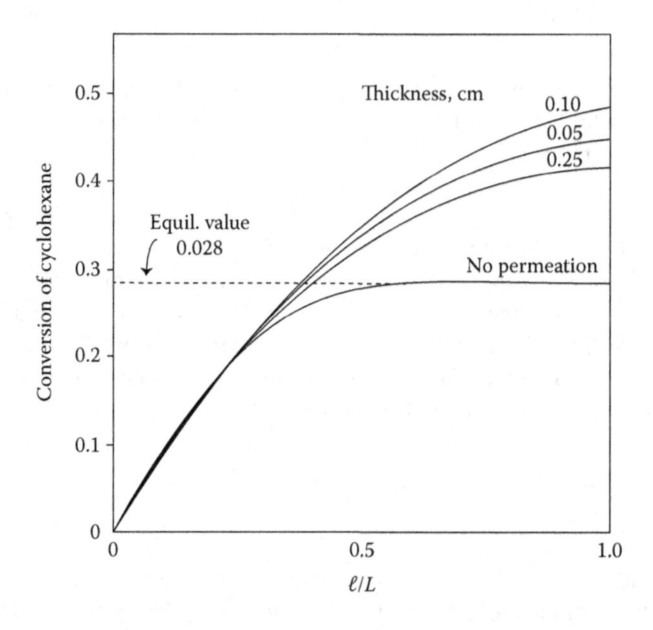

Figure 13.5 Dehydrogenation of cyclohexane: effect of membrane thickness on conversion. (Adapted from the results of Itoh, N. et al., *Int. Chem. Eng.*, **25**(1), 139, 1985.)

A SIMPLIFIED APPROACH

A simplified approach would be to straight away postulate that only hydrogen gas permeates through the membrane and that the rate of permeation is given by the simple relation

$$\Pi_H a = k_{per}[H], \text{ mol/m}^3 \text{ s} \qquad \text{(E13.1.12)}$$

Then, using the material balance of Equation 13.1, and restricting the analysis to the inner tube (thus dropping the superscripts *T* and *S*), the following equations can be readily written:

$$\frac{dF_A}{dV} = -(-r_A), \quad \frac{dF_H}{dV} = 3(-r_A) - \Pi_H a$$

or

$$\begin{aligned} \frac{dF_A}{d\bar{t}} &= -(-r_A)Q_0 \\ \frac{dF_R}{d\bar{t}} &= r_R Q_0 = (-r_A)Q_0 \\ \frac{dF_H}{d\bar{t}} &= 3(-r_A)Q_0 - Q_0 k_{per}[H] \end{aligned} \qquad \text{(E13.1.13)}$$

where $\bar{t} = V/Q_0 = L/u_0$. The initial conditions are

$$\bar{t} = 0, \quad F_A = F_{A0}, \quad F_R = F_H = 0, \quad Q = Q_0 \qquad \text{(E13.1.14)}$$

Noting that concentration = F/Q, these equations can be solved to obtain the exit flow rates F_A, F_R, and F_H as functions of the residence time $\bar{t}$.

Extension to consecutive reactions It seems likely that the use of IMR-P for a consecutive reaction can have beneficial effects on selectivity. Indeed, using the partial oxidation of methane to formaldehyde as the test reaction, the selectivity was found to be higher than in a PFR when the IMR-P was operated in the conventional way with the usual assumptions: sweep gas in the shell side under isothermal plug-flow conditions (Agarwalla and Lund, 1992; Lund, 1992).

Fluidized-bed inert selective membrane reactor (IMR-F)

A useful configuration is the IMR-F sketched in Figure 13.2d. An advantage of the fluidized-bed reactor, inadequately emphasized in the literature, is the fact that in reactions such as dehydrogenation of ethylbenzene to styrene, the bubbles of hydrogen act as "natural membranes" which remove the hydrogen formed in the dense phase.

The use of IMR-F not only improves the selectivity for the desired product in a complex reaction, but can even also raise it beyond that of the fixed bed. For instance, in the dehydrogenation of ethylbenzene to styrene given by the scheme (Sheel and Crowe, 1969)

$$C_6H_5CH_2CH_3 \leftrightarrow C_5H_5CHCH_2 + H_2 \tag{R2.1}$$

$$C_6H_5CH_2CH_2 \rightarrow C_6H_5 + C_2H_4 \tag{R2.2}$$

$$C_6H_5CH_3 + H_2 \rightarrow C_6H_5CH_3 + CH_4 \tag{R2.3}$$

$$2H_2O + C_2H_4 \rightarrow 2CO + 4H_2 \tag{R2.4}$$

$$H_2O + CH_4 \rightarrow CO + 3H_2 \tag{R2.5}$$

$$H_2O + CO \rightarrow CO_2 + H_2 \tag{R2.6}$$

removal of hydrogen from the scene of reaction in the dense phase both by transport of hydrogen bubbles away into the dilute phase (where little reaction occurs) and by the permselective action of the membrane tube wall (Figure 13.2d) prevents the conversion of styrene to toluene by reaction R2.3, thus improving styrene selectivity.

Catalytic selective membrane reactor (CMR-E)

In analyzing the performance of a CMR-E (Figure 13.6), we make the same assumptions as we did for the IMR-P. There is, however, a major difference in the equations. As the reaction occurs entirely within the membrane, it is bounded by the internal surface of the membrane wall, corresponding to $r = R_1$, for flow in the inner tube, and by the outer surface of the same wall, corresponding to $r = R_2$, for flow in the shell side. The resulting material balance equations for reaction R1 are formulated below for flow in the two tubes and for reaction in the membrane.

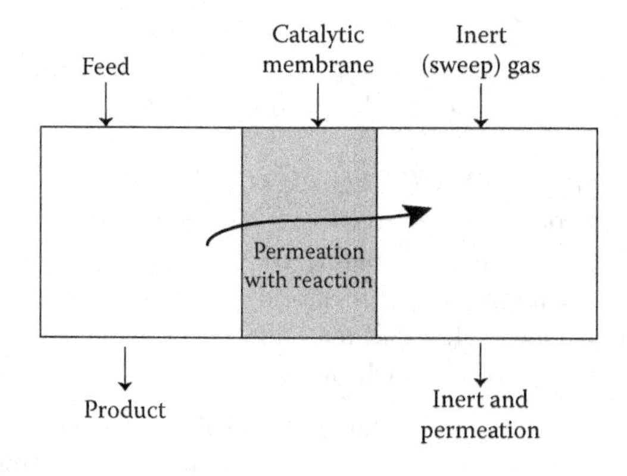

Figure 13.6 Catalytic selective membrane reactor.

Model equations

Inner (feed) tube:

$$\frac{dF_i^T}{dz} = \frac{2\pi D_i}{R_g T}\left[r\left(\frac{dp_i}{dr}\right)\right]_{r=R_1} \tag{13.11}$$

with

$$Z = 0, \quad F_i^T = F_{i0}^T = Q_0^T\left(\frac{y_{i0}^T P_t^T}{R_g T}\right) \tag{13.12}$$

Outer (shell side) tube:

$$\frac{dF_i^S}{dz} = -\frac{2\pi D_i}{R_g T}\left[r\left(\frac{dp_i}{dr}\right)\right]_{r=R_2} \tag{13.13}$$

with

$$z = 0, \quad F_i^S = F_{i0}^S = Q_0^S\left(\frac{y_{i0}^S P_t^S}{R_g T}\right), \quad Q_I^S = \alpha Q_{A0}^T \tag{13.14}$$

where α is the sweep ratio given by Equation 13.4.

Inside the membrane (with reaction)

$$\frac{D_i}{R_g T}\frac{1}{r}\frac{d}{dr}\left[r\left(\frac{dp_i}{dr}\right)\right] = -\nu_i r_i = -\nu_i k\left(p_A - \frac{p_R p_s}{K}\right) \tag{13.15}$$

The boundary conditions are

$$\begin{aligned} r = R_1, \quad p_i &= y_i^T P_t^T = y_i^T P^T \\ r = R_2, \quad p_i &= y_i^S P_t^S = y_i^S P^T \end{aligned} \tag{13.16}$$

The last terms in the above conditions are the result of the assumption $P_t^T = P_t^S$ constant (P_t) along the length of the reactor.

Main features of the CMR-E Equations 13.11 and 13.16 can either be solved as such, or nondimensionalized and then solved. Solutions can be obtained by the IMSL subroutine DBVFD along with a third-order Runge–Kutta technique. Experimental data on the dehydrogenation of ethane (Champagnie et al., 1992) reasonably uphold the predicted ethane profiles both in the tube and shell sides.

Further experimental data on the same reaction confirm trends typical of these reactions. For example, Figure 13.7 (Tsotsis et al. 1993a) shows the variation of conversion with sweep ratio α for an inner tube pressure

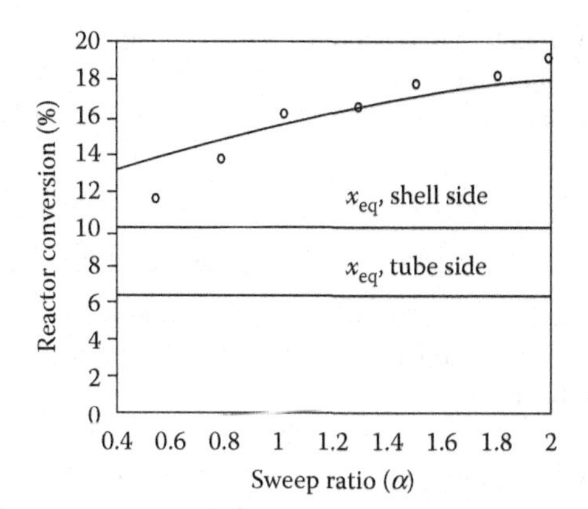

Figure 13.7 Conversion versus sweep ratio in the dehydrogenation of ethane at fixed values of other parameters. (Redrawn from Tsotsis, T.T. et al., *Sep. Sci. Technol.*, **28**, 397, 1993a.)

of 2 atm and outer shell pressure of 1 atm. It can be seen that the actual conversion obtained is significantly higher than the equilibrium conversion. Note that, in conformity with the nature of the reaction, that is, increase in volume, the equilibrium conversion on the shell side (which is at a lower pressure) is higher than in the inner tube. It increases with increasing α because the concentration of R on the shell side is reduced with increasing α, resulting in an increase in the rate of depletion of R in the reaction tube. This in turn leads to a further increase in the rate of reaction on the catalyst in the inner tube.

A vexing problem with the inert sweep gas is the need to separate it from the product. In reactions such as, for example, dehydrogenation of butane (Gobina and Hughes, 1996) and of cyclohexane (Wang et al., 1992), this has been overcome by using oxygen in admixture with an inert (such as CO) after eliminating potential formation of an oxide layer by prior high-temperature reduction with hydrogen. Such reaction-assisted transport can substantially raise conversions.

As in the case of IMR, here too the thickness of the membrane is an important adjustable parameter that must be optimized. It appears as the boundary values R_1 and R_2, the difference $(R_2 - R_1)$ being the membrane thickness. Experimental results on ethane dehydrogenation (Champagnie et al., 1992) clearly indicate the existence of an optimum thickness for best performance.

Packed-bed catalytic selective membrane reactor (CMR-P)

This reactor obviously has two catalyst zones: the packed bed and the membrane itself. As already stated, although theoretically daunting, it offers the best configuration for a complete analysis of membrane reactors. Experimental studies on the dehydrogenation of ethane showed considerable enhancement over both tube and shell side equilibrium conversions (Tsotsis et al., 1992). Further improvement was possible with increase in sweep ratio.

Catalytic nonselective membrane reactor (CNMR-E)

This reactor, sketched in Figure 13.2g, was developed specifically to prevent slip in reactions required to be strictly stoichiometric. Modeling of CNMR-E has been attempted both for fast irreversible and reversible reactions (Sloot et al., 1990, 1992; Zaspalis et al., 1991; Veldsink et al., 1992), notably the Claus reaction $2H_2S + SO_2 \rightarrow (3/8)S_8 + 2H_2O$. This concept can also be used in partial oxidations in organic technology, for example, partial oxidation of ethylene to acetaldehyde (Harold et al., 1992). The stoichiometry of this reaction can be represented as

$$A + \nu_B B \rightarrow \nu_{R1} R$$

$$A + \nu_{R2} B \rightarrow \nu_S S \qquad \text{(R3)}$$

where R is the desired product. Experiments in which the active side of the membrane was exposed to a $(C_2H_4 + He)$ mixture and the support side to a $(O_2 + He)$ mixture gave significantly higher selectivities than in experiments in which the active layer was exposed to a $(C_2H_4 + O_2 + He)$ mixture and the support side to pure He.

Catalytic nonselective hollow membrane reactor for multiphase reactions (CNHMR-MR)

A multiphase reactor design, very similar to the trickle-bed reactor, is the tubular multiphase hollow membrane wall reactor sketched in Figure 13.2h. In a regular trickle-bed reactor, the liquid flows over a partially wetted pellet as a thin film and supplies the liquid phase reactant to the catalyst pores. This action, however, has the effect of hindering pore access to the gas, thus lowering the reaction rate. On the other hand, in the multiphase membrane reactor, the liquid-filled membrane is directly accessible to the gas flowing in the inside tube. Thus, mass transfer in this reactor is considerably more efficient than in the conventional trickle-bed reactor.

It is also possible to have the liquid side fully mixed. This would mean that the external surface of the hollow tube would be exposed to a liquid of uniform composition.

One can develop governing equations for the two cases mentioned above and compare their performances with the conventional trickle-bed reactor modeled as a string of suspended spherical pellets contacted by cocurrent flow of gas and liquid. Based on such a comparison, the following observations can be made (Harold and Cini, 1989; Harold et al., 1989; Cini et al., 1991):

a. Direct supply of gas to the catalyst pores without an intervening liquid film greatly improves the reactor performance for very active catalysts.
b. For catalysts of low or moderate activity, the tube walls should (and can) be made as thin as possible to improve internal transport. Such a flexibility does not exist in the conventional trickle-bed reactor since the particle size can only be reduced at the cost of increased pressure drop.

Immobilized enzyme membrane reactor

It is easy to extend the concept to immobilization on the walls of a membrane tube. What is even more practical is to immobilize the enzyme in the usual manner on solid particles such as silica and encapsulate the particles in a ribbed sheet of a microporous plastic such as PVC. This sheet can then be rolled in a jelly-roll configuration inside a spiral reactor (Figure 13.2i). The consequent large surface area of immobilized enzyme available per unit volume of reactor space makes such a spiral reactor an attractive choice.

A reactor of such a configuration has been used in the clarification of fruit juice by elimination of pectin by the enzyme pectinase. The pectin, which is present in colloidal form, aggregates in the presence of the enzyme and settles down, leading to easy physical clarification (Shao et al., 1989).

Operational features

Combining exothermic and endothermic reactions

A particularly useful feature of a membrane is that it integrates reaction and separation into a single process, thereby increasing the conversion beyond equilibrium. If this concept can be extended to an integration of two reactions, one exothermic and the other endothermic, carried out on the opposite sides of the membrane, then we would have a thermally self-sustaining reaction. To expand this concept further (Figure 13.8), if hydrogen from an endothermic dehydrogenation reaction in the inner tube is permselectively transported to the shell side where, instead of being physically swept from the system, it is oxidized (exothermally) by oxygen over a catalyst present there, we would have an interactive thermal effect superimposed on the membrane's predisposition to separation. In other words, the heat liberated during oxidation in this "separation side" flows through the membrane into the "reaction side," thus providing the heat required for this endothermic reaction.*

* It would seem more appropriate to call these as "reaction 1 side" and "reaction 2 side" instead of reaction and separation sides; but to avoid conflict with the existing literature, we continue with the original nomenclature.

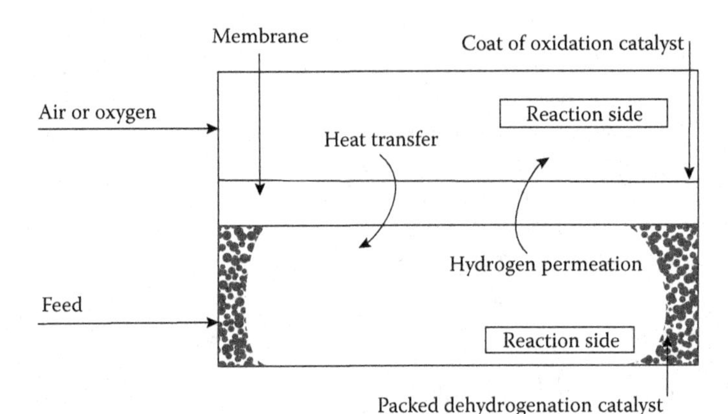

Figure 13.8 Coupling of endothermic (dehydrogenation) and exothermic (oxidation) reactions.

Dehydrogenation and hydrogenation are obvious choices for such coupling. The main disadvantage of this is that there is seldom a complete energetic, kinetic, or thermodynamic matching of the two reactions (Basov and Gryaznov, 1985). Thus, it may often be necessary to supplement the "reactant" generated on the one side or heat on the other by direct additional supply of the deficient quantity to the opposite side.

One can set up equations based on this model for transport of mass as well as energy axially through the reactor which also includes transport across the membrane. But we restrict our treatment here to a few important qualitative observations (see Itoh and Govind, 1989, for a quantitative model) when the reactor is operated adiabatically.

The reaction side temperature (solid lines in Figure 13.9) registers a fall in the vicinity of the entrance at low values of the heat transfer rate ($\Gamma = UA/C_{pA}F_{A0}$), as indicated by curve A_1. This fall disappears as the heat transfer rate rises and reaches a very high value within the first small fraction of the reactor length (curves A_2 and A_3). In the case of the separation side (broken lines), because of the exothermicity of the reaction there can be no minimum in temperature, but the trends beyond the initial region are similar to those for the reaction side (the two curves A_3 and B_3, corresponding to a very high transfer rate, almost completely coinciding with each other). Similar trends are observed for the conversion (with no dip at any point in its value) (Figure 13.9).

Controlled addition of one of the reactants in a bimolecular reaction using an IMR-P There are several instances of industrial organic reactions that are bimolecular and exothermic. An important example is the production of chloromethanes. The temperature rise can be controlled by axially distributed addition of chlorine at several discrete points into a packed bed, fluidized bed, or empty tube reactor through which methane is passed (Doraiswamy et al., 1975). The membrane reactor

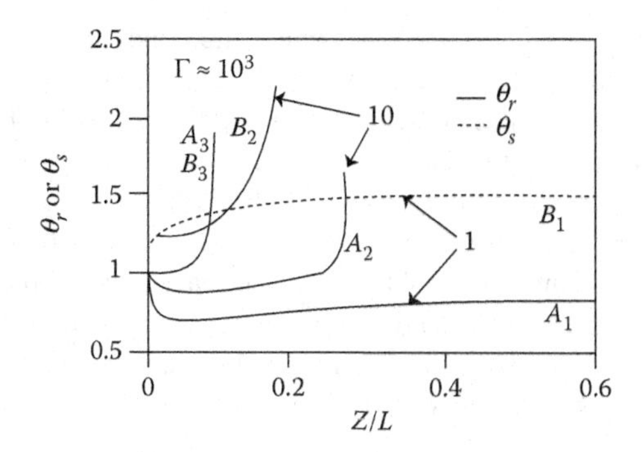

Figure 13.9 Temperature profiles in adiabatic operation of coupled reactions. $\theta_s = T^s/T_0$, $\theta_r = T^T/T_0$, $\Gamma = UA_m/F_AC_{pA}$. (Redrawn approximately from Itoh, N. and Govind, R., *AIChE Symp. Ser.*, **85**, 10, 1989.)

would appear to be an ideal choice for such reactions since it can now be allowed to permeate over the entire length of the membrane from the shell side into the inner tube or vice versa.

Experimental results on the oxidative dehydrogenation of ethane to ethylene (Tonkovich et al., 1995) clearly demonstrate the superior performance of IMR-P in relation to that of PFR. With ethane in the tube and air in the outer shell, the ratio of ethane to air (β) is an important parameter in determining the performance of IMR-P. At high values of β, the amount of air introduced into the shell is relatively small, and hence its permeation into the inner tube is also correspondingly small. Thus, the contact time of ethane with the catalyst does not change significantly through the reactor. Since plug-flow conditions can be assumed to prevail within the reactor, and permeation does little to alter the situation, the performance of IMR-P will be similar to that of PFR. This is brought out in the high-β runs plotted in Figure 13.10.

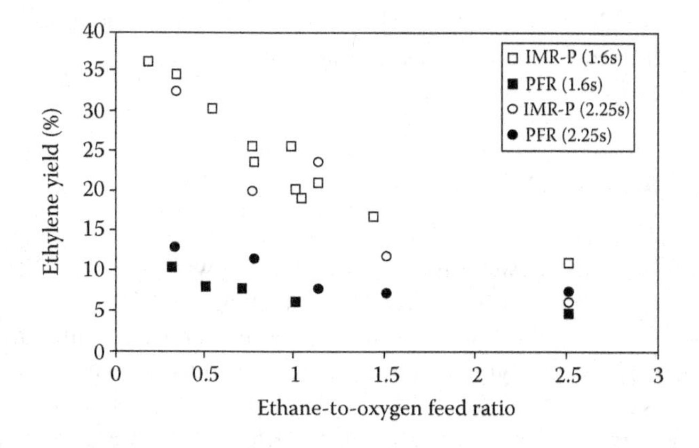

Figure 13.10 Ethylene yields in IMR-P and PFR runs. (Redrawn from Tonkovich, A.L.Y. et al., *Separation Sci. Technol.*, **30**, 1609, 1995.)

For the case of low β, on the other hand, the conditions at the entrance correspond to long contact times, which are not favorable for high selectivity of the intermediate (ethylene); but the low value of β favors high selectivity. As ethane moves downstream, the situation is reversed, that is, the rate of permeation decreases (and hence β increases), while at the same time the contact time decreases. It will, therefore, be seen that the residence time changes favorably down the reactor, while the ethane–air ratio changes unfavorably. The overall effect appears to be a very large enhancement in selectivity over the PFR values at low ratios. In fact, there is a three-fold enhancement over PFR at $\beta \approx 0.3$, as shown in the figure.

The plug-flow reactor is generally accepted as the most favorable with respect to intermediate selectivity in series–parallel reactions. The results of this example clearly show that the membrane reactor can significantly outperform the PFR.

Effect of tube and shell side flow conditions

In the developments presented above, co-current plug flow was assumed both in the tube and shell sides of the reactor. It would be instructive to analyze the effect of countercurrent flow as well as different combinations of plug and mixed flow on the two sides of the membrane. Countercurrent flow can be achieved merely by changing the direction of sweep gas flow. However, this results in a split boundary-value problem since the conditions in the shell side, unlike those in the tube side, are specified at the outlet instead of at the inlet. Substitution of mixed flow for plug flow is straightforward since one has only to use uniform concentrations everywhere in the region.

Five models are possible, as listed in Figure 13.11. Results of simulation using these possible models have shown that (Itoh et al., 1990) model (a), the countercurrent plug flow model, is clearly the best, while model (e), the mixing-mixing model, gives the poorest performance. Changes in parameter values do not change the order of these extreme models, but do alter the sequence of the intermediate models.

Comparison of reactors

It would be instructive to compare the performances of IMR-P and CMR-E not only between themselves, but also with those of PFR and MFR. We shall do so by making the following assumptions: reaction is isothermal, and fluids on each side of the membrane are fully mixed. The latter is a simplifying assumption that permits one to assign a specific pressure to each side. In comparing the performances of these reactors, we must consider the effects due to difference in pressure between the tube and shell sides [effect (1)] and of selective separation [effect (2)] (Sun and Khang, 1990).

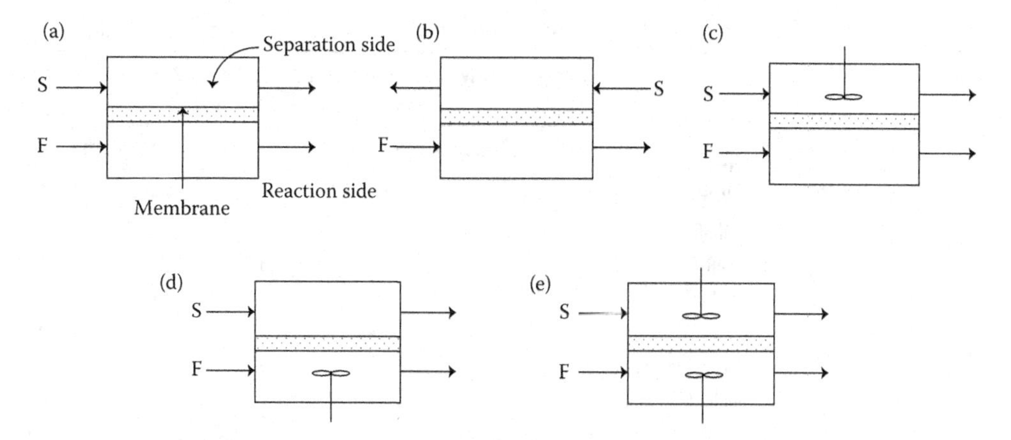

Figure 13.11 Schematic representation of ideal flow configurations. F, feed; S, sweep gas. (a) cocurrent (plug–plug) model, (b) countercurrent (plug–plug) model, (c) plug–mixing model, (d) mixing–plug model, and (e) mixing–mixing model.

Effect (1)

Since reaction occurs simultaneously with permeation, the equilibrium conversion on the separation side will depend on whether P^S is lower than, equal to, or higher than P^T. This in turn will depend on whether there is a decrease, no change, or an increase in the number of moles. Thus, we have: case 1, $\Delta v > 0$; case 2, $\Delta v = 0$; case 3, $\Delta v < 0$.

Effect (2)

This is the characteristic separation effect of the membrane.

Combined effect

A qualitative grading of the performances of IMR-P and CMR-E, along with those of PFR and MFR operated both at the tube and shell side pressures, is presented in Table 13.1. This table may be used as a preliminary guide to reactor selection.

Examples of the use of membrane reactors in organic technology/synthesis

Many of the laboratory-scale studies reported are on hydrogenation, dehydrogenation, and hydrogenolysis reactions involving medium-to-large-volume chemicals. Most of the modeling studies considered in the earlier sections were also based on these reactions. The majority of these investigations have used dense platinum-based membranes. Extensive lists of these studies (many as patents with inadequate information) are presented by Shu et al. (1991), Hsieh (1991), and Saracco and Specchia

Table 13.1 Order of Performance of Different Reactors on a Scale of 1–6

Reactor Type	Reaction: Any Reaction such as $v_A A + v_B B \rightarrow v_R R + v_S S$, $\Delta v = (v_R + v_S) - (v_A + v_B)$					
	Low Space Time			High Space Time		
	$\Delta v > 0$	$\Delta v = 0$	$\Delta v < 0$	$\Delta v > 0$	$\Delta v = 0$	$\Delta v < 0$
IMR-P	4	2	2	4	1	1
CMR-E	6	4	6	3	2	4
PFR (P^T)	3	1	1	5	3	2
PFR (P^S)	1	—	4	1	—	5
MFR(P^T)	5	3	3	6	4	3
MFR (P^S)	2	—	5	2	1	6

Source: Adapted from Sun, Y.M. and Khang, S.J., *Ind. Eng. Chem. Res.*, **29**(2), 232, 1990.

(1994). Studies on small- and medium-volume chemicals were briefly reviewed in Doraiswamy (2001). Details of two processes, for vitamin K and linalool, are outlined below.

Small- and medium-volume chemicals

Vitamin K Conventional production of vitamin K consists of four steps: hydrogenation of 2-methylnaphthoquinone-1,4 to 2-methylnaphthohydroquinone-1,4 in a solvent in the presence of Raney nickel; separation of the product from the catalyst by filtration; evaporation of the solvent; and boiling with acetic anhydride. Since the anhydride is highly corrosive, it tends to attack the nickel, and hence complete separation of the catalyst is necessary. On the other hand, use of a palladium alloy membrane reactor eliminates corrosion and makes it possible to complete the whole process in a single step (Gryaznov et al., 1986), the overall reaction being

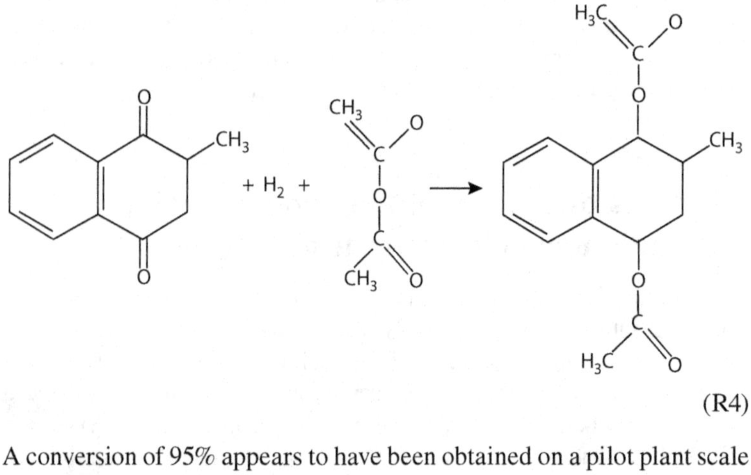

(R4)

A conversion of 95% appears to have been obtained on a pilot plant scale compared to 80% by the conventional process.

Linalool (a fragrance) Another successful application of palladium-based membranes is the hydrogenation of an acetylenic alcohol to an ethylenic alcohol in the synthesis of the perfume linalool in 98% yield. A palladium–ruthenium catalytic membrane was used with high selectivity toward triple bond-to-double-bond hydrogenation. Membranes can give higher selectivities than even the best hydrogenation catalysts in straight catalysis. Production of linanol as shown below (Gryaznov et al., 1982, 1983) is a pointer to further industrial applications of catalytic membranes.

$$CH_3-\underset{\displaystyle CH_3}{\underset{|}{C}}=CH-CH_2-CH_2-\overset{\displaystyle OH}{\overset{|}{\underset{\displaystyle CH_3}{\underset{|}{C}}}}-C\equiv CH + H_2 \longrightarrow CH_3-\underset{\displaystyle CH_3}{\underset{|}{C}}=CH-CH_2-CH_2-\overset{\displaystyle OH}{\overset{|}{\underset{\displaystyle CH_3}{\underset{|}{C}}}}-CH=CH_2 \qquad (R5)$$

Membrane reactors for economic processes (including energy integration)

One of the ways of using a membrane reactor is to selectively feed hydrogen from a hydrogen-rich stream into the reactor. Hydrogen-rich gases are available from refineries, ammonia, and many large-scale dehydrogenation plants. These cannot be used directly to produce high-purity chemicals such as pharmaceuticals and fragrances in view of the often unwanted impurities present in them. Hence, expensive pure hydrogen is almost always used. Membrane reactors make it possible to use these hydrogen-rich gases directly.

Reactions used to convert triple C≡C bonds to double C–C bonds, nitro groups to amino, quinones to hydroquinones, and aldehydes to acids, as well as those used to replace multistep synthesis, have been performed with better results than without membranes (Gryaznov, 1986). An attractive feature of many of these reactions, for example, production of vitamin K, is that corrosive conditions are avoided. Often in single-step reactions, the same catalysts have been used as in the corresponding conventional processes. A reaction such as selective hydrogenation of butynediol to butenediol, for which new catalysts are under constant development, can benefit greatly by use of membrane reactors. In fact, most reactions involving selective saturation of a triple bond to a double bond and partial oxidation can be promising candidates.

Another major advantage of membrane processes is the energy integration that can be introduced by combining an exothermic reaction on one side of the membrane with an endothermic process on the other. Shu et al. (1991) and Saracco and Specchia (1994) give a number of examples of such combined reactions, for example, dehydrogenation of borneol to

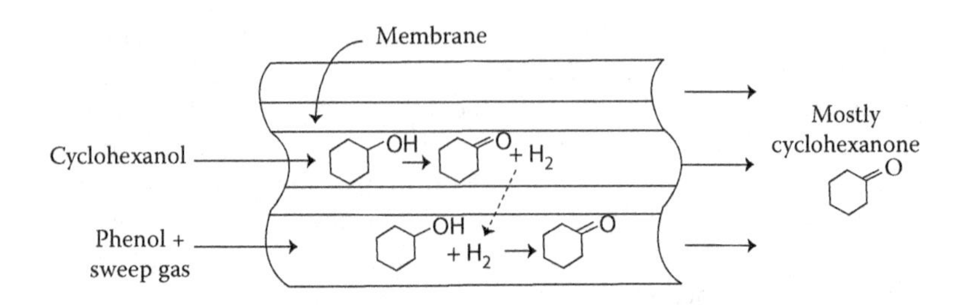

Figure 13.12 Coupling of reactions to produce the same product (cyclohexanone).

camphor combined with hydrogenation of cyclopentadiene to cyclopentene (Smirnov et al., 1981a,b); dehydrogenation of isopropanol to acetone combined again with hydrogenation of cyclopentadiene to cyclopentene (Mikhalenko and coworkers, 1977, 1978; Gryaznov et al., 1981); and dehydrogenation of cyclohexanol to cyclohexanone combined with hydrogenation of phenol to cyclohexanone (Basov and Gryaznov, 1985). The last example is particularly attractive as it produces the same product, cyclohexanone, in both the reactions (Figure 13.12).

References

Agarwalla, S. and Lund, C.R.F., *J. Membr. Sci.*, **70**, 129, 1992.

Basov, N.L. and Gryaznov, V.M., *Membr. Katal.*, 117, 1985.

Champagnie, A.M., Tsotsis, T.T., Minet, R.C., and Wagner, E., *J. Catal.*, **133**, 713, 1992.

Cini, P., Blaha, S.R., Harold, M.P., and Venketeraman, K., *J. Membr. Sci.*, **55**, 199, 1991.

Doraiswamy, L.K., *Organic Synthesis Engineering,* Oxford University Press, New York, 2001.

Doraiswamy, L.K., Krishnan, G.R.V., and Sadasivan, N., *National Chemical Laboratory (India) Report*, 1975.

Gobina, E. and Hughes, R., *Appl. Catal. A*, **137**, 119, 1996.

Gryaznov, V.M., *Plat. Met. Rev.*, **30**(2), 68, 1986.

Gryaznov, V.M., Smirnov, V.S., Mischenko, A.P., and Aladyshev, S.I., British Patent 1,342,869, 1974a.

Gryaznov, V.M., Smirnov, V.S., Mischenko, A.P., and Aladyshev, S.I., U.S. Patent 3,849,076, 1974b.

Gryaznov, V.M., Smirnov, V.S., and Ülin'ko, M.G., *Stud. Surf. Sci. Catal.* **7**, 224, 1981.

Gryaznov, V.M., Karavanov, A.N., Belosljudova, T.M., Ermolaev, A.M., Maganjuk, A.P., and Sarycheva, I.K., Br. Patent 2,096,595, 1982.

Gryaznov, V.M., Karavanov, A.N., Belosljudova, T.M., Ermolaev, A.M., Maganjuk, A.P., and Sarycheva, I.K., British Patent 2,096,595, 1982; U.S. Patent 4,388,479, 1983.

Gryaznov, V.M., Mishchenko, A.P., Smirnov, V.A., Kashdan, M.V., Sarylova, M.E., and Fasman, A.B., French Patent 2,595,092, 1986.

Harold, M.P. and Cini, P. *AIChE Symp. Ser.*, **85**, 26, 1989.

Harold, M.P., Cini, P., Patanaude, B., and Venkatraman, K., *AIChE Symp. Ser.*, **85,** 26, 1989.

Harold, M.P., Zaspalis, V.T., Keizer, K., and Burggraaf, A.J., *5th NAMS Meeting*, Lexington, KY, May 1992.
Hsieh, H.P., *Catal. Rev. Sci. Eng.*, **33,** 1, 1991.
Itoh, N. and Govind, R., *AIChE Symp. Ser.,* **85**, 10, 1989.
Itoh, N., Shindo, Y., Haraya, K., Obata, K., Hakuta, T., and Yoshitome, H., *Int. Chem. Eng.*, **25**(1), 139, 1985.
Itoh, N., Shindo, Y., and Haraya, K., *J. Chem. Eng. Jpn.*, **23**, 420, 1990.
Lund, C.R.F., *Catal. Lett.*, **12**, 395, 1992.
Mikhalenko, N.N. and Tabares, C., *Sovrem. Zadachi V Tochn. Naukakh,* **126**, 1977.
Mikhalenko, N.N., Khrapova, N.E.V., and Gryaznov, V.M., *Neftekhimiya*, **18**, 354, 1978.
Oertel, M., Schmitz, J., Welrich, W., Jendryssek-Neumann, D., and Schulten, R., *Chem. Eng. Technol.*, **10**, 248, 1987.
Saracco, G. and Specchia, V., *Cat. Rev. Sci. Eng.*, **36**, 302, 1994.
Shao, X., Xu, S., and Govind, R., *AIChE Symp. Ser.*, **268**, 1, 1989.
Sheel, H. P. and Crowe, C. M., *Can. J. Chem. Eng.*, **47**, 183, 1969.
Shu, J., Grandjean, B.P.A., Van Neste, A., and Kaliaguine, S., *Can. J. Chem. Eng.*, **69**(10), 1036, 1991.
Sloot, H.J., Versteeg, G.F., and van Swaaij, W.P.M., *Chem. Eng. Sci.*, **45**(8), 2415, 1990.
Sloot, H.J., Smolders, C.A., van Swaaij, W.P.M., and Versteeg, G.F., *AIChE J.*, **38**, 887, 1992.
Smirnov, V.S., Gryaznov, V.M., Ermilova, M.M., Orekhova, N.V., Roshan, N.R., Polyakova, V.P., and Savitskii, E.M., German Patent 3,003,993, 1981a.
Smirnov, V.S., Gryaznov, V.M., Ermilova, M.M., Orekhova, N.V., Roshan, N.R., Polyakova, V.P., and Savitskii, E.M., Soviet Patent 870,393, 1981b.
Sun, Y.M. and Khang, S.J., *Ind. Eng. Chem. Res.*, **29**(2), 232, 1990.
Tonkovich, A.L.Y., Secker, R.B., Reed, E.L., Roberts, G.L., and Cox, J.L., *Separation Sci. Technol.*, **30**, 1609, 1995.
Tsotsis, T.T., Champagnie, A.M., Vasileiadis, S.P., Ziaka, Z.D., and Minet, R.C., *Chem. Eng. Sci.*, **47**(9–11), 2903, 1992.
Tsotsis, T.T., Champagnie, A.M., Vasileiadis, S.P., Ziaka, Z.D., and Minet, R.C., *Sep. Sci. Technol.*, **28**, 397, 1993a.
Tsotsis, T.T., Minet, R.G., Champagnie, A.M., and Liu, P.K.T., In *Computer-Aided Design of Catalysts* (Eds. Becker, E.R. and Pereira, C.J.), Marcel-Dekker, New York, 1993b.
Veldsink, J.W., Damme, R.M.J., Versteeg, G.F., and van Swaaij, W.P.M., *Chem. Eng. Sci.*, **47**, 2939, 1992.
Wang, A.W., Reich, B.A. Johnson, B.K. and Foley, H.C., *Symp. Octane and Cetain Enhanc. Proc. Red. Emissions Motor Fuels, Am. Chem. Soc.*, San Francisco Meeting, 5–10 April, 1992.
Zaspalis, V.T. and Burggraaf, A.F., In *Inorganic Membranes Synthesis, Characteristics and Applications* (Ed. Bhave, R.R.), Van Nostrand Reinhold, New York, 1991.

Chapter 14 Combo reactors

Distillation column reactors

This chapter is primarily concerned with the distillation–reaction combo reactor (or distillation column reactor (DCR) as it is usually called). We shall also briefly consider the case where reaction is imposed on a difficultly separable mixture to achieve complete separation. Membrane reactors are already treated separately in Chapter 13.

As mentioned in Interlude I, combo reactors can be (i) reaction oriented or (ii) separation oriented. Irrespective of whether reaction or separation is of primary concern, three types of combo reactors are commonly used: *reaction–extraction*, *reaction–distillation*, and *reaction–crystallization* (Doraiswamy, 2001).

Crystallization is almost always used for separation and seldom for enhancing a reaction. A notable exception is when one of the reactants is a sparingly dissolving solid and the size of the crystallizing solid is less than the thickness of the film surrounding the reactant. Any phase with particles of such small sizes is often referred to as a microphase. This crystallizing microphase enhances the rate of dissolution of the reactant solid by getting inside the film and disturbing it. The result is that the rate of dissolution (and therefore of reaction) is enhanced. As this is a particular novel strategy for enhancing the rate of a reaction, and has so far not been industrially exploited, we describe a potentially important example of it below.

The manufacturing process for citric acid involves fermentation followed by downstream purification. A common method of purification is to treat the products of fermentation with lime slurry followed by reaction of the solid calcium citrate formed with aqueous sulfuric acid to give an aqueous solution of pure citric acid and a precipitate of calcium sulfate. Note that the final purification step represents a system in which a sparingly soluble solid (calcium citrate) reacts with a liquid product (citric acid) and a solid precipitate (calcium sulfate). Experimental data of Anderson et al. (1998) plotted as conversion versus time showed a sudden rise in conversion at 85%, the exact point at which precipitation started. This clearly suggests "autocatalytic" action by the precipitating solid.

Distillation column reactor

The equipment used in the case of reaction enhanced by separation is often referred to as a DCR. The chief advantage of this method is that the reactants can be used in stoichiometric quantities, with attendant elimination of recycling cost. It is also possible to use this strategy to suppress undesirable chemical reactions, such as in the alkylation of isobutane by butene in the manufacture of isooctane. In the presence of butene, isooctane can undergo further alkylation, thus reducing the selectivity. Use of reactive distillation removes isooctane continuously from the column, thus enhancing the selectivity.

Even for highly exothermic reactions where the heat release can significantly affect the conversion rate, by conducting the reaction in a reactive distillation column, the heat of reaction can be used to remove the product continuously from the reaction mixture. Thus, chemical reactions that exhibit either an unfavorable reaction equilibrium or significant heat of reaction can benefit from reactive distillation column technology.

Enhancing role of distillation: Basic principle

Consider a reversible reaction of the type

$$A + B \leftrightarrow R + S \tag{R1}$$

If one of the products can be continuously removed by carrying out the reaction simultaneously with distillation, then the reaction will be driven further, thus increasing the conversion. The reaction can be carried out either in a simple batch reactor or in a continuous distillation column. The continuous column can be either a plate column (with variations in design) or a packed column. Several studies have been reported on the modeling of such units [see, e.g., the comprehensive reviews by Malone and Doherty (2000) and Chopade and Sharma (1997)].

We consider the following aspects of these reactors:

1. Batch reactor with continuous removal of one of the products
2. Packed-column reactor where the packing is also the catalyst
3. The *residue curve map* (RCM) and its use as the basis for design (the most important strategy)

Batch reactor with continuous removal of product

Let reaction R1 be carried out in a batch reactor with an attached column for separating R. We assume that stoichiometric quantities of A and B are present initially and consider two cases: (1) there is no accumulation of S because it is vaporized as soon as it is formed and (2) there is an accumulation of S because only a fraction of it is vaporized.

Case 1: Accumulation of S

$$N_A = N_{A0}(1 - X_A) \tag{14.1}$$

$$N_S = N_{A0}X_A - \int_0^t F_S\,dt \tag{14.2}$$

The rate equation may be written as

$$-r_A = k_2[A][B] - k_{-2}[R][S] \tag{14.3}$$

where

$$[A] = [B] = \frac{N_{A0}(1 - X_A)}{V}, \quad [R] = \frac{N_{A0}X_A}{V} \tag{14.4}$$

thus

$$-r_A V = N_{A0}\frac{dX_A}{dt} = \frac{k_2 N_{A0}^2(1 - X_A)^2}{V} - \frac{k_{-2}N_{A0}X_A}{V}\left(N_{A0}X_A - \int_0^t F_s\,dt\right) \tag{14.5}$$

An overall mass balance gives

$$\frac{d(\rho V)}{dt} = -F_s M_s \tag{14.6}$$

which, for constant density, becomes

$$\frac{dV}{dt} = -\left(\frac{M_s}{\rho}\right)F_s = -\alpha' F_s \tag{14.7}$$

where M_s is the molecular weight of S and $\alpha' = M_s/\rho$.

Because it is assumed that the product S evaporates as quickly as it is formed

$$N_{A0}X_A = \int_0^t F_s\,dt \tag{14.8}$$

giving

$$F_s = N_{A0}\frac{dX_A}{dt} = \frac{k_2 N_{A0}^2(1 - X_A)^2}{V} \tag{14.9}$$

Substituting this in Equation 14.7, we obtain

$$\frac{dV}{dt} = -\alpha' N_{A0}\frac{dX_A}{dt}$$

or

$$\frac{dV}{dX_A} = -\alpha' N_{A0} \tag{14.10}$$

giving

$$V = V_0(1 + \varepsilon_{L1} X_A) \tag{14.11}$$

where

$$\varepsilon_{L1} = -\alpha'[A]_0 \tag{14.12}$$

Combining Equations 14.9 and 14.11, we obtain

$$\frac{dX_A}{dt} = \frac{k_2 N_{A0}^2 (1 - X_A)^2}{V} = k_2 [A]_0 \frac{(1 - X_A)^2}{(1 + \varepsilon_{L1} X_A)} \tag{14.13}$$

This equation can be solved analytically to give the time required for a specific conversion X_A:

$$t = \frac{1}{k_2 [A]_0} \left[\frac{(1 + \varepsilon_{L1}) X_A}{1 - X_A} - \varepsilon_{L1} \ln \left(\frac{1}{1 - X_A} \right) \right] \tag{14.14}$$

Case 2: S is not completely vaporized
Let us assume that a fraction of product S formed is lost by vaporization. This fraction β' *depends* on the vapor–liquid equilibrium and the heating policy used.

$$N_s = N_{A0} X_A (1 - \beta') \tag{14.15}$$

where

$$\beta' = \text{Reaction} - \text{Separation parameter} = \frac{\int_0^t F_s\, dt}{N_{A0} X_A} \tag{14.16}$$

The parameter β' can be physically interpreted as the ratio of the moles of S removed by vaporization to the moles formed by reaction and may therefore be regarded as a reaction–separation parameter. Equation 14.11 for this case will be modified as follows:

$$V = V_0 - \alpha' \int_0^t F_s\, dt = V_0 - \alpha' \beta' N_{A0} X_A$$

$$= V_0 (1 + \varepsilon_{L2} X_A) \tag{14.17}$$

$$\varepsilon_{L2} = \alpha' \beta' [A]_0 = \frac{M_s \int_0^t F_s\, dt}{\rho N_{A0} X_A} [A]_0 \tag{14.18}$$

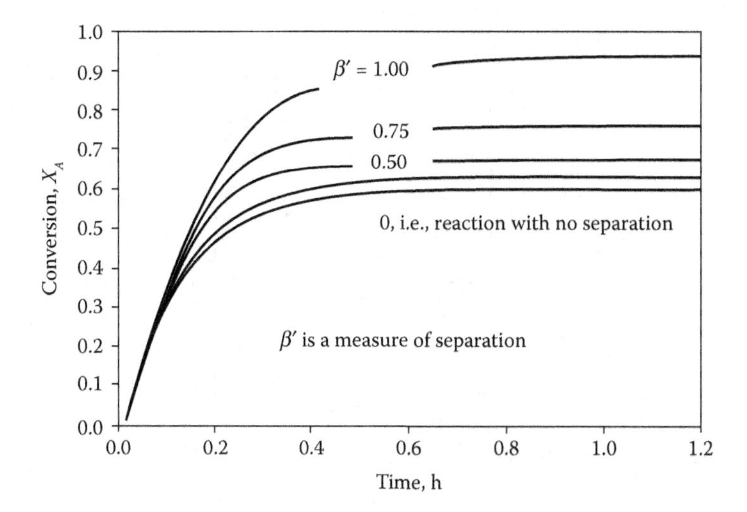

Figure 14.1 Performance of a DCR for the reaction $A + B \rightarrow R + S$. Conversion–time profiles for different values of the extent of product (S) removal β'.

Then, by combining Equations 14.13 through 14.16, we finally obtain

$$\frac{dX_A}{dt} = \frac{k_2[A]_0(1 - X_A)^2}{1 + \varepsilon_{L2}X_A} - \frac{k_{-2}[A]_0 X_A^2}{1 + \varepsilon_{L2}X_A}(1 - \beta') \tag{14.19}$$

Equation 14.19 can be solved numerically for different values of β' corresponding to the extent to which the product is removed. Some results of numerical integration are shown in Figure 14.1. Clearly, the conversion increases with an increase in β'. The maximum conversion is obtained when the product is instantaneously removed from the reaction mixture. At $\beta' = 0$, the conversion approaches the value corresponding to the limiting condition of reaction with no separation.

We now consider the design of industrial DCRs. First, we consider a simple design without going into details of the role of distillation boundaries for both straight distillation and reactive distillation. This will be followed by a mathematical analysis of these factors.

Packed DCR

A packed DCR has the advantage that it speeds up the reaction in the column and also supplies a packing surface for mass transfer at the vapor–liquid interface. As in any packed-bed reactor, the principal aim of modeling a packed DCR is to obtain the concentration (or mole fraction) profiles of the different components along the reactor. A basic requirement for doing this is an equation for mass flux at the surface that incorporates the effect of chemical reaction. This is not needed in modeling plate columns. In addition, the packing usually also acts as

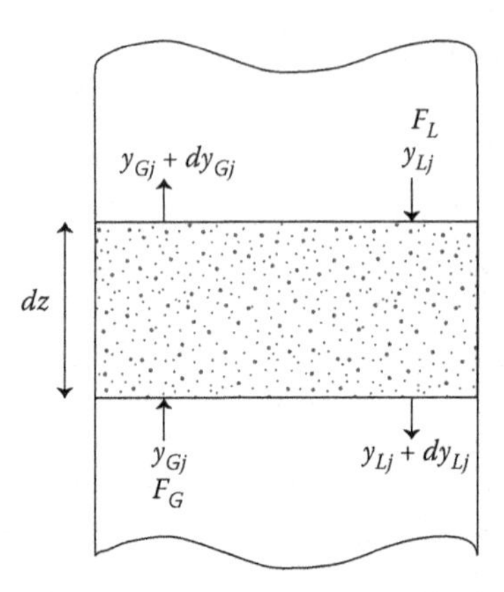

Figure 14.2 Flows in a differential element of a packed DCR.

a catalyst. Hence, the large numbers of models developed for the plate column are not applicable to the packed column.

Consider Figure 14.2, which represents a differential element of a packed-column reactor. The following component mass balance can be readily written as

$$d(F_G y_{Gj}) = -N_j' a A_c dl = d(F_L y_{Lj}) - r_j h_0 A_c dl \tag{14.20}$$

where F_G and F_L are the gas and liquid molal flow rates, respectively, A_c is the cross-sectional area of the column, h_0 is the liquid holdup in the packing per unit volume of the packed bed, N_j' is an overall mass flux that includes the effect of reaction, and a is the area per unit volume. The following expression for N_j' can be used (Sawistowski and Pilavakis, 1979, 1988):

$$N_j' = \left[\frac{1}{k_{Gj}^0} + \left(\frac{K_j}{k_{Lj}^0}\right)\left(\frac{1+\varepsilon_G}{1+\varepsilon_L/2}\right)\right]^{-1}\left[\left(\frac{1+\varepsilon_G+\varepsilon_L/2}{1+\varepsilon_L/2}\right)y_{Gj}^* - y_{Gj}\right] \tag{14.21}$$

where K_j is the equilibrium constant for gas–liquid equilibrium at the surface (i.e., $Y_{Gj,s} = y_{Lj,s}$), and ε_G and ε_L are constants defined by

$$\varepsilon_G = \frac{q_s/\Delta H_v}{k_{Gj}^0}, \quad \varepsilon_L = \frac{q_s/\Delta H_v}{k_{Lj}^0} \tag{14.22}$$

where the k's represent the mass transfer coefficients of species j in the gas and liquid phases (mol/m² s) for pure distillation. The term q_s essentially corresponds to the heat of reaction (since the heat of dilution and heat loss are negligible in comparison) and can be calculated from

$$q_s = \frac{r_j h_0 (\Delta H)}{a} \tag{14.23}$$

Based on Equation 14.20, the following overall and component derivatives with respect to column height can be written

$$\frac{dF_G}{dl} = -\left(\frac{q_s}{\Delta H_v}\right)aA_c$$

$$\frac{dF_L}{dl} = -\left(\frac{q_s}{\Delta H_v}\right)aA_c$$

$$\frac{dy_{Gj}}{dl} = \frac{\left[-y_{Gj}\frac{dF_G}{dl} - N_j aA_c\right]}{F_G} \qquad (14.24)$$

$$\frac{dy_{Lj}}{dl} = \frac{\left[r_j h_0 A_c - y_{Lj}\frac{dF_L}{dl} - N_j aA_c\right]}{F_L}$$

where r_j is the rate of reaction of species j, mol/m^3 s.

The set of Equations 14.24 can be solved provided the following information is available: vapor–liquid equilibrium data, for example, the ternary equilibrium data for a typical esterification reaction; mass and enthalpy balances around the feed point, reflux inlet, and reboiler to account for the flow rates, compositions, and thermal conditions of the external streams; mass transfer coefficients in the absence of reaction (either by experimental determination or by estimation from available correlations); liquid holdup (usually from available correlations); and an expression for the reaction rate. Then, the equations can be solved by any convenient method, preferably the Runge–Kutta routine, to obtain the mole fraction of each component as a function of height.

Overall effectiveness factor in a packed DCR

An important feature of packed DCRs is the need to pack the catalyst in a special way to ensure good flow, mass transfer, and contact characteristics. An example of this is the use of an ion-exchange resin catalyst (Amberlyst 15) in methyl tertiary butyl ether (MTBE) manufacture. The bed consists of bags made in the form of a cloth belt with narrow pockets sewn across it (Figure 14.3). The pockets are filled with catalyst granules, and the belt is twisted into a helical form, referred to as a bale (see Smith, 1980, for details). Clearly, each pocket represents a closely packed bed of unconsolidated particles, and the pocket and the individual particle exhibit, respectively, their own distinctive macro- and microdiffusional features. This is broadly similar to the particle–pellet model of a catalyst pellet (see Chapter 8) but with distinctly different "pellet" behavior. Therefore, it is necessary to define an overall effectiveness factor that takes this unique feature into account. An attempt to do this was reported (Xu et al., 1995) for the MTBE reaction, for which

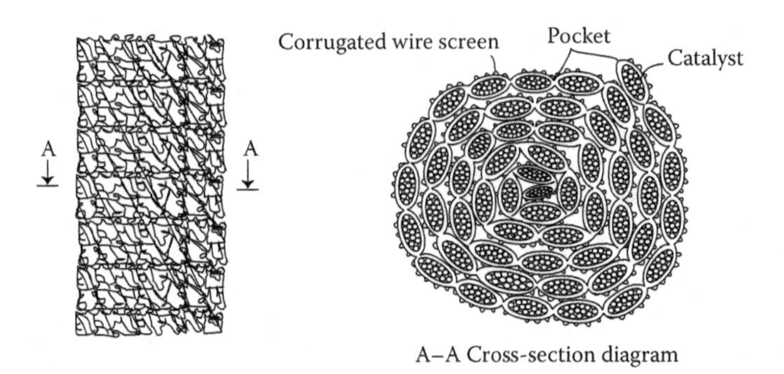

Figure 14.3 Structure of the catalyst bale used in the MTBE reactor. (Adapted from Xu, X., Zheng, Y., and Zheng, G., *Ind. Eng. Chem. Res.*, **34**, 2232, 1995.)

the overall effectiveness factor was found to vary from 0.2 to 0.9 depending on the conditions of operation.

Residue curve map (RCM)

We now describe a procedure where we begin with distillation and add reaction to it, first with an equilibrium reaction and then a reaction with a finite forward rate constant (in other words, we will explore the effect of reaction kinetics on distillation behavior). A common example of distillation–reaction is esterification, such as Eastman Kodak's process for methyl acetate:

$$\text{Methanol} + \text{acetic acid} \leftrightarrow \text{methyl acetate} + \text{water} \qquad \text{(R2)}$$

in which the conversion is limited by thermodynamics. By continuously removing water or methyl acetate from the reaction mixture, the reaction equilibrium can be forced completely to the product side without using an excess of any reactant. The methods presented previously for accomplishing this notwithstanding, the best theoretical approach is to base the analysis on the concept of RCM for pure distillation and then examine the consequences of introducing a chemical reaction under both equilibrium and nonequilibrium conditions. In the latter situation, reaction kinetics will also be involved. We start by defining RCM.

Let us consider a mixture of three components A, B, and R. Of special importance are the so-called residue curves, which represent the liquid residue compositions with time in a simple batch distillation. Different curves are obtained for different starting compositions, and a collection of these curves for a given ternary system is called an RCM. A typical RCM is shown in Figure 14.4a. Some important features of these curves should be noted: If the direction of the curve is assumed to be from the starting composition to the ending composition, then the arrow on each curve points from the lower-boiling component or azeotrope (if one is formed) to the higher. The presence of azeotropes can create boundaries

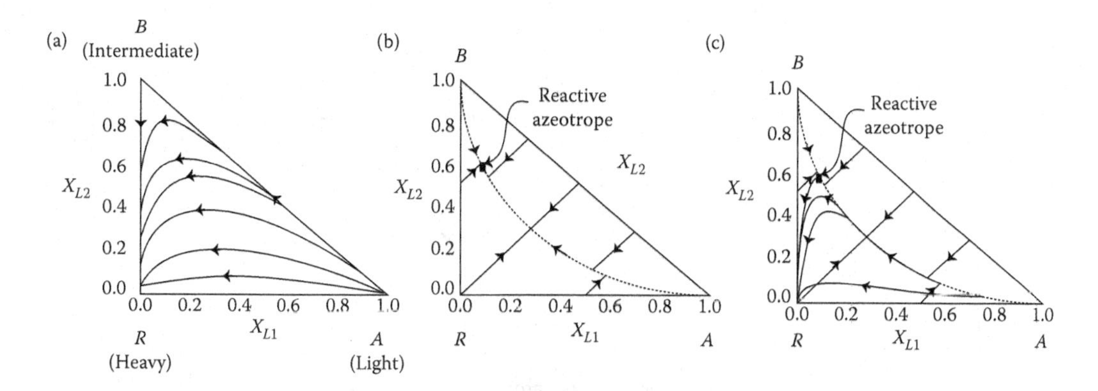

Figure 14.4 RCMs for different situations in a DCR. (a) Nonreactive mixture of A, B, and R; (b) at Da = 100, for a reactive mixture $A + B \rightarrow 2R$; (c) RCM for the same mixture using Equation 4.33. In all these cases, the relative volatility of A and B relative to R are 5 and 3, respectively. (From Venimadhavan, G. et al., *AIChE J.*, **40**, 1814, 1994.)

for the residue curve. These curves merge with the reaction equilibrium curve, thus establishing a new distillation boundary on which the light or starting residue composition is the lower-boiling pure component or azeotrope and the heavy or ending composition residue is the higher pure component or azeotrope.

Design methodology The design of DCRs has developed into a highly specialized area, and it is not our intention to cover this in great detail here. However, some of the theoretical foundations are presented, based essentially on the studies of Barbosa and Doherty (1987a,b, 1988a,b,c), Buzad and Doherty (1994), and Venimadhavan et al. (1994).

The design equations would include, in addition to the usual heat and mass balances and vapor–liquid equilibria, equations for chemical equilibria alone or with the forward rate equation. The occurrence of a chemical reaction can severely restrict the allowable ranges of temperatures and phase compositions by virtue of the additional equations for chemical equilibrium/kinetics. As already explained, this effect can be quantitatively analyzed by constructing an RCM. We illustrate this by considering the reaction

$$A + B \leftrightarrow 2R \qquad \text{(R3)}$$

in which the three components form an ideal liquid mixture. The rate of reaction per mole of mixture is given by

$$r_A = r_B = -\frac{r_C}{2} = -k_f\left(y_{L1}y_{L2} - \frac{y_{L3}^2}{K}\right) \qquad (14.25)$$

where components 1, 2, and 3 correspond to A, B, and R, respectively, K is the thermodynamic equilibrium constant, k_f is the forward reaction

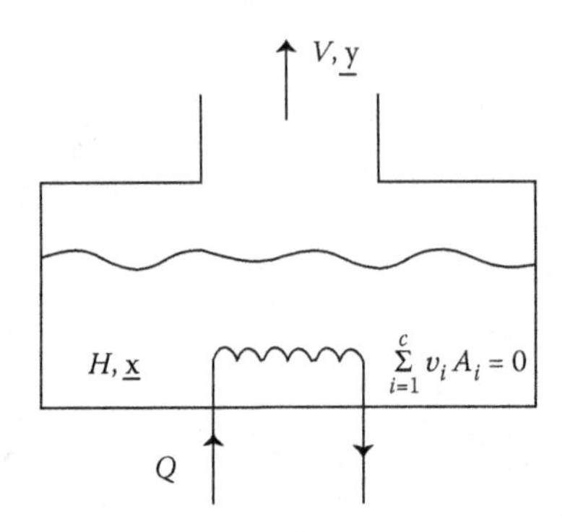

Figure 14.5 System parameters for the design methodology.

rate constant, and x is the mole fraction. We use mole fractions (instead of concentrations) because then the rate constant can be expressed in units of reciprocal time regardless of the order of the reaction, allowing for a simple universal definition of the dimensionless Damköhler number (which, in this case, is the ratio of reaction to separation rates) for all reaction orders. As this number represents the rate of reaction relative to the rate of product removal, it is a direct measure of the effectiveness of reactive distillation.

Referring to Figure 14.5, the overall and component material balances for simple distillation are

$$\frac{dH}{dt} = -V \tag{14.26}$$

H: holdup

$$\frac{dHy_{Li}}{dt} = -Vy_{Gi} - Hk_f\left(y_{L1}y_{L2} - \frac{y_{L3}^2}{K}\right), \quad i = 1,2 \tag{14.27}$$

$$H\frac{dy_{Li}}{dt} - Vy_{Li} = -Vy_{Gi} - Hk_f\left(y_{L1}y_{L2} - \frac{y_{L3}^2}{K}\right), \quad i = 1,2 \tag{14.28}$$

which, by simple manipulation, leads to

$$\frac{dy_{Li}}{d\xi} = y_{Li} - y_{Gi} - Da\left(y_{L1}y_{L2} - \frac{y_{L3}^2}{K}\right)\frac{H}{H_0}\frac{k_f}{k_{f,\min}}\frac{V_0}{V}, \quad i=1,2 \tag{14.29}$$

where

$$d\xi = \left(\frac{V}{H}\right)dt \quad \text{dimensionless time} \tag{14.30}$$

and Da is the Damköhler number given by

$$Da = \frac{H_0 k_{f,\min}}{V_0} \tag{14.31}$$

$k_{f,\min}$ is the forward reaction rate at the temperature of the lowest-boiling pure component or azeotrope. For simplicity, we assume that k_f is constant and equal to k. Expressing t in its dimensionless form (Equation 14.30) and integrating [with $H(\xi = 0) = H_0$]

$$\frac{H}{H_0} = e^{-\xi} \tag{14.32}$$

Substituting Equation 14.32 in Equation 14.29 and recalling that $k_f/k_{f,\min} = 1$ (assumed)

$$\frac{dy_{Li}}{d\xi} = y_{Li} - y_{Gi} - Da\left(y_{L1}y_{L2} - \frac{y_{L3}^2}{K}\right)\frac{V_0}{V}e^{-\xi}, \quad i = 1,2 \tag{14.33}$$

We now run into a problem in that Equation 14.33 cannot be solved without first specifying the ratio V/V_0, which represents the rate of vapor generation. In other words, a heating policy has to be defined. Venimadhavan et al. (1994) defined it mathematically as

$$V = V_0\, e^{\epsilon\xi} \tag{14.34}$$

where ϵ is a general parameter that determines the vapor rate policy. Substituting this in Equation 14.33 gives

$$\frac{dy_{Li}}{d\xi} = y_{Li} - y_{Gi} - Da\left(y_{L1}y_{L2} - \frac{y_{L3}^2}{K}\right)e^{-(1+\epsilon)\xi}, \quad i = 1,2 \tag{14.35}$$

We will not consider this equation as such for it has multiple solutions (the curious student is encouraged to explore this further). We restrict the treatment to a relatively simple case by assuming $\epsilon = -1$, which amounts to assuming a decreasing vapor rate policy giving $V/V_0 = H/H_0$. With this assumption, Equation 14.35 becomes

$$\frac{dy_{Li}}{d\xi} = y_{Li} - y_{Gi} - Da\left(y_{L1}y_{L2} - \frac{y_{L3}^2}{K}\right), \quad i = 1,2 \tag{14.36}$$

This is an important equation in spite of some simplifying assumptions, for it defines the effect of chemical reaction with a finite rate on the distillation behavior of the system. We now use this equation to calculate the RCMs.

Generating residual curve maps

Generating residual curve maps We go back to reaction R3 and assume a temperature-independent value of $K = 2$. We also assume that A has a constant volatility relative to R of 5, and B a relative value of 3. The first step is to generate the RCM for the nonreactive mixture of A, B, and R. This is just a triangular plot of the binary liquid phase compositions AB, BR, RA. The procedure consists of the following steps:

1. Construct this ternary diagram with the low-boiling component at the right vertex, the high-boiling component at the left vertex, and the intermediate component at the top vertex.

2. Draw arrows on the edges of the triangle in the direction of decreasing temperature for each binary.
3. Each vertex representing a pure component is called a saddle. The highest boiling component (or azeotrope) in a region is the stable node and the lowest boiling component is the unstable node.
4. Pure components and azeotropes that have a boiling point between the stable and unstable nodes are called saddles. These are characterized by residue curves that move toward and then away from the saddles.
5. Draw the residue curves. Each such curve represents the liquid residue composition with time in the still during a simple one-stage batch distillation.
6. Mark on the diagram the direction of the residue curve as from the starting composition to the ending composition. When this is done, the arrow on each curve will point from a lower-boiling component (or azeotrope) to a higher-boiling component (or azeotrope).
7. The presence of azeotropes can create distillation boundaries, and (as so clearly explained in Widagdo and Seider (1996)), these boundaries cannot be crossed by a residue curve. They represent the cases of a light or starting residue being a lower-boiling pure component or azeotrope, and the heavy or ending residue being a higher-boiling pure component or residue.
8. These distillation boundaries divide the map into different distillation regions in such a way that separation of two pure components from different regions is not possible using conventional distillation.

saddle
stable node
unstable node

Figure 14.4a represents the RCM for the ideal mixture of pure A, B, and R in the absence of reaction. It will be noted that the arrows on the map conform to the requirements of an RCM as brought out above. Let us now impose reaction R3 on the system and ensure that it reaches equilibrium almost instantaneously. We also assume that the rate of reaction is much faster than the rate of product removal, thus conforming to a high value of the Damköhler number. Figure 14.4b represents this situation with $Da = 100$. The central curve in the figure is the reaction equilibrium curve. It will be seen that under the conditions assumed, a reactive azeotrope is formed, and the components from both ends move toward it. All lines from the three sides of the triangle stop at the equilibrium curve, thus defining the distillation boundary for this situation (distillation with equilibrium reaction). Let us now remove the equilibrium restriction and assume that the reactions move forward at a finite rate, again at $Da = 100$. In other words, we consider the effect on distillation of reaction R3 proceeding at a very fast forward rate. This situation is depicted in Figure 14.4c and clearly shows that the residue curves are all confined to the region left of the equilibrium curve, and at long times the chemical equilibrium curve (the right side boundary) is recovered.

Meaning of azeotropy in reactive distillation

New coordinates for reactive distillation It is clear from Figure 14.4 that the condition for azeotropy is *not* equality of compositions in the liquid and vapor phases, as in straight distillation. Instead, the following equality has been proposed (Barbosa and Doherty, 1988a,b,c):

$$\frac{y_{L1} - y_{G1}}{v_1 v_t y_{L1}} = \frac{y_{Li} - y_{Gi}}{v_i v_t y_{Li}}, \quad i = 2,\ldots,c, \; v_t = \sum_0^{\infty} v_i \tag{14.37}$$

where y_{Li} and y_{Gi} are the mole ratios of component i in the liquid and gas phases, respectively, and c is the number of components.

Equation 14.37 shows that composition is not a convenient measure of azeotropy in reactive distillation. For equilibrium reactive mixtures, it is more convenient to use transformed variables that represent the equivalent amounts of reactants present in the equilibrium mixture. Thus, for the reaction

$$v_A A + v_B B \leftrightarrow v_R R \tag{R4}$$

these variables are defined as

$$\begin{aligned} Y_{LA} &= \frac{y_{LA} - (v_A/v_R)y_{LR}}{1 - (v_t/v_R)y_{LR}} \\ Y_{LB} &= \frac{y_{LB} - (v_A/v_R)y_{LR}}{1 - (v_t/v_R)y_{LR}} \end{aligned} \tag{14.38}$$

The transformed composition variables have two convenient properties: they have the same numerical value before and after reaction, and their sum is unity:

$$Y_{LA} + Y_{LB} = 1 \tag{14.39}$$

The condition for reactive azeotropy when expressed in terms of these transformed variables takes the familiar form of compositional equality and may be written as

$$Y_{Li} = Y_{Gi} \tag{14.40}$$

Reference may be made to the original publication of Barbosa and Doherty on the use of these transformed variables.

Distillation–reaction

We now take a brief look at the situation where reaction is used as an aid in separating difficultly separable mixtures. We examine a case where distillation, reaction, and extraction are simultaneously involved. In other words, two separation processes are combined with reaction.

Dissociation–extractive distillation

As mentioned at the beginning, it is also possible to impose a chemical reaction on the distillation of two closely boiling compounds to achieve effective separation of the components. This method is based on exploiting the dissociation processes involved in the acid–base reactions imposed on separation by distillation of the components of two closely boiling liquids. This is referred to as *dissociation–extractive distillation.*

Distillation combined with reaction has been successfully used for the separation of close-boiling mixtures. When used in this separation mode, the technique is frequently referred to as dissociation–extractive distillation. It can also be used in the reaction mode by continuous separation of the products of reaction from the reactants. The equipment used in the latter case is often referred to as the DCR. The chief advantage of this method is that the reactants can be used in stoichiometric quantities, with attendant elimination of recycling costs.

Basic principle The mole fraction of the ith component (y_{Gi}) in the vapor phase in equilibrium with a liquid mixture with a mole fraction y_{Li} at a total pressure P (not much higher than atmospheric) is given by

$$y_{Gi} = \frac{\gamma_i y_{Li} P_i}{P_T} \tag{14.41}$$

where γ_i is the activity coefficient of i. By definition, the relative volatility of a mixture of components i and j is given by

$$\beta_{ij} = \frac{y_{Gi}/y_{Li}}{y_{Gj}/y_{Lj}} \tag{14.42}$$

or

$$\frac{y_{Gi}}{y_{Gj}} = \frac{\gamma_i y_{Li} P_i}{\gamma_j y_{Lj} P_j} \tag{14.43}$$

It is evident that the relative volatility can be manipulated through the γ's by the addition of suitable reacting/complexing agents.

Theory Acid–base reactions, which are typically very fast and reversible, can take advantage of steric and/or acidity differences of the components of a mixture, usually isomers. Consider a mixture of two bases B_1 and B_2 that are to be separated by imposing a chemical reaction with an acid introduced in stoichiometric deficiency. The reactions involved are

$$B_1 + AH \overset{K_{B_1}}{\leftrightarrow} B_1H^+A^- \tag{R5}$$

$$B_2 + AH \overset{K_{B2}}{\leftrightarrow} B_2H^+A^- \tag{R6}$$

The complexes formed are assumed to be nonvolatile. In other words, the chemical reactions are assumed to be confined to the liquid phase. The following competitive reaction now occurs between the bases and the complexes:

$$B_1H^+A^- + B_2 \overset{K_{12}}{\leftrightarrow} B_2H^+A^- + B_1 \tag{R7}$$

where

$$K_{12} = \frac{[B_2H^+A^-]_L[B_1]_L}{[B_1H^+A^-]_L[B_2]_L} = \frac{K_{B_2}}{K_{B_1}} \tag{14.44}$$

Also, liquid–vapor equilibrium is established between the free bases, giving

$$\beta_{12} = \frac{[B_1]_G[B_2]_L}{[B_1]_L[B_2]_G} \tag{14.45}$$

Two steps are involved in the protonation of the bases. These are ion pair formation and dissociation of the ion pair into charged species as shown below:

$$B_i + AH \overset{K_{PT}}{\leftrightarrow} B_iH^+A^- \overset{K_D}{\leftrightarrow} B_iH^+ + A^- \tag{R8}$$

where K_{PT} is the protonation constant and K_D the ion pair dissociation constant. The solvent dielectric constant, ε_D, plays a role in the ion pair dissociation step (the attractive force between ions is inversely proportional to ε_D). Thus, with weak acids in nonaqueous media, the formation of free ions should be very low. On the other hand, the higher stability of the anion A^- of a strong acid allows ion dissociation, in which case the free ions should be formed in greater amount if the dielectric constant of the solvent is high. Moreover, the K_{12} value should be higher with free ions than with ion pairs. Normally, K_{12} is dependent on the difference between the pK_a's of the acid and bases involved in this acid–base reaction except for orthosubstituted compounds because of the steric effects. The value of K_{12} must be known from experiments in both the presence and absence of the reacting components.

If the entire amount of acid is consumed in the complexation because of stoichiometric deficiency, then one can assume

$$N = [B_1H^+A^-]L + [B_2H^+A^-]L \tag{14.46}$$

We now define an apparent relative volatility as

$$\beta_a = \frac{\left[[B_1]_G/([B_1]_L + [B_1H^+A^-]_L)\right]}{\left[[B_2]_G/([B_2]_L + [B_2H^+A^-]_L)\right]}$$

$$\beta_a = \left[\frac{[B_1]_G}{[B_2]_G}\frac{[B_2]_L}{[B_1]_L}\right]\frac{\left[(1 + [B_2H^+A^-]_L)/[B_2]_L\right]}{\left[(1 + [B_1H^+A^-]_L)/[B_1]_L\right]}$$

$$\beta_a = \beta_{12}\frac{\left[(1 + [B_2H^+A^-]_L)/[B_2]_L\right]}{\left[(1 + [B_1H^+A^-]_L)/[B_1]_L\right]} \tag{14.47}$$

Elimination of $[B_1H^+A^-]$ and $[B_2H^+A^-]$ through Equations 14.44 and 14.46 leads to the following expression for relative volatility:

$$\beta_a = \beta_{12}\left[1 + \frac{N(\delta'\delta + 1)(K_{12} - 1)}{N(\delta' + 1) + K_{12}[B_{12}]_L + \delta'[B_{12}]_L}\right] \tag{14.48}$$

where δ' and $[B_{12}]_L$ are two new parameters defined as

$$\delta' = \frac{[B_2]_L}{[B_1]_L} \tag{14.49}$$

$$[B_{12}]_L = [B_1]_L + [B_2]_L \tag{14.50}$$

The value of K_{12} is dependent on the relative strengths of the two bases. For $K_{12} > 1$, β_a will always be greater than β_{12}.

Clearly, the presence of competitive reactions can increase the relative volatility of the mixture. For most close-boiling substances, especially isomers, β_{12} is close to unity. Thus, the increase in relative volatility is because of the bracketed quantity, which can be manipulated by varying the reacting component, thereby changing K_{12}. Figure 14.4 shows the variation in β_a/β_{12} with K_{12} for a set of selected parameter values.

Examples

An acid–base reaction can be used to enhance separation by taking advantage of the difference in the pK_a values of the components to be separated (Duprat and Gau, 1991). For example, in the close-boiling 3-/4-picolines mixture, addition of trifluoroacetic acid in stoichiometric deficiency results in preferential complexation of 4-picoline with a selectivity of about 2 in formamide as solvent. As the pyridinium salts are nonvolatile, the vapor phase is enriched with respect to 3-picoline. A near-complete separation can be expected by repetition of this enrichment in countercurrent staging. 3-Picoline will leave the column as the distillate at the top, while 4-picoline will leave as a liquid phase complex with the acid as the bottom product. 4-Picoline can be regenerated from the complex by addition of a stronger base.

Acid–base reactions can also be exploited by taking advantage of the steric and/or acidity differences of alcohol isomers by conducting the distillation of these mixtures in the presence of amines (Gassend et al., 1985). More recently, organic bases have been used in the reactive distillation of close-boiling phenolic substances (Mahapatra et al., 1988).

References

Anderson, J.G., Larson, M.A., and Doraiswamy, L.K., *Chem. Eng. Sci.,* **53**, 2459, 1998.
Barbosa, D. and Doherty, M.F., *Chem. Eng. Sci.,* **43**, 529, 1988a.
Barbosa, D. and Doherty, M.F., *Chem. Eng. Sci.,* **43**, 541, 1988b.
Barbosa, D. and Doherty, M.F., *Chem. Eng. Sci.,* **43**, 1523, 1988c.
Barbosa, D. and Doherty, M.F., *Proc. R. Soc.,* **A413**, 443, 1987a.
Barbosa, D. and Doherty, M.F., *Proc. R. Soc.,* **A413**, 459, 1987b.
Buzad, G. and Doherty, M.F., *Chem. Eng. Sci.,* **49**, 1947, 1994.
Chopade, S.P. and Sharma, M.M., *React. Funct. Polym.,* **32**, 53, 1997.
Doraiswamy, L.K. *Organic Synthesis Engineering,* Oxford University Press, New York, 2001.
Duprat, F. and Gau, G., *Can. J. Chem. Eng.,* **69**, 1320, 1991.
Gassend, R.G., Duprat, F., and Gau, G., *Nouv J. Chim.,* **19**, 703, 1985.
Mahapatra, A., Gaikar, V.G., and Sharma, M.M., *Sep. Sci. Technol.,* **23**, 429, 1988.
Malone M.F. and Doherty M.F., *Ind. Eng. Chem. Res.,* **39**, 3953, 2000.
Sawistowski, H. and Pilavakis, P.A., *Inst. Chem. Eng. Symp. Series,* **56**, 42, 1979.
Sawistowski, H. and Pilavakis, P. A., *Chem. Eng. Sci.,* **43**, 355, 1988.
Smith, L.A. Jr., U.S. Patent 4,242,530, 1980.
Venimadhavan, G., Buzad, G., Doherty, M.F., and Malone, M.F., *AIChE J.,* **40**, 1814, 1994.
Widagdo, S. and Seider, W.D., *AIChE*, **42**, 96, 1996.
Xu, X., Zheng, Y., and Zheng, G., *Ind. Eng. Chem. Res.,* **34**, 2232, 1995.

Chapter 15 Homogeneous catalysis

Introduction

General

Catalysis by soluble complexes of transition metals is a rapidly gaining mode of catalysis in organic synthesis. These metals form bonds with one or more carbons in an organic reactant, resulting in complexes that are known as organometallic complexes. Catalysis by these complexes is often referred to as homogeneous catalysis. Among the important applications of homogeneous catalysis in organic synthesis are isomerization of olefins, hydrogenation of olefins (carried out using Wilkinson-type catalysts), hydroformylation of olefins to aldehydes with CO and H_2 (the *oxo* process), carbonylation of unsaturated hydrocarbons and alcohols with CO (and coreactants such as water), and oxidation of olefins to aldehydes, ketones, and alkenyl esters (Wacker process).

Formalisms in transition metal catalysis

Uniqueness of transition metals

Transition metals with partially filled *d* shells

Transition metals have the distinctive property (not shared by other metals) that their *d* shells are only partially filled with electrons. This gives them the unique ability to exist in several oxidation states. As a result, the use of organic complexes of transition metals can provide pathways for an extraordinary range of reactions.

A typical transition metal atom has nine valence shell orbitals: one *s*, three *p*, and five *d*, in which it accommodates valence electrons that bond with other moieties known as *ligands* (usually represented by the letter L) to form two types of bonds, *covalent* and *coordinated.* In a typical covalent bond, one or more electrons are shared between two atoms. However, the electrons that constitute a bond need not be equally shared; in fact, all of the electrons can come from just one of the atoms. Although this kind of bond is also a covalent bond, it is usually regarded as a subcategory of the covalent bond and is referred to as a coordinate bond. This ability to form these two types of bonds with a number of ligands is responsible for the unique catalytic properties of the transition metals and their complexes.

Example 15.1: The hydride complex $ReH_7(PMe_2Ph)_2$ is known to be catalytically very active. Why is this so?

The hydride complex contains phosphine as the ligand (Figure 15.1) (see Bau et al., 1978). There are seven Re–H bonds and two P–Re bonds. The seven metal–hydrogen bonds are covalent bonds formed by pairing of the lone electrons from seven hydrogen atoms with valence electrons of rhenium, one from each of its seven orbitals. They are metal–ligand bonds similar to any C–H bond of an organic molecule. On the other hand, the two P $\rightarrow$ Re bonds are coordinate bonds formed by the donation of a complete electron pair from just one partner, which then is shared by the other. The donor component in this case is the phosphine ligand.

covalent bond

coordination

The metal bonds with nine ligands in two different ways, seven covalently and two coordinatively. This ability of the transition metal to bond with such a large number of ligands makes it unique. The changes that this coordination introduces in the electron distribution in the complex greatly enhance the reactivity of the ligand molecule, leading to its unique catalytic action.

Another concept that is fundamental to an understanding of homogeneous catalysis is that of the oxidation state of a complex. Although commonly used in general chemistry, a less puristic definition of this state that may be regarded as a formalism more suited to transition metal complexes has become the cornerstone of homogeneous catalysis.

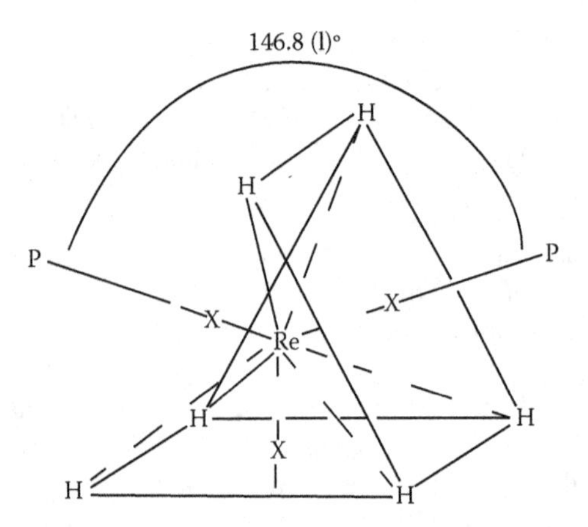

Figure 15.1 A postulated structure of H_7Re (P-Me_2Phh). (From Bau, R. et al., In *Transition Metal Hydrides, Am. Chem. Soc. Ser.,* Vol. 167, Washington, DC, 1978.)

Indeed, the application of transition metal catalysis in organic technology is built around much such formalism. In addition to the formal oxidation state, these include *coordinative unsaturation, coordination number,* and *coordination geometry; hydride formalism;* and the *18-electron* and *16–18-electron rules*, also referred to as *electron bookkeeping.*

coordination

electron bookkeeping

Oxidation state of a metal

According to Collman et al. (1987), the *oxidation state* is defined as "the charge left on the central atom when the ligands are removed in their normal closed-shell configurations." A particularly important fact in organometallic chemistry is that both hydrogen and carbon are more electronegative than the transition metal (recall that H is electropositive in acids and C in organic molecules). Thus, the M–H group is considered as M^+H^-. This is commonly referred to as the *hydride formalism.* Oxidation states can be similarly assigned to any ligand if its normal closed-shell configuration is known.

The formal oxidation state of a metal is related to its *d* electron configuration as shown in Table 15.1 (Collman et al., 1987). This table is constructed on the assumption that the outer electrons are all *d* electrons. This is usually a good approximation. Thus, Co(O) is d^9, Re(IV) is d^3, Ir(II) is d^7, and so on. A metal complex consists of the metal and a ligand. Thus, the formal ligand charge should be known before we can assign a formal oxidation state to the metal. Collman et al.'s table of ligand charges (1987) is reproduced in Table 15.2. Example 15.2 illustrates the procedure for obtaining the oxidation states of metals in various complexes (see Kegley and Pinhas, 1986; Collman et al., 1987, for more examples).

Table 15.1 Relationships between Oxidation State and d^n in Transition Metals

Group Number		**IVa**	**Va**	**VIa**	**VIIa**		**VIIIa**		**Ib**
First row	3d	Ti	V	Cr	Mn	Fe	Co	Ni	Cu
Second row	4d	Zr	Nb	Mo	Tc	Ru	Rh	Pd	Ag
Third row	5d	Hf	Ta	W	Re	Os	Ir	Pt	Au
Oxidation state	0	4	5	6	7	8	9	10	—
	I	3	4	5	6	7	8	9	10
	II	2	3	4	5	6	7	8	9
	III	1	2	3	4	5	6	7	8
	IV	0	1	2	3	4	5	6	7

Source: Collman, J.P. et al., *Principles and Applications of Organotransition Metal Chemistry,* 2nd ed, University Science Books, Mill Valley, CA, 1987.

Table 15.2 Charges and Corresponding Coordination Numbers for Typical Ligands

Ligand	Charge[a]	Coordination[a] Number
X (Cl, Br, I)	−1	1(2)
H	−1	1(2,3)
CH_3	−1	1(2)
Ar	−1	1
RCO	−1	1(2)
Cl_3Sn	−1	1
R_3Z (Z = N, P, As, Sb)	0	1
R_2E (E = S, Se, Te)	0	1
CO	0	1(2,3)
RNC	0	1(2)
RN	0(−2)	1(2)
R_2C	0(−2)	1(2)
N_2	0	1(2)
$R_2C{=}CR_2$	0(−2)	1(2)
$RC{\equiv}CR$	0(−2)	1(2)
CN	−1	1
η^4-Cyclobutadiene	0	2
$CH_2{=}CHCH_2{-}$	−1	1
η^3-Allyl	−1	2
η^6-Benzene (C_6H_6)	0	3(2,1)
η^5-Cyclopentadienyl (C_5H_5)	−1	3
η^7-Cycloheptadienyl (C_7H_7)	+1	3
NO[b]	+1(−1)	1(2)
ArN_2[b]	+1(−1)	1(2)
O	−2	2
O_2	−2(−1)	2(1)

Source: Collman, J.P. et al., *Principles and Applications of Organotransition Metal Chemistry*, 2nd ed, University Science Books, Mill Valley, CA, 1987.

Note: The superscript on η implies that all carbon atoms interact with the metal.

[a] Less common or alternative formulation in parentheses.

[b] Noninnocent ligand (two or more discrete bonding modes).

Example 15.2: Determine the oxidation state of the metal in each of the following complexes: $[Fe(CO)_4]^{2-}$, $[Ni(CO)_4]$, $[Cu(NH_3)_4]^{2+}$, $[(\eta^5\text{-}C_5Me_5)IrMe_4]$, and $(PMe_3)_2Pd(\eta^3\text{-}C_3H_3)]^+$

Note from Table 15.2 that the carbonyl ligands are formally neutral. Because the charge on the iron carbonyl is 2^-, the charge on iron is $(-2 - 0) = -2$. Hence, the oxidation state of iron in this complex is −2, which is represented as Fe(−II).

In $[Ni(CO)_4]$, because the charge on the molecule is zero and the carbonyl ligands are also neutral, the charge on the nickel is also zero. Thus, the oxidation state of nickel in nickel carbonyl is Ni(O).

In the complex $[Cu(NH_3)_4]^{2+}$, note from Table 15.2 that the ammonia ligand is also neutral. Hence, the charge on Cu is $(2 - 0) = 2$, that is, the oxidation state of copper in the complex is Cu(+II).

In the iridium complex $[(\eta^5\text{-}C_5Me_5)IrMe_4]$, the charges on CH_3 and $\eta^5\text{-}C_5Me_5$ are −1. Hence, the oxidation state of Ir in the complex is $[0 - (-5)] = 5$, that is, Ir(+V).

The ligand $(\eta^3\text{-}C_3H_3)$ in the complex $[(PMe_3)_2Pd(\eta^3\text{-}C_3H_3)]^+$ has a charge of −1, and PMe_3 has a charge of 0. Because the complex itself has a charge of 1, the oxidation state of Pd is $[1 - (-1)] = 2$, that is, Pd(II).

Coordinative unsaturation, coordination number, and coordination geometry

coordinative unsaturation

The presence of a vacant coordination site (analogous to an active site in heterogeneous catalysis) is an important prerequisite for homogeneous catalysis and is termed as *coordinative unsaturation.* When the total number of electrons in metal–ligand binding is 18, the complex is considered to be coordinatively saturated. If the electron count is 16 or less, the metal ion possesses at least one vacant coordination site and is said to be coordinatively unsaturated. For example, the dissociation of PPh_3 (triphenylphosphine) from $RhCl(PPh_3)_3$ (Wilkinson's complex) is an important step in forming the coordinatively unsaturated reactive species $RhCl(PPh_3)_2$ in the hydrogenation of olefins. This kind of vacant coordination site can either be in-built in a catalyst or created *in situ* by appropriate design of the metal ion–ligand–solvent system.

A single metallic ion is usually surrounded by a variable number of ligands resulting in different structures. The number of ligands involved in each structure is known as the *coordination number* (n_c) of the complex and is an important parameter of transition metal catalysis.

Ligands and their role in transition metal catalysis

Another important formalism implicit in the definition of the oxidation state considered previously is with respect to the classification of ligands. We formally define a ligand as "an element or a combination of elements which form(s) chemical bond(s) with a transition element" (Masters, 1981). Essentially, two types of ligands have been formally recognized: ionic, such as Cl^-, H^-, OH^-, CN^-, alkyl-, aryl-, and $COCH_3^-$, and neutral, such as primary, secondary, and tertiary phosphines, CO, alkene, amine, and so on. Such a clear-cut distinction can be misleading to the purist in coordination chemistry. Thus, the "ionic" ligands such as H^- and CH_3 are nearly neutral in terms of metal–ligand charge separation in the classical sense, and the "neutral" ligands such as tertiary phosphines are known to violate their neutrality (Chatt and Leigh, 1978).

Ligands can be bonded to metal through single coordination sites or through more than one coordination site. These are known as *unidentate* and *multidentate ligands*, respectively. Examples of unidentate ligands are the halides, NH_3, and H_2O. CO can form both unidentate and multidentate bonds as shown below:

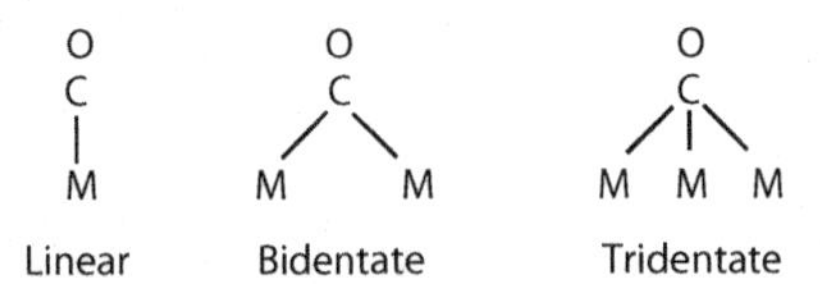

The iodide ligand is particularly important in homogeneous catalysis. The phosphine ligands are even more important; typical examples are unidentate phosphine, unidentate triphenylphosphine (TPP), bidentate diphosphine (DIPHOS), and bidentate chiral diphosphine (DIPAMP).

Multidentate ligands are not commonly used in industry because of the heavy costs associated with their manufacture. It is also important to note that hydride (H^-) and alkyl (R^-) are two of the most commonly used ligands in homogeneous catalysis.

A practically more useful classification can be made in terms of the actual presence or otherwise of a ligand in the final product(s) of reaction (see Masters, 1981). In some cases, the ligands are carried over intact in the products of catalytic cycles, such as CO in a catalytic carbonylation or hydroformylation, for example

$$RCH{=}CH_2 + CO + H_2 \xrightarrow{CO_2(CO)_8} RCH_2CH_2CHO + RCH(CHO)CH_3 \quad (R1)$$

Such ligands are termed *participative ligands*. On the other hand, there are ligands that promote a catalytic cycle but do not end up in the product(s), such as in the preparation of dimethyl maleate from acetylene, CO, and MeOH, catalyzed by palladium chloride in the presence of thiourea (NH_2CSNH_2) and a trace of oxygen:

$$CH{\equiv}CH + CO + MeOH \xrightarrow{PdCl_2/NH_2CSNH_2 + O_2\ (\text{trace})} MeO_2CCH{\equiv}CHCO_2Me \quad (R2)$$

Note that chlorine and thiourea do not appear in the product. Such ligands are known as *nonparticipative* ligands and can be usefully employed by exploiting their structural properties.

Electron rules ("electron bookkeeping")

It should be clear from the earlier sections that the basic requirement in applying the principles of organometallic chemistry is counting electrons associated with a complex. The formalisms arising out of this have been combined into two powerful rules of electron bookkeeping: the *18-electron rule* and its corollary the *16–18-electron rule.*

18-electron rule This rule arises from the assumption that the valence shell electrons of the metals are all in the Nd shell, where N is the principal quantum number. From quantum theory considerations, the d levels are usually associated with the highest energies and are hence the most amenable to the exchange of electrons. Because the number of electrons is related to the oxidation state of a metal, it is clear that the number of d electrons, denoted by d^n, determines this state. This bookkeeping function of the d orbitals in determining the oxidation state of the complex is best illustrated by its usefulness in formulating a rule for the maximum allowable number of ligands for each d^n. The rule in its final form, known as the *18-electron rule*, is

$$d^n + 2n_{c,\max} = 18 \tag{15.1}$$

where $n_{c,\max}$ is the maximum coordination number. This rule is limited to cases where the complex has only one metal atom (i.e., is mononuclear), and d^n is an even number with all electrons paired (diamagnetic). Stated simply, the rule asserts that for a metal complex to be stable, the nine outer orbitals of a transition metal must accommodate 18 electrons.

Exceptions to this rule are not many but do exist. Thus, more than 18 valence electrons can be accommodated in some cases. The more important cases are those where the number is less than 18, leading to what are known as *coordinatively unsaturated complexes.* Some of the latter (the 14- and 16-electron complexes) are particularly useful in catalysis, resulting in the important rule described later. A procedure for determining the stabilities of complexes based on this rule is illustrated below for two complexes (see Kegley and Pinhas, 1986, for details).

Example 15.3: Determining the stability of complexes

Are the complexes $Ru(PPh_3)_2(CO)_2$ and $Cp_2NbH(C_2H_4)$ stable and observable?

To answer this question, it is convenient to construct a table of ligand coordination numbers and electrons for the different components of each complex as shown in Table 15.3.

In preparing the table, the total coordination number for any ligand (column 3) is calculated as (n_c) (number of ligand units), n_c being obtained from Table 15.2. Thus, for CO in complex 1,

Table 15.3 Ligands, Coordination Numbers, and Electrons

Complex	Ligand(s)	Total Coordination Number	Number of Electrons	Charge
1. $Ru(PPh_3)_2(CO)_2$	2 CO	$1 \times 2 = 2$	4	Neutral
	2 PPh_3	$1 \times 2 = 2$	4	Neutral
	Ru(0)	—	8	Neutral
		4	16	
2. $Cp_2NbH(C_2H_4)$	Cp^-	$3 \times 2 = 6$	12	−2
($Cp = \eta^5\text{-}C_5H_5$)	H^-	$1 \times 1 = 1$	2	−1
	C_2H_4	$1 \times 1 = 1$	2	Neutral
	Nb(III)	—	2	+3
		8	18	

the value is $(1 \times 2) = 2$. The number of electrons for any ligand (column 4) is calculated as (number of electrons for each unit of a ligand)(the number of ligand units). Thus, for CO, the value is $(2 \times 2) = 4$. Where all the *d* electrons are paired, the number of electrons is twice the coordination number. Recall that our treatment is restricted to such systems. Column 4 gives directly the number of *d* electrons in the outer shell.

The conclusions from column 4 for the two compounds are: compound 1 is an unstable 16-electron complex, while compound 2 is a stable and observable 18-electron complex.

16–18-electron rule When a complex is coordinatively saturated, any subsequent ligand substitution by nucleophilic attack can only occur by a mechanism in which the nucleophile does not appear in the rate-determining step. In other words, the mechanism is governed solely by transformations within the complex in which the 18-electron compound would dissociate to give an unsaturated complex, which can then bind (i.e., associate) with other potential ligands. Such a mechanism corresponds to the S_N1 mechanism and has given rise to the so-called *dissociation–association* or *16–18-electron rule*, first proposed by Tolman (1972). A useful interpretation of this rule is that in a series of steps in a reaction, no step is possible in which the number of valence electrons changes by more than 2.

Operational scheme of homogeneous catalysis

It is clear from the basic principles (or, more correctly, the formalisms) of homogeneous catalysis just outlined that the essence of homogeneous catalysis lies in the formation of a transition metal complex with a coordination sphere that offers an environment conducive to chemical change. These transformations may involve rearrangement or migration of ligands already present and their elimination as product(s) or insertion of an external ligand within the coordination sphere to form a product that is then eliminated from the complex.

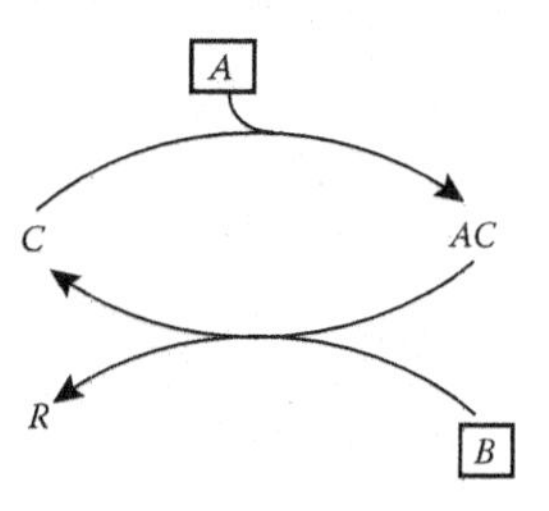

Figure 15.2 A typical catalytic cycle. Reaction $A + B \rightarrow R$.

The second essential feature of any homogeneous catalytic process is that the catalyst must be regenerable. Clearly, this involves a *catalytic cycle* as shown in Figure 15.2 for a simple hypothetical case. Each step constitutes a fundamental reaction of organometallic chemistry. Justifiably, therefore, the critical economic issue that determines the success of a technology is its ability to regenerate the costly catalyst (say Pd) at the expense of cheap raw materials (such as oxygen in oxidation). Based on these observations, the complete methodology for developing a process using homogeneous catalysis is described in the chart shown in Figure 15.3.

Basic reactions of homogeneous catalysis

The chief basic reactions of transition metal chemistry are reactions of ligands, coordination, and addition reactions (sometimes called the *elementary* or *activation* steps), and insertion and elimination reactions (the main reactions). We consider each of these here with examples.

Reactions of ligands (mainly replacement)

The replacement of one ligand by another in a transition metal complex is a common basic reaction in homogeneous catalysis. In a reaction of this kind where two reactants are involved, the kinetics can be described by the S_N1 or S_N2 mechanism. Thus, for $Cr(CO)_6$, the replacement of one CO by a triphenylphosphine ligand occurs by the S_N1 mechanism. In other words, it is influenced only by the local environment within the coordination sphere of the complex. Similarly, one can also think of

S_N1, S_N2: Nucleophilic substitution reactions

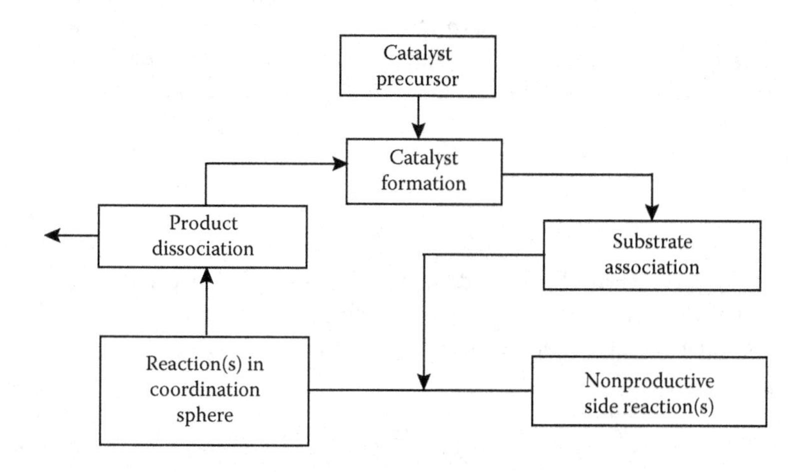

Figure 15.3 Overall methodology of homogeneous catalysis.

ligands that conform to the S_N2 mechanism by actively participating in the displacement of an existing ligand. Examples of these are the substitution reactions of $Mo(CO)_6$ and $W(CO)_6$. All of these examples belong to even-electron-count systems, that is, diamagnetic complexes.

Examples of odd-electron-count (i.e., paramagnetic) intermediates are also known, but are very rare and much less understood. Therefore, our focus will continue to be the even-electron-count complexes considered so far.

Elementary reactions (or activation steps)

Essentially, two broad types of elementary reactions occur in homogeneous catalysis: coordination and addition. Although these are reactions, and we designate them so, they should more appropriately be viewed as steps in a catalytic cycle that "activate" the substrate prior to reaction.

Coordination reactions When CO or an alkene coordinates with a metal center, it appears in the final product without losing its integrity. Consider, for example, the reaction (Masters, 1981)

CH$_2$=CH$_2$ / Pd / OAc ⟶ Pd–CH$_2$CH$_2$OAc (R3)

Here, the ethylene does not lose its integrity in the reaction with acetate in the coordination sphere of Pd. Reactions of this type are known as *coordination reactions* (comparable to nondissociative adsorption in heterogeneous catalysis).

Addition reactions In *addition reactions*, the substrate splits and each fraction bonds separately with the catalyst (metal) center. These reactions can be classified as *heterolytic addition*, *homolytic addition*, and *oxidative addition*. They are distinguished by the nature of the change in the oxidation state upon addition of a substrate to a metal center.

Heterolytic addition

$$M^nL_y + XY \rightarrow M^nL_{y-1}X + Y^+ + L^- \qquad \text{(R4)}$$

This is characterized by the addition of a substrate XY to a single metal center (through X or Y) with no overall change in the oxidation state or coordination number of the metal.

Homolytic addition

$$2M^nL_y + XY \rightarrow 2M^{n+1}(X)(Y)_{L_y} \qquad \text{(R5)}$$

Here, a substrate *XY* adds on to two metal centers in such a way that the formal oxidation state of each metal center increases by 1.

Oxidative addition

$$M^n L_y + XY \rightarrow M^{n+2}(X)(Y)L_y \qquad \text{(R6)}$$

In this addition, the formal oxidation state of the metal is increased by 2. The reverse of this reaction is known as *reductive elimination.* An example is the following oxidative addition in an Rh complex (Forster, 1979):

$$[I_2Rh(I)(CO)_2] + CH_3I \longrightarrow [I_2(CH_3)Rh(III)(CO)_2] \qquad \text{(R7)}$$

Rh(I) ⟶ Rh(III)

Main reactions

After the complex has been formed, that is, after the reactant molecule has been activated, the next step is the main reaction leading to the formation of the products. These reactions generally occur either by insertion or elimination.

Insertion

Single-metal center

$$\underset{|}{X}\atop{M—Y} \rightleftharpoons \text{M-X-Y or M-Y-X} \qquad \text{(R8)}$$

Multiple metal centers

$$(X)M_1—M_2(Y) \longrightarrow (XY)M_1—M_2 \qquad \text{(R9)}$$

As can be seen from both these general schemes, what is really involved is migration of a ligand within the molecule. Insertion is the key step in several catalytic cycles such as those associated with hydrogenation, oligomerization, and hydroformylation. For example, in hydroformylation reactions using $HRh(CO)(PPh_3)_3$ as the catalytic complex, insertion of precoordinated CO into the alkyl group by ligand migration within the coordination sphere of the complex is an important step in the catalytic cycle (Evans et al., 1968a,b):

$$(CH_3CH_2)Rh(CO)_2(PPh_3)_2 \rightarrow (CH_3CH_2CO)Rh(CO)(PPh_3)_2 \qquad \text{(R10)}$$

Elimination

Reductive elimination When $Cp(CO)_2Fe{-}CH_3$ is reacted with HBF_4, a complex is formed from which CH_4 is eliminated:

$$[Cp(CO)_2Fe(CH_3)H]^+[BF_4]^- \xrightarrow{\text{Reductive elimination}} [Cp(CO)_2Fe]^+[BF_4]^- + CH_4 \quad \text{(R11)}$$

In this step, the two ligands H and CH_3 combine to form the product CH_4, which then leaves the coordination sphere of the metal. The formal oxidation state of the metal in this reductive elimination step is reduced by two units, which may be regarded as the reverse of oxidative addition.

β-Hydrogen elimination This is the reverse of insertion in which the β-hydrogen of an alkyl group migrates to the metal atom and further strengthens the description of the elimination reaction as a ligand migration reaction. α-Elimination is also possible, but is much less understood.

Main features of transition metal catalysis in organic synthesis: A summary

The special features of transition metal complexes that are responsible for their remarkable attractiveness in organic synthesis may be summarized as follows.

1. General readiness to bond with a large number of metals in the periodic table and with just about any organic molecule.
2. Ability to activate (through coordination) a variety of industrially available feedstocks such as CO, H_2, olefins, and alcohols or their derivatives.
3. Ability to stabilize unstable intermediates like metal hydrides and metal alkyls in relatively stable but kinetically reactive complexes.
4. Accessibility to different oxidation states and ability to move from one oxidation state to another during the course of a reaction.
5. Ability to assemble and orient various reactive components within the coordination sphere (the *template effect*).
6. Ability to accommodate within the coordination sphere both participative and nonparticipative ligands; this is useful in modifying the steric and electronic properties important in determining catalytic activity and selectivity.

A typical class of industrial reactions: Hydrogenation

Based on the activation and elementary steps outlined, a variety of catalytic reactions can be better understood, for example, isomerization, hydrogenation, carbonylation, hydroformylation, oxidation, and metathesis. We illustrate this by considering the hydrogenation reaction with Wilkinson's homogeneous catalyst.

Hydrogenation by Wilkinson's catalyst

Wilkinson's catalyst

The three types of activation characterized by R12, R13, and R14 can be written specifically for hydrogen activation by Ru, Co, and Rh as follows:

$$Ru^{2+} + H_2 = Ru^{2+}H + H^+ \quad \text{(heterolytic splitting)} \quad (R12)$$

$$2Co^0 + H_2 = 2Co^{1+}H \quad \text{(homolytic splitting)} \quad (R13)$$

$$Rh^+ + H_2 = Rh^{3+}H_2 \quad \text{(oxidative addition)} \quad (R14)$$

Of these, oxidative addition is a particularly important class, with several examples of reactions catalyzed by complexes of iridium(I) and rhodium(I). The following reaction scheme for iridium(I) is illustrative:

$$Ir^{I}(CO)(PPh_3)_2Cl + XY \rightarrow Ir^{III}(Cl)(X)(Y)(CO)(PPh_3)_2 \quad (R15)$$

where the substrate XY can be

$$X\text{–}Y \Rightarrow H\text{–}H, \quad H\text{–}Cl, \quad CH_3CO\text{–}Cl, \quad Cl\text{–}Cl$$

Because the oxidation state increases, activation by oxidative addition is more effective for metals in a low state of oxidation. Therefore, noble metals of Group VIII, which satisfy this requirement, are the most suited for this purpose. (See Chaloner et al., 1994, for a detailed treatment of homogeneous hydrogenation.)

Wilkinson's catalyst

Ruthenium, iridium, and rhodium are the most frequently used metals. The best known is the rhodium catalyst $RhCl(PPh_3)_3$ first discovered by Wilkinson* and bearing his name.

* G. Wilkinson and E.O. Fischer were awarded the Nobel Prize in 1973 for their work, which started a burst of activity in homogeneous catalysis.

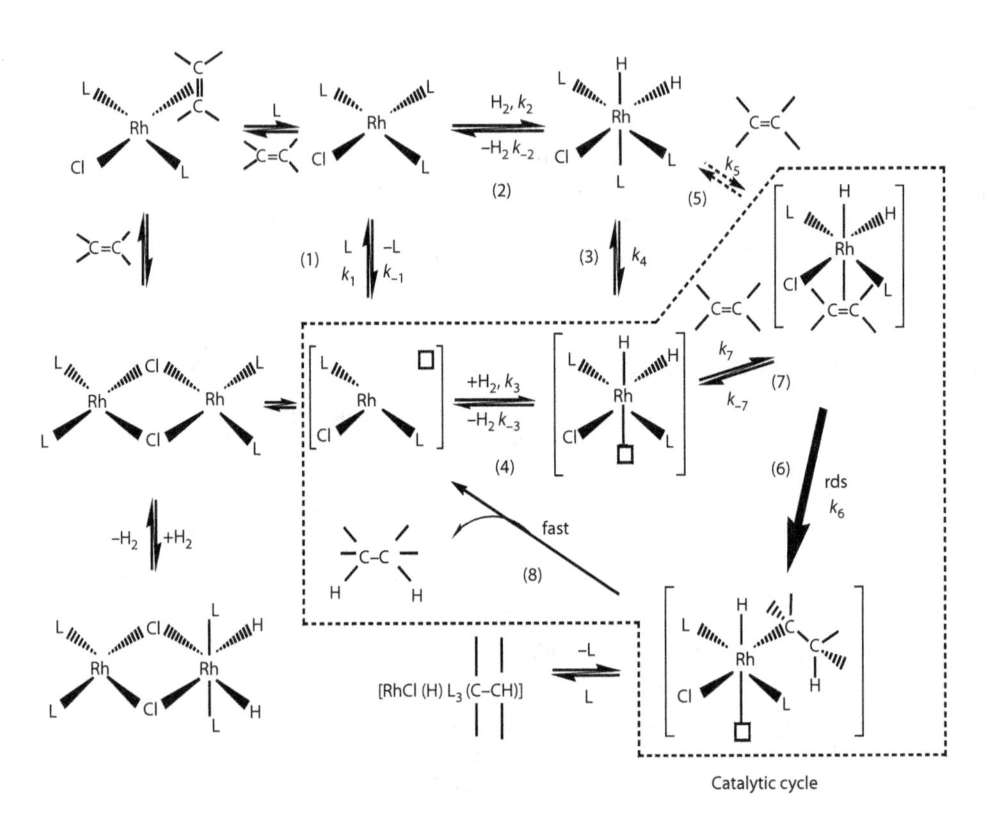

Figure 15.4 Catalytic cycle for hydrogenation of olefins using a Wilkinson's catalyst. L = PPh_3, 0 = vacant coordination site, rds = rate-determining step. (Redrawn from Halpern, J. et al., *J. Mol. Catal.*, **2**, 65, 1976.)

The catalytic cycle The mechanism of Wilkinson hydrogenation is briefly outlined here, mainly to demonstrate a procedure for formulating appropriate catalytic cycles and formulating rate models. The complete cycle for Wilkinson hydrogenation is given in Figure 15.4, based on which the following conclusions are important:

Step 6 in the catalytic cycle (boxed) represents substrate insertion and is invariably the rate-controlling step. Thus, let us consider the hydrogenation of cyclohexene by Wilkinson's catalyst. The catalytic cycle in Figure 15.4 shows that all steps other than step 6 are characterized as fast or having reached equilibrium. Hence, the rate-determining step of the cycle is step 6.

Kinetics and modeling The kinetics of this reaction can be formulated by the method outlined in Chapters 5 and 7 (see also Gates, 1992, for this specific reaction). Let A = cyclohexene, $B = PPh_3$, $C = RhCl(PPh_3)_2H_2A$, and $D = RhCl(PPh_3)_3H_2$. Then, the rate of disappearance of cyclohexene is given by the rate of step 6:

$$r_A = -\frac{d[A]_b}{dt} = \frac{d[C]_b}{dt} = k_6[C]_b \qquad (15.2)$$

where subscript b represents the liquid bulk. The equilibrium of the previous step is

$$K_5 = \frac{[B]_b[C]_b}{[A]_b[D]_b} \tag{15.3}$$

Because both the complexes C and D are saturated octahedral complexes, we can use the total concentration of the rhodium complex

$$[T]_b = [C]_b + [D]_b \tag{15.4}$$

as a readily measurable correlating parameter to eliminate the unknown concentration $[C]$. Thus, from Equations 15.3 and 15.4,

$$[C]_b = \frac{K_5[A]_b[B]_b}{K_5[A]_b + [B]_b} \tag{15.5}$$

Combining this with Equation 15.3 leads to

$$-r_A = \frac{k_6 K_5[A]_b[T]_b}{K_5[A]_b + [B]_b} \tag{15.6}$$

It will be noticed that the final rate equation given by Equation 15.6 contains only quantities that are readily measurable (with the possible exception of K_5).

A general hydrogenation model

A similar analysis can be made for other reactions also. However, because detailed information on the catalytic cycle for a reaction at hand would not normally be available, the value of the equilibrium constant corresponding to K_5 would not be known. From a practical (reaction engineering) point of view, therefore, Equation 15.6 may be recast as a more general two-parameter rate equation of the form

$$-r_A = \frac{k_1[A]_b[T]_b}{K_5[A]_b + [B]_b} \tag{15.7}$$

and the constants k_1 and K_5 determined from simple statistical methods as described in Chapter 7. Then, the constant k_6 can be obtained from $k_6 = k_1/K_5$. The equation can be generalized by letting K_5 be any constant k_2.

It is interesting to note that the nature of the ligand that is bonded to the metal and that is not a reactant is important in determining the reactivity of the complex. In fact, the reactivity of a Wilkinson's hydrogenation catalyst can be enhanced several fold merely by changing the ligand.

General kinetic analysis

Intrinsic kinetics

Recall that in the majority of reactions using homogeneous catalysts in the liquid phase, a gas phase is also present, mainly hydrogen and/or carbon monoxide. This diffusion of gas in liquid can falsify the kinetics. We consider in this section the modeling of gas–liquid reactions in the absence of diffusional effects.

The development of a mechanistic model for a homogeneous reaction requires construction of a catalytic cycle, which is quite difficult. On the other hand, simple kinetic expressions both of the power law and hyperbolic types can be readily derived. These are usually adequate for purposes of reactor design. Thus, in the analysis of homogeneous catalysis involving a gas–liquid reaction, the following general hyperbolic form of the rate equation may be used:

$$-r_A = \frac{k[A]^*[B]_b[C]_b}{(1 + K_A[A]^* + K_B[B]_b)^n} \tag{15.8}$$

or

$$-r_A = \frac{k[A]^*[B]_b[C]_b}{(1 + K_E[E]^*)^n} \tag{15.9}$$

where A, B, and C represent, respectively, the gas and liquid phase reactants and the catalyst. The second form, where E represents carbon monoxide, is particularly useful for hydroformylation reactions. The value of n varies depending on the substrate used. For hydroformylation of hexene, for example, $n = 2$, and for that of allyl alcohol, $n = 3$.

Multistep control In the foregoing examples, rate equations were developed on the basis of a single rate-determining step. It is possible that many steps of a cycle would be simultaneously controlling, as in the Wacker process. The rate equation for such a reaction tends to be more complicated, but can be developed by the methods discussed in Chapter 5. Thus, for the oxidation of triphenylphosphine with a Pt complex, a rate equation can be developed based on the catalytic cycle shown in Figure 15.5 (Halpern and Birk et al., 1968a,b; Pickard, 1970):

$$\text{Rate} = \frac{k_1 k_2[\text{Pt complex}][\text{PPh}_3][\text{O}_2]}{k_1[\text{O}_2] + k_2[\text{PPh}_3]} \tag{15.10}$$

where k_1 and k_2 are the rate constants for the steps marked 1 and 2 in the cycle, that is

$$\text{Pt(PPh}_3)_3 + \text{O}_2 \xrightarrow{k_1} \text{Pt(PPh}_3)_2\text{O}_2 + \text{PPh}_3 \tag{R16}$$

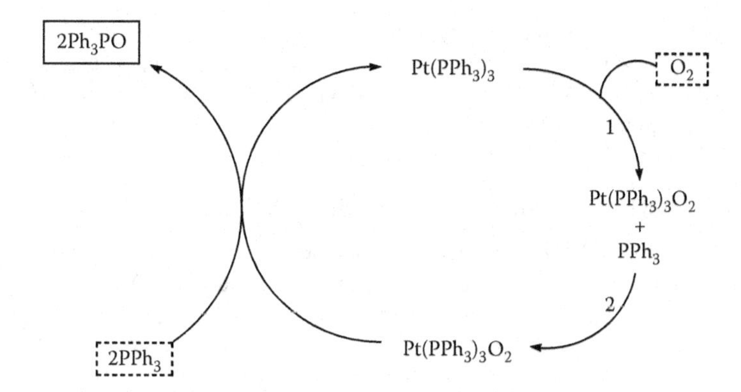

Figure 15.5 Catalytic cycle for the oxidation of triphenylphosphine. (Drawn from reactions given by Halpern, J. and Pickard A.L., *Inorg. Chem.*, **9**, 2798, 1970.)

$$Pt(PPh_3)_2O_2 + PPh_3 \xrightarrow{k_2} Pt(PPh_3)_3O_2 \qquad (R17)$$

Role of diffusion

Recall that there are a number of reactions where homogeneous catalysis involves two phases, liquid and gas, for example, hydrogenation, oxidation, carbonylation, and hydroformylation. The role of diffusion becomes important in such cases. In Chapter 6, we considered the role of diffusion in solid catalyzed fluid-phase reactions and gas–liquid reactions. The treatment of gas–liquid reactions makes use of an "enhancement factor" to express the enhancement in the rate of absorption due to reaction. A catalyst may or may not be present. If there is no catalyst, we have a simple noncatalytic gas–liquid heterogeneous reaction in which the reaction rate is expressed by simple power law kinetics. On the other hand, when a dissolved catalyst is present, as in the case of homogeneous catalysis, the rate equations acquire a hyperbolic form (similar to LHHW models discussed in Chapters 5 and 6). Therefore, the mathematical analysis of such reactions becomes more complex.

Complex kinetics—Main issue

Hyperbolic equations were used in Chapter 6 to represent reactions catalyzed by solid surfaces. They are referred to as LHHW models and they can be empirically extended to homogeneous catalysis in liquid phase reactions. The actual rate equation to be used for a given reaction will depend on the regime of that reaction. Methods of discerning the controlling regimes for catalytic gas–liquid reactions described in the gas–liquid chapter were based on simple power law kinetics. Extension of these methods to gas–liquid reactions catalyzed by homogeneous catalysts involves no new principles, but the mathematics becomes more complicated because of the hyperbolic nature of the rate models. In this section, we consider reactions involving one gas and a liquid (e.g., carbonylation, hydrogenation, oxidation) and two gases and a liquid (e.g., hydroformylation) (Table 15.4).

Reactions involving one gas and one liquid

Regimes 1 and 2 The reaction is controlled by the chemical kinetics in this regime, and hence diffusional limitations are absent. Depending on the reaction, the rate equation can have any of many hyperbolic forms, such as those presented in Chapter 5.

Regime 2 corresponds to film diffusion control and hence is independent of the kinetics. Thus, the developments outlined in Chapter 6 for gas–liquid reactions with simple-order kinetics are equally valid for reactions with LHHW kinetics.

Regime between 1 and 2 (reaction in bulk) In this regime, the rate is controlled by both film diffusion and reaction kinetics. Assuming that the reaction is represented by a typical LHHW equation, the following equations must be solved simultaneously for the enhancement factor:

Diffusion

$$-r_A = k_L' a_L([A]^* - [A]_b) \tag{15.11}$$

Reaction

$$-r_A = \frac{k[A]_b[B]_b[C]_b}{1 + K_A[A]_b + K_B[B]_b} \tag{15.12}$$

where C represents the catalyst. Because the rates are equal at steady state, the unknown $[A]_b$ can be eliminated and the following expression obtained for the enhancement factor:

$$\eta = \frac{(-r_{Aa})}{k_L' a_L [A]^*}$$

$$= \frac{(1 + k_a k_b + M_{LH}\gamma) - \sqrt{(1 + k_a + k_b + M_{LH}\gamma)^2 - 4k_a M_{LH}\gamma}}{2k_a} \tag{15.13}$$

where

$$k_a = K_a[A]^*,\ k_b = K_B[B]_b,\ M_{LH} = \frac{\sqrt{D_A k[B]_b[C]_b}}{k_l'},\ \gamma = \frac{k_L'}{D_A a_L} \tag{15.14}$$

The condition to be satisfied is

$$M_{LH} \leq 0.3\text{–}0.8 \tag{15.15}$$

The value of γ is always less than or equal to unity.

Table 15.4 Summary of Kinetic Models Used in Homogeneous Catalysis

S. No.	Reaction System	Catalyst	Rate Equation, $-r_A$ mol/cm³s	Reference
1.	Hydrogenation of cyclohexene	$RhCl(PPh_3)_3$	$\frac{k[A]^*[B]_b[C]_b}{1+K_A[A]^*+K_B[B]_b}$	Osborn et al. (1966)
2.	Hydrogenation of allyl alcohol	$RhCl(PPh_3)_3$	$\frac{k[A]^*[B]_b[C]_b}{1+K_A[A]^*+K_B[B]_b^2}$	Wadkar and Chaudhari (1983)
3.	Oxidation of cyclohexane	$Mn(OAc)_2$	$\frac{k[B]_b[C]_b}{(k_1+k_2[C]_b)}$	Kamiya and Kotake (1973)
4.	Oxidation of ethylene	$RCl(PPh_3)_3$	$\frac{k[A]^*[B]_b[C]_b}{(1+K_A[B]^*)}$	Chaudhari (1984)
5.	Carbonylation of methanol	$RhCl_3$ HI solution	$k[P]_b[C]_b$	Roth et al. (1971)
6.	Hydroformylation of propylene	$HCo(CO)_4$	$\frac{k[A]^*[A]^*[C]_b}{[E]^*}$	Natta et al. (1952)
7.	Hydroformylation of diisobutylene	$CO_2(CO)_8$	$\frac{k[A]^*[P]_b[C]_b}{[E]^*+K_A[A]^*}$	Martin (1954)
8.	Hydroformylation of hexene and allyl alcohol	$HRhCO(PPh_3)_3$	$\frac{k[A]^*[B]_b[C]_b}{(1+K_E[E]^*)^n}$, $n=2$ for hexene, 3 for allyl alcohol	Deshpande and Chaudhari (1983)
9.	Hydroformylation of allyl alcohol	$HRhCO(PPh_3)_3$	$\frac{k[A]_s^{*1.52}[E]^*[B]_b[C]_b}{(1+K_E[E]^*)^3(1+K_B[B]_b)^2}$	Deshpande and Chaudhari (1989)
10.	Hydroformylation of 1-decene	$HRh(CO)(PPh_3)_3$	$\frac{k\,p_H p_{CO}[A]^*[C]_b}{1+K_1p_{CO}+K_1K_2p_{CO}[A]^*+K_1K_2K_3p_{CO}^2[A]^*+K_1K_2K_3K_4p_{CO}^3[A]^*}$	Divekar et al. (1993)

Source: Adapted from Mills, P.L. et al. *Reviews in Chemical Engineering*, **8**, 1, 1992..

Note: $[A]^*$ = interfacial concentration of O_2, H_2, CO; $[B]^*$ = interfacial concentration of gaseous substrate; $[B]_b$ = concentration of liquid phase substrate; $[C]_b$ = concentration of catalyst; $[E]^*$ = interfacial concentration of CO; $[P]_b$ = concentration of promoter.

Regime 3 (reaction in film) It was pointed out earlier that regime 3 is controlling (for simple power law kinetics) if $M \gg 1$. It can similarly be shown (Chaudhari, 1984) that regime 3 is controlling for LHHW models if

$$M \gg 1 \tag{15.16}$$

If there is to be no depletion of B, the following additional condition must be satisfied:

$$M_{LH} \ll \left(1 + \frac{[A]_b}{[B]_b}\right) \tag{15.17}$$

Then, the enhancement factor for such a reaction can be expressed in terms of a generalized Hatta number defined as

$$\begin{aligned} M'_{LH} &= \frac{\sqrt{2D_A}}{k'_L[A]^*}\left[\int_0^{[A]}(-r_A)d[A]\right] \\ &= M_{LH}\left\{\frac{2}{k_a^2}\left[k_a + (1+k_b)\ln\frac{1+k_b}{1+k_a+k_b}\right]\right\}^{1/2} \end{aligned} \tag{15.18}$$

Equation 15.13 for the enhancement factor is equally valid for the present system, but the definition of the Hatta number is replaced by Equation 15.18:

$$\eta = \frac{M'_{LH}}{\tanh M'_{LH}} \tag{15.19}$$

Note that the generalized Hatta number depends on the Hatta number M'_{LH} in addition to the other parameters of the system. Because M'_{LH} depends on the catalyst concentration, the enhancement factor given by Equation 15.19 also depends on the catalyst concentration (and k_a and k_b). This is illustrated in Figure 15.6, which is a solution of this equation for different values of k_a and for a fixed value of k_b.

An inspection of the figure leads to two major conclusions: (1) an increase in catalyst concentration (i.e., in M'_{LH}) leads to a shift in regime from chemical to diffusion control (as expected) and (2) an increase in the diffusion parameter k_a raises the enhancement factor in the kinetic regime but lowers it in the diffusion regime.

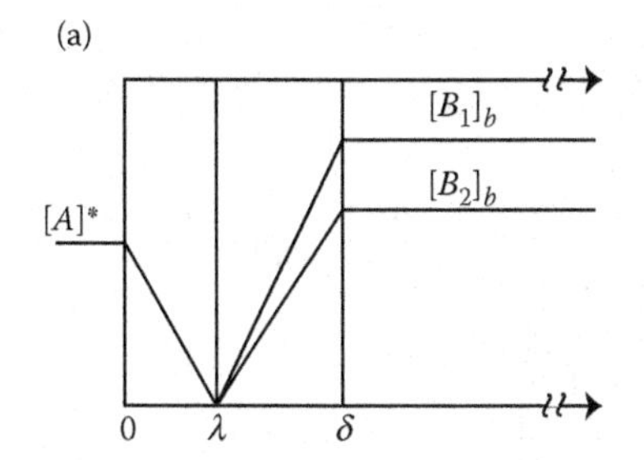

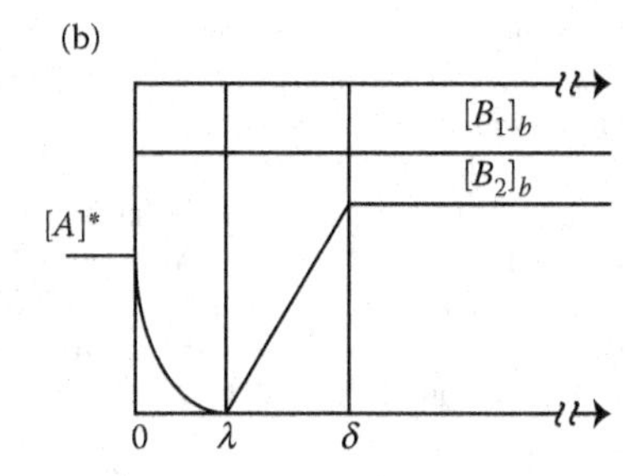

Figure 15.6 Absorption and reaction of a gas with two liquid phase reactants, second step instantaneous. (a) Both steps instantaneous and (b) first step in regime 2–3 and second step instantaneous.

Two gases and a liquid

The most important class of reactions in this category is hydroformylation. A number of kinetic and modeling studies on the hydroformylation of selected substrates have been reported (see Doraiswamy, 2001). The role of diffusion in these reactions was also studied and equations for the enhancement factor were proposed (Bhattacharya and Chaudhari, 1987). It was shown that multiple solutions can exist under certain conditions. Also, the fact that the reaction is first order in H_2 and negative order in CO leads to results that contradict accepted trends. Mainly, agitation seems to have no effect under conditions of significant mass transfer limitation.

Liquid phase oxidation of gaseous substrates with O_2, such as the oxidation of ethylene to acetaldehyde (Wacker process), is another example of this class of reactions. A mathematical model for a bubble column reactor for this reaction, assuming plug flow of gas and mixed flow of liquid, was developed (Rode et al., 1994).

References

Bau, R., Carroll, W.E., Hart, D.W., Teller, R.G., and Koetzle, T.F., In *Transition Metal Hydrides* (Ed., Bau, R.), *Am. Chem. Soc. Ser.* Vol., 167, Washington, DC, 1978.

Bhattacharya, A. and Chaudhari, R.V., *Can. J. Chem. Eng.*, **65**, 1018, 1987.

Birk, J.P., Halpern, J., and Pickard, A.L., *J. Am. Chem. Soc.*, **90**, 4491, 1968a.

Birk, J.P., Halpern, J., and Pickard, A.L., *Inorg. Chem.*, **7**, 2672, 1968b.

Chaloner, P.A. et al., *Homogeneous Hydrogenation,* Kluwer Academic Publishers, Dordrecht, 1994.

Chatt, J., and Leigh, G.J., *Angew. Chem., Int. Ed. Engl.*, **17**, 400, 1978.

Chaudhari, R.V., In *Frontiers in Chemical Reaction Engineering*, Vol. I (Eds. Doraiswamy, L.K. and Mashelkar, R.A.), Wiley Eastern, New Delhi, 1984.

Collman, J.P., Hegedus, L.S., Norton, J.R., and Finke, R.G., *Principles and Applications of Organotransition Metal Chemistry*, 2nd ed, University Science Books, Mill Valley, CA, 1987.

Deshpande, R.M. and Chaudhari, R.V., *Ind. I. Tech.*, **21**, 351, 1983.

Deshpande, R.M. and Chaudhari, R.V., *J. Catal.*, **115**, 326, 1989.
Divekar, S.S., Deshpande, R.M., and Chaudhari, R.V., *Catal. Lett.*, **21**, 191, 1993.
Doraiswamy, L.K., *Organic Synthesis Engineering,* Oxford University Press, New York, 2001.
Evans, D., Osborn, J.A., and Wilkinson, G., *J. Chem. Soc. A.,* 3133, 1968a.
Evans, D., Yagupsky, G., and Wilkinson, G., *J. Chem. Soc. A.,* 2660, 1968b.
Forster, D., *Adv. Organomet. Chem.,* **17**, 255, 1979.
Gates, B.C., *Catalytic Chemistry,* Wiley, New York, 1992.
Halpern, J. and Pickard A.L., *Inorg. Chem.,* **9**, 2798, 1970.
Halpern, J., Okamoto, T., and Zakhariev, A., *J. Mol. Catal.,* **2**, 65, 1976.
Kamiya, Y. and Kotake, M., *Bull. Chem. Soc. Jpn.*, **46**, 2780, 1973.
Kegley, S.E. and Pinhas, A.R., *Problems and Solutions in Organometallic Chemistry,* University Science Books, Mill Valley, CA, 1986.
Martin, A.R., *Chem. Ind. (London)*, 1536, 1954.
Masters, C., *Homogeneous Transition-Metal Catalysis,* Chapman & Hall, New York, 1981.
Mills, P.L., Ramachandran, P.A., and Chaudhari, R.V. *Reviews in Chemical Engineering*, **8**, 1, 1992.
Natta, G., Ercoli, R., Castellano, S., and Barbieri, F.H., *J. Am. Chem. Soc.*, **76**, 4049, 1952.
Osborn, J.A., Jardine, F.H., Young, J.F., and Wilkinson, G., *J. Chem. Soc. A*, 1711, 1966.
Rode, C.V., Gupte, S.P., Chaudhari, R.V., Pirozhkov, C.D., and Lapidus, A.L., *J. Mol. Catal.,* **91**, 195, 1994.
Roth, J.F., Craddock, J.M., Hershman, A., and Paulik, F.E., *Chemtech*, 600, 1971.
Tolman, C.A., *Chem. Soc. Rev.,* **1**, 337, 1972.
Wadkar, I.G. and Chaudhari, R.V., *J. Mol. Catal.*, **22**, 105 1983; *Ind. I. Tech.*, **21**, 351, 1983.

Chapter 16 Phase-transfer catalysis

Introduction

Phase-transfer catalysis (PTC) is an important tool in the organic chemist's armory to conduct reactions involving two mutually insoluble solvents. Chemical engineers were almost completely unconcerned with PTC till around mid-1970s. Then, after a period of lukewarm interest for over a decade (when some basic advances were made), it gained a respectable position in the 1980s and has since been experiencing a steady growth. This period (starting from around 1970) also roughly coincided with CRE becoming more inclusive in that large-volume chemicals were no longer its single main focus, such as the petroleum refining, petrochemical, and heavy organic and inorganic chemical industries. Areas of general interest in small- and medium-volume chemicals, such as pharmaceutical engineering and PTC, moved increasingly from technology to engineering science orientation.

A good understanding of chemistry is obviously one of the underpinnings of CRE. This is particularly so in certain areas. PTC is one such area where the role of the catalysts used and ionic interactions in the liquid phases are important considerations. It is worth remembering that PTC always involves at least one liquid phase. These facts will become evident as we proceed further in this chapter. For a deeper understanding of the more chemistry-based areas of CRE, reference to the book *Organic Synthesis Engineering* (Doraiswamy, 2001) is recommended.

What is PTC?

We have already given a relatively detailed introduction to PTC in the first Interlude. Here, we will start by remembering that, PTC involves reactions in heterogeneous liquid–liquid or solid–liquid systems in which inorganic anions (or organic anions generated through deprotonation with an aqueous phase base) react with organic substrates through the mediation of a phase-transfer catalyst. The reactive anions are introduced into the organic phase in the form of lipophilic ion pairs or complexes that they form with the PT catalyst (Reaction R1a). This ion-pair partitions between the organic and aqueous phases due to its lipophilic nature and once the anion is transferred to the organic phase, it can react with the organic substrate, yielding the desired product (Reaction R1b), with high yields and often with high selectivity. In the absence of the PT catalyst, the anions and the organic phase cannot react as they are physically isolated from each other in two different mutually immiscible phases.

$$\begin{aligned} &Q^+Y^- + M^+Y^- \leftrightarrow M^+Y^- + Q^+X^- \quad \text{(a) Ion exchange reaction} \\ &Q^+Y^- + R\text{–}X \leftrightarrow R\text{–}Y + Q^+X^- \quad \text{(b) Organic phase reaction} \end{aligned} \qquad \text{(R1)}$$

In addition, PTC has the advantage that easily recoverable solvents such as dichlormethane, toluene, and hexane can be used rather than polar solvents. All these benefits lead to enhanced productivity with higher safety and lower environmental impact. Also, it must be mentioned that although a PT catalyst is mainly used to enhance reaction rates and yield, it can be a useful tool in many cases to selectively synthesize one product or significantly reduce an undesired by-product. One of the main concerns with PTC that has been a significant barrier to industrial adoption, especially in the pharmaceutical and food additives industry, is the issue of catalyst recovery from the final product stream.

Fundamentals of PTC

Classification of PTC systems

PTC reactions can be classified broadly into soluble and insoluble PTC reactions (Figure 16.1). Soluble PTC is further categorized based on the phases involved in the reaction into liquid–liquid (LLPTC) and solid–liquid (SLPTC), which are discussed in detail in the following sections. Some of the other variants of PTC are discussed below.

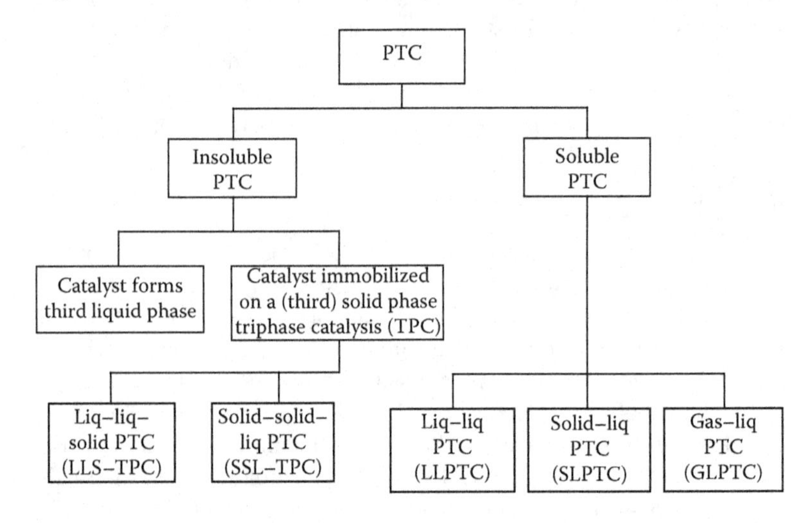

Figure 16.1 Classification of PTC systems.

Gas–solid–liquid PTC is a particularly interesting variation due to the absence of an organic solvent and the possibility of continuous operation in a plug-flow reactor packed with a solid such as inert alumina spheres. Other atypical variants include inverse PTC, where an organic-soluble reagent is transported by a suitable transfer agent into the aqueous phase, for reaction to occur there. Because of the reverse direction of catalyst transfer, it is appropriately called reverse PTC. Insoluble PTC results when the PT catalyst is immobilized on a solid support and used in a traditional liquid–liquid reaction system, or a three-phase liquid–liquid–liquid (L–L–L) system is involved where the PT catalyst is concentrated in a third liquid phase.

Phase-transfer catalysts

Quaternary ammonium salts (known commonly as quats) are the most frequently used PT catalysts, followed by the corresponding phosphonium salts, macrocyclic polydentate ligands like cryptands and crown ethers, and open-chain polyethers like poly(ethylene glycols) (PEGs). The basic requirements of a typical PT catalyst are that it should be cationic (quats) or have some way of complexing with the reactive anion (crown ethers, cryptands, PEGs) and have enough organic structure to partition into the organic phase. Table 16.1 summarizes the most commonly used catalysts in terms of their activity, stability under reaction conditions, availability, costs, and recovery. Stability under reaction conditions (especially temperature conditions), ease of recovery and separation after the reaction, cost, availability, toxicity concerns (especially with crown ethers), and ease of recovery and disposal are some of the other considerations that guide catalyst choice.

The attractiveness of quaternary ammonium salts stems from their overall superiority in terms of activity, stability, availability, and costs. For

Table 16.1 Common Phase-Transfer Catalysts

Catalyst	Stability and Activity	Cost and Availability	Use and Recovery
Ammonium salts	Moderately stable under basic conditions and up to 100°C. Decomposition under basic conditions. Moderately active	Cheap and easily available	Widely used but recovery steps are relatively difficult
Phosphonium salts	More stable thermally than ammonium salts but less active under basic conditions	More expensive and less readily available than ammonium salts	Widely used but recovery steps are relatively difficult
Crown ethers and cryptands	Stable and highly active, even under basic conditions and can be used up to 150°C	Expensive and relatively less readily available in bulk quantities	In addition to cost, toxicity and handling concerns limit use
PEG and derivatives	More stable than ammonium salts but much lower activity	Very cheap and easily available in bulk quantities	Large amounts needed but relatively easy to recover due to high water solubility

certain solid–liquid systems, the more expensive but stable and highly active cryptands (commonly [2.2.2]-cryptand) and crown ethers (typically, 18-crown-6 and dibenzo-18-crown-6 for K salts, and 15-crown-5 for Na salts) are good choices, although industrial use of these compounds is limited due to higher costs and toxicity concerns. For certain applications, the less active but cheaper open-chain polyethers like PEGs and their dimethyl ethers (glymes), referred to as "a poor chemist's crown ether," may be suitable. Due to their water solubility, they are poor or at best mediocre catalysts in liquid–liquid systems. However, PEG can, under suitable conditions, form a third catalyst-rich phase and function as an active PT catalyst in a L–L–L system (Jin et al. 2003).

A higher degree of hydration of the anion indicates a higher tendency for the anion to stay in the aqueous phase. For ions that have been transferred to the organic phase, although some water can be transferred with the ion pair, the hydration shell is smaller and reactivity is higher than with the metal salt. Anion hydration is an important factor in PTC, and it is important to note that the hydration sphere is reduced or totally eliminated in SLPTC conditions, as discussed in another section.

Mechanism of PTC

Liquid–liquid PTC

Mechanistically, typical PTC reactions take place via two main mechanisms in liquid–liquid systems. The extraction mechanism (Starks, 1971) is the most commonly accepted one for simple nucleophilic substitution under neutral conditions for reactions with a range of anions (e.g., halides, cyanide, thiocyanate, sulfite, nitrite, acetate, carbonate, etc.). According to this mechanism (Figure 16.2a), the PT catalyst, denoted by Q^+X^-, is a vehicle to transfer the reactive anion Y^- of the metal salt M^+Y^- from the aqueous phase into the organic phase, where it reacts with the organic substrate, RX, to give the desired product RY and regenerating Q^+X^-, which can continue the PTC cycle. Typically, the active form of

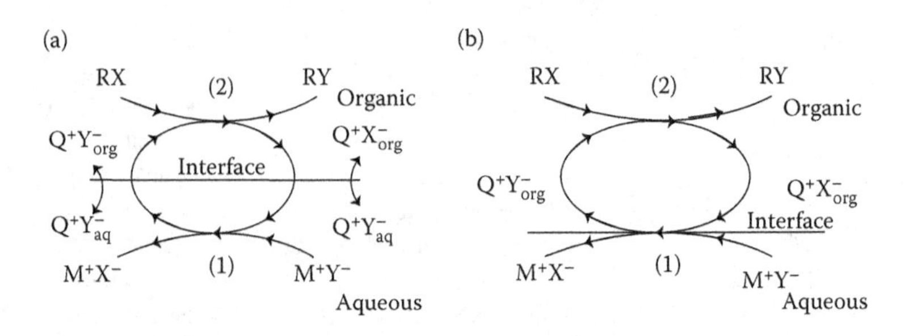

Figure 16.2 LLPTC mechanism. (a) Starks' extraction mechanism and (b) Brandström–Montanari modification.

the PT catalyst (Q^+Y^-), formed after ion exchange in the aqueous phase, partitions between the aqueous and organic phases.

If the catalyst is highly organophilic, it can be present exclusively in the organic phase and still be a good PT catalyst. Brandström–Montanari (Brandström, 1977; Landini et al. 1977) modification to the extraction mechanism is more appropriate (Figure 16.2b). Here, ion exchange takes place at the interface of the two phases and not in the aqueous phase. Thus, partitioning of the quat between the two liquid phases is not a necessary condition for effective phase-transfer action in liquid–liquid systems.

In both cases, the phase-transfer cation–anion pair Q^+Y^- is the reactive species in the organic phase. It should be noted that most ion pairs have a very low degree of disassociation in organic solvents and thus the quat species exist in the organic phase as ion pairs Q^+Y^- and Q^+X^-, whereas the aqueous phase remains in equilibrium between the ion pairs and the free ions Q^+, X^-, and Y^-.

Many PT-catalyzed reactions such as C-, N-, O-alkylations of weak acids like aliphatic alcohols, isomerizations, additions, and hydrolysis take place in the presence of a base, typically aqueous NaOH or solid K_2CO_3 and are hereafter grouped together as PTC/OH^- reactions. An advantage of carrying out reactions in the presence of a base in biphasic systems is that it prevents hydrolysis of the organic reactant since OH^- has limited solubility in the organic phase, where the organic reagent is not subjected to the alkaline conditions in the aqueous phase. Also, in the absence of PTC, these reactions would require severe anhydrous conditions, handling of corrosive and expensive reducing agents such as metal hydrides (sodium hydride), alkali metal alkoxides (potassium t-butoxide), sodium metal, or other organometallic agents.

Starks' extraction mechanism is not applicable for reactions in the presence of bases, and deprotonation of moderately and weakly acidic organic compounds at the interface is the most likely mechanism. This is a reaction that takes place even without the PT catalyst. The main role of the quat species (Makosza, 1975) then is to detach this organic carbanion formed at the interphase by forming the ion-pair Q^+OR^- and drawing it into the organic phase, where reaction with the organic substrate can continue. We illustrate this in Figure 16.3, which is a schematic of the alkylation of phenylacetonitrile via reaction with alkyl halides in the presence of a base, in which hydrogen abstraction of the phenylacetonitrile takes place at the interphase, yielding a carbanion, PhCHCN. After deprotonation, the PT catalyst draws the carbanion into the organic bulk as a $Q^+PhCHCN^-$ ion pair, which reacts with the organic substrate (the alkyl halide, R′Y), forming the desired phenylalkylacetonitrile product and regenerating the PT catalyst species, Q^+X^-. Thus, in this case, the role of the PT catalyst is to ferry the *in situ* generated anion from the interphase to the organic bulk, where the organic phase reaction can proceed.

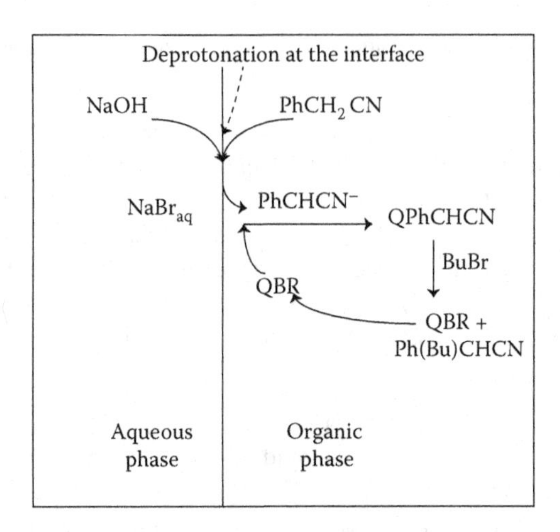

Figure 16.3 Makosza interphase mechanism for base-mediated PTC reactions.

If the organic phase reaction is very rapid, then reaction can take place very close to the interface itself.

Solid–liquid PTC

In certain cases, enhanced reaction rates as well as higher selectivity can also be obtained by switching from LLPTC conditions to SLPTC conditions. LLPTC systems suffer from slow reactions due to the hydration of the anions and can also undergo undesired hydrolysis side reactions in the presence of OH^- ions. Thus, SLPTC reactions often demonstrate higher reaction rate (due to higher anion in the absence of a significant hydration shell around the transferred anion) as well as higher selectivity than LLPTC reactions. In a solid–liquid system, the quat has to approach the solid surface and undergoes ion exchange at or very close to the solid surface (or in some cases within the solid) to form the active form of the PT catalyst. Hence, crown ethers and polyethylene glycols are found to be more effective in SLPTC reactions initially, though further studies have shown that some quaternary onium salts, with easy accessibility to the solid surface, can also be effective in SLPTC. Depending on the location and mechanism of the ion-exchange reaction and the solubility level of the solid in the organic phase, two scenarios can be anticipated. These are shown in Figure 16.4 and are referred to as homogeneous and heterogeneous solubilization mechanisms (Melville and Goddard, 1988).

Homogeneous solubilization takes place when the solid has some finite solubility, albeit very low, in the organic phase and involves dissolution of the inorganic solid salt into the organic phase, followed by the ion exchange in the liquid phase very close to the solid surface. In this case, Q^+X^- does not interact directly with the solid surface but picks up the Y^- anion from a thin film near the solid surface (since the ion-exchange

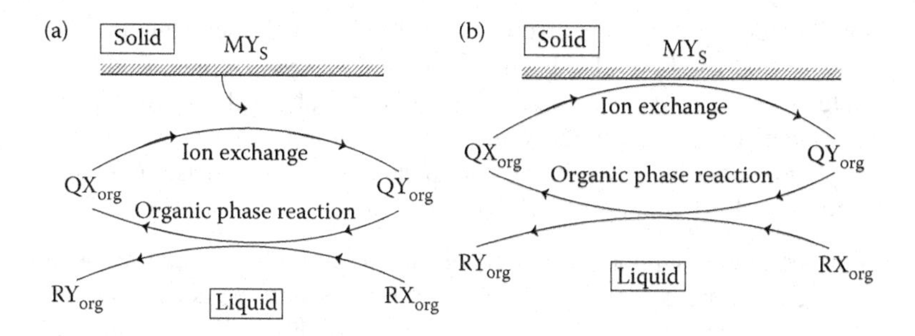

Figure 16.4 Mechanisms of SLPTC: (a) homogeneous solubilization; (b) heterogeneous solubilization.

reaction in the liquid phase is usually very fast) and ferries the Y^- ion into the organic bulk in the form of Q^+Y^-. Depending on the rate of the organic phase reaction, this reaction can take place completely in the liquid film itself (very fast organic reaction) or in the organic bulk (slower organic reaction). However, when the solid does not have a finite solubility in the organic phase, the PT catalyst has to approach the surface of the solid crystalline lattice to pick up the Y^- ion. In this case, the ion-exchange reaction can be the slow and rate-determining step. Small amounts of water (typically 0.5–2.0% of the solid salt) have been found to enhance the reaction rate significantly in such cases. According to a theory proposed by Liotta et al. (1987), traces of water enhance the rate of the PTC cycle due to solid dissolution in a thin aqueous film (called the omega phase) that coats the solid particles.

Solid-supported PTC or triphase catalysis (TPC)

The recovery and removal of the PT catalyst from the organic phase can be a cumbersome process that limits its adoption in many industrial applications. However, if the PT catalyst can be restricted in a third phase (be it a third insoluble liquid phase or supported on a solid phase), then separation can be easily achieved via phase separation or a simple filtration and centrifugation process. However, the introduction of a third phase introduces new interfaces, which come with new diffusion and transfer resistances that can slow down the reaction rate. In fact, higher costs and lower reactivity (than the soluble analogs) have greatly limited the use of supported PTC.

The most commonly studied resin support is polystyrene crosslinked with divinylbenzene (DVB) (1–2% crosslinking) in its microporous form, though it has also been used in its macroporous and popcorn forms. The PT catalyst can be physically adsorbed or chemically bound on the support with or without a spacer chain between the support and the PT catalyst. These supported catalysts are active in a variety of reactions but find limited commercial applications due to lower reactivity than the soluble analogs, which is mainly due to diffusional limitations in the solid support phase. Higher costs and physical and chemical stability for repeated use have also been a concern. However, simple supported systems where quat salts are adsorbed on organophilic clays have proved to be inexpensive and robust options and even found to be more active than corresponding polymer-supported catalysts in some cases (Lin and Pinnavaia, 1991; Akelah et al., 1999).

Similarly, quats supported on high-surface-area silica and aluminas also give reaction rates that are significantly higher than the corresponding soluble analogs. This is probably due to the highly polar environment of the alumina, which alters the microenvironment of the reaction site and provides a highly favorable condition for substitution reactions. As explained by Desikan and Doraiswamy (2000), a positive "polymer" effect is created that compensates for the retardation of the reaction due to diffusional limitations.

However, examples of superior activity of supported catalysts are more the exception than the rule. Overall, it is fair to say that the disadvantages have overwhelmed the obvious advantages of a supported catalyst, namely ease of separation of the catalyst after reaction, ease of recycle and reuse of catalyst, and adaptation to continuous commercial processes. However, with ever-increasing raw material prices and the high costs of energy required in catalyst recovery steps, the differential in terms of higher costs is getting narrowed.

Modeling of PTC reactions

We shall now apply the methods developed in the previous chapters to model PTC reactions in liquid–liquid and solid–liquid systems, including solid-supported systems. For a more detailed account of these methods, reference may be made to the articles, among others, of Naik and Doraiswamy (1998), and Yadav and collaborators (1995, 2004). The rate of the overall PTC cycle is dependent on the relative rates of the different steps in the PTC cycle. Thus, when the basic conservation equations for mass balance are written for a PTC system, the individual steps that comprise the PTC cycle must be accounted for. These steps are: the ion-exchange reaction, interphase mass transfer of both inactive and active forms of the phase transfer (PT) catalyst, partitioning of the catalyst between the two phases (in liquid–liquid systems), and the main organic phase reaction. When these are considered, the normal assumption of pseudo-first-order kinetics (Equation 16.1) is no longer valid.

$$-r_{RX} = k_{obs}[RX]_{org} = k_2[QY]_{org}[RX]_{org} \tag{16.1}$$

It is important to note that the concentration of Q^+Y^- in the organic phase, $[QY]_{org}$, is critical in determining the rate of the PTC cycle, since it governs the rate of product formation through the organic phase reaction.

It is now fairly well recognized that the Q^+Y^- concentration varies with time under various conditions, including some in which a pseudo-first-order rate constant fits the experimental data.

In general, differential–algebraic equations can be written for the mass and energy balances of the chemical species involved in the reaction, typically using thermodynamic and phenomenological system constants as model parameters. The interphase species transport is defined in terms of an overall mass transfer coefficient, which is a combination of the individual local mass transfer coefficients for each phase; the interface itself is assumed to provide no resistance to species transport. PTC enhances not only slow reactions (kinetic regime), where mass transfer considerations have no effect on the overall reaction rate, but also fast reactions, where reaction and mass transfer rates are of comparable magnitudes and the reaction occurs partially or entirely in a diffusion film close to the phase interphase. The relative rates of convection, diffusion, and reaction in PTC systems can be defined, like in other multiphase systems, in terms of dimensionless numbers like the Damköhler number (*Da*) or Thiele modulus and Péclet number (*Pe*), which characterize the relative importance of diffusion vs. reaction and convection vs. diffusion, respectively. Compositions in each phase can be expressed using equilibrium constants or partition coefficients, which are typically assumed to be independent of bulk phase composition.

A typical process model requires the estimation of several physicochemical properties of the reagents as well as transport and kinetic parameters. It is better if these are independently obtained and not obtained by parameter regression, where raw experimental data is force fitted to a model. Of the different systems listed in Table 16.1, we shall consider below the two most general and important systems, LLPTC and SLPTC.

LLPTC models

The following factors must be considered in any LLPTC model:

- Intrinsic kinetics of ion-exchange reaction
- Intrinsic kinetics of organic phase reaction
- Mass conservation of species
- Overall mass conservation
- Interphase mass transfer
- Intraphase mass transfer
- Catalyst loading and dissociation kinetics
- Equilibrium partitioning of catalyst between phases
- Location of reaction—organic, aqueous, or interface

In addition, reactor dynamics such as flow patterns, heat transfer, and so on can also affect the reaction profiles. Typically, it is safe to assume that the aqueous side film resistance is negligible and the ion-exchange reaction is fast in liquid–liquid systems. The ion-exchange reaction is usually a reversible reaction, while the organic phase reaction is usually irreversible in the case of nucleophilic substitution reactions (though depending on the chemistry involved, this reaction can also be reversible).

In general, the following model parameters need to be independently determined to develop a thorough understanding of LLPTC kinetics: *dissociation constant in aqueous phase; mass transfer coefficient for QX and QY; and intrinsic kinetics of the ion-exchange and organic phase reactions; and anion selectivity ratio of the quat.*

SLPTC models

SLPTC reactions can proceed essentially under two conditions: homogeneous and heterogeneous. In the homogeneous case, the solid first dissolves in the liquid, after which all steps proceed as in LLPTC. The only difference is that all steps involved in solid dissolution should also be considered. In heterogeneous solubilization, on the other hand, there is no dissolution involved, and reaction within the solid must be considered, as described in Chapters 6 and 9 for fluid–solid reactions. A useful way of developing a framework for modeling is to propose the simplest model to begin with, based only on the kinetics of the reaction involved and then upgrade it progressively by adding one step at a time till all the steps are included (the opposite, starting with the most complex model, is perhaps more common). When such a development is attempted, as shown in Figure 16.5, four models result. Model D in the figure is the most complete and incorporates all possible sources of resistance, namely solid dissolution, ion exchange, mass transfer of the quat species to and away from the solid surface, and the organic phase reaction.

The model equations (see Doraiswamy, 2001) can be nondimensionalized in terms of a Thiele prameter and a Biot number to account for the mass transfer steps. As can be seen in Figure 16.6, the assumption of a constant QY concentration in the organic phase and subsequent pseudo-first-order reaction rate (Model A) can lead to a gross overestimate of the expected conversion in a given time. Owing to the QY concentration gradually building up with time, conversion is much slower when the contributions of the other steps in the SLPTC cycle, namely solid dissolution, ion exchange, and mass transfer of the quat species, are accounted for.

For heterogeneous solubilization (Figure 16.7), the PTC cycle is made up of diffusion through the liquid film (external mass transfer), diffusion steps within the reactive solid, adsorption–desorption steps at the solid surface (if any), surface ion exchange reaction, and the liquid phase organic reaction. Transient conditions prevail within the solid and the controlling regime can change due to structural changes within it during the course of the reaction. As with gas–solid reactions

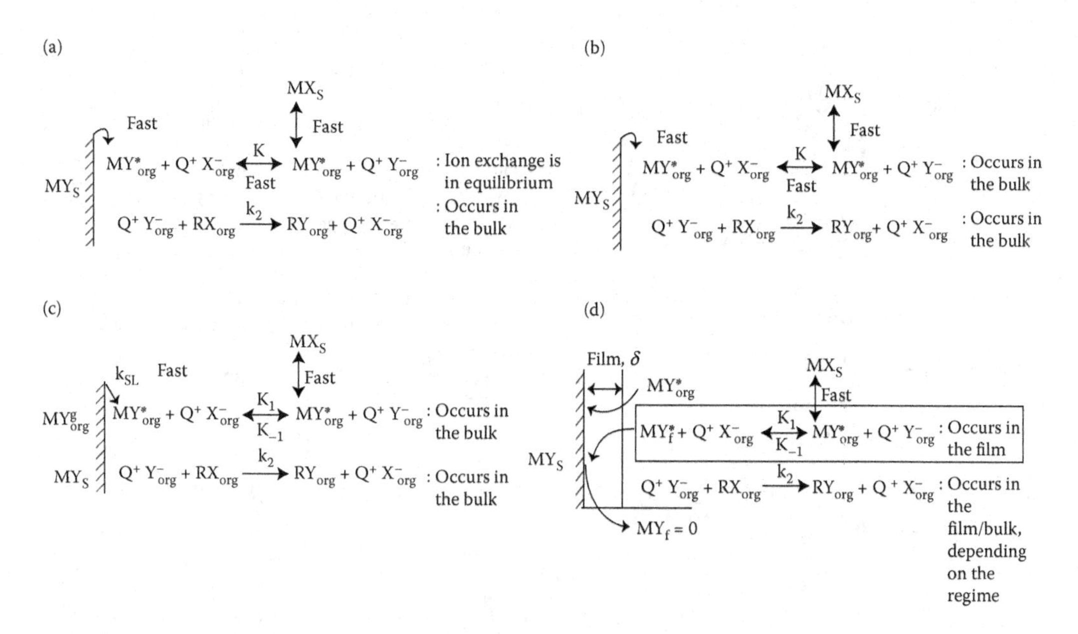

Figure 16.5 Models for homogeneous solubilization. (a) Model A, (b) Model B, (c) Model C, (d) Model D. (Adapted from Naik, S.D. and Doraiswamy, L.K., *AIChE J.*, **44**, 612, 1998.)

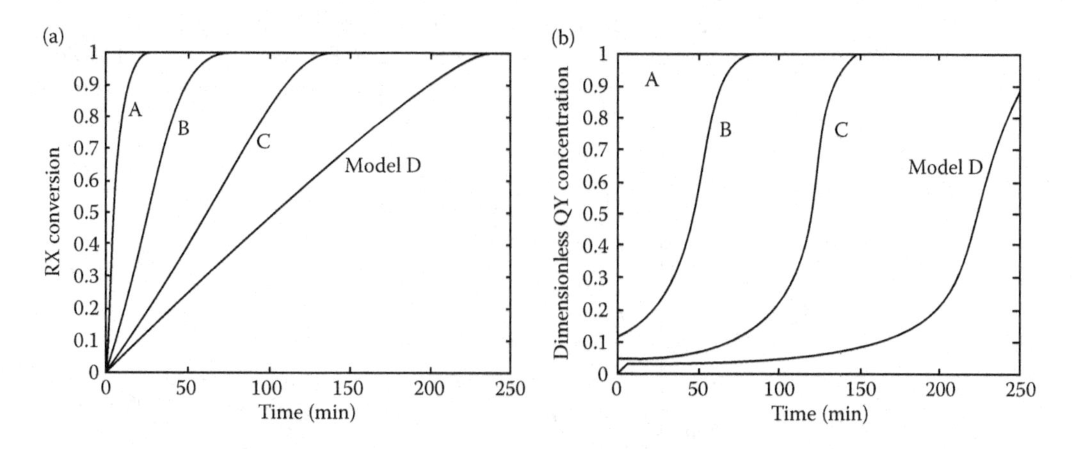

Figure 16.6 Model simulations for homogeneous solubilization: (a) RX conversion; (b) quat concentration in the organic phase. (Adapted from Naik, S.D. and Doraiswamy, L.K., *AIChE J.*, **44**, 612, 1998.)

considered in Chapter 6, depending on the porosity of the solid, a shrinking core or a volume reaction model can be considered for solid–liquid PTC reactions. In this case, ion exchange can be one of the rate-controlling steps due to limited access of the quat cation to the anions in the lattice structure and the deposition of the product MX on the solid surface. And this can lead to a contradictory interpretation of the role of diffusion.

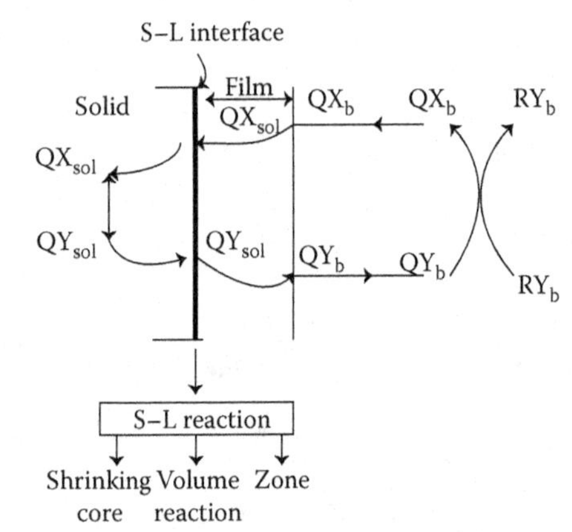

Steps involved in the PTC cycle are:

1. Liquid phase mass transfer: diffusion of Q^+X^- to the solid surface
2. Diffusion of Q^+X^- within the solid
3. Possible adsorption of Q^+X^- on the solid surface
4. Reaction (ion exchange) at the solid surface, forming Q^+Y^-
5. Desorption of Q^+Y^-
6. Diffusion of Q^+Y^- out of the solid
7. Liquid phase mass transfer: diffusion of Q^+Y^- to the liquid bulk
8. Organic reaction of Q^+Y^- with RX, regenerating Q^+X^-

Figure 16.7 Model schematic for heterogeneous solubilization. (Adapted from Naik, S.D. and Doraiswamy, L.K., *AIChE J.*, **44**, 612, 1998.)

Interpretation of the role of diffusion: A cautionary note Once again, assuming constant values for diffusivity and solid–liquid mass transfer coefficients for QX and QY, the model equations can be nondimensionalized in terms of the Thiele parameter (ϕ^2) and the Biot number (Bi_m). An important observation from the simulation analysis is with respect to the effect of the solid phase on the conversion of the organic substrate in the organic phase. As shown in Figure 16.8, under conditions of low diffusional limitation (low ϕ^2), overall conversion is lower than that under

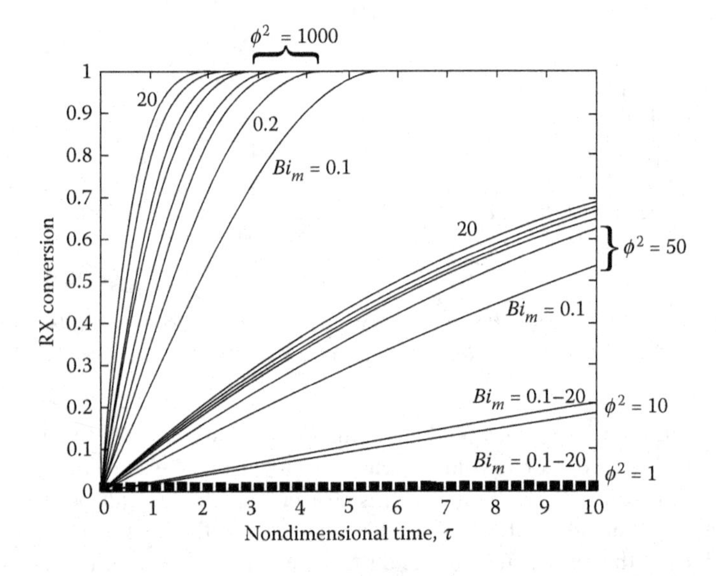

Figure 16.8 Model simulation for heterogeneous solubilization. (Adapted from Naik, S.D. and Doraiswamy, L.K., *AIChE J.*, **44**, 612, 1998.)

higher diffusional limitation conditions. This is the exact opposite of the diffusion effect in traditional heterogeneous catalysis. The reason for this is that the situation is quite different here, for what is involved is not the effect of solid phase diffusion on reaction within the solid phase, but the effect of diffusion (on ion-exchange reaction) within the solid phase on the reaction in the surrounding liquid phase. Thus, in the SLPTC system, higher values of ϕ^2 correspond to the case with high diffusional limitation and a fast rate of the ion-exchange reaction. Ion exchange takes place on the surface of the solid and generates significant amounts of QY at the solid interface for consumption in the organic phase. On the other hand, when the relative rate of the ion-exchange reaction is low compared to diffusion (low ϕ^2), sufficient amounts of QY are not generated and the overall conversion in the organic phase is also low. Also, film-transfer limitations, as expressed by the Biot number, have an effect at higher values of ϕ^2.

It should be noted that both homogeneous and heterogeneous solubilization can occur simultaneously in an SLPTC system. However, a general comprehensive model combining the features of both mechanisms has not yet been developed. Instead, the overall rate can perhaps be obtained by merely combining the rates of reaction predicted by the two models. It is the heterogeneous model that largely contributes to the overall reaction rate in most cases, though the contribution of the homogeneous model can also be significant in cases where the solid phase has a finite solubility in the organic phase. Also, mechanistic differences occur in the presence of small traces of water in SLPTC systems, involving the dissolution of the solid in a thin aqueous film, the so-called omega phase that coats the solid particles. This aspect of the problem has yet to be rigorously explored.

Supported PTC (TPC)

In general, supported PT catalysis (also called triphase catalysis by many groups) is slower than the soluble analogs, mainly due to the diffusional resistance of the solid support. The complex interactions between the three phases involved in a supported PTC system are difficult to understand but important in tailoring the support structure for improved reactivity and stability. Support macrostructure is determined by properties such as the degree of crosslinking, the number of anchoring sites, the size of spacer chains that separate the anchored quat species from the support surface, the size of the support particles, and the macroporosity of the support, and plays an important role in determining the activity of the supported catalyst. These factors also affect access to the active site due to their effect on swelling of the resin in the presence of a solvent, which, for polymeric resins, is a function of degree of crosslinking and solvent type. The presence of spacer chains (typically 8–12 carbon atoms) moves the reaction center away from the polymer substrate and facilitates access to the active sites. In addition to macrostructure of the support, the microenvironment within the support is crucial as it affects the interactions of the aqueous and organic phases with the PT catalyst immobilized on the support phase. While soluble PTC requires the transfer of one of the reagents to the phase of the second reactant, supported PT systems require both reagents to diffuse to the active PT catalyst sites on the support. Thus, diffusion to the surface of the support and inside the porous support structure for both the aqueous and organic phases becomes critical. Equally important is the distribution dynamics of the two liquid phases within the solid.

Kinetic mechanism of TPC systems

A liquid–liquid–solid triphase reaction system consists of an organic liquid phase containing a substrate (typically the dispersed phase), an aqueous liquid phase containing a reagent (typically the continuous phase), and a solid-supported catalyst. The mechanism is similar to Starks' extraction mechanism for homogenous PTC systems, with one major difference. In Starks' mechanism it is assumed that the PT catalyst moves freely between the organic and aqueous phases, whereas in a TPC system the catalyst movement is restricted and the organic and aqueous reagents must be brought to the catalyst cation. The immobilization of a PT catalyst to a solid support also introduces characteristics in the reaction system, which is typical of heterogeneous catalysis. For instance, instead of considering a planar phase boundary through which the catalyst transports the anions as assumed in classical two-phase systems, one will need to consider a volume element which contains the active catalytic sites as well as the continuous and dispersed bulk phases. For example, as shown in Figure 16.9, a polymer-supported TP catalyst contains the hydrophobic

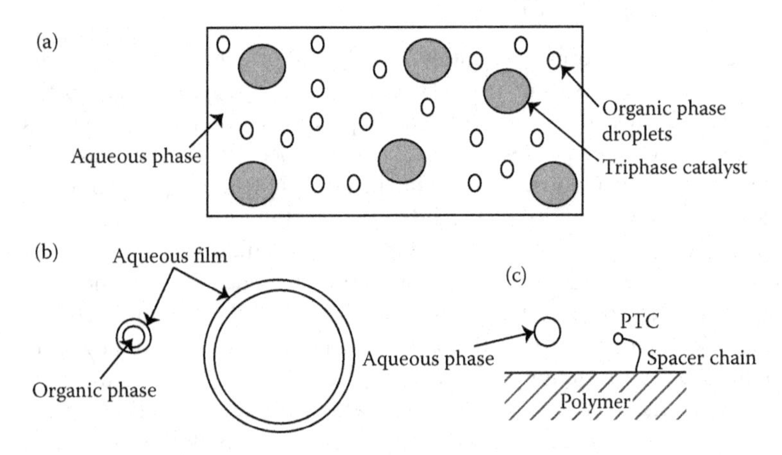

Figure 16.9 A schematic diagram of reaction mechanism for TPC system. (a) Reaction system, (b) polymer (catalyst) particle, and (c) polymer (catalyst) pore.

polymer backbone solvated by the organic solvent and an aqueous phase that contains water and the inorganic nucleophile. It is likely that the TP catalyst sites involved in the reactions are the ones that are present on the interface between the phases. Given the dynamic nature of the interface in this system, the catalyst sites can alternate between being available and unavailable for the reactions. From studies on ion-exchange resins, it is commonly suggested that the ion-exchange reaction is very fast. This enables one to assume that the ion-exchange reaction is always in equilibrium. Therefore, the overall reaction rate in an L–L–S triphase reaction is determined by the following kinetic steps: (1) mass transfer of reactants, (2) intraparticle diffusion of reactants, and (3) intrinsic organic reaction rate at the active sites. These kinetic steps show similarities of TPC to traditional heterogeneous catalysis. While traditional heterogeneous catalysis involves diffusion of reactants through a single gaseous or liquid phase, transport of reactants from two liquid phases involves a diffusion–reaction scenario that is much more complicated.

Another point that adds to the complexity of triphase catalytic systems is the determination of phase continuity within the solid support. The continuities would be determined by the volume ratio of the phases as well as to the lipophilicity/hydrophilicity of the polymer support. It is most likely that the lipophilic polymer support imbibes the organic solvent, thus making the organic phase the continuous phase with the dispersed aqueous phase droplets being transported through it to come in contact with the immobilized catalytic sites. Other factors, such as the type of solvent, reactant concentration, and the type of the inorganic anion, organic leaving group, and catalyst cation, also have a role to play. Modeling the phase distribution, by itself, is a very challenging task if not an impossible one. To date, no kinetic models on TPC which consider the complicated dynamics of phase distribution within the catalyst particle are available.

Methodology for modeling solid-supported PTC reactions

The methods used for modeling-supported PTC systems are all based on the standard equations developed for porous catalysts in heterogeneous catalysis (Chapter 6). These are expressed in terms of an overall effectiveness factor that accounts both for the mass transfer resistances outside the supported catalyst particles (film diffusion resistance, expressed as a Biot number) and within them (intraparticle diffusional resistance, expressed in terms of a Thiele modulus). Then, for any given solid shape, the catalytic effectiveness factor ε_c can be derived as a function of the Thiele modulus ϕ_j. Thus, for a spherical support solid, we have

$$\varepsilon = \frac{3\phi\coth(3\phi) - 1}{3\phi^2} \tag{16.2}$$

where

$$\phi = \frac{R_c}{3}\sqrt{\frac{k_2 M_c \lambda_C}{V_{\text{cat}} D_q}} \tag{16.3}$$

M_c is the mass, V_{cat} the volume, λ_C the concentration, D_q the diffusivity of the quat species in the organic phase, and R_c the radius of the catalyst support particle. The overall effectiveness factor ε_o can then be derived by incorporating the film diffusion term (in terms of Biot number, Bi_m) in addition to the interparticle diffusion term (accounted for by ε_c).

$$\varepsilon_o = \frac{\varepsilon_c}{1 + \phi^2 / Bi_m} \tag{16.4}$$

However, this model considers only the organic phase explicitly and the kinetics of the ion-exchange reaction and any resistances to transport of aqueous phase reagents from the aqueous bulk to the surface of the supported catalyst and diffusion within the support are ignored. In other words, the analysis accounts for steps 4–6 listed above but assumes the rates of steps 1–3 too fast to affect the overall rate of the PTC cycle. These restrictions were subsequently removed (Wang and Yang, 1991a,b, 1992) and the model was further generalized by accounting for reversibility of the aqueous phase ion-exchange reaction (Desikan and Doraiswamy, 1995), which led to a lower effectiveness factor than when the reverse reaction was neglected.

Supported PTC with LHHW kinetics

As TPC involves the reaction between a solid and a fluid (liquid), rigorous modeling of the reaction should be possible by invoking the LHHW model for the reaction. We have already described in Chapter 5 the principles underlying LHHW modeling of reactions involving a solid catalyst. The various steps involved in a typical PTC cycle were explained above. We now illustrate through a fully worked out example the complete modeling of a triphase catalytic reaction involving a nucleophilic substitution reaction.

Example 16.1

Consider the reaction system: octyl bromide (RBr) dissolved in toluene, reacting with potassium acetate (K^+OAc^-) dissolved in the aqueous phase, to yield octyl acetate (ROAc) and potassium bromide (K^+Br^-) in their respective phases. The overall reaction can be expressed as

$$RBr_{org} + (K^+OAc^-)_{aq} \rightarrow ROAc_{org} + (K^+Br^-)_{aq} \tag{E16.1.1}$$

The following assumptions can be made: reactions take place at the catalyst cations located at the interphase in the catalyst particle;

phase distribution within the catalyst is constant and not affected by phase composition changes; diffusion coefficients of reactants are constant; there is no change in phase volumes during reaction; organic reactant and product are insoluble in the aqueous phase; organic and aqueous phase bulks are well mixed; isothermal conditions prevail throughout the course of reaction; extraction at the interface is in equilibrium.

Construct an LHHW model for the reaction and identify the various constants of the model. Then, assuming that all the experiments have been carried out and the required data are available, test your model using these data (given at the end of the problem).

SOLUTION

The steps that are involved in the reaction are

Step 1: Transfer of the nucleophile (acetate ion, OAc^-) from the bulk phase into TP catalyst particle. The transfer involves the diffusion of OAc^- in the aqueous phase in the bulk and within the catalyst particle.

Step 2: Ion-exchange reaction between the nucleophile and the leaving anion (bromide anion, Br^-) attached to the catalyst cation Q^+ to form active sites Q^+OAc^-.

Step 3: Transfer of the organic substrate (octyl bromide, RBr) from the bulk phase into the catalyst particle. The transfer involves the diffusion of the substrate in the organic phase in the bulk and within the catalyst particle.

Step 4: Reaction between the substrate and the nucleophile at the active sites of the catalyst located at the interface to form the organic product (octyl acetate, ROAc) and to reform the attached leaving anion on the catalyst cation.

Step 5: Transfer of the leaving anion and the organic product from the catalyst particle to the bulk aqueous phase and organic phase, respectively.

At reaction conditions where the rates of mass transfer are much higher than the rate of reaction, the mass transfer steps can be ignored. Thus, we assume that the reaction mechanism consists of an ion-exchange reaction step between OAc^- and Q^+X^- (for the first ion-exchange cycle, $X = Cl^-$ and for all subsequent cycles, $X = Br^-$) to form an active site, Q^+OAc^-, followed by reaction of RBr at this site to form the final product ROAc and an inactive site Q^+Br^-. These steps may be expressed as follows:

Ion-exchange step:

$$(Q^+Br^-)_s + (O^+Ac^-)_{aq} \leftrightarrow (Q^+OAc^-)_s + (Br^-) \quad \text{(E16.1.2)}$$

Organic phase reaction step:

$$(Q^+OAc^-)_s + (RBr)_{org} \leftrightarrow (Q^+Br^-)_s + (ROAc)_{org} \quad \text{(E16.1.3)}$$

The overall reaction can be seen as analogous to an Eley–Rideal modification of the LHHW model that involves a reaction between an adsorbed reactant with an unadsorbed reactant from the bulk phase (see Chapter 5). We treat the ion-exchange step as the adsorption of the first reactant on the inactive catalyst sites to form active sites, and the organic phase reaction step as the reaction of the second reactant with the adsorbed reactant at the catalyst sites.

Note that the definition of "inactive" and "active" sites for a TP catalytic system is different from that for traditional heterogeneous catalysis. In the latter, an active site is a catalyst site at which the adsorption and reaction steps take place. Reaction does not take place at an inactive site. An active site can be either vacant or occupied depending on whether it contains an adsorbed atom/molecule/complex or not. In TPC, a site is meant to be the catalyst cation, that is, Q^+. It is considered inactive when it is attached to the catalyst's original anion or to the by-product anion (in our case Cl^- or Br^-). On the other hand, a site is active if it is attached to the inorganic nucleophile, that is, OAc^- anion. The organic phase reaction occurs at this activated site.

The reversible ion-exchange reaction step may be compared to the Langmuir–Hinshelwood adsorption/desorption mechanism. Using the traditional notation of heterogeneous catalysis, we can express the reversible ion-exchange reaction E16.1.3 as

$$OAc^- + S^+Br^- \leftrightarrow Br^- + S^+OAc^- \qquad \text{(E16.1.4)}$$

where S^+ is the triphase catalyst's cation. Let us now assume that a transitional site $Br^-S^+OAc^-$ is formed between the forward and reverse reaction steps, that is,

$$OAc^- + S^+Br^- \leftrightarrow Br^-\ S^+OAc^- \leftrightarrow Br^- + S^+OAc^- \qquad \text{(E16.1.5)}$$

We now come to the main feature of the example: Split the steps into two separate equilibrium attachment/detachment steps. The forward reaction step can be seen as the "attachment/detachment" of the OAc^- anion on the inactive site $S^+\ Br^-$, that is,

$$OAc^- + S^+Br^- \leftrightarrow Br^-\ S^+OAc^- \qquad \text{(E16.1.6)}$$

Similarly, the reverse reaction step can be seen as the "attachment/detachment" of the Br^- anion on an active site S^+OAc^-, that is,

$$Br^- + S^+OAc^- \leftrightarrow Br^-S^+OAc^- \qquad \text{(E16.1.7)}$$

Assuming the rates of attachment and detachment to be in equilibrium, we obtain Equations E16.1.8 and E16.1.9 for reactions E16.1.6 and E16.1.7, respectively.

$$\theta_{BrOAc} = K_{OAc}[OAc^{-1}]_{aq}(1 - \theta_{OAC} - \theta_{BrOAC}) \qquad \text{(E16.1.8)}$$

$$\theta_{BrOAc} = K_{Br}[Br^-]_{aq}(1 - \theta_{Br} - \theta_{BrOAC}) \qquad (E16.1.9)$$

where K_{OAc} and K_{Br} are the equilibrium attachment/detachment constants for OAc^- and Br^- anions, respectively; $[OAc^-]_{aq}$ and $[Br^-]_{aq}$ are the concentrations in the aqueous phase of OAc^- and Br^- anions, respectively; and θ_{OAc}, θ_{Br-}, and θ_{BrOAc} are the fractions of the total numbers of TPC cations attached to OAc^-, Br^- and both OAc^- and Br^- anions, respectively. It is postulated that transition sites Br^- S^+ OAc^-, once formed, are transformed instantaneously to either active sites S^+ OAc^- or inactive sites S^+ Br^-. Thus, Equations E16.1.8 and E16.1.9 can be written as

$$\theta_{OAc} = K_{OAc}[OAc^-](1 - \theta_{OAc} - \theta_{Br}) \qquad (E16.1.10)$$

$$\theta_{Br} = K_{Br}[Br^-](1 - \theta_{OAc} - \theta_{Br}) \qquad (E16.1.11)$$

Combining the expressions for θ_{Br} and θ_{OAc}, we obtain a hyperbolic equation for the fraction of active TPC sites as

$$\theta_{OAc} = \frac{K_{OAc}[OAc^-]_{aq}}{1 + K_{OAc}[OAc^-]_{aq} + K_{Br}[Br^-]_{aq}} \qquad (E16.1.12)$$

or in terms of catalyst concentration:

$$[S^+ \cdot OAc^-] = [s^+]_{tot} \frac{K_{OAc}[OAc^-]_{aq}}{1 + K_{OAc}[OAc^-]_{aq} + K_{Br}[Br^-]_{aq}} \qquad (E16.1.13)$$

where $[S^+]_{tot}$ and $[S^+ OAc^-]$ are the total concentration of catalyst and the concentration of catalyst attached to OAc^- anions, respectively. Finally, combining Equations E16.1.8 and E16.1.13, we obtain the following expression for the rate of the organic reaction:

$$-r_{org} = k_{org}[RBr]_{org}[s^+]_{tot} \frac{K_{OAc}[OAc^-]_{aq}}{1 + K_{OAc}[OAc^-]_{aq} + K_{Br}[Br^-]_{aq}} \qquad (E16.1.14)$$

The model assumes that the conversion rate is a linear function of the total concentration of the catalyst. To account for a possible nonlinearity between catalyst concentration and the conversion rate, the model can be modified as

$$-r_{org} = k_{org}[RBr]_{org}[s^+]_{tot}^{\alpha} \frac{K_{OAc}[OAc^-]_{aq}}{1 + K_{OAc}[OAc^-]_{aq} + K_{Br}[Br^-]_{aq}} \qquad (E16.1.15)$$

where α is the power-law exponent on the concentration of catalyst.

Model verification:

Values of the parameters k_{org}, K_{OAc}, and K_{Br} are needed to validate the model. Satrio et al. (2000) determined these values from independent sets of experiments.

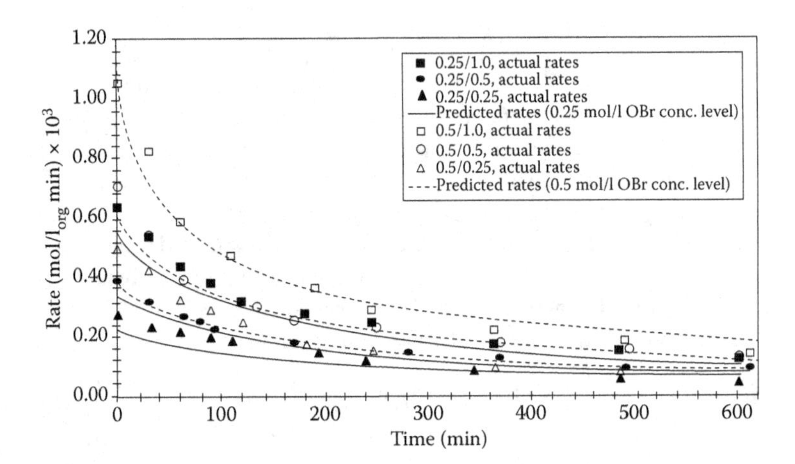

Figure 16.10 Predicted and actual reaction rate plots for the octyl bromide-potassium acetate system in the presence of polymer-supported TBMAC catalyst at 95°C and at 0.25 and 0.5 mol/l$_{org}$ octyl bromide concentration levels. (Adapted from Satrio, J.A.B., Glatzer, H.J., and Doraiswamy, L.K., *Chem. Eng. Sci.*, **55**(21), 5013, 2000.)

Figure 16.10 compares the rates predicted from Equation E16.1.9 with the experimental rates. It will be noticed that there is good agreement between the two.

Conclusions:

Two conclusions stand out.

1. The scope of the LHHW models has been expanded to include a situation where the same site becomes alternately active and inactive.
2. The model withstands the rigorous test of comparison with experimental data, *based on independent measurements of the kinetic and equilibrium model constants.* This constitutes a powerful confirmation of the model. It will be recalled (Chapter 7) that the usual practice is to postulate a number of plausible models and simultaneously extract all the parameter values for a given model from a *statistical analysis of the same raw kinetic data for the reaction.* Based on certain criteria, the parameter values should meet, the *best* model is chosen, which is not necessarily the *correct* model (see Chapter 7 for the limitations of this procedure).

"Cascade engineered" PTC process

An interesting possibility, labeled "*cascade engineered* PTC process," has been proposed for multi-step synthesis (Yadav, 2004). In this process, instead of conducting several steps of a multistep process, with intermittent separation and purification, the same catalyst is used for a number of reactions without separation. By choosing appropriate reaction conditions with different modes of operation (S–L, L–L, L–L–L, etc.), an optimum choice of catalyst and solvent can be made for a series of reactions, resulting in improved atom economy for the reagents used and reduced waste from the overall process. It is also possible that all the steps of a multistep process cannot be consolidated into a single step, in which case two- or more-step consolidations may have to be accommodated.

Before accepting the PTC route (indeed any *a priori* favored route) for a product, it is desirable to carry out a comparative assessment of competing routes, including cascade engineering of steps. When this is done, it may turn out that the favored route is not necessarily the best one. We present two case studies in the final interlude following this chapter: one involving cascade engineering of all steps into a single consolidated step, and the other in which several options are considered, including more than one consolidation.

References

Akelah, A., El-Borai, M.A., AbdEl-Aal, M.F., Rehab, A., and Abou-Zeid, M.S., *Macromol. Chem. Phys.*, **200**, 955, 1999.

Brandström, A. Principles of phase transfer catalysis by quarternary ammonium salts, in *Advances in Physical Organic Chemistry*, Academic Press, London and NY 1977.

Desikan, S. and Doraiswamy, L.K., *Ind. Eng. Chem. Res.*, **34**, 3524, 1995.

Desikan, S. and Doraiswamy, L. K., *Chem. Eng. Sci.*, **55**, 6119, 2000.

Doraiswamy, L.K., *Organic Synthesis Engineering*, Oxford University Press, New York, 2001.

Jin, G., Morgner, H., Ido, T., and Goto, S., *Catal. Lett.*, **86**, 207, 2003.

Landini, D., Maia, A., and Montanari, F., *J. Chem. Soc. Chem. Comm.* 112, 1977.

Lin, C.L. and Pinnavaia, T.J., *Chem. Mater.*, **3**, 213, 1991.

Liotta, C.L., Burgess, E.M., Ray, C.C., Black, E.D., and Fair, B.E., *Am. Chem. Soc. Ser. No. 326*, Washington, DC, 15, 1987.

Makosza, M., *Pure Appl. Chem.*, **43**, 439, 1975.

Melville, J.B. and Goddard, J.D., *Ind. Eng. Chem. Res.*, **27**, 551, 1988.

Naik, S.D. and Doraiswamy, L.K., *AIChE J.*, **44**, 612, 1998.

Satrio, J.A.B., Glatzer, H.J., and Doraiswamy, L.K., *Chem. Eng. Sci.*, **55**(21), 5013, 2000.

Starks, C.M., *J. Am. Chem. Soc.*, **93**, 195, 1971.

Wang, M.L. and Yang, H.M., *Chem. Eng. Sci.*, **46**, 619, 1991a.

Wang, M.L. and Yang, H.M., *Ind. Eng. Chem. Res.*, **30**, 2384, 1991b.

Wang, M.L. and Yang, H.M., *Ind. Eng. Chem. Res.*, **31**, 1868, 1992.

Yadav, G.D. *Top. Catal.* **29**, 145, 2004.

Yadav, G.D. and Mistry, C.K., *J. Mol. Catal. A Chem.*, **102**, 67, 1995.

Chapter 17 Forefront of the chemical reaction engineering field

Objective

The objective of this chapter is to give the reader a perspective of foresight for the future.

- What are the future trends in the world?
- How should the chemical community respond to that?
- How much of the present knowledge base in the CRE field is sufficient to tackle these problems?
- What are the new tools needed in a chemical reaction engineer's toolbox?

Introduction

The increasing world population and increasing standards of living have brought about resource economy and sustainability to the forefront of the everyday discussions. As one of the key players, the chemical process industries around the world are moving toward more economical methods of chemical synthesis. Future chemical reaction engineers are responsible for designing new and novel methods of chemical synthesis, with broader emphasis toward chemistry as well as engineering of the process. Process intensification is one tool of the trade that we have to learn. Of the many tools of process intensification, use of nontraditional reactors is increasingly being practiced. We gave brief descriptions about such reactors in Part III. The coverage was not comprehensive, but rather based on our personal biases. In this closing chapter, we will try to be more comprehensive about the future trends in chemical process industries and address the changing face of process optimization.

Resource economy

The development of ammonia synthesis technology was an economically driven situation, with its own wars in the nineteenth century in South

America. Today, the realities in terms of the oil-dependent cultures are the everyday discussions and part of the political agenda of every country. The responsibility of the chemical process industries is to find cheaper methods of producing commodity chemicals as well as fine and specialty chemicals and find alternative raw materials for chemical synthesis.

Carbon and hydrogen

For the chemical process industry, the fundamental raw material is carbon and hydrogen. Thus, finding cheap and abundant sources of carbon and hydrogen is very important. The ultimate source of carbon is CO_2, and technologies are needed to process CO_2 with the efficiency that plant cells process. The ultimate source of hydrogen is water, and an energy-efficient technology to obtain hydrogen from water is the holy grail of the present-day research. Once these very fundamental raw materials are obtained, then the synthesis of the variety of chemicals is possible even with present-day existing technologies.

The $CO_2 + H_2O$ technology has been readily available in the photosynthetic cells for about 3.5 billion years. The chemical industry is gradually learning to look at the successful examples from nature, and photosynthesis is one of them. In the photosynthetic cell, the reactors are compartmentalized for water splitting and carbon dioxide hydrogenation, that is, they take place in separate locations. Water-splitting reaction demands solar energy, while carbon dioxide reduction reaction proceeds as long as hydrogen atoms are available. The carbon chains with glucose as building blocks are used to synthesize oil to store energy or synthesize starch and cellulose. Using these molecules as carbon source and converting them into traditional commercial hydrocarbons, such as gasoline or diesel, is the greater challenge of chemical engineering academia and industry simultaneously.

Catalysis and CRE developments, hand in hand, can bring about exciting solutions where CO_2 transforms from being a greenhouse gas to a desired raw material. To develop selective CO_2 utilization processes such as those available in plant cells, a mere understanding of the chemical details of the reactions taking place is not enough. It is also necessary to understand transport processes as well as the chemical and phase equilibria of the processes across lipid membranes comprising the substructures of a plant cell with the eyes of a chemical engineer. The field of CRE, with its built-in expertise of the reactions, transport processes, and optimization skills coupled with the material science and chemistry background of the catalysis domain is ready to tackle the complex world of chemical transformations in biological systems.

Bio-renewables

Economy-driven selection rules point toward finding local sources of carbon and hydrogen. Ligno-cellulosic materials appear as successful candidates among many. However, efficiently converting them into marketable fuels is the challenge.

The bio-renewable domain of the chemical process industries is rapidly emerging as an alternative for its fossil fuel-driven counterpart. In this field, there are mainly two pathways:

i. Use of biological resources for transportation fuel manufacture
ii. Use of biological resources for chemical manufacture

Conversion of the carbon in the cellulosic structure to olefins serves as for route (i), while route (ii) requires complex conversion routes.

The brute force approach to the problem is using the expertise in coal or solid fuel technology developed so far to produce syn-gas which can be further used as the raw material for a variety of the processes. The next generation of the engineers will be developing gentler and more efficient technologies for the next generation of chemical process industries.

The resource economy requires that the synthesis to be carried out with, if possible, no by-products. Therefore, an in-depth understanding of the chemistry is utterly needed to design systems without many by-products. In Chapters 14 and 15, we have given a brief description of homogeneous catalysis and phase-transfer catalysis, two of the most commonly used methodologies for fine and specialty chemical syntheses. The developments in asymmetric synthesis will add to our ability to synthesize a pure enantiomer, which in turn will vastly improve the resource economy as well as the energy (used for product purification) economy.

While the traditional realm of chemical engineering is scaling-up of processes, the resource economy demands that we should be able to scale down and design at the micro- or at the nano-scale. Micro-scale manufacturing is emerging parallel to developments in microelectronic manufacturing. On the other hand, the accumulated knowledge basis in chemistry and biology is now driving the field toward abilities in manufacturing in nanoscopic reactors, where one molecule per turnover is manufactured.

Energy economy

The global energy challenges and the demands on energy no longer allow for luxuries of process optimization disregarding process energy demands. It must be borne in mind carefully that the second law of thermodynamics hints toward slower processes for minimum entropy generation and therefore higher energy efficiencies. Thus, we discover one aspect of the scaling down of the processes.

Oil refining, as one classical process example, is a high-energy demand large-scale process. The refining process itself is imbued with the heat exchange units, optimization of which adds tremendously to the process economics. In oil refining, optimizations on many fronts, such as market-driven product distribution, the variations in the raw material qualities, in addition to energy takes place.

From the most simplistic point of view, if we make the same quality (or better) product at lower temperatures and pressures, the process investment and operational costs as well as the process energy demand would be lower. Thus, the chemical process industry energy economy resided in finding better catalysts that can operate at lower temperatures and pressures with higher yields. The catalysis field is still an empirical science. The prospect for the future is the ability to integrate the molecular understanding with the large-scale processing.

Heat integration in microreactors

At smaller sizes and high surface-to-volume ratios provided by the microfluidic systems, heat transfer efficiency increases and it is possible to integrate the endothermic and exothermic reactions in one process unit. Emerging examples in this area indicate high promises. Though scale-up remains an issue.

Sonochemical reaction triggering

It is possible to locally excite a part of the reactor instead of the large structure by using focused sound waves. This offers some unique and unexplored capabilities of catalytic or noncatalytic reactions. The local hot spots enable very high temperatures, whereas the smallness of the heated area makes it possible to use technique in heat-sensitive systems as well. Furthermore, the delivery of energy to where it is needed is a great leap toward energy savings in chemical processing.

Photochemical or photocatalytic systems

Another emerging chemical/catalytic area is using photons, especially in the visible spectral range to trigger chemical conversions. The reaction system can be light-sensitive and driven by itself, or light-harvesting molecules or sensitizers may be needed to drive the reaction. The best example in the area is photosynthesis, offering many challenges and solutions for chemical conversions using light.

Electrochemical techniques

While the use of electrochemical methods to selectively synthesize chemicals is attractive, the electrochemical route is becoming more and more popular for energy generation, such as in fuel cells, and in energy storage in batteries. A better understanding of electrochemical routes is evidently necessary for future generations of chemical reaction engineers who are required to model chemical reactions along with mass transfer and electron transport problems.

Microwaves

So far, we have discussed the interaction of matter with electromagnetic radiation of visible light, sound waves, and electrons, and here we will briefly touch upon the interaction of the matter with microwaves. Similar to the technologies we have discussed so far, microwaves also offer the potential for local hot spots. Depending on the resonance frequency of the matter at hand, increased absorption efficiencies can lead to unique phase and chemical transformations.

Chemical reaction engineer in the twenty-first century

The process optimization engineer in the twenty-first century must have in his/her armory the tools of information and computational technologies, in addition to the insight and intuition (s)he brings along with the chemical engineering training. The process engineer of the twenty-first century must incorporate the following:

A priori reactor design for nonideal surfaces The length- and time-scale integration problem is one of the challenges of the chemical engineer. Efficient processes require an improved understanding of the molecular phenomena and better integration of the molecular phenomena in the design heuristics.

Biomimetics in chemical conversions Has to be carefully integrated in the design heuristics. It must be remembered that biological systems have benefited from billions of years of evolution triggered by a multitude of optimization constraints. They offer unique process alternatives within themselves.

Ant colony optimization is a tool that will be attractive for process optimizations where probabilistic models are needed. The computational optimization technique mimics the complexity arisen from simplicity in ant colonies for process optimizations and control.

Formulation engineering is another area that a well-equipped chemical reaction engineer can bring in the hard skills of mixing, size distribution control, and solids handling to manufacturing drugs or other types of specialty chemicals.

Crystal engineering also requires strong skills of CRE starting from modeling a chemical reaction at the molecular level to the control of heat and mass transfer to synthesize and process high-purity crystals. The major consumer of such skills is the microelectronic industry.

Concept of personal reactors is another area that the dreamers must follow. The ability to manufacture the demanded chemical, at the exact amount that is needed, will transform the chemical manufacturing cultures in the same way that the personal computer has transformed the present-day society.

In Closing

In these limited pages, you have found a perspective toward CRE which is inevitably biased from our experiences. The responsibility of the seasoned learner is to use this text, along with many others, as his/her compass to explore the vast universe.

> They (students) come in believing textbooks are authoritative but eventually they figure out that textbooks and professors don't know everything, and then they start to think on their own. Then, I begin learning from them.
>
> **— Theodore (Ted) Henry Geballe**

As quoted in the autobiography of Stephen Chu, in Gösta Ekspong (ed.), *Nobel Lectures: Physics 1996–2000* (2002).

Subject Index

A

Activated complex, *see* Transition state theory
Activity
  Coefficients, 133
  medium and substituent effects, 132
Addition reactions, 462
Additivity rules, 135
Adiabatic reactor, 266
  conversion–temperature relationship in, 266
  fixed bed, 254, 266, 277
Adsorption, 139
  Henry's law, 141
  Inhomogeneities in, 142
  isotherms of two-dimensional equations of state, 143
  Langmuir isotherm, 141, 159
  Thermodynamics, 139
Age of the fluid, 101
Alternative fixed-bed designs, 276
Ammonia synthesis, 276
  kinetics, 149–152
Axial dispersion model, 89, 99, 246, 279
  elimination of, 231

B

Backmixing, 275
Batch reactor, 13–19, 56–69
  Comparison to PFR, and MFR, 29
  Design for complex reactions, 48
  Laboratory reactors, 220
  non-isothermal operation of, 16–18
  optimum operating policies for, 18–19
  rate parameters from, 221
Bed voidage, 246
Bio-renewables, 498
Biot number, 209
Bodenstein approximation, 162
Bodenstein reaction, 158
Boundary layer resistance, 337
Brinkmann-Foechheimer-Darcy equation, 262
Bronsted relations, *see* Polanyi and Bronsted relations
Bubble column slurry reactors, 390
  Gas holdup in, 392–393
  Minimum velocity for complete solids suspension in, 392
  Regimes of flow in, 391
Bubble rise velocity, 293
Bulk diffusion, 190

C

Caking, 299, 318
Calculation of reactor volume
  for gas-liquid reactors, 361
  for liquid-liquid contactors, 377
  Mixed gas, batch liquid, unsteady state, 364
  Mixed gas, mixed liquid, steady state, 364
  Plug Gas, mixed liquid, steady state, 364
  Plug gas, plug liquid, and cocurrent steady state, 363
  Plug gas, plug liquid, and countercurrent steady state, 363
Carbon and hydrogen, 498
Cascade engineered PTC process, 494

Catalysis
  homogeneous (*see* homogeneous catalysis)
  phase transfer (*see* phase transfer catalysis)
Catalyst
  deactivation, *see* deactivation control
  dilution for temperature uniformity, 231
  flicker, 280
  Nonuniform distributions between tubes, 275
  packing in a tubular reactor, 251
  Staging of, 305
  Wilkinson's homogenous, 465–467
Catalytic wire-gauze reactors, 280
Cell model, 257
Characteristic timescales, 102, 185
Chemical equilibrium
  constant, 124
  Derivation of relationships for simple reactions, 144
  for diffusion,reaction, and mass transfer, 185
  for reactions in gas phase, 146
  for reactions in liquid phase, 146
  Medium and substituent effects on
Chemical Reaction Engineering
  The essential minimum of, 2
  in the twenty-first century, 501
Chlorination
  of ethylene, 300
  of methane, 296
Choice between NINA-PBR and A-PBR, 274
Circulation systems, 309
Coal gasification, 320
Collection of laboratory data
  Experimental methods for, multiphase reactions, 398
  for three-phase catalytic reactions, 398
  for three-phase noncatalytic reactions, 401
  Interpretation of, 399
Collision theory, 154
Combo reactors, 77–78, 249, 435; *see also* Distillation column reactors
Complex equilibria, 128–131
Complex kinetics
  in homogeneous catalysis, 469
Complex reactions, 33–83
  analytical solutions for, 40–41
  batch reactor design, 48–49
  continuous stirred tank reactor design, 53–55
  determination of rates in, 39
  equilibria, 128–131
  extent of reaction in, 38–39
  in CSTR, 52
  in PFR, 52
  independent reactions in (*see* independent reactions), 36–38
  Mathematical representation of, 35
  mathematical representation of, 35–36
  multistep reactions in, 46–47
  paralel reactions, 56–59
  paralel-consecutive reactions, 59–60
  plug-flow reactor design, 52–53
  Rate equations for, 38
  reactor choice for maximizing yields/selectivities, 56–60
  reactor design for (*see* reactor design for complex reactions)
  Reactor design for, 48
  Reduction of, 34
  selectivity and yield in, 39–40
  Stoichiometry of, 34
Constrictivity, 179
Contact time, 184
Continuity equation, 188
Continuous reactors
  for multiphase reactors, 385
Continuous stirred tank reactor (CSTR), *see* mixed-flow reactor (MFR)
Conversion concentration relationships, 10
Coordination reactions, 462
Coordinative unsaturation, coordination number, and coordination geometry, 457
Criteria
  for energy minimization in gas-liquid reactions, 372–375

for volume minimization in gas-liquid reactions, 369–372
CSTR
Adiabatic, 27
Complex reactions, 53

D

Damköhler number, 90,185, 444
Deactivation control, 310
Basic equation of, 313
Fixed-bed reactor for, 314
Fluidized-bed reactor for, 315
Heat transfer issues for, 312
Moving bed reactor for, 315
Reactor choice for, 312
Dean flow as a static mixer, 115
Dean number, 116
Decay time, 313
Defluidization of bed: Sudden death, 299, 318
Degree of heterogeneity, 189
Design methodologies for NINA-PBR, 256
Cell model, 257
Homogeneous, 257
pseudo-homogeneous, 257
heterogeneous models, 257
Models based on the pseudohomogeneous assumption, 257
Quasi-continuum models, 257
DFT (density functional theory), 158
Differential reactors, 226
Diffusion
accompanied by an irreversible reaction of general order, 350
and reaction in series, 350
Configurational, 190
Disguises, 222, 229
Multicomponent, 199
resistance, 252
role in homogeneous catalysis, 469
the role in phase transfer catalysis: A cautionary note, 486
Diffusion and reaction in fluid-fluid reactions
fast reactions, 352
instantenous reactions, 3
Measurement of mass transfer coefficients for, 354
very slow and slow reactions, 350
Diffusivity, 178
Effective, 179
in gases, 178
in liquids, 179
Multicomponent, 199
Dilute bed region, 297
Dissociation–extractive distillation, 448
Distillation column reactor, 436–447
Batch reactor with continuous removal of product, 436–439
Design methodology of, 443
Overall effectiveness factor in a packed, 441
Packed, 439–441
Residue curve map (RCM) for, 442–447
Distillation–reaction, 447–451
Di-*t*-butyl peroxide decomposition, 224
Dynamic programming, 255, 266

E

Effective diffusivity, 179, 191, 186–196, 260, 265
Estimation in pellets, 192
external, 208
in gases, 178
in liquids, 179
internal-external, 189
miscellaneous effects, 199
modes, 190
multicomponent, 199
Nonisothermal, 197
resistance elimination, 230
slow reactions, 205
Effectiveness factor, 191, 194–196
Effects of fines, 318
Elastic turbulence, 116
Electrochemical techniques, 500
Electron rules ("electron bookkeeping") 459
18-electron rule, 459
16–18-electron rule, 460
Eley-Rideal mechanism, 165

Eliminating
axial dispersion effects, 231
film mass transfer resistance, 229
pore diffusion resistances, 230
transport disguises, 229
Elimination reactions, 464
Emulsion gas flow, 287
End region models, 297, 298
Dilute bed region, 298
Gird (Set) region, 298, 291
Energy economy, 499
Engulfment-deformation diffusion model, 104
Enhancement factor, 206, 351
Enthalpy of formation, 122–123
Equilibrium compositions
Extension to a nonideal system, 129
for complex reactions, 128
in condensed phase(s), 126
in gas phase reactions, 126
Equilibrium constant, 124–128, 155
conversion expressions, 126
medium and substituent effects, 132
Equivalent diffusion time, 183
Ergun equation, 262
Ethylation of aniline, 300
Extent of reaction, 38, 50, 129
External
diffusion, *see* diffusion
effectiveness factor, 208
mass and heat transfer, 208
Extrathermodynamic approach, 134
Group contributions or additivity principle, 135
relationships between rate and equilibrium parameters, 136
to selectivity, 138

F

Falling film, 183
Fast fluidized-bed reactor, 289, 307
Bottom dense zone, 307
Choking, 289
Fick's Laws, 180
Film theory, 181–182
Film diffusion, 246
of mass and heat transfer for fluid–fluid reactions, 185
Fine particles or fines, 318
Fixed-bed reactor design PPP
for solid catalyzed fluid-phase reactions, 251
isothermal operation, 261
Momentum balance, 261
Fixed-bed reactor, 340, 251–283
Axial flow, 278
Gas solid catalytic, 243
Gas–solid noncatalytic, 243, 340
Radial axial flow, 279
Radial flow, 276–77
Fixed-bed versus fluidized-bed reactors, 319
Flow reactors for testing gas–solid catalytic reactions, 225
accessing the rate information in, 229
Differential versus integral, 226
Fluidization, 285
Archimedes number, 287
basics of, 286
Bubble phase, 289, 303
bubbling bed, 289–291
Defluidization of bed: Sudden death, 318
Effects of fines on, 318
Emulsion phase, 289, 303
Froude group, minimum, 286
Geldart's classification of, 287
Gulf streaming, 318
Heterogeneous (aggregative), 285
Homogeneous (particulate), 285
Incipient, 289
Minimum bubbling velocity, 301
Minimum fluidization velocity, 286, 301
Minimum velocity of, 286
(Minimum) slugging velocity, 291, 301
Slugging, 318
Two-phase theory of, 287
Wake, 289
Fluidized-bed reactor design
for solid catalyzed fluid-phase reactions, 285–325
Fluidized-bed reactors
Bed behavior in, 288
Bubble growth, 292
Bubble Radius, 293
Bubble rise velocity, 293
Bubble size distribution, 292
Bubbling bed diameter, 292

Bubbling bed model of, 292
Bubbling fluidized bed, 289, 292
Bulk flow velocity, 293
Calculation of conversion in, 297
Circulation systems for, 309
Classification of, 288
Classification of, 288
Cloud (thickness), 293
Complete modeling of, 291
Complete modelling, 291
Conversion in, 297
Dilute bed region of, 297
Dilute phase region, 291
End region models of, 297
Estimation of bed properties, 295
Fast fluidized-bed, 307
Fluid mechanistic modelling, 291
Free rise velocity, 293
Gas–solid noncatalytic, 345
Grid or jet region of, 298
Heat transfer in, 295
Incipient, 289
Main fluid-bed region, 291
Packed, 303
Packing void, 290
Performance, 302
Presssurized, 320
Reactor models for packed, 303
Recommended scale-up procedure for, 300
Solids distribution, 294
Start-up of, 319
Strategies to improve performance of, 302
Transport (or pneumatic), 308
Turbulent bed, 307
Velocity limits, 289,290
Zones, 290
Fluidized-bed versus fixed-bed reactors, 319
Forced convection
axial flow, 279
radial flow, 279
Forefront of the chemical reaction engineering field, 497
Fowler Guggenheim isotherms, 142–144
Free energy change, 124
medium and substituent effects, 132
minimization, 130–132
Fugacity, 125, 129

G

Gas holdup, 390
Gas–liquid and liquid–liquid reactions and reactors, 347–381
Gas–liquid contactors, 358–375
based on manner of contact, 358
based on manner of energy delivery, 359
calculation of reactor volume, 361–365
Classification of, 358, 359
Mass transfer coefficients and interfacial areas of, 359
Role of backmixing in, 359
Gas liquid reactions, 243
in a slab, 204
Instantaneous reactions, 206
Slow reactions, 205
Two-film theory, 205
Gas-solid reactions
catalytic, 243
non-catalytic, 200, 243
Gas solid non-catalytic reaction models
Computational, 326
Discrete, 326
Grain-micrograin, 326
Nucleation, 339
Particle-pellet (grain), 326, 332, 334, 339
Percolation, 326
Reaction zone, 326, 331
Two zone, 331
Shrinking core, 326, 327
Ash diffusion, 328
Volume reaction, 326, 329, 339
Gas–solid noncatalytic reactors
fixed-bed, 339, 340
fluidized-bed, 339, 345
moving-bed, 339, 343
Gas–solid noncatalytic reactions and reactors, 325
Bulk-flow or volume-change effects in, 334
Effect of temperature change on, 335
Modeling of, 326
Models that account for structural variations for, 336

Gas–solid reaction models, 326
Extensions to the basic models, 334
A general model that can be reduced to specific ones, 338
Other models, 334
Effect of reaction, 336
Shrinking core model, 327
Effect of sintering, 338
The particle–pellet or grain models, 332
Volume reaction model, 329
Zone models, 331
Gatterman-Koch reaction, 127
Geldart's
Classification, 287
Particles, 286
Gibbs free energy, 122
Gibbs-Helmholtz relationship, 125
Gibbs adsorption isotherms, 139
General kinetic analysis, 468
Gradientless reactor, 221, 231
Grid or jet region, 298
Group contributions rules, *see* Additivity rules
Gulf streaming, 299, 318

H

Hammett relationship for dissociation constants, 137
Hatta number, *see* Hatta modulus
Hatta modulus, 185, 206, 351, 436
Heat integration
in microreactors, 500
Heat of combustion, 123,124
formation, *see* enthalpy of formation
standard, 123
Heat transfer, 295
coefficients, 186
role on diffusion, 210
Helmholtz free energy, 122
Henry's Law, 141
Heterogeneous, 253, 257
reaction, 18–19, 122, 187
Homogeneous, 253, 257
catalysis, 249
reactions, 187
Homogeneous catalysis, 453–474
Basic reactions of, 461–464
coordination reactions in, 462
Elementary reactions (or activation steps), 462
kinetic analysis of, 468–473
Operational scheme of, 460
Hydrogenation model, 467

I

Ideal reactors, 13–24
Batch reactor, 13–18
CSTR, 23–24
Nonidealities defined with respect to, 88
PFR, 20–21
two limits of, 86
Independent reactions, 36
Industrial reactions using homogeneous catalysis
Hydrogenation, 465
Hydrogenation by Wilkinson's catalyst, 465
Insertion reactions, 463
Instantaneous reactions, 206
Integral reactors, 226
Interaction by exchange with a mean, 104
Internal and external diffusion, 210
Combined effects, 209
Gas phase reactants, 210
Liquid phase reactants, 211
Relative roles of mass and heat transfer, 210
Internal energy, 122
Intraparticle resistance, 337
Intrinsic
kinetics in homogeneous catalysis, 468
rate constant, *see* Effectiveness factor

J

Joint PDF, 105

K

Keto-enol tautomerization of benzoyl camphor, 134
Kinetics, 153
in relation to chemical reaction engineering, 118
mechanism of TPC systems, 488

mechanism reduction methodology, 170
Proposing a kinetic model, 158
Kinetic regime diagnosis, 231
Knudsen pump, 115
Diffusion, 190
Kolbe-Schmitt carbonation, *see* three-phase noncatalytic reactions
Koros–Nowak criterion, 231
Kulkarni-Ramachadran-Doraiswamy criterion, 286
Kunii and Levenspiel's model, 286

L

Laboratory reactors: Collection and analysis of the data, 217
Gradientless reactors, 231
Lagrange multipliers, 130
Langmuir isotherm, 141, 159
LHHW (Langmuir-Hinshelwood-Hougen-Watson) models, 159–165, 235–239
Lewis and Whitman film theory, *see* film theory
Ligands
and their role in transition metal catalysis, 457
Reactions of (mainly replacement), 461
Liquid phase reactions, 124
activity coefficient, 132
Partial molar properties, 131
Solvent and solute operators, 133
Thermodynamics, 131
Liquid–liquid contactors, 375–379
calculation of reactor volume, 377–380
Classification of, 375
classification of, 375–376
interfacial areas in, 376
mass transfer coefficients in (*see* mass transfer)
stirred tank reactor for, 379
Liquid–liquid PTC, 478
models for, 483
Loop slurry reactors, 396–398
Mass transfer in, 397
Types of, 396

M

Macromixing
partial, 99–100
Marangoni effect, 354
Mars-van Krevelen mechanism, 166
Mass balance, 279
Mass transfer
across interfaces, 187
between bubble and emulsion, 294
coefficient, 181–184
film, 181
gas-liquid contactors, 359
governing equations, 187
in bubble column slurry reactors, 391–392
in loop slurry reactors, 396–398
in three phase reactors, 389
in trickle bed reactors, 398
liquid-liquid contactors, 376
measurement of, in gas-liquid systems, 354–355
phenomenological coefficient, 180
resistance elimination, 229
Mass transfer coefficients and interfacial areas
of some common fluid-fluid contactors, 359
for liquid liquid contactors, 376
Maxwell-Boltzmann velocity distribution, 154
Mean field theory, 111
Mechanically agitated slurry reactors, 389
Controlling regimes in, 390
Membrane reactors, 78, 248
Catalytic nonselective (CNMR-E), 424
Catalytic nonselective hollow for multiphase reactions (CNHMR-MR), 424
Catalytic selective (CMR-E), 421–423
Combining exothermic and endothermic reactions in, 425
Controlled addition of one of the reactants in a bimolecular reaction in, 80, 426
coupling of reactions in, 80

Membrane reactors (*Continued*)
  Effect of tube and shell side flow conditions in, 428
  equilibrium shift in, 79
  exploitable features of, 79
  Fluidized-bed inert selective (IMR-F), 412, 420
  for economic processes (including energy integration), 431
  hybridization in, 80
  Immobilized enzyme, 425
  in organic technology/synthesis, 429
  in small- and medium-volume chemicals manufacture, 430
  inorganic, 78
  Major types of, 411–414
  Modeling of, 414–425
  Operational features, 425–428
  Packed-bed catalytic selective (CMR-P), 423
  Packed-bed inert selective with packed catalyst (IMR-P), 414
  reactant slip in, 80
Membrane-assisted reactor engineering, 411–432
Metals
  transition (*see* transition metals)
Michelis-Menten mechanism, 168
Microfluidic devices, 247, 354–355
  for fluid-fluid reactions, 354
  heat integration in, 500
Micromixing,
  partial, 99–100
  policy, 98–99
Microkinetic analysis, 169
  Determining the kinetic parameters, 171
  Postulating a mechanism, 171
Micro-macro random pore model, 191
Microwaves, 501
Minimization of free energy, 130
Minimum speed for complete suspension, 390
Mixed flow reactor (MFR or CSTR), 23–26
  basic (and design) equations for, 23–24
  nonisothermal operation of, 24–26
Mixer
  Dean flow as static, 115
  Elastic turbulence as, 116
  Slug flow as, 115
Mixing, 95–97, 113, 115
  Concept of, 95
  Engulfment-deformation diffusion, 104
  Fully segregated flow, 97
  interaction by exchange with a mean, 104–105
  Micromixing policy, 98
  Models for partial micromixing, 100
  Models for, 99
  partial, 99–100
  Passive devices for, 115
  Practical implications in chemical synthesis, 106
  practical implications of, 106
  Regions of, 95
  regions of, 95–97
  segregated flow, 97–98
  surface mixing, 111
  turbulent mixing model, 101
Models for
  PTC reactions, 482
  membrane reactors, 414
  methodology for solid-supported PTC reactions, 489
  of gas–solid reactions, 326
  partial micromixing, 100
  partial mixing, 99
Momentum balance, 261, 279
Moving-bed reactors
  Gas–solid noncatalytic, 343
Multiple bed reactor, 266, 270
Multicomponent diffusion, 199
  Extension to complex reactions, 200
Multiphase reactions and reactors, 383–410
  See three-phase catalytic reactors; three-phase non-catalytic reactions
Multiphase reactions
  in homogeneous catalysis, 470–473
Multiple steady states, 26–29
  in a CSTR, 26
  Stability of the steady states, 28

Multiple-bed reactor, 270
Multistep reactions, 46
  Yield versus number of steps in, 47

N

Non-adiabatic, 254, 277
Noncatalytic gas–solid reactions, 200
Nonideal reactor analysis, 85
  of MFR, 90
  of PFR, 88–90
Nonisothermal operation, 16, 254
  of a SBR, 71
Nonisothermal, nonadiabatic, and adiabatic reactors, 254
NINA-PBR, 254–256, 274
Non-uniform packing, 275

O

Optimum temperatures/temperature profiles
  Consecutive reactions, 73
  extension to batch reactor, 73
  in a CSTR, 71–72
  in a PFR, 72
  In SBR, 71
  Parallel reactions, 72
Oxidation
  of ammonia, 280
  of ethylene, 280
  of methanol, 280
  zinc sulfide, 331

P

Packed bed reactor (*see* fixed bed reactor)
Packed fluidized-bed reactors, 303
Paradox of heterogeneous kinetics, 169
Parallel path model, 191
Parallel reactions, 72–73
  consecutive reactions, 59
  Effect of reaction order in nonreacting products, 56
Partial molar properties, 131, *see* thermodynamics
PCA (Principal component analysis), 172–174
Peclet number, 90, 265, 275
Pellet diameter, 276
Penetration theory, 181, 183
Perfectly Mixed Flow Reactors (MFR), 23
  Basic equation, 23
  Nonisothermal operation, 24
  See also CSTR
Plug Flow Reactor, 19–23, 52
  Basic equations for, 21–23
  Design equations, 21
  Nonisothermal operation of, 21
  Optimal operating policies, 19
  with recycle, 60–64, 86–87
Phase transfer catalysis, 81, 249, 475
  Classification of systems, 476
  Fundamentals of, 476–478
  liquid-liquid, 478–480
  modelling, 482–488
  solid-liquid, 480–481
  solid-supported, 481–482
Phase-transfer catalysts, 81–82, 477
Photochemical or photocatalytic systems, 500
Pneumatic reactor, *see* transport reactor
Polanyi and Brønsted relations, 136
Pore
  diffusion resistance, 201, 246
  network model, 191
  size distribution, 336
Porosity, 179
Probability factor, 155
Process intensification, 246
Pseudoequilibrium hypothesis, 161
Pseudo-homogeneous

Q

Quasi-continuum models, 257

R

Radial dispersion, 246, 279
Radial-flow reactors, 276
Random pore model, 336
Rate
  and equilibria, 121
  Basic equation, 8–9
  constants, 157
  data analysis for, 235
  determining step, 160

Rate (*Continued*)
Different definitions of, 6–7
from batch reactor data, 221
From concentration data, 221
From pressure data, 223
of reaction, 5–9
Rate equation
Stoichiometry of the, 9
Basic relationships of, 9
Rate-determining step, 160
Rates and equilibria
thermodynamic and extrathermodynamic approaches, 121
Reaction coordinate
Reactions
with volume change, 14
without volume change, 14
Reactions and reactors, 5–31, 76
See batch reactor; mixed-flow reactor; plug-flow reactor; semibatch reactor
Continuous in 3-phase catalytic reactions, 385, 387
Reactions with an interface: Mass and heat transfer effects, 177
Reactive distillation, 77
Reactor choice
Volume minimization criterion, 369
Energy minimization criteria, 372
Reactor design
for catalytic reactions, *see* fixed-bed reactors, fluidized-bed reactors, gas-liquid contactors, three- phase reactors
for complex reactions
CSTR, 52
PFR, 52
for fluid-fluid reactions, 355
A generalized form of equation for all regimes, 356
for very slow reactions, 356
for fast reaction, 357
for instantenous reaction, 357
reactor efficiency, *see* three-phase catalytic reactors
Reactor model
Reactor, 12
height, 276
Microfluidic, 114
nonadiabatic with no axial diffusion, 262
Short contact time, 113
Recycle Flow reactor (RFR) 62; *see also* PFR with recycle regimes of control, 190
Optimal design of, 62
Use in complex reactions, 64
Reduction of ilmenite, 329
Regimes of control, 211
Regimes of operation, 211, 234
Relative volatility, 448
Residence time distribution, 91
Characteristic timescales, 102
the age of the fluid at a point, 101
Turbulent models of101
Types of, 93
Residue curve map (RCM), 442
Resource economy, 497
Role of diffusion
in homogeneous catalysis, 469
Catalyst effectiveness, 189
in a spherical catalyst, 191
in pellets, 189

S

Scale-up considerations, 276
of fluidized bed reactors, 299
Segregated flow, *see* mixing
Selectivity, 39
Analytical solutions, 40
Extrathermodynamic approach, 138
general expression for, 39–40
in multistep reactions, 46–47
Maximizing, 43
Optimal temperature profiles to maximize, 71
Reactor choice for maximizing, 56
Semibatch reactors, 46–47
for multiphase (1,0)- and (1,1)-order reactions, 384
Constant volume, 64
General expression for multiple reactions, 70
Non-isothermal operation, 71
Variable volume, 66–70
with constant rate of inflow, 66
with constant rate of outflow, 67

Sharp interface model, *see* Shrinking core model
Shrinking core model, 200, 327
Simple reactions
  Mathematical representation of, 35
  Stoichiometry of, 34
Single-bed reactor, 268
Sintering, 338
Site-energy distribution, 142
Slug flow as a mixer, 115
Slugging, 318
Solid catalyzed fluid reactions, 189
Solid catalyzed reactions, 235
  Influence of surface nonideality, 239
  LHHW models, 236
  Selection of the most plausible model, 236
Solid–liquid PTC, 480
  models for, 484
Solid packing, 252
Solid-supported PTC or triphase catalysis (TPC), 481
Solvation effects, 122
Solvent and solute operators, 133
Sonochemical reaction triggering, 500
Spreading pressure, 140
Staging of catalyst, 305
  Countercurrent, 305
  Cross-current, 305
  Efficiency, 306
  Vertical, 305
Stark's mechanism, 488
Start-up, 319
Statistical experimental design, 218
Steric factor, *see* probability factor
Stokes-Einstein equation, 179
Stirred tank reactor (STR), practical considerations, 379–380
Strategies for heat exchange, 273
Styrene dehydrogenation, 278
Supported PTC (TPC), 487
Surface
  coverage, 161
  diffusion, 190
  nonideality, 168
Surface renewal theory, 181, 184
surface reaction mechanisms, 159
Langmuir–Hinshelwood–Hougen–Watson models, 159
  Eley–Rideal mechanism, 165
  Influence of surface nonideality, 168
  Mars–van Krevelen mechanism, 166
  Michelis–Menten mechanism, 168
Surface renewal theory, 184
Suspension, 390

T

Tanks-in-series model, 87, 100
The particle–pellet or grain models, 332
Thermodynamics
  of adsorption, 139
  consistency tests, 158
  in relation to chemical reaction engineering, 118
  Basic relationships and properties, 122
Thiele modulus, 185, 193–196, 335
Three-phase reactors, 383
  catalytic, 383
  Design of, 383
  Types of, 387
Three-phase catalytic reactors, 383–398
  continuous, 385–387
  dispersion model of, 387
  laboratory data: interpretation of, 398–401
    effect of temperature on, 398
  mixed-flow model of, 387
  plug-flow model of, 387
  reactor efficiency, 385–386
  semibatch, 384–385
Three-phase noncatalytic reactions, 401–407
  Kolbe-Schmidt carbonation of β-naphthol, 404–407
  solid insoluble, 404
  solid-slightly soluble, 402–404
Timescales
  characteristic, 102–104
  for engulfment, 102–103
  of molecular diffusion, 102
  of reaction, 102
  reactor time constant, 103
  of recirculation, 104
  of turbulence, 103
Topochemical, 331, 334

Tortuosity, 179
Transition metal catalysis, 453–460
16–18-electron rule, 460
18-electron rule, 459
coordination geometry, 457
coordination number, 457
coordinative unsaturation, 457
Formalisms in, 453
Ligands and their role in, 457
multidentate ligands, 458
nonparticipative ligands, 458
Oxidation state of a metal, 455–457
participative ligands, 458
unidentate ligands, 458
Transition state theory, 155–157
Transport reactor, 308
Transport disguises, 229, 231, 234
in perspective, 231
Guidelines for eliminating or accounting, 234
Transport phenomena
between phases, 180
in relation to chemical reaction engineering, 118
Trickle Bed Reactors (with cocurrent downflow), 397
Controlling regimes in, 398
Mass transfer in, 398
Regimes of flow in, 397
Trickle-bed reactors, 397–398
controlling regimes in, 398
mass transfer in (*see* mass transfer)
mimicking of, 80
regimes of flow in, 397
Tube diameter, 276
Tubular reactors
Axial dispersion model, 89, 275
Nonidealities in, 88
See also PFR or Plug flow reactors
Turbulent bed reactor, 307
Two film theory, 185, 205
Tyn-Calus correlation, 179

U

UBI-QEP (Unity bond index-quadratic exponential potential), 171

V

Vacant sites, 160
van't Hoff relationship, 125
Variable volume reactors, 66–68
Variable-density reactions, 11
Voidage, 336
Volmer's isotherm, 142–144
Volume change modulus, 335
Volume reaction model, 200, 329

W

Wall effect, 246
Weisz modulus, 196
What is PTC?, 475
Wilkinson's catalyst, 465
Wire gauze reactors, 276

Y

Yield, 39
Analytical solutions, 40
in multistep reactions, 46–47
vs. number of steps in a multistep scheme, 47–48
Optimal temperature profiles to maximize, 71
Reactor choice for maximizing, 56

Z

Zone model, 105
Zone models, 331

Author Index

A

AbdEl-Aal, M.F., 482
Abou-Zeid, M.S., 482
Adachi, T., 390
Agarwalla, S., 420
Akelah, A., 482
Akita, K., 392, 393
Akiyama, S.J., 78
Aladyshev, S.I., 414
Amundson, N.R., 286, 336
Anderson, J.G., 435
Anzai, H., 78
Aris R., 18, 36, 48, 190, 191, 196, 198, 200, 266, 271
Arrhenius, S., 9
Asai, H., 392, 393
Aslan, M.Y., 150
Asperger, S., 175
Aubin, J., 355
Avidan, A., 306
Avrami, M., 334

B

Baeyens, J., 318
Bagajewicz, M., 343
Baldi, G., 398, 399
Baldyga, J., 104, 106
Barbosa, D., 443, 447
Barnard, J.A., 162
Basov, N.L., 426, 432
Batteas, J.D., 165
Bau, R., 454
Beekman, J.W., 191
Beenackers, A.A.C.M., 90, 390
Behie, L.A., 298
Bell, A.T., 165
Bellman, R., 266
Belosljudova, T.M., 431
Belton, D.N., 165
Benson, S.W., 135
Bern, L., 389
Berty, J.M., 227
Beveridge, G.S.G., 334
Bhatia, S., 326, 338
Bhatia, S.K., 334, 336, 338
Bhattacharya, A., 473
Birk, J.P., 468
Bischoff K.B., 71, 191, 238
Bisio, A., 374
Black, E.D., 481
Blaha, S.R., 424
Blenke, H., 374, 397
Bodenstein, M., 162
Bohmer, K., 374
Bolthrunis, C.O., 306
Botterill, J.S.M., 296
Botton, R., 374
Bourdart, M., 169, 239
Bourne, J.R., 102, 103, 104, 106, 107
Bouzek, K., 169, 239
Bowen, J.H., 331
Box, G.E.P., 218
Brandström, A., 478, 479
Briens, C.L., 306
Briggs, R.A., 329
Broadbelt, L.J., 165
Brown, H.C., 138
Brown, L.F., 191
Buchholz, R., 374
Bukur, B.D., 286
Burgess, E.M., 481
Burggraaf, A.J., 79, 80, 424
Buss, J.H., 135
Butt, J.B., 175, 191, 200
Buzad, G., 443
Buzzelli, D.T., 59

C

Cadle, P.J., 191
Calderbank, P.H., 359
Calus, W.F., 179
Campbell, R.R., 335
Cankurt, N.T., 286
Carberry J.J., 64, 191, 200, 209–212, 227, 275
Carroll, W.E., 454
Čatipović, N., 289
Cermak, J., 365, 369, 370, 373, 374
Chaloner, P.A., 465
Champagnie, A.M., 422, 423
Chan, S.F., 338
Chandrasekhar, B.C., 343
Chang, C.L., 390
Charpentier, J.C., 397, 398
Chatrand, C., 271
Chatt, J., 458
Chaudhari, R.V., 384, 393, 394, 396, 397, 471, 472, 473
Cheng, C.K., 331
Chitester, D.C., 286
Cholette, A., 100
Chopade, S.P., 436
Chorkendorff, I., 175
Chou, T.S., 398
Choudhary, V.R., 227, 238
Chrostowski, J.W., 336
Chuang, K.C., 308
Chunningham, R.S., 191
Cini, P., 80, 424
Cloutier, L., 100
Collman, J.P., 455, 456
Cortez, D.H., 280
Couderc, J.P., 286
Cox, J.L., 427
Cresswell, D., 286
Crowe, C.M., 271, 421
Cussler, E.L., 183

D

Damme, R.M.J., 424
Danckwerts, P.V., 91, 94, 100, 181, 354
Davidson, J.F., 286, 291, 294, 299, 318
Davidson, J.M., 318
Davis, M.M., 134
de Croocq, D., 308
de Lange, R.S.A., 80
de Mello, A.J., 115
Deckwer, W.D., 359, 361, 392
Degaleesan, T.E., 359, 376
del Broghi, 330
DeLasa, H.I., 286
Denbigh, K.G., 60
Desikan, S., 482, 490
Dilke, M.H., 127
DiMaggio, C.L., 165
Dixon, A.J., 262
Doğu, G., 212
Doherty, M.F., 436, 443, 447
Doraiswamy, K.L., 14, 19, 72, 78, 143, 144, 169, 191, 199, 200, 212, 227, 234, 235, 238, 239, 254, 257, 259, 260, 271, 272, 286, 290, 296, 299, 300, 307, 309–313, 316, 317, 319
Doraiswamy, L.K., 326, 330, 343, 348, 352, 354, 359, 360, 361, 391, 400, 404, 406, 407, 426, 430, 435, 473, 475, 482, 484, 485, 486, 490, 493, 494
Douglas, J.M., 22
Dovi, V.G., 238
Drese, K.S., 116
Dudukovic, M.P., 330, 336
Dumesic, J.A., 169, 171
Dumez, F.J., 238
Duprat, F., 450, 451
Duque, R., 249
Dutta, S., 244, 248, 320, 340

E

Eagleton, L.C., 22
Edwards, M., 306
Edwards, W.M., 280
Ehrfeld, W., 248
El-Borani, M.A., 482
Eldridge, J.W., 88
Eley, D.D., 127
Elnashaie, S.S.E.H., 286
El-Rahaiby, S.K., 334
Ergun, S., 262
Ermilova, M.M., 78, 432
Ermolaev, A.M., 431
Erofeev, B.V., 334
Ertl, G., 112, 113, 143
Erwin, M.A., 280
Evans, M.G., 155

Evans, D., 463
Evans, J.W., 341, 342
Eyring, H., 155

F

Fair, B.E., 481
Fan, L.T., 90, 297
Fasman, A.B., 78, 80, 430
Favier, M., 397, 398
Feng C.F., 191
Ferrero F., 150
Finke, R.G., 455, 456
Foley, H.C., 423
Forni, L., 150
Forni, G.F., 36
Fowler, R.H., 147
Fox, R.O., 91, 94, 96, 97, 105, 106
Franckaerts, J., 238
Froment, G.F., 71, 191, 238
Fukushima, S., 398, 399
Fuller, E.N., 192
Furusaki, S., 286, 298

G

Gad-el-Hakk, M., 115
Gaikar, V.G., 451
Gardin, D.E., 165
Garvrilidis, A., 200
Garza-Garza, O., 336
Gassend, R.G., 451
Gates, B.C., 466
Gau, G., 450, 451
Gavalas, G.R., 79
Geankoplis, J., 191
Geldart, D., 286, 287, 293, 318
Georgakis, C., 336
Ghali, E., 78
Gillespie, B.M., 64
Gillespie, G.R., 280
Gilliland, E.R., 178, 286
Glasstone, S., 175
Glatzer, H.J., 493, 494
Gobina, E., 423
Godbole, S.P., 359, 392
Goddard, J.D., 480
Godfrey, J.H., 354
Goldie, P.J., 334
Gorring, R.L., 335
Goto, S., 385, 387, 478
Govind, R., 413, 414, 425
Gower, R.C., 334
Grace, J.R., 286, 289, 291
Grandjean, B.P.A., 78, 429, 431
Grunwald, E., 136, 137
Gryaznov, V.M., 78, 80, 414, 426, 430, 431, 432
Gualy, R., 244, 248, 320, 340
Gugenheim, E.A., 147
Gunn, D.J., 295, 296
Gunther, A., 115
Gupta, J.S., 326, 338
Gupte, S.P., 473

H

Hakuta, T., 417, 418, 419
Halpern, J., 466, 468, 469
Hammett, L.P., 134
Haraya, K., 417, 418, 419, 428
Hardt, S., 116
Harold, M.P., 80, 424
Harrison, D., 286, 291, 318
Harrison, D.P., 338
Hart, D.W., 454
Hegedus, L.S., 455, 456
Hegner, B., 359
Heijnen, J.J., 374
Heine, H., 373
Henzler, H.J., 374
Hessel, V., 115, 248
Hetzer, J.H., 134
Hickmann, D.A., 113, 114
Hicks, J.S., 198, 199
Higbie, R., 181
Hikita, H., 392, 393
Hilal, N., 295, 296
Hills, A.W.D., 335
Himmelblau, D.M., 238
Hirner, W., 397
Holmes, M.H., 112
Horio, M., 286, 289
Hougen, O.A., 162, 236
Hsieh, H.P., 429
Huet, S., 238
Hughes, R., 423
Huis in't Veld, M.B.H.J., 79
Hung, J., 374

I

Ido, T., 478
Ishida, M., 329, 331
Ishikawa, H., 392
Itoh, N., 417, 418, 419, 428

J

Jazayeri, B., 306
Jensen, K.A., 115
Jhunjhunwala, M., 115
Jiang, F., 116
Jin, G., 478
Johnson, M.L.F., 191
Johnson, B.K., 423
Johnston, H.F., 287
Joo, Y.L., 116
Joshi, J.B., 234, 254, 290, 309, 312, 326, 343, 348
Juvekar, V.A., 365

K

Kabel, R.L., 374
Kaliaguine, S., 78, 429, 431
Karavanov, A.N., 78, 431
Kashdan, M.V., 78, 80, 430
Kastaněk, F., 359, 361, 365, 369, 370, 373, 374
Kasuka, K., 398, 399
Kato, K, 303, 304, 305
Kegley, S.E., 455, 459
Keizer, K., 79, 80, 424
Kelbar, B.G., 392
Kelkar, B.G., 359
Kenney, C.N., 398
Kenson, R.E., 280
Khang, S.J., 428
Khrapova, N.E.V., 432
Kikukawa, H., 392
Kim, K.K., 338
Kiperman, S.L., 169, 239
Kitao, M., 392, 393
Kittrell, J.R., 238
Koetzle, T.F., 454
Koros, R.M., 231, 235
Kozicki, F., 104, 107
Kramers, H., 371, 372
Kramers, H.A., 157
Kratochvil, J., 359, 365, 369, 370, 373, 374
Krishna, R., 355, 358, 376
Krishnan, G.R.V., 426
Kubota, H., 272
Kulkarni, A.A., 209
Kulkarni, B.D., 257, 260, 286, 326, 330, 352
Kumar, B., 108
Kumbilieva, K.E., 169
Kunii, D., 286, 289, 291, 293, 296, 298, 307, 312, 317
Küpper, M., 116
Kurten, H., 359, 374
Kusaka, K., 398, 399
Kusakabe, K., 78
Kuznecov, A.M., 373
Kwauk, M., 286

L

Laddha, G.S., 359, 376
Ladhabhoy, M.E., 366
Laidler, K.J., 157, 158
Lamba, H.S., 330
Landini, D., 479
Languasko, J.M., 212, 234
Lapidus, A.L., 473
Larson, M.A., 435
Lazic, Z.R., 218
Lee, C., 80
Lee, H.H., 200
Lee, K.U., 266, 271
Leenars, A.F.M., 79
Leffler, J.E., 136, 137
Leigh, G.J., 458
Leuteritz, G., 396, 397
Levenspiel, O., 63, 93, 266, 273, 286, 289, 291, 293, 294, 296, 298, 307, 312, 317, 354, 363
Lewis, W.K., 181, 286, 297, 303
Li, L.F.C., 278, 279
Li, Y., 286
Li, Z.Y., 78
Lidefelt, J.O., 389
Lin, 482
Lin, Y.S., 79
Liotta, C.L., 481
Liu, P.K.T., 81
Lombardo, S., 165
Louise, A., 1
Löwe, H., 115
Lund, C.R.F., 420
Luss D., 191, 280, 336

M

Maeda, H., 78
Maestri, M., 170
Maganjuk, A.P., 431
Mahapatra, A., 451
Mahoney, D.J., 336

Makosza, M., 479
Malone, M.F., 436, 443
Mandler, J., 227
Mangartz, K.H., 392
Mann, R., 374
Mantri, V.B., 331
Masel, R.I., 175
Masters, C., 457, 458, 462
Mathis, J.F., 303
Matsubara, M., 387
McCabe, R.W., 162, 164, 165
McDaniels D.H., 138
Mears, D.E., 275
Megiris, C.E., 79
Meister, D., 374
Melville, J.B., 480
Mhadeswar, A.B., 158, 170, 172–174
Mhaskar, R.D., 387
Miglio, R., 36
Mikhailov, A.S., 143
Mikhalenko, N.N., 432
Mills, P.L., 397, 471
Minet, R.C., 422, 423
Mishchenko, A.P., 78, 80, 414, 430
Mistry, C.K., 482
Mitchell, D.S., 162
Miyauchi, T., 297, 298
Montanari, F., 478, 479
Montgomery, D.C., 218
Morbidelli, M., 27
Morgner, H., 478
Mori, S., 298
Morooka, S., 78
Muchi, I., 296
Mullet, G.M., 126
Mutasher, E.I., 341

N

Nagel, O., 359, 374
Naik, S.D., 482, 485, 486
Nam, S.W., 79
Narasimhan, G., 335
Nauman, E.B., 48, 93, 99, 100
Nejemeisland, M., 262
Neuberg, H.J., 334
Neurock, M., 157, 158
Ng, K.Y.S., 165
Noddings, C.R., 126
Norton, J.R., 455, 456
Nowak, E.J., 231, 235

O

Obata, K., 417, 418, 419
Okamoto, T., 466
Okufi, S., 390
Oldshue, J.Y., 359
Onken, U., 374
Oran, U., 166
Orekhova, N.V., 78, 432
Osborn, J.A., 463

P

Papa, G., 303
Paraskos, J.A., 387
Patanaude, B., 80, 424
Patel, P.V., 191
Perez de Ortiz, E.S., 390
Perlmutter, D.D., 334, 336, 338
Pernicone, N., 150
Peters, M.S., 224
Petersen, E.E., 191
Petrov, L.A., 169
Phadtare, P.G., 404, 406, 407
Pickard, A.L., 468, 469
Pilavakis, P.A., 440
Pilhofer, T.H., 392
Pinhas, A.R., 455, 459
Pinnavaia, 482
Piret, E.L., 88
Pirozhkov, C.D., 473
Plath P., 159, 235
Pohorecki, R., 355
Polanyi, M., 155
Polyakova, V.P., 432
Post, T., 374
Powell, R., 280
Prasad, K., 317
Prasannan, P.C., 331, 339
Prausnitz, J.M., 147
Pyzhev, V., 149

R

Rader, C.G., 280
Raghavan, N.S., 238
Rahse, W., 202
Rajadhyaksha, R.A., 212, 234
Ramachandran, P.A., 286, 326, 330, 336, 338, 384, 393, 394, 471
Ranade, P.V., 338
Rao, Y.K., 334
Rase, H.F., 276-278
Ravindranath, K., 107

Ray, C.C., 481
Reed, E.L., 427
Rehab, A., 482
Reich, B.A., 423
Reid, R.C., 179
Rekoske, J.E., 165
Reylek, M., 374
Rihani, D.N., 231
Roberts, D., 280
Roberts, G.W., 227
Roberts, G.L., 427
Rode, C.V., 473
Rose, L.M., 380
Roshan, N.R., 432
Rossetti, I., 150
Rothermund, H.H., 143
Roušar, I., 169, 239
Rovind, R., 413, 414
Ruckenstein, E., 334
Runger, G.C., 218
Russel, T.W.F., 59
Ruthven, D.M., 142
Rys, P., 104, 107

S

Sacco, A., 329
Sadana, A., 312, 316, 317
Sadasivan, N., 426
Sahimi, M., 326
Samenkov, V.V., 373
Sampath, B.S., 343
Sandler, S.I., 17, 125, 144
Sano, Y., 390
Saracco, G., 429, 431
Sarycheva, I.K., 431
Sarylova, M.E., 78, 80, 430
Satrio, J.A.B., 493, 494
Satterfield, C.N., 175, 191, 227, 280
Savitskii, E.M., 432
Sawistowski, H., 390, 440
Schmidt, L.D., 29, 113, 114, 280
Schmidt, M.A., 115
Schöfeld, F., 115, 116
Schoon, N.H., 389
Schügerl, K., 359
Secker, R.B., 427
Sedriks, W., 398
Segawa, K., 392, 393
Seider, W.D., 446
Serebryannikova, O.S., 78
Serov, Y.M., 78
Shah, M.J., 238
Shah, M.A., 280
Shah, Y.T., 359, 361, 387, 392, 396
Shao, X., 413, 414, 425
Shaqfeh, E.S.G., 116
Sharma, M.M., 72, 191, 200, 212, 234, 235, 259, 271, 272, 307, 312, 317, 319, 330, 354, 359, 360, 361, 366, 391, 400, 436, 451
Sheel, H.P., 421
Shendye, R.V., 169, 239
Shindo, Y., 417, 418, 419, 428
Shorter, J.A., 138
Shu, J., 78, 429, 431
Shustorovich, E., 165, 171
Sie, S.T., 358
Singh, A.K., 279
Sinke, G.C., 128
Sinn, R., 359, 374
Sisak, C., 374
Sitting, W., 373
Skorpinski, E.J., 224
Sloot, H.J., 424
Smirnov, V.A., 78, 80, 414, 430, 432
Smith, J.M., 191, 336, 338, 385, 387
Smith, L.A., 441
Sobieszuk, P., 355
Sofekun, O.A., 231
Sohn, H.J., 334, 335
Sohn, H.Y., 334, 335, 343
Somorjai, G.A., 165
Song, S., 341, 342
Sotirchos, S.V., 341
Specchia, V., 398, 399, 429, 431
Squires, A.M., 286, 289, 306
Starks, C.M., 81, 478
Steinemann, J., 374
Steward, P.S.B., 318
Stewart, W.E., 191
Stoltze, P., 169
Stull, D.R., 128
Sun, Y.M., 428
Szekely, J., 326, 343

T

Tabares, C., 432
Talmor, E., 398
Tanigawa, K., 392, 393
Taylor, R., 376
Teller, R.G., 454

Temkin, M.I., 149
Thaller, L.H., 238
Thalmann, M., 115
Thiel, P.A., 165
Thodos, G., 238
Thoma, S., 107
Tolman, C.A., 460
Tonkovich, A.L.Y., 427
Toomey, R.D., 287
Treybal, R.E., 375
Tsepetonides, J., 374
Tsotsis, T.T., 422, 423
Turner, J.C.R., 60
Tyn M.T., 179

U

Uhlhorn, R.J.R., 79
Uku, J., 392
Ülin'ko, M.G., 432
Ulrichson, D.L., 336
Uner, D.O., 163, 166

V

van Baten, J.M., 355
Van de Vusse, J.G., 60, 64
van Deemter, J.J., 289
Van den Bleek, C.M., 231
Van Heerden, C., 27
van Hove, M.A., 165
van Neste, A., 429, 431
van Santen, R.A., 157, 158
van Swaaij, W.P.M., 90, 390, 424
van't Riet, K., 374
Varghese, P., 203
Varma A., 27, 200
Vasileiadis, S.P., 422, 423
Vavanellos, T., 334
Veldsink, J.W., 424
Venimadhavan, G., 443
Venkatraman, K., 80, 424
Venkitakrishanan, G.R., 19
Verma, R.K., 108
Versteeg, G.F., 424
Viesturs, U.E., 373
Vilermaux, J., 104
Vlachoz D.G. 158, 170, 172–174
Voetter, H., 60
Volk, W., 300
von Oertzen, A., 143
Vortmeyer, D., 343

W

Wagner, E., 422, 423
Wakao, N., 191
Walas, S.M., 147
Wang, A.W., 423,
Wang, M.L., 490
Warwick, G.C.I., 375
Watabe, S., 387
Watson, C.C., 303
Watts, D., 238
Waugh, K., 169
Weekman, V.W., 312
Weekman, W.W., 335
Weinberg, W.H., 165
Weisz, P.B., 198, 199
Weller, S., 169, 239
Weller, S.W., 280
Wen, C.Y., 90, 287, 298, 329, 331
Wendel, M., 275
Werther, J., 286, 289
Westerterp, K.R., 90, 180, 371, 372
Westrum, E.F. Jr. 128
Whitman, W.G., 181
Widagdo, S., 446
Wiedeskehr, H., 389
Wilkinson, G., 463
William, E.D., 165
Wirges, H.P., 202
Wong, C.J., 162, 164, 165
Wong, C.W., 390
Worley, F.L., 398
Wu, H., 27
Wu, J.C.S., 81

X

Xu, S., 413, 414, 425
Xu, X., 441, 442

Y

Yadav, G.D., 482, 494
Yadav, N.K., 309, 319
Yagi, H., 389
Yagupsky, G., 463
Yamaguchi, N., 390
Yang, K.H., 162, 236
Yang, H.M., 490
Yang, W.C., 286
Yates, J.T.Jr., 165
Yerushelmi, J., 286

Yoshida, F., 389, 392, 393
Yoshitome, H., 417, 418, 419
Yu, Y.H., 287
Yung, C.N., 390

Z

Zabrodsky, S.S., 296
Zahradnik, J., 359, 365, 369, 370, 373, 374
Zakhariev, A., 466
Zarkanitis, S., 341
Zaspalis, V.T., 80, 424
Zenz, F.A., 303
Zhang, Y., 355
Zheng, G., 441, 442
Zheng, L., 355
Zheng, Y., 441, 442
Ziaka, Z.D., 422, 423
Zwietering, T.N., 94, 100

Printed in the United States
by Baker & Taylor Publisher Services